ELECTRICAL MACHINES
Modelling and Analysis

MRITTUNJAY BHATTACHARYYA

Emeritus Professor
Electrical Engineering Department
St. Thomas College of Engineering and Technology
Kolkata

Formerly

Director, Institute of Technology
Banaras Hindu University, Varanasi

Vice-Chancellor
Assam (Central) University, Silchar

Vice-Chancellor
West Bengal University of Technology, Kolkata

Founder Vice-Chancellor
Kalinga Institute of Industrial Technology (Deemed University)
Bhubaneswar

PHI Learning Private Limited

Delhi-110092
2016

₹ 595.00

ELECTRICAL MACHINES—Modelling and Analysis
Mrittunjay Bhattacharyya

ISBN-978-81-203-5177-6

Published by Asoke K. Ghosh, PHI Learning Private Limited, Rimjhim House, 111, Patparganj Industrial Estate, Delhi-110092 and Printed by Mohan Makhijani at Rekha Printers Private Limited, New Delhi-110020.

To

My revered parents Late Manibala and Satish Chandra
and
All my teachers and well-wishers

Contents

Foreword

A question often arises is how best the undergraduate students learn about electrical machinery?

Since the days of invention of Faraday's Law of Electromagnetic Induction, Electromagnetic machines have emerged in various diverse forms—direct current machines, and alternating current machines—asynchronous and synchronous, and others. There arrived Linear machines and then Plasma machines. There evolved methods for the steady-state analysis of machines, which takes each device in a peacemeal way, and as the issues of interconnection arose which demanded analysis under transient and dynamic conditions, the Generalised theory and the Unified Theory were formulated.

The Generalised study, developed in terms of circuit equations on the basis of Park's two-axis model or Doherty and Nickle's symmetrical component model, is an excellent approach. Performance equations are derived in the form of impedance and connection matrices. The approach is extremely user-friendly, probably because of its basis on circuit conception, which is a continuity of a student's earlier training on circuit analysis. But this has a limitation of non-applicability to certain problems such as, machines with more than one saliency.

The Unified approach, on the other hand, is based on energy principles which are applicable to all kinds of electromagnetic devices. The machine is considered as part of a set of linked electrical, magnetic and mechanical systems, and provides a physical basis of generalised approach.

This book, entitled 'Electrical Machines—Modelling and Analysis' by Prof. Mrittunjay Bhattacharyya, attempts the students to be introduced to the Unified Theory, and in a lucid way, presents sound understanding of all normal electrical machines based on their common physical principles to help in the building up of advanced concepts.

It would be reasonable to expect a graduate student to have an understanding of everything covered in the book. Practicing engineers, desiring an introduction in modern methods of analysis of electrical apparatus, should find the book useful.

I wish the author a rewarding experience with the book.

SANKAR KUMAR SEN
Formerly, Prof. and Head, Electrical Engineering,
Bengal Engineering College, Shibpur, and
Vice Chancellor, Jadavpur University

Preface

I still, fondly, remember my revered teacher, Professor B. Ravindranath, who introduced me to the parallel operation of d.c. generators. He made me realise the importance of appreciating the understanding and stressing on the physical aspects of practical technological problems. I also remember the continual encouragement I received from Professor S.K. Sen in my career growth. Professor Sen, formerly Head of Department of Electrical Engineering in the erstwhile Bengal Engineering College, Shibpur, and Vice-Chancellor, University of Jadavpur, Kolkata, is an eminent educationist of this region. He achieved the special glory in his distinguished career by successfully tackling the problem of nagging power-outage in West-Bengal during the last decade of the twentieth Century in his capacity as the then Minister-in-charge of Power and Non-conventional Energy of the Government of West Bengal.

The present book is the culmination of the enthusiasm gathered from the pioneering works of Dr. C.V. Jones, Professor A.J. Ellision, Professors D. O'Kelly and S. Simmons, Professor W. Leonhard and his team and Dr. P. Vas. The need for mathematical modelling of electrical machines assumed special significance with the introduction of modern techniques in machine control and increasing requirement of the need for viewing the multiple classes of machines under the same broad head.

The present book starts with a review of the process of energy conversion involved in electro-mechanical devices and a broad recapitulation of the circuit elements involving basic electrical circuits. The importance of dimensional analysis of different quantities encountered in the deliberations has been stressed upon for physical appreciation of the problem involved.

Chapter 2 reviews the mathematical tools used in the book, and the basic principles involving the circuit analysis, together with a review of the concept of energy associated with electrical circuits. Chapter 3 deals with the concept of energy storage in magnetic fields and goes on to develop force/torque in electro-mechanical transducers. Here, the concept of singly- and multiply-excited machines are introduced for the first time. Energy and co-energy concepts associated with magnetic circuits are introduced. Mechanism of force/torque production in translating/revolving types of electrical machines is indicated. This leads to generalised equations of torque developed in rotating electrical machines under different conditions of movements, leading to the development of the dynamic equations, including production of voltages in the machine circuits.

Chapter 4 is exclusively dedicated to the forms of electrical circuits encountered in electrical machines and their classification, including the effect of coil-distribution and coil-spanning. Mathematical representation of electrical machine parameters is also highlighted in this chapter.

Chapter 5 records the general voltage developed with special reference to distributed type of windings, stressing that the voltages developed by transformer-induced effect and the speed-induced effect are the same for a constant rotating flux in the air-gap. The need for the effect of transformer-induced and speed-induced voltages has been separately considered and superimposed for time-varying flux in the air-gap under special conditions.

Chapter 6 deals with three-phase systems with distributed windings and stresses on the concept of transformation used extensively to electro-mechanical energy converters, thus, leading to the concept of Unified Theory of Electrical Machines. In this chapter, the quantitative relationship among self-, mutual- and leakage-reactance has been physically discussed.

Chapters 7 and 8 deal with the analysis of actual standard machines like the d.c. machines of different configurations, the induction machine and the synchronous machine with both non-salient and salient-pole configurations.

Chapter 9 stresses on the transient and dynamic operation of machines with special reference to standard machines, and shows how this aspect has assumed special relevance with requirement of fast control of drives.

Chapter 10 is devoted to unbalanced operation of induction motor and motors with lack of symmetry. It also explains why a synchronous machine starting on damper-bars with its field short-circuited often locks at half-synchronous speed—a phenomenon often encountered by students in laboratories. The last chapter highlights the concept of space-vectors and briefly introduces the concept of decoupled speed control of a.c. drives.

I enjoyed the unique privilege of close interaction with a number of stalwarts in the field of electrical engineering like Professor G.H. Rawcliffe, Dr. W. Fong, and Professor E.R. Laithwaite, together with Dr. R.F. Burbidge and Dr. Alex Broadway of the University of Bristol in the United Kingdom. I still remember the important lesson received from the Professor (Late) Rawcliffe of the University of Bristol, who said:

..."if there is a system that physically works, there will always be a mathematical formulation to justify its working, but developing a mere mathematical relationship does not ensure that a physical system will work".

I shall fail in my duties if I do not include my acknowledgement to Dr. Krishnendu Chakrabarty, Principal, Government Engineering College, Kalyani, and Professor Pramathes Das for their effort for improving the manuscript. I also recognise the contribution of my two daughters, Amrita and Bishakha, who provided encouragement.

I also record my sincere appreciation for the PHI Learning team for typesetting, drawing diagrams and editing of the manuscript.

Mrittunjay Bhattacharyya

List of Symbols

Chapter 1

F	Force in newton experienced by an elemental carrying conductor
L	Length of conductor
I_1, I_2	Conductor current
r	Separating distance in metre
B	Flux density
H	Magnetic force or magnetic field intensity
r, r_1, r_2	Proportionality constant connecting current flow in conductors and potential difference—that is resistance
v	Potential difference or voltage
T_1, T_2	Absolute temperature in kelvin = 234.5 + t
t_1, t_2	Temperature in degrees celsius
A	Cross-sectional area of material
L	Inductance parameter of coil
S	Reluctance of the magnetic circuit
$F = Ni$	Magneto-motive force or mmf
N	Number of turns of the coil
Q	Charge between two conductors
v_c	Potential difference
C	Capacitance
E	Electric field intensity in volt/m
μ	Magnetic permeability
μ_r	Relative permeability
μ_0	Absolute permeability of free-space or vacuum
Þ	Resistivity in ohms per metre
ϕ	Flux
€	Permittivity or the specific di-electric of the material between conducting surfaces
μ	Permeability of the material
μ_0	Permeability of free space or vacuum

Chapter 2

a, r	Number of rows
c, d	Number of columns
R	Resistance parameter
X	Inductive reactance parameter
M	Mutual inductance/reactance parameter
$[A]$	A general matrix
$[A_t]$	Transpose of a general matrix
$a, b, ...$	Elements of a general matrix
$[A^*]$	Conjugate of a general matrix
$[A]^{-1}$	Inverse of a general matrix
$[V']$	New column voltage matrix in compound form
$[I']$	New column current matrix in compound form
$[Z']$	New impedance matrix in compound form
p	Mean power
I^*	Conjugate of current
V_t^*	Conjugate of voltage
$[V]_t$	Transpose of voltage matrix
$[I]_t$	Transpose of current matrix
$[Z]_t$	Transpose of impedance matrix

Chapters 3 and 4

F	mmf
F	Mechanical force
x, X	Distance
v	Voltage
V	Velocity
L	Length of wire
r	Radius of pick-up coil
N	Number of turns of coil
$[G]$	G-matrix coupled to
Ω_r	Rotor speed
M	Mutual inductance between two windings carrying current
L	Self-inductance of a winding
λ	Air-gap permeance of mutual flux and the stator self flux respectively
θ_r	Angle between the stator and the rotor respectively in mechanical degrees
T_e	Electro-magnetic torque in doubly-fed maching
W_{elect}	Power flowing from electrical inputs
W_{le}	Power lost electrically
W_{fld}	Power stored in magnetic field
W_{mech}	Power in mechanical form
W_{mt}	Power lost mechanically
W_{lm}	Power lost friction and windage losses
W_{Sm}	Mechanical power remains stored in rotational systems

Ψ_r	Rotor coil flux-linkage
M_{sr}	Stator to rotor mutual inductance
B_d, B_q	Fundamental components of special distribution of flux density along *d*-axis and *q*-axis
B	Spatial distribution of flux density
F	Spatial distribution of mmf
F_r	Crest value of F
B_r	Crest value of B in rotor
B_{rd}, B_{rq}	Crest values of B_r in rotor along the *d*-axis and the *q*-axis
ψ	Phase spread
K_d	Distribution/spread/layer factor
$K_p / K_b / K_c$	Pitch factor/breadth factor/coil-span factor
K_w	Winding factor
d	Diameter
l	Length
Z	Number of conductors per radian where specified
P	Total number of poles
F_{ad}, F_{aq}	mmf components acting along a point displaced by an angle a from *d*- and *q*-axis
Λ_d, λ_q	Permeance of gap along *d*- and *q*-axis for full-pitch coil span
L_{r1}, L_{r2}	Self-inductance component of rotor as specified
L_s	Self-inductance of stator winding
L_r	Self inductance of rotor winding
F_{max}	Maximum value of mmf distribution in rotor
A, B, C, A', B', C'	Beginnings and ends of phase *A, B, C* phase belts
K	Constant harmonic coefficients
r_S	Resistance of stator winding
r_r	Resistance of rotor winding
a	Angle of separation from *d*-axis
β	General coil-span as indicated

Chapter 5

θ	Angle of separation between axes of coils
ω_r	Angular velocity of rotor in radians/sec and for a 2-pole arrangement
L_s	Self-inductance of stator
L_r	Self-inductance of rotor
M_{sr}	Mutual inductance of stator and rotor
λ_s	Air-gap permeance of stator as a function of rotor coil
λ_r	Air-gap permeance of rotor
λ_{sr}	Permeance between stator and rotor
ψ	Phase spread
¥	Space angle
σ	Angular separation between stator axis and rotor coil axis
M	Coil number per phase

m	Number of turns per coil
v (*new*)	Angle of brush separation
E_{brush}	r.m.s emf across brushes
$_{max}E_{brush}$	Maximum amplitude of emf between brushes
f_r	Rotor frequency corresponding to motion of rotor with respect to stator
f_s	Frequency in the stator
T	Total number of turns
a	Pairs of parallel paths
ϕ_m	Maximm amplitude of flux
B_m	Maximum amplitude of flux-density
P	Number of pole-pairs in machine
C	Number of coils
E_S	Aggregate emf

Chapter 6

R	Resistance
L	Coefficient of self-inductance
M	Coefficient of mutual inductance
X	Reactance due to coefficient of self-inductance
X_m	Reactance due to coefficient of mutual inductance
F_a, F_b, F_c	Crest values of *a, b, c* phasor mmfs of 3-phase system which are orthogonal equivalent
d, q	*d*- and *q*-axis
F_d, F_q	Crest values of mmf along *d*- and *q*-axis of an equivalent 2-phase system
$[I']$	New current column matrix
$[I]$	Original column matrix of old currents
$[V']$	New voltage column matrix
$[V]$	Original column matrix of old voltages
K	Arbitrary constant introduced to maintain orthogonality
v_+	$K(v_a + jv)$, the positive sequence component of voltage
v_-	$K(v_a - jv)$, the negative sequence component of voltage
L	Coefficient of self-inductance of individual phase winding
l	Leakage inductance of the individual phases in winding used in estimating X_0

Chapter 7

R_s	Resistance of the stator winding
L_s	Self inductance of the stator winding
M_{sr}	Mutual inductance between the stator and the rotor winding and vice versa
P	Heaviside's differential operator
P	Number of pole-pairs when especially specified
I_a	Armature current
R_f	Resistance of field winding
I_f	Current in the field

M_a	Mutual inductance between the armature and the field
θ_r	Instantaneous time angle between the armature and field axes
T_e	Torque developed on the armature
A_{ds}, B_{dr}, C_{qr}, D_{ds}	Coil nomenclature of primitive machine
M_{dsqr}, M_{srqs}	Mutual inductance coefficients between stator *d*-axis and rotor *q*-axis coils and vice versa
M_{dsdr}, M_{qsqr}	Mutual inductance coefficients between stator and rotor coils along direct and quadrature axis and vice versa
L_{ds}, L_{qs}	Self inductance coefficients between stator coils along *d*- and *q*-axis
L_{dr}, L_{qr}	Self inductance coefficients between rotor pseudo-stationary coils along *d*- and *q*-axis
R_{rh}	Resistance of external field rheostat
R_a	Resistance of armature winding
S_1	Terms related to separately-excited or shunt field winding
S_2	Terms related to series field winding
V_S	Voltage appearing across the series field winding
s	Per-unit slip of induction motor

Chapter 8

l	Leakage inductance per phase of winding
$\bar{L}_1$	Self-inductance per phase*
$\bar{M}_1$	Mutual inductance between phases* winding inductance with a 1-phase
$\bar{L}_L$	Voltage applied to the phase* Self-inductance for 1-phase excitation
$\bar{L}_A$, $\bar{L}_B$, $\bar{L}_C$	For a 3-phase system*
L_1	Self-inductance of stator winding
$\bar{M}_1$	Stator-to-stator mutual inductance between phases
L_{01}	Stator leakage inductance of symmetrical 3-phase stator winding
$\bar{L}_2$	Self-inductance of symmetrical 3-phase rotor winding per phase
L_{02}	Rotor leakage inductance of symmetrical 3-phase rotor winding
L_2	Self-inductance of rotor winding
$\bar{M}_2$	Rotor-to-rotor mutual inductance between phases
$\bar{M}_3$	Mutual-inductance component between phases for third harmonic
Z_{11}, Z_{12}, Z_{21}, Z_{22}	Matrix elements of old compound matrix as indicated in Equation (8.2)
Z_{11}', Z_{12}', Z_{21}',Z_{22}'	Matrix elements of the new compound matrix as indicated in Equations (8.6) and (8.9)
θ_r	Angular displacement between rotor 'a' and stator 'A' phase for mutual inductsnce estimation
$\theta_2 = \theta_r + 2\pi/3$	Angular displacement between rotor 'b'/phase and stator 'A' phase for mutual inductance estimation
$\theta_3 = \theta_r + 4\pi/3$	Angular displacement between rotor 'c' phase and stator 'A' phase for mutual inductance estimation

M_{aA}, M_{bA}, M_{cA}, etc.	Mutual inductance between stator and rotor decided as per suffices
ψ_{la}	Leakage flux linkage with of rotor phase 'a'
ψ_a	Flux linkage of phase 'a' of rotor
ψ_{ga}	Gap flux linkage of phase 'a'
λ_l	Permeance seen by leakage flux
λ_d	Permeance seen by *d*-axis of phase 'a'
λ_q	Permeance seen by *q*-axis of phase 'a'
ψ_{ba}	Flux linkage of phase '*b*' by total flux of rotor
ψ_{la}	Flux linkage of leakage flux of phase 'a'
L_{aF}	Mutual inductance between the field and the rotor phase 'a'
L_{aD}	Mutual inductance between the *D*-axis damper and the rotor 'a' phase
L_{aQ}	Mutual inductance between the *Q*-axis damper and rotor 'a' phase
M_{rD}	Maximum mutual inductance between *D*-axis damper and 'a' phase
M_{rQ}	*Q*-axis damper and 'a' phase
M_{aF}	Maximum mutual inductance between field and rotor phase 'a'
D, d, Q, q	*d*-axis/*q*-axis for the equivalent 2-phase machine
δ	Load angle
δ_T	Torque angle
K	Effective turns ratio of secondary to primary
X_S	Synchronous reactance
Ψ	Power factor angle on the operation chart
I_{aD}, I_{aQ}	Direct- and quadrature-axis components of rotor current
X_{aD}, X_{aQ}	*D*- and *Q*-axis components magnetizing reactance
X_l	Leakage reactance component
X_d, X_q	*D*- and *Q*-axis synchronous
X_{ad}, X_{aq}	*D*- and *Q*-axis synchronous reactance
X_l	Leakage reactance component
X_d, X_q	*d*- and *q*-axis synchronous reactance components
F	Armature MMF of the machine of the *q*-axix
F_d, F_q	*d*-axis and *q*-iaxis components of armature mmf
δ	load angle
ψ_{phase}	Flux-linkage matrix of phase
L_{ph}	Inductance matrix of phase
V_{phase}, I_{phase}	Voltage/current matrix of phase windings
λ_θ	Permeance of air-gap at angle θ from a general rotor position or *d*-axis of motor

Chapter 9

$\dot{\omega}_m$	Angular velocity in rads/sec
P	Number of poles
Z	Total number of armature conductors
a	Number of parallel paths in armature
K_a	Generator constant
Φ_f	Flux per pole

K_f	Field constant
τ_f	Field time-constant
$G(S)$	Gain factor of the open-loop control of armature
V_f	Voltage applied to field current
M	Steady-state gain of separately-excited dc motor
G_{ss}	Steady-state gain of separately-excited generator
H	Transfer function of feedback circuit
R_l, L_l	Resistance/inductance of load
R_a, L_a	Resistance/inductance of armature
R, L	Total resistance/inductance of armature and load
$T(s)$	Transfer function in *s*-plane
A	Amplifier gain
R_b	Effective brush resistance
R_a	Armature resistance
V_t	Applied voltage to the motor armature
I_a	Motor armature current
Φ_f	Flux per pole interacting with armature
R_a	Resistance of armature circuit
$K_0 = \dfrac{PZ}{2.a}$	Constant of the motor
J	Moment of inertia of the motor rotor
F	Damping constant of the motor rotor
D	Spring constant of the motor rotore
$K = K_a K_f$	Mtor constants
K_t	Tacho-generator constant in volts/rad/sec
V_R	Reference voltage
P_b	Balance of power ($P_m - P_e$) on the shaft
M	Mass of the pendulum bob
L	Length of suspension
θ	Displacement of the bob from its natural equilibrium position at a general time, *t*
$\dot{\theta}$	Angular velocity at a general time instant, *t*
θ_0, $\dot{\theta}_0$	Respective initial values of and at time $t = 0$
$\ddot{\theta}$	Acceleration due to gravity
$\Omega = \sqrt{\dfrac{g}{L}}$	Angular velocity at a given time instant, *t*
v	Linear velocity of the pendulum bob
A	Amplitude of the swing
ϕ	Phase angle

Chapter 10

x_s	Leakage reactance per phase of the stator winding
R_r'	Resistance per phase of the rotor referred to stator

X_r'	Leakage reactance per phase of the rotor referred to the stator
i_{s1}	Positive sequence current in stator referred to Phase A
i'_{r1}	Positive sequence current of rotor referred to Phase A
e_{s1}	Positive sequence line-to-neutral voltage of stator referred to phase A
s	Per-unit slip
s_2	Per unit negative sequence slip
P_{shaft}	Total shaft power output
T_{shaft}	Net torque at the shaft
E_{s1}	Positive sequence voltage applied to stator
E_{g1}	Positive sequence voltage induced in stator
E_{r1}	Rotor terminal voltage due to positive sequence

Chapter 11

$I_0(t)$	Zero-sequence current
$I_A(t)$, $i_B(t)$, $i_C(t)$	Current input to the stator 3-phase winding
N_{SE}	Effective number of turns in stator per phase
N_S	Actual number of turns in stator phase winding
K_{WS}	Winding factor of the stator winding
$F_S(\theta, t)$	Resultant mmf distribution
θ	Angle around the periphery taking phase A axis
$I_S(t)$	Space-phasor of stator current in the complex plane with reference to statury reference frame of the stator
1, a, $a2$	Special operation similar to time operation in 3-phase system used in symmetrical components
$\|i_S\|$	Modulus of stator space phasor
a_S	Phase angle with reference to real axis of the stationary reference frame fixed to the stator
$i_S(t)$	General stator current
I_S	Maxixum value of stator current which is a constant
$I_{SD}(t)$, $i_{SQ}(t)$	General d-axis and q-axis stator current
K	Arbitrary constant
$RE(I_S)$	Real axis component of stator current per phase
$IM(I_S)$	Imaginary axis component of stator current per phase
N_{rE}	Effective turns per phase in rotor winding
N_r	Actual turns per phase in rotor
K_{wr}	Winding factor of rotor
P	Number of pole-pairs
$[C_{12}]$	Commutator transformation matrix
F	Magneto-motive force, mmf
I_{ra}, i_{rb}, i_{rc}	Instantaneous values of rotor current in rotor phases *ra*, *rb* and *rc* respectively
I_{rd}, i_{rq}	Instantaneous values of d-axis and q-axis rotor currents expressed in the stator reference frame

i_{ra}, i_{ra}	Instantaneous values of *d*-axis and *q*-axis rotor current components expressed in rotor reference frame
i_{rx}, i_{ry}	Instantaneous values of the *d*-axis and *q*-axis rotor current expressed in the general special reference frame
i_{SA}, i_{SB}, i_{SC}	Instantaneous values of stator current in stator phase SA, SB, SC respectively
i_{SD}, i_{SQ}	Instantaneous values of *d*-axis and *q*-axis stator current components with respect to stationary reference frame
i_{Sd}, i_{Sq}	Instantaneous values of *d*-axis and *q*-axis stator current expressed with respect to rotor reference frame
i_{Sx}, i_{Sy}	Instantaneous *d*-axis and *q*-axis stator current components expressed to a general or special reference frame of *J* flux and torque producing stator current comp.
I_{S0}	Instantaneous value of zero-sequence stator current comp *J* polar mass moment of inertia in kg.m^2
L	Dynamic inductance
L_m	Magnetizing inductance
L_S, L_{Sl}	Self and leakage inductance of stator
L_e, L_{rl}	Self and leakage inductance of rotor
p	Heaviside's differential operator (*d/dt*)
P_S, P_r	Instantaneous power of stator and rotor
P	number of pole-pairs
R_S, R_r	Resistance of stator and rotor phase windings
t	time
T_e	Instantaneous electro-magnetic torque
T_m	Mechanical torque on motor shaft in N-m equivalent
V_{dx}, V_{dy}	Instantaneous *d*-axis and *q*-axis decoupling voltage comps. Expressed in special reference frame
V_S	Space phasor of stator voltage expressed in terms of stator reference frame
V_{Sg}	Space phasor of stator voltage with reference to general reference frame
V_{Sm}	Space phasor of stator voltages expressed reference to the Magnetizing flux-oriented reference frame
$V_{S\psi S}$	Space phasor of stator voltage with reference to the stator flux-oriented reference frame
$V_{S\psi r}$	Space phasor of the stator voltages expressed as rotor flux-oriented reference frame
V_{SA}, V_{SB}, V_{SC}	Instantaneous stator voltages in phase *SA*, *SB* and *SC*
V_{SD}, V_{SQ}	Instantaneous *d*- and *q*-axis stator expressed with reference to stationary reference frame
V_{Sx}, V_{Sy}	Instantaneous *d*- and *q*-axis stator voltage components Indicated in general special reference frame
V_r	Space phasor of rotor voltage with reference to rotor reference frame
V_r'	Space phasor of rotor voltage expresses as reference to stator reference frame
V_{rg}	Space phasors of the rotor voltages expressed in general reference frame

$V_{r\phi S}$	Space phasor of the rotor voltage expressed with reference to stator flux-oriented reference frame
V_{rd}, V_{rq}	Instantaneous voltages of the *d*-axis and *q*-axis rotor voltage expressed as function of stator reference frame
V_{rx}, V_{ry}	Instantaneous *d*- and *q*-axis rotor voltages expressed in the general or special reference frame
V_{ra}, V_{r}	Instantaneous *d*- and *q*-axis rotor voltages with respect to rotor reference frame
V_{r0}	Instantaneous zero-sequence rotor voltage
W_{mech}	Instantaneous mechanical output energy
δ	Load angle
θ_r	Rotor angle
P_m	Phase angle of the magnetizing flux-linkage space phasor with reference to *d*-axis stator reference frame
ρ_r	Phase angle of the rotor flux-linkage space phasor with *d*-axis stator reference frame
ρ_S	Phase angle of the stator flux-linkage space phasor with *d*-axis stator reference frame
σ	Resultant leakage constant
σ_r	Rotor leakage constant
ψ_f	Excitation flux linkage
ψ_m'	Space phasor of the magnetizing flux-linkage expressed with reference to stator reference frame
$I_{mS}(i_{mS})$	Space phasor of the stator magnetizing currents expressed in the magnetizing flux-oriented reference frame
$I_{mr}(i_{mr})$	Space phasor of the rotor magnetizing currents expressed in the magnetizing flux-oriented reference frame
$I_r(i_r)$	Space phasor of the rotor currents expressed in the rotor reference frame
$i_r'(i_r')$	Space phasor of the rotor currents expressed in the stator reference frame
$I_{rg}(i_{rg})$	Space phasor of the rotor currents expressed in the general reference frame
$i_{rm}(i_{rm})$	Space phasor of the rotor currents expressed in the magnetizing flux-oriented reference frame
$i_{r\psi S}$	Space phasor of the rotor currents expressed in the stator flux oriented reference frame
$i_S(i_S)$	Space phasor of the stator currents expressed in the stator reference frame
$i_S'(i_S')$	Space phasor of the stator currents expressed in the rotor reference frame
i_{sg}	Space phasor of the stator currents expressed in the general reference frame
i_{Sm}	Space phasor of the stator currents expressed in the magnetizing flux-oriented reference frame
$i_{S\psi r}$	Space phasor of the stator currents expressed in the rotor flux-oriented reference frame
$i_{S\psi S}$	Space phasor of the stator currents expressed in the stator flux-oriented reference frame
i	Spatial operation similar to time operation in 3-phase system

CHAPTER 1

Introduction to Electro-mechanical Energy Conversion and Basic Circuit Elements

1.1 ELECTRO-MECHANICAL ENERGY CONVERSION

Electro-mechanical energy conversion, the basic process that governs the functioning of electric motors and generators, uses *electro-magnetic field* as a coupling interface to effect interchange of energy between its electrical and mechanical energy handling components. Figure 1.1 shows a schematic representation of energy flow in a typical electro-mechanical energy converter.

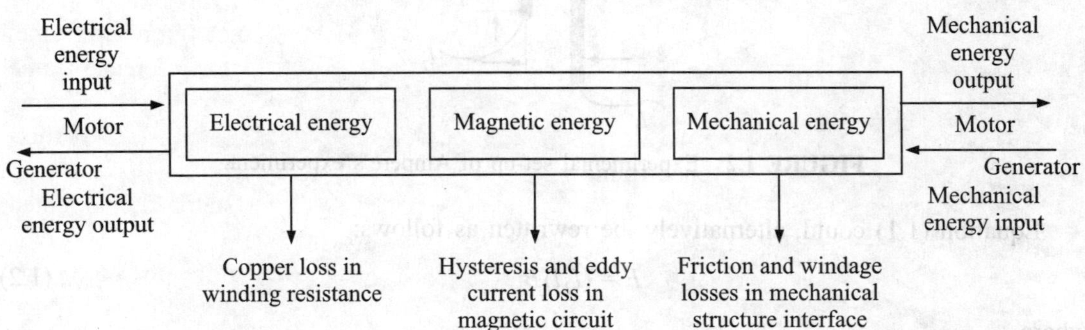

FIGURE 1.1 Schematic of energy-flow and distribution in electro-mechanical energy conversion arrangement.

An understanding of the electro-magnetic process, leading to energy conversion, demands a clear appreciation of *Ampere's Law*. This law is the starting point in comprehension of the nature of forces associated with interaction of magnetic fields between mutually coupled current-carrying conductors.

1.1.1 Ampere's Law

According to the experimental findings of Ampere, force between two coplanar current-carrying conductors is in direct proportion of the magnitudes of involved currents and the magnetic property (called the *magnetic permeability*) of the medium in which the current-carrying conductors are located. And the above force is in inverse proportion to the distance separating the concerned current elements. It envisages that the current-carrying conductor creates around itself a special area of influence called *electro-magnetic field*, in which invisible lines of influence are developed in the form of forces exerted on a current element, called the *magnetic flux-density* and is denoted by B. It has the physical dimension of force per current length.

The following quantitative representation of Ampere's Law of electro-magnetic induction leads to appreciation of its physical meaning in the context of magnetic flux produced by current-carrying conductors. According to Ampere's Law:

$$F = \frac{\mu I_1}{2\pi r}(I_2 l) \text{ newton} \quad \text{...(1.1)}$$

where, F denotes the force in newton experienced on an elemental current-carrying conductor of length l, and carrying a current of I_2, while it is placed under the influence of another current-carrying conductor with a current of I_1 through it. The medium of influence has an absolute magnetic permeability denoted by μ, and the two conductors are taken to be coplanar, separated by a distance r, as shown in Figure 1.2.

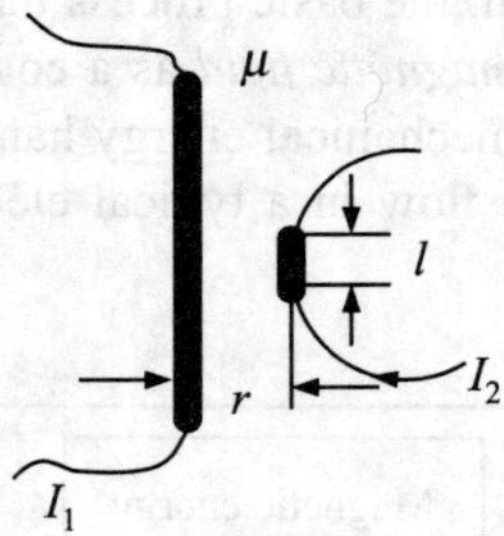

FIGURE 1.2 Experimental set-up of Ampere's experiment.

Equation (1.1) could, alternatively, be rewritten as follows:

$$F = (I_2 l) B \quad \text{...(1.2)}$$

where,

$$B = \frac{\mu I_1}{2\pi r} \quad \text{or} \quad B = \frac{F}{I_2 l}$$

which has a physical dimension of $\left[\dfrac{\text{Newton}}{\text{Ampere-metre}}\right]$.

And since

$$\text{Force} = \frac{\text{Energy}}{\text{Length}}$$

dimensionally speaking,

$$\text{Newton} = \frac{\text{Watt-second}}{(\text{Metre})}$$

which leaves,

$$B = \frac{\text{Volt-second}}{(\text{Metre})^2}$$

Since, dimensionally speaking,

$$\text{Voltage} = \frac{\text{Magnetic flux}}{\text{Time}}$$

$$\therefore \quad B = \frac{\text{Magnetic flux}}{(\text{Metre})^2}$$

B, therefore, signifies the flux-density in the medium of influence of the current-carrying conductors. The physical dimensions of μ, the magnetic permeability of the medium, can be similarly worked out. Thus,

$$\mu = \frac{\text{Newton}}{(\text{Ampere})^2} = \frac{\text{Volt-ampere second}}{(\text{Ampere})^2\,(\text{Metre})}$$

$$= \frac{\text{Volt-second}}{\text{Ampere-metre}}$$

But, $\dfrac{\text{Volt-second}}{\text{Ampere}}$ dimensionally represents inductance in henry, presented by the electrical circuit (see Section 1.2.2). Hence, μ has the unit of henry/metre. The value of the permeability, μ for free-space or vacuum (as it is sometimes indicated), is denoted by μ_0, and its value in MKS system is:

$$\mu_0 = 4\pi \times 10^{-7}$$

To account for any chance in respect of medium of influence to any medium other than free-space, a concept of relative permeability, μ_r is found useful, so that the effective permeability, μ, for a medium assumes the form:

$$\mu = \mu_r \mu_0$$

where, μ_r is a dimensionless entity.

It is now possible for us to develop a simplistic picture of the actual scenario obtaining two sets of current-carrying conductors when placed under the influence of each other. It is possible to visualize that whenever current passes through an electrical circuit, magnetic flux will produced as a consequence. The magnitude and nature of distribution of flux depend on the magnitude and distribution of the current and the nature of the surrounding medium. It is convenient to theorize that the current produces a *magnetic force*, H, which will has a magnitude and direction at every point in the medium where the current-carrying conductor is placed. The presence of magnetic force, H, at a given point is manifested by the development of a corresponding value of magnetic flux density B, whose magnitude and direction of influence are dependent on those of H, and the magnetic permeability of the medium. H is popularly identified as *magneto-motive force per unit length of the flux-path* or *magnetic field intensity* at a point. In quantitative terms:

$$B = \mu H$$

Distribution of magnetizing force and the resulting flux density in a medium are always in closed loops. There is always a value of H at every point along the line of force, suitably directed along the line. If we integrate H with respect to the length of the path over the entire length of the closed loop of the magnetic circuit, we obtain the total magneto-motive force (mmf) required to maintain the resulting flux-density in the magnetic circuit. The mmf, which is distributed along the entire loop, may not be uniformly distributed along the entire loop due to different media with multiple values of μ usually constituting the magnetic circuit of electro-mechanical energy converters. The magnitude of the mmf is actually equal to the total value of the electric current encircled by the loop. This is expressed as follows:

$$\oint H.dl = I \quad ...(1.3)$$

1.2 BASIC CIRCUIT ELEMENTS—RESISTANCE, INDUCTANCE AND CAPACITANCE

We wish to review the nature and role assumed by the basic circuit elements—the resistance, the inductance and the capacitance—associated with electric circuits.

1.2.1 Resistance Parameter

Resistance is one of the basic parameters of electric circuit, and every circuit element presents some resistance. But in case of inductors and capacitors and also in case of connecting wires among different circuit elements, we wish to underplay the role of resistance and usually ignore it or treat it as a lumped external component. In a general way, resistance can be looked upon as the property or manifestation of a circuit element opposing the flow of current through it. In the process, it converts electrical energy into heat energy. According to Ohm's law, it is introduced as a proportionality constant r, independent of the magnitude and direction of the current flow in the circuit element and the potential difference across the terminals of the element.

Originally, the free electrons associated with the atoms of conducting materials undergo random motion due to thermal energy in them, but it does not contribute a regular net current flow through them. If a voltage source is connected to the terminals of the system, it produces an electric field in the body of the conducting system. This field imparts a directed velocity component to the above random motion. This velocity is primarily directed in opposition to the electric field, and the randomly moving electrons meet with frequent collisions with the atoms in the lattice structure of the conducting materials and produce irreversible heat loss and consequent irretrievable loss of energy. This loss is termed as *Copper loss*.

Circuit Viewpoint of Resistance Element

A circuit element is usually described from circuit viewpoint as its relationship between its associated current and the potential difference. From Ohm's law, we have

$$v = r\,i, \quad \text{or} \quad r = \frac{v}{i} \qquad \text{...(1.4)}$$

Thus, Ohm's law indicates that the potential difference between the two points: A and B—in Figure 1.3 is connected to the ensuring current by the proportionality constant r, which, as has already been pointed out earlier, is independent of the magnitude and direction of flow of the current. This is a depiction of resistance element. It clearly indicates that the resistance is the slope of a straightline, when plotted between the current flowing through the element at a given temperature and the potential difference developed across the element: AB—as indicated in Figure 1.4.

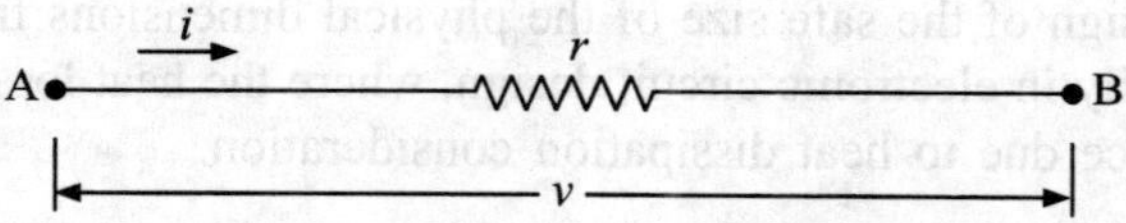

FIGURE 1.3 Current flow through resistive element.

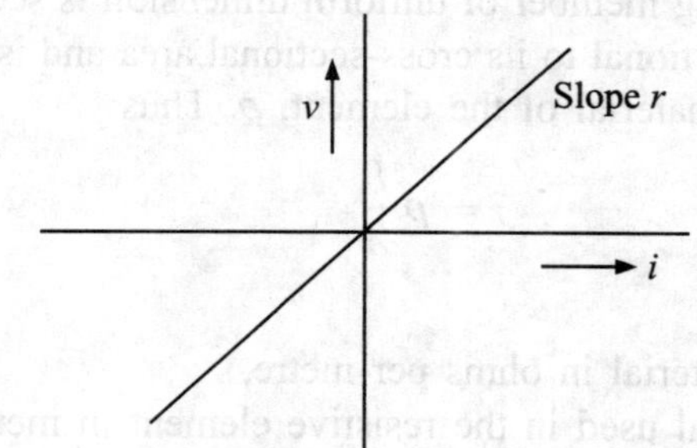

FIGURE 1.4 *v–i* characteristic of a resistive element.

Energy Viewpoint of Resistant Element

Resistance parameter is inexorably linked with its property of converting electrical energy into heat energy. The characteristic equation for the power (P) associated with the current flowing in a resistor is given by Joule's law, which states that the number of electrons colliding with atoms in a resistive environment to produce heat is proportional to the square of the current:

$$P = vi = (ir)\,i = i^2 r \qquad \text{...(1.5)}$$

The associated energy (W) converted into heat in the time interval ($dt = t_2 - t_1$), can be estimated as follows:

$$W = Pdt = i^2 r\,(t_2 - t_1) \qquad \text{...(1.6)}$$

The current-carrying conductors and resistance element incur losses proportional to i^2r, which manifests in the form of heat. The heat so produced is, at first, absorbed in the body of the system. Once the temperature rises to the ambient value, the unit starts dissipating heat into the surrounding medium based on the difference of temperature. A stage is finally reached when the temperature stops rising and, at this stage, the rate of heat generation equals that of heat dissipation. Experiment shows that in almost all types of metallic conductors, the resistance varies with temperature. The mathematical relationship denoting the increase of resistance as a function of temperature can be shown as follows:

$$r_2 = r_1[1 + \sigma(T_2 - T_1)] \quad ...(1.7)$$

where, r_1 is the resistance of the system at a absolute temperature T_1, and r_2 is the resistance of the same body at absolute temperature T_2, σ being the resistance–temperature coefficient, a property of the material of the resistor over a limited temperature range.

Since the physical shape and size of the resistor decide the capacity of the element to dissipate the heat generated, it becomes important to indicate the 'power rating' of a resistive element in its specification, in addition to its current rating. Thus, although the power rating of a resistor is of not much significance in the choice of the circuit element, the information is significant in the design of the safe size of the physical dimensions from safety and security considerations, especially in electronic circuit design, where the heat loss in the circuit elements assume great importance due to heat dissipation consideration.

Physical Significance of the Resistance Element

The resistance of a conducting member of uniform dimension is seen to be in direct proportion to its length, inversely proportional to its cross-sectional area and is determined by the physical conduction property of the material of the element, ρ. Thus:

$$r = \rho \frac{l}{A} \quad ...(1.8)$$

where

ρ = resistivity of the material in ohms per metre,
l = length of the material used in the resistive element in metres and
A = cross-sectional area of the material used in the resistive element in (metre)^2.

1.2.2 Inductance Parameter

A coil formed by a conducting material with an enclosure having enclosed space about which the coil circumscribes, when carrying a current, is known to produce a magnetic flux along its longitudinal axis. The concept of inductance associated with such a coil was first introduced by Michael Faraday, based on his classical experiments in 1831, which ultimately led to the discovery of *electro-dynamics* and *electro-mechanical energy conversion*. The property of inductance can be characterized as that of a circuit element which enables it to store energy in its magnetic flux field.

The speciality of inductance is that it is effective in a circuit only when the *current flowing in the coil is changing with time*. The current–voltage relationship can be expressed in terms of the inductance parameter of the circuit by the following equation:

$$v = v_L + v_r$$

where

v = net voltage across the coil carrying current.

v_L = drop of voltage due to inductive reactance of the coil and

v_r = drop of voltage due to resistance component of the coil.

Both resistance and inductance parameters of the circuit parameters are *distributed parameters*, and the coil formed to act as an inductance element usually comprises low resistance. Besides, a circuit element is usually denoted by its equivalent representation in which resistance and inductance components are treated as lumped parameters. Further, it is customary to neglect the resistance of the coil to begin with, so that as a first degree of approximation:

$$v = v_L \text{ for an inductance coil}$$

Thus,
$$v = v_L = L\left(\frac{di}{dt}\right) \qquad \text{...(1.9)}$$

where

v_L is the voltage appearing across the coil at an instant,

L is its inductance parameter, and

$\left(\frac{di}{dt}\right)$ is the rate of chance of current flowing through it at the particular instant in question.

Examination of Figure 1.5 indicates that if the current entering terminal A is positive and increasing [that is, $\left(\frac{di}{dt}\right)$ is positive], then the voltage appearing across terminal A is positive as per the above equation, or the voltage developed across the terminals A – B is directed by the head of the arrow as indicated. In other words, v_L acts *in opposition* to an applied voltage v, and so $(v - v_L)$ leads to flow of current i.

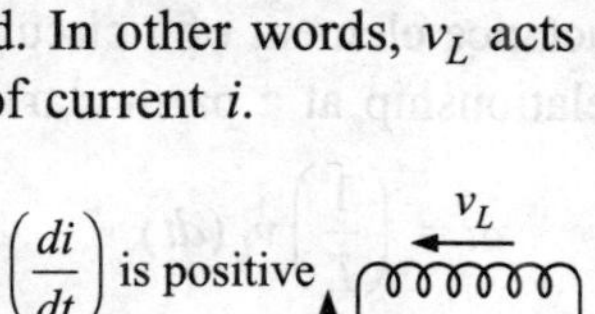

FIGURE 1.5 Schematic of an inductance parameter with negligible resistance, indicating current flow and induced voltage directions.

In the idealistic system, therefore, with the coil resistance neglected, we have

$$L = \frac{v_L}{\left(\frac{di}{dt}\right)} \qquad \text{...(1.10)}$$

The potential difference at a given time instant across the inductive element and the corresponding derivative of the current–time function define the inductance parameter, measured by units of *henry*. For a linear inductance element, *in which the inductance is not dependent on the magnitude of current in the circuit*, it indicates that the magnetic flux produced along the axis of the coil is strictly in proportion to the current carried by the coil—a situation available for low values of magnetic flux with no magnetic saturation and where the flux develops along a magnetic circuit containing air-gaps. For such a case, the inductance of a linear inductance element can be depicted as the slope of a straightline passing through the origin when plotted with v_L and $\frac{di}{dt}$ along the axes of the plot as shown in Figure 1.6.

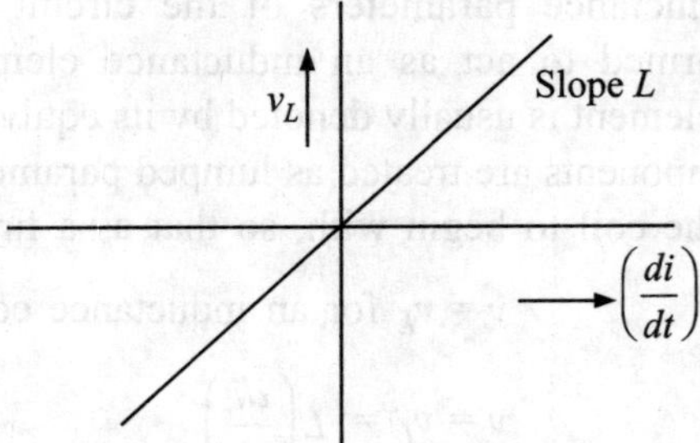

FIGURE 1.6 v_L–$\left(\frac{di}{dt}\right)$ characteristic of an inductance element without magnetic saturation.

In this manner, a similarity exists between the resistance parameter, *r*, as depicted in Figure 1.4, and inductance parameter.

When a magnetic core is associated with an inductance element, an inductance parameter may not remain constant due to magnetic saturation of the magnetic circuit, if we do not confine the operation to low values of flux. In such cases, the inductance parameter becomes 'non-linear' and it becomes dependent on the circuit current. But it is customary to assume the inductance as a constant parameter (over a particular range of current magnitudes) by manipulating the saturation level in practice. We may reconsider Eq. (1.9) to have a clear understanding of the nature and phenomenon of the inductance element of a circuit. Let us rewrite the same equation in specifying the voltage/current relationship at a particular instant of the operation as follows

$$di = \left(\frac{1}{L}\right) v_L (dt) \qquad \text{...(1.11)}$$

Written in integral form:

$$\int_{i_1}^{i_2} di = \left(\frac{1}{L}\right) \int_0^t v_L \, dt$$

or

$$i_2 = \left(\frac{1}{L}\right) \int_0^t v_L \, dt + i_1 \qquad \text{...(1.12)}$$

where

i_1 represents the initial value of current at $t = 0$, and i_2 is the value of the current at a time t, during which period the circuit current was changing.

Equation (1.12) reveals that the current in the inductance element is a function of the voltage developed across the terminals and the value of the initial current at the start of the consideration. We can, therefore, conclude by visual scrutiny of Eqs. (1.11) and (1.12) that:

(i) The current through the inductance element cannot change abruptly in zero time interval, as a finite change in current in zero time shall require that an infinite value of voltage shall be required to be available across the terminals.
(ii) In 'zero-time', the contribution to the current through the inductance element is zero, so that the value of current immediately prior to or succeeding the appearance of terminal voltage across the impedance element remains the same.

Hence, the property of inductance parameter can be likened to *inertial effect of electrical circuits*.

Circuit Viewpoint of Inductance Element

We shall, now include the resistance parameter of the coil acting with the inductance element. The arrangement, with an air-gap along its magnetic circuit, in shown in Figure 1.7.

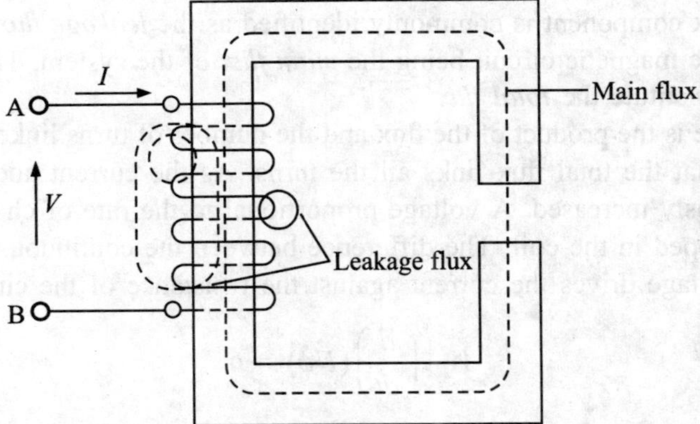

FIGURE 1.7 Flux-distribution diagram in the magnetic core of an inductance coil.

In electrical engineering, it is a common practice to depict an actual system by its equivalent circuit consisting of its circuit model, which behaves like the actual system under identical terminal conditions. Different models of a device may be required for different conditions of representation, but a model is valid for a range of terminal conditions. The difference between the actual system and its model is imperceptible and presents identical behaviour for all practical purposes. The lumped parameter representation shown in Figure 1.8 is one such example for calculation purposes.

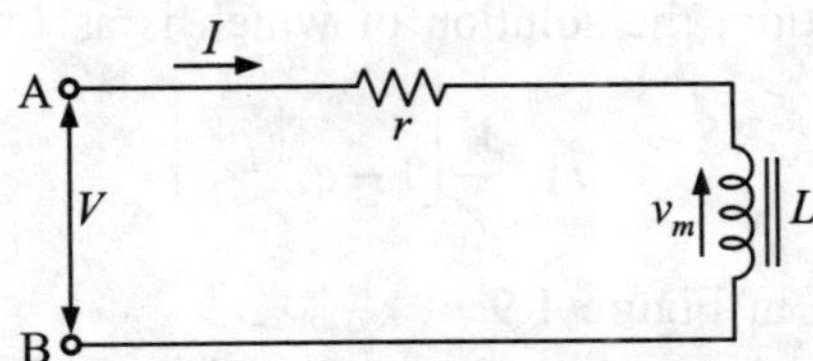

FIGURE 1.8 Equivalent circuit representation of an inductance coil on lumped-parameter basis.

In actual inductor, both resistance and inductance are mixed together along its length.

Direct voltage applied across the inductor coil: We are ready to examine what shall happen if we apply direct voltage across the terminal A–B of the inductance coil by connecting a battery in Figure 1.7. The circuit model in Figure 1.8 shall now carry a current i which will rise from $i = 0$ at $t = 0$, to $i = I$ at a time t, the entire duration of the period t experiencing a changing current. The inductance coil shall present a constant inductance L for the above period because of the air-gap in the magnetic circuit. If the coil has an effective number of turns, N, the current shall produce a magneto-motive force Ni and, therefore, a flux ϕ along its axis. At any time, the flux is given by:

$$\phi = \frac{F}{S}$$

where, $F = Ni$, is the mmf and S the reluctance of the magnetic circuit. The mmf will distribute the resulting flux along all possible paths. Although most of the flux will pass along the core, i.e., along the low reluctance magnetic circuit, a small portion may take air paths as shown in Figure 1.7. This flux component is commonly identified as the *leakage flux*, the remaining flux established along the magnetic circuit being the *main flux* of the system. The sum of these two flux components constitute the *total flux*.

The flux-linkage is the product of the flux and the number of turns linked by it, and is given by ϕN, assuming that the total flux links all the turns. As the current and flux rise, the flux-linkage is continuously increased. A voltage proportional to the rate of change of current-flow in the coil is developed in the coil. The difference between the continuous battery voltage (V) and the induced voltage drives the current against the resistance of the current.

$$V - \left(\frac{d}{dt}\right)(N\phi) = ir$$

or

$$V - N\left(\frac{d\phi}{dt}\right) = ir$$

As discussed in section 1.2.2, the inductance L of the coil can be written as:

$$L = \frac{N\phi}{i}$$

So that we can write the voltage–current equation of the inductance L as follows:

$$V - L\frac{di}{dt} = ir$$

It is a linear differential equation, the solution of which is as follows:

$$i = \frac{V}{r}\left[1 - e^{-\left(\frac{r}{L}\right)t}\right] \qquad \text{...(1.13)}$$

The plot of which is depicted in Figure 1.9.

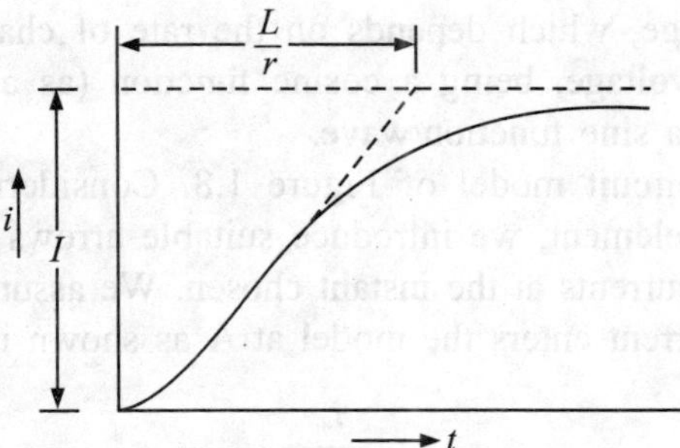

FIGURE 1.9 Exponential rise of current through the inductance coil, when a constant battery voltage is applied across it.

It is possible to further refine the lumped parameter model of Figure 1.8 by bifurcating the total impedance element L into two components, L_l and L_m as inductance components of 'leakage reactance', and 'mutual reactance', and that of main-flux reactance, so that:

$$L = L_l + L_m$$

This modification in the model is depicted in Figure 1.10.

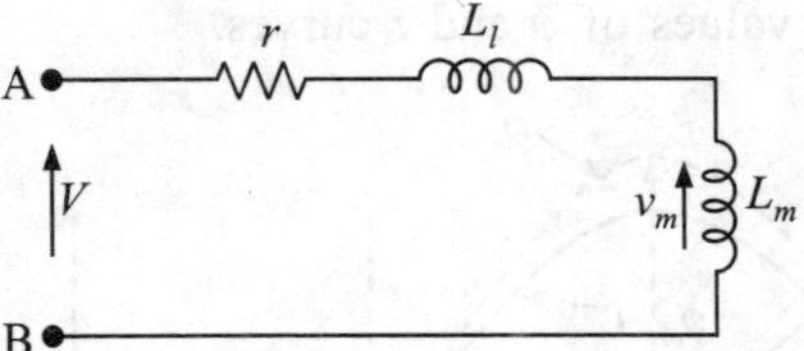

FIGURE 1.10 Modified circuit model, with leakage and main-flux reactances divided.

Both the models depicted in Figure 1.8 and Figure 1.10 may be used for estimating the current–voltage relationships with the provisio that such changes do not occur rapidly. The parameters are constant only if the current (and hence, the flux linking the inductance coil) is not large, as operation under large currents shall cause saturation of its magnetic circuit (causing L_m to vary with current magnitude) but L_l will remain virtually constant, the flux-path being dominated by air, v_m is the voltage drop across the mutual reactance component.

Operation of reactance parameter when connected to alternating voltage source: If, now, we consider that an alternating voltage source:

$$v = V_{\max} \cos \omega t \qquad ...(1.14)$$

is connected across the terminals A – B of the depiction in Figure 1.8, an alternating current, i shall now flow through the coil set-up. The value of the current will not be much affected by ignoring the coil resistance after the coil functions under study-state conditions, following the lapse of the initial transient period. But, we expect that the steady-state current (being a function of t now) shall be entirely dominated by the inductance parameter of the circuit. We can, therefore, conveniently neglect the effect of resistance being negligibly small. The value of the current shall, now, be determined by the fact that the induced voltage produced by the continuously changing flux-linkage must, for all practical purposes, be equal to the applied voltage. The induced voltage, which depends on the rate of change of current, shall act in opposition to the applied voltage, being a cosine function (as assumed). The current must, therefore, be the nature of a sine function/wave.

Let us first take the circuit model of Figure 1.8. Considering modification introduced by ignoring the resistance element, we introduce suitable arrows along the assumed positive directions of voltages and currents at the instant chosen. We assume the terminal A to be at a higher potential, so that current enters the model at A as shown in Figure 1.11(a):

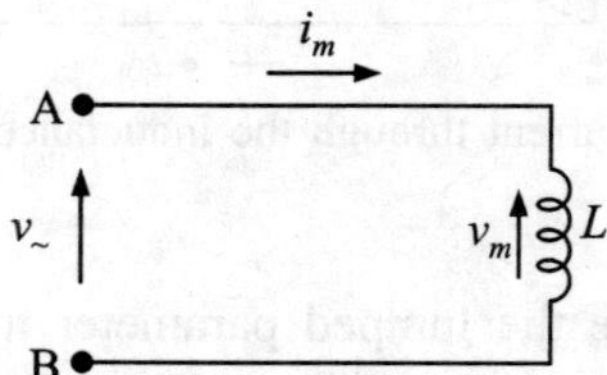

FIGURE 1.11(a) Depiction of voltage developed, v_m, when an a.c. applied voltage, $v_\sim$ is applied across A – B, and current i is flowing, so that at each point, $v_\sim = v_m$.

Let us, now, indicate the instantaneous values of the current and voltages as functions of time in Figure 1.11(b) and their phasor relationships by the phase-diagram in Figure 1.11(c). ϕ_m, i_m are the maximum values of ϕ and i curves.

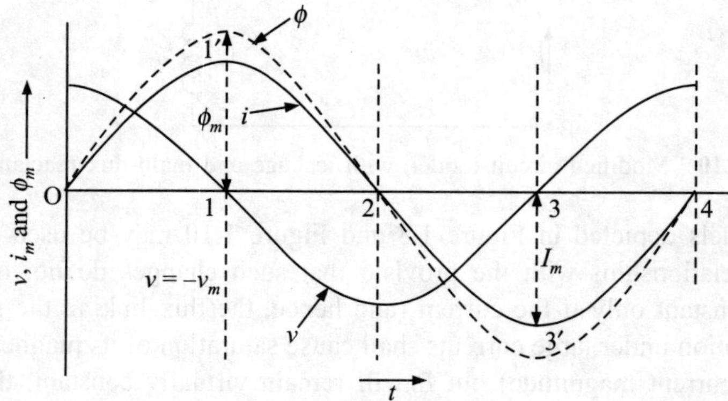

FIGURE 1.11(b) Instantaneous time diagram of voltages, current and flux produced in the equivalent circuit model of inductance element, with alternating voltage applied across its terminals.

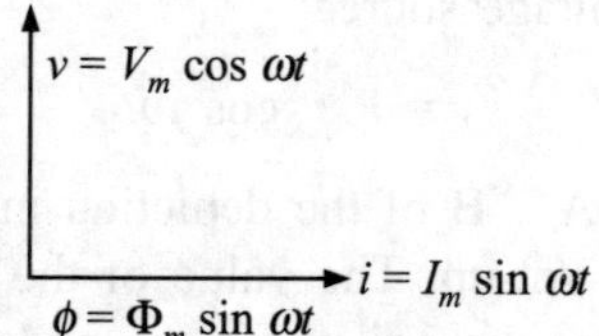

FIGURE 1.11(c) Phasor diagram showing phasor positions of v—the induced voltage, i—the circuit current and ϕ—the magnetic flux.

A critical examination of Figure 1.11(b) indicates that the instantaneous values of the induced voltage is zero when the rate of change of current is zero [at points 1 and 3 in the diagram in Figure 1.11(b)] and is maximum at points 2 and 4, when the rate of change assumes maximum value. Further, the direction of the induced emf is dependent on the nature of slope of the current curve. It is negative during the entire period for instants 0 to 2, commensurates with the negative slope of the current wave, and positive during the entire period from 2 to 4, as the slope of the current wave is positive during this interval.

As indicated in Figure 1.11(c), the phasor relationships show that voltage V leads the current by 90°. In Eq. (1.14), the applied voltage was indicated as a phasor rotating at an angular velocity, $\omega = 2\pi f$ radian per second for a supply frequency f Hz. Then the current i can be denoted by a phasor with mathematical representation of the form:

$$i = I_{\max} \cos\left(\omega t - \frac{\pi}{2}\right) \quad ...(1.15)$$

and, the net energy input to the inductance coil over one cycle of operation is given by:

$$\begin{aligned} p &= vi \\ &= V_{\max} \cos \omega t \, I_{\max} \cos\left(\omega t - \frac{\pi}{2}\right) \\ &= 2V_{\max} I_{\max} \cos \omega t \sin \omega t \\ &= V_{\max} I_{\max} \sin 2\omega t. \end{aligned}$$

The energy flow in the inductive coil, therefore, oscillates at double the supply frequency, the net energy input being zero over one cycle of operation. In other words, for half the cycle, during which the voltage and current have the same signs [periods indicated by durations 0 to 1′ and 2 to 3′ in Figure 1.11(b)], energy flows into the magnetic circuit/field, and during the remaining half, during which voltage and current have opposite signs [typified by intervals 1′ to 2 and 3′ to 4 in Figure 1.11(b)], the magnetic field returns the energy to the supply system.

The value of $I_{\max}$ indicates that the maximum mmf is required to produce the maximum flux, $\Phi_{\max}$ in the magnetic circuit. Since,

$$v = N\frac{d\phi}{dt}$$

$$\therefore \quad \begin{aligned} \phi &= \left(\frac{1}{N}\right)\int v \, dt \\ &= \left(\frac{1}{N}\right)\int V_{\max} \cos \omega t \, dt \\ &= \frac{V_{\max}}{\omega N} \sin \omega t \end{aligned}$$

Hence, $$\Phi_{\max} = \frac{V_{\max}}{\omega N} = \frac{V_{\max}}{2\pi f N}$$

We can, now, express the rms value of voltage (V) in terms of f, N and ϕ_{max} as,

$$V = \frac{2\pi f N}{\sqrt{2}} \phi_{max}$$

or $$V = 4.44\, fN\phi_{max} \text{ volt} \qquad \text{...(1.16)}$$

under steady-state conditions in an a.c. circuit.

While considering the equivalent circuit model of the inductance element, the current required to produce the flux at every instant is called the magnetizing current. This is denoted by i_m in Figure 1.11(a).

Since we assumed that the magnetic core made up of magnetic material is also present, the alternating flux associated with the model for use in alternating current systems, shall have an additional continuous power loss component, due to hysteresis and eddy-current effects in the core—a loss that we did not have to consider in the model used with direct voltage applied. These losses are termed as *iron losses* and should be additionally incorporated in the model for completeness. It is a common practice in electrical modelling representations, that power losses are indicated by introducing a resistance element in the circuit such that the copper loss associated with this resistance element when carrying a suitable current through it equals the loss in question. Hence, we need to introduce modification in the model by inserting a resistance in the circuit model of Figure 1.11(a), to include the effect of iron losses in the inductance core. But both the main flux and the iron-loss associated with the core have direct bearing on each other. Hence, the resistance should be introduced *in parallel* with the inductance element representing the main flux producer, in which the voltage v_m is produced. This is the desired modification in the equivalent circuit for the inductance element working with alternating current supply system. This is shown in Figure 1.12.

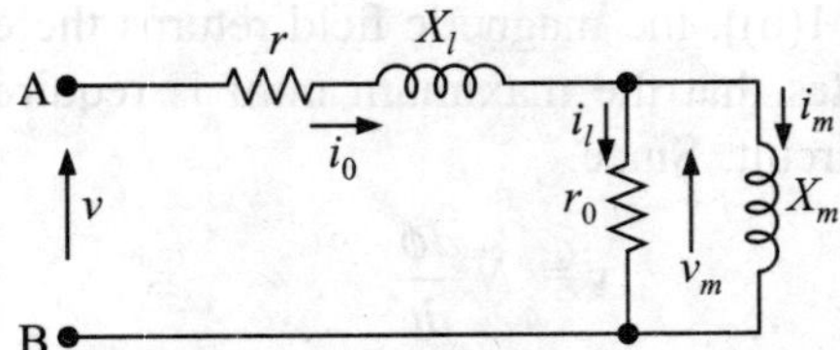

FIGURE 1.12 Modified equivalent circuit model incorporating the iron loss in the magnetic core, when working on a.c. supply.

Introduction of this additional resistance in parallel, causes an increase in the total current flow from the supply. Further, we also observe that the magnetizing current i_m, flowing in the inductance branch lags behind voltage v_m by 90°. The current in the resistance branch, r_0 has the same voltage v_m effective across it and the current i_l flowing through it is co-phasal with v_m. Hence, the model takes a total input current i_0 from the supply, such that i_0 is the phasor sum of i_m and i_l, as indicated in Figure 1.13. It must be emphasised that this modified model of Figure 1.12 is only an approximation. It is, at best, good for, only small values of i_0, and for a particular value of the applied voltage. For other values of applied voltage, the iron loss shall not be strictly equal to the copper loss in the resistance r_0. Besides, the model is effective only in case of the applied voltage v and the induced emf v_m can be approximated to be equal. Also, if the applied voltage is increased to a value high enough

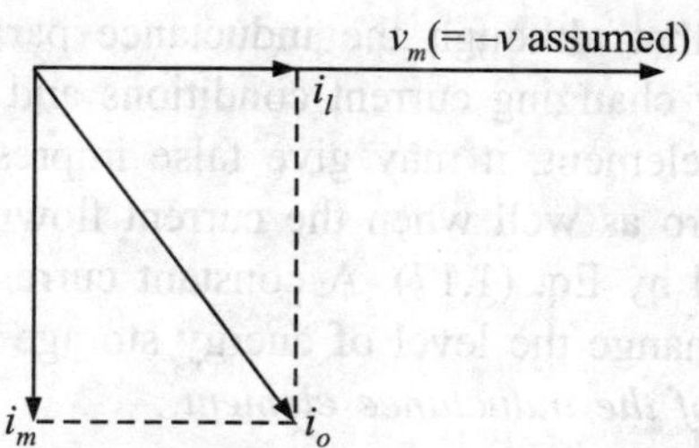

FIGURE 1.13 Phasor diagram showing current i and its phasor relations with i_l and i_m.

to effect saturation of the core, L_m shall not remain constant but will decrease in value and bring in an increased anomaly to the above model representation.

But this model finds extensive application in the study of electro-mechanical energy converters due to simplicity of its representation, keeping the limitations of the model in mind.

Energy Viewpoint of the Inductance Element

We shall continue to assume that inductance element has negligible internal resistance. If the initial current through the element is 'zero', and if a current i is made to flow through it such that the induced voltage v_m appears across it due to changing-flux linked by itself in time t ($t = 0$ at $i = 0$), the total energy (W) flowing to the element from the supply is given by:

$$W = \int_0^t v_m i \, dt \text{ joule}$$

This leads to the following relationship:

$$W = \int_0^t \left(L \frac{di}{dt} \right) i \, dt = \int_0^t Li \, di$$

$$= \frac{1}{2} Li^2 \text{ joule} \qquad \text{...(1.17)}$$

Equation (1.17) indicates that the inductance element absorbs an amount of energy proportional to the inductance parameter L and square of the instantaneous current, this energy is stored in the element in its magnetic field in the form of *magnetic energy*. This energy becomes zero, whenever the current becomes zero. Because the energy associated with the inductance parameter increases or decreases with the current flowing through it, we can conclude that the inductor has the property of being capable of returning the stored energy kept as magnetic energy in its field, to the supply source from which it received it. Equation (1.17) can be rearranged in the form:

$$L = \frac{2W}{i^2} \text{ henry} \qquad \text{...(1.18)}$$

Thus, an alternative way of defining the inductance parameter is in terms of the amount of energy stored in its magnetic field corresponding to an instantaneous current. This is the energy description of the inductance parameter.

It is interesting to note that although the inductance parameter has a voltage induced across the element only under changing current conditions and a constant current results in a zero voltage drop across the element, it may give false impression that the energy stored in its magnetic field becomes zero as well when the current flowing through it is constant. This is not correct as demonstrated by Eq. (1.17). A constant current only leads to a fixed energy stored in it. Any attempt to change the level of energy storage is resisted by the system. *This is, in fact, the inertial effect of the inductance element.*

Physical viewpoint of the Inductance Element

We had seen under section 1.2.2 that the voltage drop is given [Eq. (1.9)] by the following equation:

$$v_m = L\left(\frac{di}{dt}\right)$$

However, the same voltage drop may be described by Faraday's law in terms of the flux produced by the current, that is:

$$v_m = N\left(\frac{d\phi}{dt}\right)$$

which leads to the relation:

$$L = N\left(\frac{d\phi}{di}\right)$$

In case, when ϕ is directly proportional to i, for all values of i (that is for a 'linear' inductance), we can rewrite the above expression in the form:

$$L = \frac{N\phi}{i}, \text{ in weber-turn per ampere.}$$

If we substitute,
$$\phi = \frac{\text{mmf}}{\text{magnetic reluctance}} = \frac{Ni}{R}$$

where, the mmf denotes the magneto-motive force which produces the flux ϕ in the magnetic circuit having a reluctance of R, we can attain a more useful expression in terms of the physical dimensions of the magnetic circuit as will be indicated in expression (1.19) presently.

The schematic diagram in Figure 1.14, shows the arrangement of an inductance coil. The core, which carries a total flux ϕ, is assumed to have a coil of N turn surrounding its cross-section, and has an enclosed circuit of length l and a cross-section area of A.

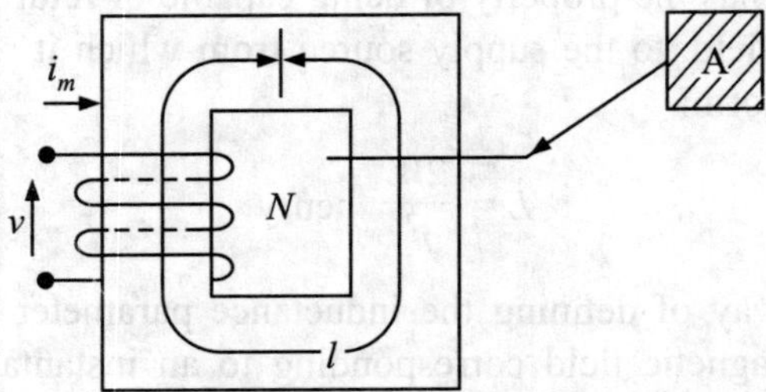

FIGURE 1.14 Core arrangement of an inductance coil having N turns with magnetic core-length of l, and a cross-section area of A.

Then, $$R = \frac{l}{\mu A}$$

where, μ is the magnetic permeance of the material, l is the length of the flux path and A is the cross-sectional area of the core. The inductance of the element can be expressed in its physical dimensions as follows:

$$L = \frac{N^2 \mu A}{l} \qquad \text{...(1.19)}$$

Examination of the above relation reveals that similar to resistance parameter, the inductance parameter is also dependent on the physical dimensions and the magnetic properties of the medium of the magnetic circuit—a feature not emerging when considering the parameter under *circuit* and *energy* viewpoints. This is significant because it indicates what needs to be done to change the value of the inductance parameter. Thus, the inductance can be changed by any of the following four manipulations or a combination of them:

1. By changing the number of turns of the coil
2. By changing the magnetic circuit material
3. By modifying the cross-sectional area of the flux-path through the inductance core
4. By changing the length of the flux-path

as is evident from the Eq. (1.19).

1.2.3 Capacitance Parameter

This is the third basic parameter in an electric circuit. Capacitance can be characterized as that basic property of a circuit element in which energy is capable of being stored in the electric field of the element. Its influence in an electric circuit is present only when there exists a charging potential difference across the terminals of the circuit element.

Circuit viewpoint of Capacitive Element

The concept of capacitance is introduced as a proportionality constant term, relating the charge between two metallic surfaces (or conductors) to the corresponding potential difference between them. Thus:

$$Q = Cv_c$$

where, Q denotes the charge, v_c gives the potential difference and C is the capcitance. The definition of capacitance from circuit viewpoint requires the concept of current in the element. General expression of current is defined from the concept of charge as follows:

$$i = \left(\frac{d}{dt}\right)Q = C\left(\frac{d}{dt}\right)v_c \text{ ampere} \qquad \text{...(1.20)}$$

The current flowing through the capacitance parameter is related to the potential difference appearing across it by Eq. (1.20). Thus, any circuit element in which the current flowing in it is directly proportional to the rate of change of potential difference across the element, is called a capacitive element.

A capacitance is made with a large metallic surface, separated by a small distance between the two surfaces. Definition of a capacitance can, then, be summed up by the following mathematical relationship:

$$C = \frac{i}{\left(\dfrac{dv_c}{dt}\right)} \text{ farad} \qquad \text{...(1.21)}$$

Dimensionally, C is expressed as ampere-second per volt or coulomb per volt. Equation (1.20) can be manipulated to show that the voltage across the terminals of the capacitor element, in terms of current flowing through it, gives:

$$dv_c = \left(\frac{1}{C}\right) i\, dt$$

and, integrating both sides, we get

$$\int_{v_c(0)}^{v_c(t)} dv_c = \left(\frac{1}{C}\right)\int_0^t i\, dt + v_c \quad (\text{at } t = 0)$$

or

$$v_c(\text{at } t) = \frac{1}{C}\int_0^t i\, dt + v_c \quad (\text{at } t = 0)$$

where, v_c is the initial voltage appearing at $t = 0$ in the plates of the capacitance element. If this voltage is zero, then:

$$v_c = \frac{1}{C}\int_0^t i\, dt \qquad \text{...(1.22)}$$

Figure 1.15 shows a graphical representation of Eq. (1.22).

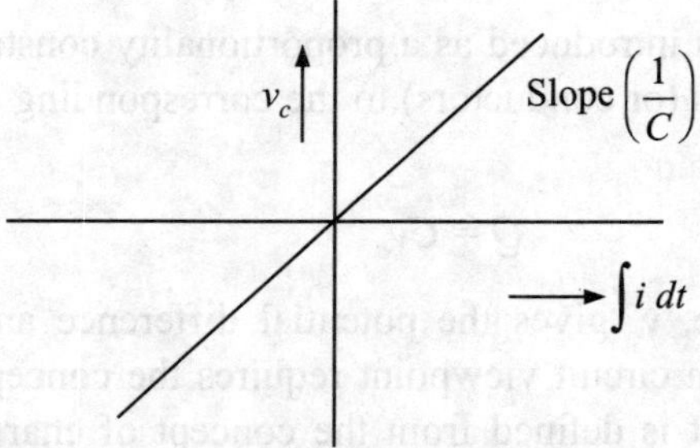

FIGURE 1.15 A plot of the potential difference between the plates of a capacitance element and the current in the element.

A straightline relationship is maintained between the variation of v_c and $\left[\int i\, dt\right]$, the slope of the straightline indicates the inverse of the capacitance, that is, $\left[\frac{1}{C}\right]$. If we compare

Figure 1.15 with Figure 1.4 and Figure 1.6, we observe similarity in behaviour among all circuit elements, when plotted against different values of voltage along dependent axes. Figure 1.16 shows the diagrammatic representation of the circuit with capacitance element. For the direction of current shown in the diagram, the potential at the upper plate increases, with the upper plate becoming more positive. Electrons shall now be transferred from the upper plate to the lower plate through the external circuit. The arrow-head representing the direction of v_c is pointed upwards. If an alternating voltage source is connected across the capacitance element, the flow reverses and the lower plate gains positive potential during the negative half-cycle operation.

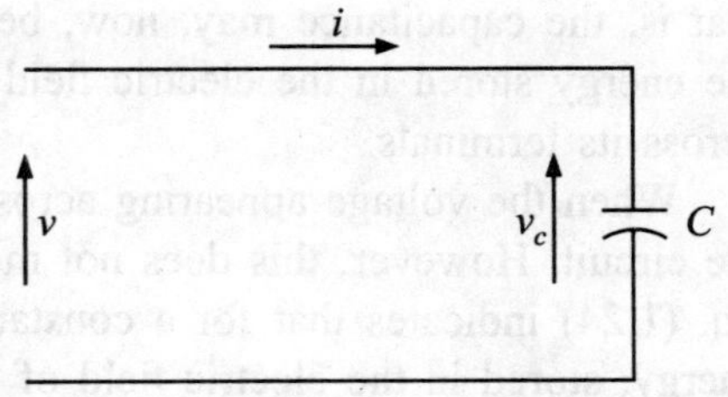

FIGURE 1.16 Circuit model of capacitance element.

A study of Eqs. (1.20) and (1.22) shall reveal that the voltage across the capacitance element cannot change disjointedly. A finite change in v_c in zero-time has to have a value of infinity for $\left(\frac{dv_c}{dt}\right)$, that is, a capacitance current i has to be infinite—a physical impossibility. On the other hand, for a finite change in capacitor current—however, large—the integral contribution in zero time must be zero.

Energy Viewpoint of Capacitance Element

Let us consider that the capacitance shown in Figure 1.16 has zero initial voltage across it, and a current i is flowing in the circuit for a time interval dt. The energy (W) delivered to the capacitance over this time is given by:

$$W = \int_0^t v_c i \, dt \qquad ...(1.23)$$

Incorporating Eq. (1.20) in Eq. (1.23), we get

$$W = \int_0^t v_c \left(C \frac{dv_c}{dt}\right) dt = \int_0^{v_c} C v_c dv_c$$

or

$$W = \frac{1}{2} C v_c^2 \text{ joule} \qquad ...(1.24)$$

Thus, the capacitance absorbs an amount of energy proportional to the capacitance parameter and the square of the instantaneous value of voltage appearing across the capacitor terminals. The absorbed energy is stored by the element in its electric field existing between the two plates. Now, as the capacitor voltage increases so does the energy stored and as v_c decreases in magnitude, the associated energy also decreases. Thus, like an inductance element, the capacitance element also has the capacity of interchanging energy with the supply system. As before [in Eq. (1.7) for the case of inductance element], we can manipulate Eq. (1.24) to define capacitance from energy viewpoint. We can write:

$$C = \frac{2W}{(v_c)^2} \text{ farad} \qquad ...(1.25)$$

that is, the capacitance may, now, be identified in terms of the instantaneous value of twice the energy stored in the electric field and the corresponding value of the potential difference across its terminals.

When the voltage appearing across the capacitance is constant, there is no current flow in the circuit. However, this does not mean that the energy in the element is not stored. In fact, Eq. (1.24) indicates that for a constant capacitance voltage, there exists a finite and constant energy, stored in the electric field of the capacitance element.

Physical viewpoint of Capacitive Element

The amount of charge that accumulates on the metallic surface of a capacitance element, can be expressed as:

$$Q = Cv_c \text{ coulomb} \qquad \text{...(1.26)}$$

By Gauss' flux-theorem, it is possible to express the accumulated charge in terms of the electric field quantity, E as follows:

$$Q = \in A E \qquad \text{...(1.27)}$$

In the above equation, $\in$ is the permittivity, that is, the specific dielectric constant of the material between the conducting surfaces constituting the capacitance element, A denotes the area of the conducting surfaces E being the electric field intensity with a unit of volt/metre.

For the sake of an illustration, let us consider the capacitance element to have a definite configuration, that is, two parallel plates separated by a distance d between them, and carrying a dielectric material of the same thickness (d) intervening the plates. Then, for this system:

$$E = \frac{v_c}{d} \qquad \text{...(1.28)}$$

Hence, Eq. (1.27) assumes, after substitution of Eq. (1.28) in it:

$$Q = \in A\left(\frac{v_c}{d}\right)$$

which leads to:

$$Cv_c = \in\left(\frac{A}{d}\right)v_c$$

or

$$C = \in\left(\frac{A}{d}\right) \qquad \text{...(1.29)}$$

Although, we derived this expression for capacitance for a particular geometry of the element for a parallel-plate configuration, we can conclude that the capacitance parameter, based on physical dimensions can, in general, be derived as proportional to the dielectric constant of the material intervening the space between the charge carriers, is inversely proportional to the spacing between the charge accumulators, and is directly proportional to the metal surface areas.

Special attention of the reader is drawn in the physical viewpoints of the circuit elements, because these have direct bearing on their physical dimensions and configuration, and certain informations are not available in their discussions in *circuit* and *energy* viewpoints. In practice, the capacitance parameter does not enjoy the patronage and applications in modelling of electro-mechanical energy converters, but is included for completeness.

OBJECTIVE TYPE QUESTIONS

1.1 The following magnetic and electrical quantities are displayed on the left column and the S.I. units are shown on the right column. Match the left column quantities with the right column lists:

(i) Potential	1. Volt/metre
(ii) Current	2. Ohm
(iii) Voltage gradient	3. Weber
(iv) Current density	4. Ampere-turn/metre
(v) Specific resistivity	5. Henry/metre
(vi) Resistance	6. Henry
(vii) Magneto-motive force	7. Volt
(viii) Permittivity of free space	8. Ampere $(\text{metre})^2$
(ix) Flux	9. Farad/metre
(x) Flux-linkage	10. Weber-turn or volt second
(xi) Flux density	11. Ampere-turn/weber
(xii) Magnetic field strength	12. Coulomb
(xiii) Reluctance	13. Ampere
(xiv) Permeance	14. Ohm-metre
(xv) Permeability	15. Ampere-turn
(xvi) Charge	16. Tesla
(xvii) Capacitance	17. Weber/ampere-turn
(xviii) Inductance	18. Farad

1.2 A magnetic circuit with a relative permeability $\mu_r = 100$, has a cross-section area of 5 cm^2 and a mean length of 25 cm. The coil on the core has 100 turn and an mmf of 1000 A-T. Which of the following statements is correct?

(a) The inductance of the coil is 2.512 mH
(b) The flux-density of the core is 0.5 tesla
(c) The flux linked by the coil is 0.025 Wb-T
(d) The reluctance offered by the core is 4×10^6 A-T/Wb
(e) All statements are true.

1.3 Select the correct answer of the given alternatives provided:

(a) Inductance is the same as:

(i) Weber-turn/ampere (ii) Volt-second/ampere
(iii) Henry-turn^2 (iv) All the above

(b) Capacitance is the same as:

(i) Coulomb/volt (ii) Ampere-second/volt
(iii) Volt-second/ampere (iv) Both (i) and (ii) above

(c) Resistance is the same as:

(i) Coulomb/volt (ii) Ampere-second/volt
(iii) Volt-second/ampere (iv) None of the above

(d) Voltage gradient is equivalent to:

(i) Specific resistance × Current density
(ii) Coulomb/farad-metre
(iii) Volts/metre
(iv) All of the above

(e) Joule is equivalent to:

(i) Coulomb-volt (ii) Volt × Ampere
(iii) $(\text{Volt})^2$/ohm (iv) Current/resistance

(f) Flux-density is equivalent to:

(i) Tesla (ii) Ampere-turn × henry/$(\text{metre})^2$
(iii) Weber/$(\text{metre})^2$ (iv) All the above

1.4 (a) Distinguish between the terms resistor and resistance.

(b) Clarify the term:
'resistance is defined from the viewpoint of circuits'.

1.5 How does the value of temperature affect the resistance magnitude? Why is it important to know the value of its effect in design?

1.6 Why does the temperature of a resistance element not rise continuously even if a current is continuously passing through it?

1.7 Why is it required to declare the voltage rating of a resistor element?

1.8 Distinguish between an inductor and an inductance.

1.9 Indicate the circumstances which help to conclude that a circuit is inductive.

1.10 Explain physically why the current flowing through an inductor cannot be instantaneously changed.

1.11 Identify the important information that is obtained from the geometrical description, which cannot be obtained from analysis based on circuit view point or energy view point in inductors.

1.12 By what factor will the inductance of a coil be increased when the number of turns is increased to 1.5 times the original value, keeping all other quantities unchanged?

1.13 Elaborate the statement: 'Inductance is a function of the medium in which it operates'.

1.14 What is the main reason for the value of inductance to be implicity non-linear?

1.15 How can you show that an electrical circuit containing a single element is capacitive?

1.16 Explain why the voltage across the terminals of a capacitor is not possible to change suddenly?

1.17 Discuss capacitance of a circuit using energy and geometrical viewpoints. What important information follows from geometrical viewpoint, which cannot be obtained directly from the circuit and energy descriptions?

1.18 By what factor is the capacitance of a pair of parallel conducting plates, separated by a dielectric material altered, when the separating distance between the plates is halved and the surface area of the plates is doubled?

1.19 Explain why it is an undesirable practice to interrupt an electrical circuit with the help of a knife switch, when the circuit has a large inductance. How can you minimise the undesirable effect of using a knife switch in such circuits?

1.20 By what factor shall the inductance of a coil be increased, if the number of turns of the coil is doubled?

Answers

1.1 (i)–7; (ii)–13; (iii)–1; (iv)–8; (v)–14; (vi)–2; (vii)–15; (viii)–9; (ix)–3; (x)–10; (xi)–16; (xii)–4; (xiii)–11; (xiv)–17; (xv)–5; (xvi)–12; (xvii)–18; (xviii)–6

1.2 (e) **1.3** (a) (iv); (b) (iv); (c) (iv); (d) (iv); (e) (i); (f) (iv)

1.12 2.25 **1.18** 4 times **1.20** 4 times

NUMERICAL PROBLEMS

1.1 A constant voltage source is applied to a combination of three resistors shown in Figure 1.17.

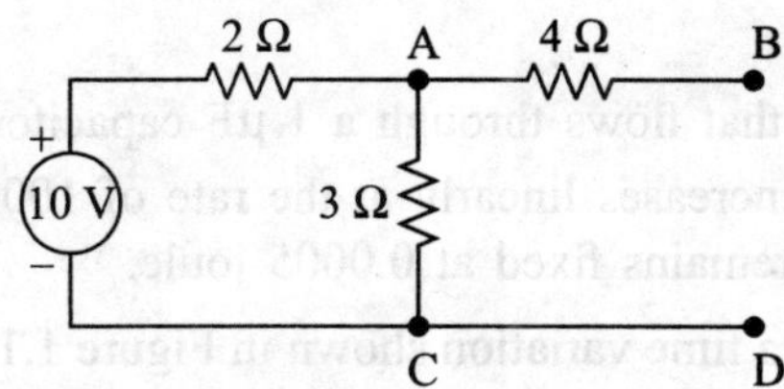

FIGURE 1.17 Constant voltage applied to a combination of three registors.

Determine V_{AC} and V_{BD}. **[Ans.** $V_{AC} = V_{BD} = 6$ V]

1.2 An electric heater with two identical elements draws 1000 watt form a 250 volt supply system with both element working normally.

(a) What power will the heater draw from a 200 volt supply, when the two elements are connected in series? **[Ans.** 0.16 kW]

1.3 An electric broiler takes a current of 10 A at 250 volt for 2 hour.

(a) If electrical energy costs ₹ 5.00 per kWh, determine the cost of operating the broiler.

(b) How much electricity in coulomb is consumed by the broiler?

(c) What is the rate at which the electrical energy is expended?

(d) What is the effective resistance of the broiler, when estimated from the energy viewpoint? **[Ans.** (a) ₹ 25; (b) 72000 C; (c) 2500 W; (d) 25 Ω]

1.4 A resistor is found to dissipate 2.5 kWh of energy in 1 hour when drawing a current of 10 A.

(a) Estimate the value of the resistance using energy approach.
(b) Obtain the resistance of the system using circuits approach. **[Ans.** 25 Ω]

1.5 A coil of standard copper wire having a resistance of 12 ohm at 20°C, is imbedded in the core of a large transformer. After the transformer is in operation for several hour, the resistance of the coil is found to be 13.4 ohm. What is the temperature of the transformer core? **[Given:** Resistance temperature coefficient of copper at 20°C is: 0.00393 Ω/°C] **[Ans.** 55°C]

1.6 A coil mounted on a continuously closed circular core of ring type, as shown in Figure 1.18, has 100 turn. The mean length of turn of the coil is 0.2 m and the cross-sectional area of the core is 0.01 m^2. The value of permeability of the core material is 10^{-3}. Determine the following:

(a) the inductance of the coil.
(b) when direct voltage is applied to the inductor, it is found to draw a current of 0.1 A. How much energy shall be stored in its magnetic field? **[Ans.** 0.5 H; 0.0025 joule]

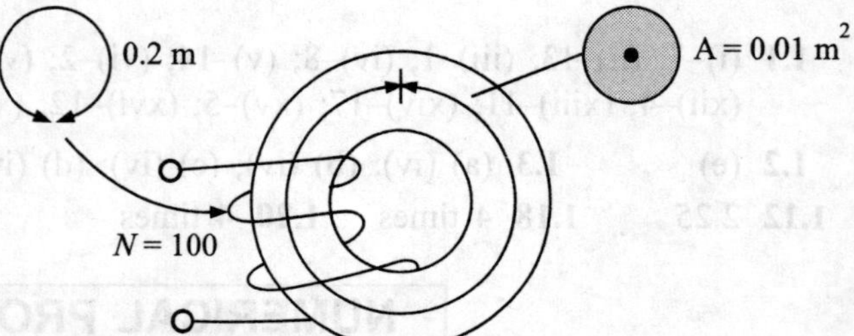

FIGURE 1.18 The magnetic circuit.

1.7 Determine the current that flows through a 1 μF capacitor:

(a) when the voltage increases linearly at the rate of 1000 volt/second, and
(b) the energy stored remains fixed at 0.0005 joule. **[Ans.** 1 mA; 0]

1.8 A voltage wave having a time variation shown in Figure 1.19 is applied to a pure inductor having a value of 2 H. Sketch the variation of the current through the inductor over a period of the first 10 second. Indicate the values on the derived current curve.

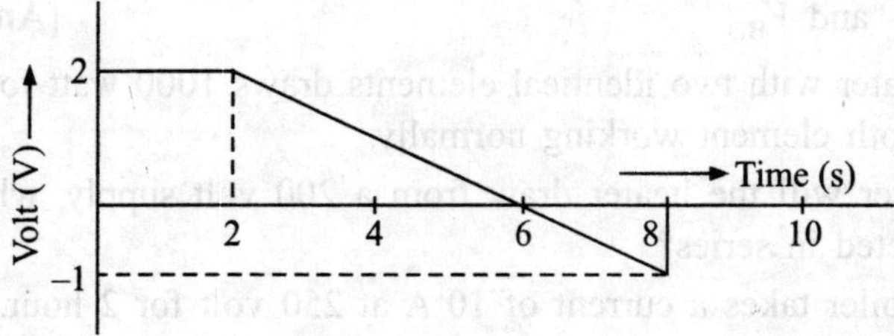

FIGURE 1.19 Voltage profile against time applied across a pure inductor.

CHAPTER 2

Mathematical and Circuit Theory Tools

2.1 INCORPORATING MATHEMATICAL TOOLS

Machine equations that we encounter in the text are integral, simultaneous, non-linear and partial differential equations. In tackling simultaneous equations, a special technique in mathematics has been evolved, called the matrix algebra. It is possible to render a much simpler treatment of the subject that is contained in mathematical text books. We confine ourselves here by indicating only aspects which are directly relevant to machine analysis.

The first step in handling a problem is to record, in a purely routine manner, the simultaneous equation relevant to the chosen problem. But before illustrating the actual details, let us acquaint ourselves with the basic definitions associated with matrix algebra first.

2.2 ELEMENTARY MATRIX ALGEBRA

A matrix is a regular array of $(r \times c)$ equations. It is arranged in r rows and c columns, each term being identified as an *element* of the matrix. The elements may be simple real numbers or algebraic symbols, complex numbers of the form $(R + jX)$ or trigonometric functions, operators like $M\left(\frac{di}{dt}\right)$ or even further matrices.

- If the number of rows of a matrix equals the number of columns, the matrix is termed a *square matrix*.
- If there is only *one column*, that is $c = 1$, the matrix is a *first order column matrix*.
- If the number of rows, $r = 1$, it is called a *first order row matrix*.
- A *zero matrix* has all its elements as zeros.
- A *unit matrix* is a square matrix in which all diagonal terms from left to right are unity and all other terms are zeros, that is:

$$\begin{bmatrix} 1 & 0 \\ 0 & 1 \end{bmatrix} \quad \text{or} \quad \begin{bmatrix} 1 & 0 & 0 \\ 0 & 1 & 0 \\ 0 & 0 & 1 \end{bmatrix}$$

- Two matrices are said to be *transpose* of each other if the rows of one matrix are changed as columns of the other matrix, that is, $1 \le ij$ term $\le n \equiv 1 \le ji$ term $\le n$ of the two matrices. Also,

$$[A_t]_t = [A]$$

where, $[A_t]$ is the transpose of $[A]$.

2.2.1 Addition and Subtraction of Two Matrices

For addition or subtraction between two matrices to be possible, they must both have the same number of rows and same number of columns. Only in such cases, each element of the sum or difference, is the sum or difference of the corresponding elements of the constituent matrices.

For example, Let $[A] = \begin{bmatrix} 1 & 2 \\ 3 & 4 \end{bmatrix}$ be a given matrix. Then, $[A_t] = \begin{bmatrix} 1 & 3 \\ 2 & 4 \end{bmatrix}$.

Therefore, $[A] + [A_t] = \begin{bmatrix} 2 & 5 \\ 5 & 8 \end{bmatrix}$ and $[A] - [A_t] = \begin{bmatrix} 0 & -1 \\ 1 & 0 \end{bmatrix}$

Examination of the sum reveals that the non-diagonal terms are identical and are of the same sign. The rows are equal to the columns. A matrix with this property is called a *symmetric matrix*. The difference, $[A] - [A_t]$ is, by contrast, called an *anti-symmetric* or *skew-symmetric*, and has the following property: The non-diagonal terms are having the same value but their signs are reversed, or, the interchanging of rows and columns changes the sign of all terms. The *cumulative* and *associative laws* are obeyed in matrix algebra, that is:

$$[A] + [B] = [B] + [A]$$

and $$[[A]+[B]]+[C] = [A]+[[B]+[C]]$$

2.2.2 Multiplication of Matrices

Multiplication by a Constant

Each element of the matrix is multiplied by the constant in question. Thus,

$$a.\begin{bmatrix} x & y \\ z & m \end{bmatrix} = \begin{bmatrix} ax & ay \\ az & am \end{bmatrix}$$

Multiplication of Two Matrices

Such a multiplication is possible only if the number of columns of the first matrix equals the number of rows of the second matrix. The resulting product is a matrix with the same number of rows as that of the first matrix being multiplied and the same number of columns as that of the second matrix being multiplied. Thus, if a matrix *A*, having *a* rows and *c* columns is denoted by ${}_a[A]_c$, and if it is multiplied by a matrix *B*, having *c* rows and *d* columns, and is depicted by ${}_c[B]_d$, then:

$${}_a[A]_c\ {}_c[B]_d = {}_a[AB]_d$$

The following points need to be noted in connection with multiplication of two matrices:

1. Multiplication may not always, or even in general, possible.
2. Multiplication may be possible in one direction but not in the other direction.
3. When multiplication is possible in both directions, the product may yield quite different results. *Thus, the cumulative law in multiplication of matrices is not obeyed.* In other words:

$${}_3[A]_2\ {}_2[B]_3 = {}_3[AB]_3$$

But,

$$\left.\begin{aligned} {}_2[B]_3\ {}_3[A]_2 &= {}_2[BA]_2 \\ \therefore \quad {}_3[AB]_3 &\neq {}_2[AB]_2 \end{aligned}\right\} \qquad \text{...(2.1)}$$

4. If the matrices are square, and there are two possible products, these two products are also of different values:

$$\left.\begin{aligned} {}_r[A]_c\ {}_c[B]_r &= {}_r[AB]_r \\ {}_c[B]_r\ {}_r[A]_c &= {}_c[BA]_c \end{aligned}\right\} \qquad \text{...(2.2)}$$

where, $r = c$

When multiplication is possible, the element in the *r*th row and the *c*th column of the product matrix is the sum of the products of the corresponding elements in the *r*th row of the first matrix and the *c*th column of the second matrix. Thus:

$$\begin{bmatrix} a & b \\ c & d \end{bmatrix}\begin{bmatrix} e & f \\ g & h \end{bmatrix} = \begin{bmatrix} (ae+bg) & (af+bh) \\ (ce+dg) & (cf+dh) \end{bmatrix}$$

But,

$$\begin{bmatrix} e & f \\ g & h \end{bmatrix}\begin{bmatrix} a & b \\ c & d \end{bmatrix} = \begin{bmatrix} (ea+fc) & (eb+fd) \\ (ga+hc) & (gb+hd) \end{bmatrix}$$

Thus, the two product matrices are different, and the cumulative law of product does not hold good in matrices.

Inversion of Matrices

$$\left.\begin{aligned} &\text{Let} && [V] = [R]\,[I] \\ &\text{or} && [R]^{-1}\,[V] = \big[[R]^{-1}\,[R]\big][I] \\ &&& \quad = [1]\;[I] = [I] \\ &\text{Therefore,} && [I] = [R]^{-1}\,[V] \end{aligned}\right\} \qquad ...(2.3)$$

It is very important to note that the order of multiplication is very important. The inverse of a matrix exists when,

1. The matrix in question is a square matrix
2. The matrix has a *non-zero determinant value*

The inverse of a matrix can be obtained by the following process:

(a) Replace the matrix by its transpose, that is, interchange the rows by corresponding column terms in a new matrix form.

(b) Replace each element by its cofactor in the determinant of the original matrix.

(c) Divide each element by the value of the determinant of the original matrix.

ILLUSTRATION 2.1 Let us find the inverse of

$$[R] = \begin{bmatrix} 12 & -6 & -4 \\ -6 & 20 & -6 \\ -4 & -6 & 10 \end{bmatrix} \qquad ...(2.4)$$

(a) In the first step, we replace the matrix by its transpose, i.e.

$$[R_t] = \begin{bmatrix} 12 & -6 & -4 \\ -6 & 20 & -6 \\ -4 & -6 & 10 \end{bmatrix} \qquad ...(2.5)$$

In this case, $[R_t]$ and $[R]$ are identical. This can be verified from Eqs. (2.4) and (2.5). In fact, all resistance matrices obtained from the application of Maxwell's cyclic current rule, detailed later in Section 2.4, to simple static networks, are always symmetrical. It is useful to note that all terms in the inverse of a static resistance matrix are positive. The work may be checked by multiplying each term in the transposed matrix by the corresponding cofactor in step (b), thus giving three values of the determinant. These must have identical values. The final check could also be made by multiplying $[R]$ with its inverse. The result is a unit matrix.

(b) In this step, we write down the cofactors of the matrix:

$$\begin{bmatrix} \oplus\begin{bmatrix} 20 & -6 \\ -6 & 10 \end{bmatrix} & \ominus\begin{bmatrix} -6 & -6 \\ -4 & 10 \end{bmatrix} & \oplus\begin{bmatrix} -6 & 20 \\ -4 & -6 \end{bmatrix} \\ \ominus\begin{bmatrix} -6 & -4 \\ -6 & 10 \end{bmatrix} & \oplus\begin{bmatrix} 12 & -4 \\ -4 & 10 \end{bmatrix} & \ominus\begin{bmatrix} 12 & -6 \\ -4 & -6 \end{bmatrix} \\ \oplus\begin{bmatrix} -6 & -4 \\ 20 & -6 \end{bmatrix} & \ominus\begin{bmatrix} 12 & -4 \\ -6 & -6 \end{bmatrix} & \oplus\begin{bmatrix} 12 & -6 \\ -6 & 20 \end{bmatrix} \end{bmatrix} \qquad ...(2.6)$$

$$= \begin{bmatrix} \oplus(200-36) & \ominus(-60-24) & \oplus(36+80) \\ \ominus(-60-24) & \oplus(120+16) & \ominus(-72-24) \\ \oplus(36+80) & \ominus(-72-24) & \oplus(240-36) \end{bmatrix}$$

[***Note:*** The sign attached to each of the cofactors at the beginning of each term of the cofactors in small circles, is for clarity.]

$$= \begin{bmatrix} 164 & 84 & 116 \\ 84 & 104 & 96 \\ 116 & 96 & 204 \end{bmatrix} \qquad ...(2.7)$$

(c) Value of the determinant, worked out on the basis of the first row:

$$= +12 \times 164 + (-6) \times 84 + (-4) \times 116$$
$$= 1968 - 504 - 464 = 1000$$

Value of the determinant, worked out on the basis of the second row:

$$= +(-6) \times 84 + (20) \times 104 + (-6) \times 96$$
$$= -504 + 2080 - 576 = 1000$$

Value of the determinant, worked aout on the basis of the third row:

$$= +(-4) \times 116 + (-6) \times 96 + (10) \times 204$$
$$= -464 - 576 + 2040 = 1000$$

Please note that the value of the determinant, when computed on the basis of the first, second and third rows, is the same, as was stated earlier.

(d) Therefore, the inverse of the matrix in Eq. (2.4) is given by:

$$[R]^{-1} = \frac{1}{1000}\begin{bmatrix} 164 & 84 & 116 \\ 84 & 104 & 96 \\ 116 & 96 & 204 \end{bmatrix} \qquad ...(2.8)$$

Please note the checks provided after each step. Finally,

$$[R]\,[R]^{-1} = \begin{bmatrix} 12 & -6 & -4 \\ -6 & 20 & -6 \\ -4 & -6 & 10 \end{bmatrix} \cdot \frac{1}{1000}\begin{bmatrix} 164 & 84 & 116 \\ 84 & 104 & 96 \\ 116 & 96 & 204 \end{bmatrix}$$

$$[RR^{-1}] = \begin{bmatrix} 1 & 0 & 0 \\ 0 & 1 & 0 \\ 0 & 0 & 1 \end{bmatrix} \qquad ...(2.9)$$

In this case, the individual element of the product, when computed, works out as follows:

$$RR_{11}^{-1} = \frac{12 \times 164 - 6 \times 84 - 4 \times 116}{1000} = \frac{1968 - 504 - 464}{1000} = 1$$

$$RR_{12}^{-1} = \frac{12\times 84 - 6\times 104 - 4\times 96}{1000} = \frac{1008 - 624 - 384}{1000} = 0$$

$$RR_{13}^{-1} = \frac{12\times 116 - 6\times 96 - 4\times 204}{1000} = \frac{1392 - 576 - 816}{1000} = 0$$

$$RR_{21}^{-1} = \frac{-6\times 164 + 20\times 84 - 6\times 116}{1000} = \frac{-984 + 1680 - 696}{1000} = 0$$

$$RR_{22}^{-1} = \frac{-6\times 84 + 20\times 104 - 6\times 96}{1000} = \frac{-584 + 2880 - 576}{1000} = 1$$

$$RR_{23}^{-1} = \frac{-6\times 116 + 20\times 96 - 6\times 204}{1000} = \frac{-696 + 1920 - 1224}{1000} = 0$$

$$RR_{31}^{-1} = \frac{-4\times 164 - 6\times 84 + 10\times 116}{1000} = \frac{-656 - 504 + 1160}{1000} = 0$$

$$RR_{32}^{-1} = \frac{-4\times 84 - 6\times 104 + 10\times 96}{1000} = \frac{-336 - 624 + 960}{1000} = 0$$

$$RR_{33}^{-1} = \frac{-4\times 116 - 6\times 96 + 10\times 204}{1000} = \frac{-464 - 576 + 2040}{1000} = 1$$

2.2.3 Compound Matrices

It can be observed that while inversion of two-by-two matrix involves no significant labour, and three-by-three matrices pose no real difficulty, the labour involved increases enormously for four-by-four and other higher order matrices. The use of the concept of compound matrices can reduce the labour involved significantly in such cases. In the analysis of electrical machines, we primarily encounter higher order matrices depending on the number of windings, but in many cases, the knowledge of the current flowing in all these windings is not of our basic interest, as some of these windings are short-circuited. Such windings cannot be ignored, however, since by mutual induction effects, they play a vital role in the operation of the machine. By a simple application of compound matrices, equations giving these currents in these windings may be eliminated while their effect on the overall performance of the machines are correctly preserved.

Let us assume a voltage matrix $[V]$ equation of the form:

$$[V] = [Z]\,[I]$$

and let us suppose that the current matrix $[I]$ has six rows while the impedance matrix $[Z]$ is a six-by-six matrix. Let the voltage and current matrices be broken into two equal parts with the help of a line drawn between the third and the fourth rows. Additionally, let the impedance matrix be separated into four equal square matrice parts, by lines drawn between the third and fourth rows and columns. This original equations can, now, be denoted by the general form:

$$\begin{bmatrix} V_1 \\ V_2 \end{bmatrix} = \begin{bmatrix} Z_{11} & Z_{12} \\ Z_{21} & Z_{22} \end{bmatrix} \begin{bmatrix} I_1 \\ I_2 \end{bmatrix} \qquad ...(2.10)$$

This is a two-by-two matrix, but each of the eight elements, V_1 ..., Z_{11} ..., I_1 ..., are themselves matrices. We can, therefore, write following equations:

$$V_1 = Z_{11}I_1 + Z_{12}I_2 \qquad ...(2.11)$$

$$V_2 = Z_{21}I_1 + Z_{22}I_2 \qquad ...(2.12)$$

Let us imagine that we wish to eliminate the matrix I_2. It enables the original six-by-six matrix to be inverted by a much simpler process of inverting three-by-three matrices from the viewpoint of computation, but it preserves the effects of I_2 matrix analytically while I_2 is eliminated.

If Eq. (2.12) is multiplied in the front by $[Z_{22}]^{-1}$, it yields:

$$[I_2] = [Z_{22}]^{-1} [V_2] - [Z_{22}]^{-1} [Z_{21}] [I_1] \qquad ...(2.13)$$

Let us, now, substitute in Eq. (2.11), the value of $[I_2]$ given by Eq. (2.13), thus leading to the development of a new equation of the form:

$$[V'] = [Z'] [I_1]$$

where,

$$[V'] = [V_1] - [Z_{12}] [Z_{22}]^{-1} [V_2]$$

and

$$[Z'] = [Z_{11}] - [Z_{12}] [Z_{22}]^{-1} [Z_{21}]$$

Then the currents $[I_1]$ may be obtained by inverting $[Z']$, and multiplying the same by $[V']$. If the eliminated current $[I_2]$, flows in the stator short-circuited meshes, then the applied voltages $[V_2]$ are all zeros, and there is simply $[V'] = [V_1]$.

2.2.4 Use of Compound Matrix in the Derivation of Well-known Network Theorems

Thevenin's and Norton's Theorems

Let us consider the general multi-mesh network of Figure 2.1(a), and concentrate our attention to the preservation of the single current $[I_1]$, flowing in mesh (1), to the right side of the separating line A–B. Let us also imagine that the complete matrix voltage equations of the network have been written down and all currents other than $[I_1]$ have been eliminated with the help of compound matrix concepts. Let $[V_n]$ be the algebraic sum of all voltages driving current in the anti-clockwise direction with respect to mesh (1), to the dividing line (A–B) or to the right of switch S, in Figure 2.1(a). Then, the cyclic voltage, $[V_1]$ in mesh (1) is given by:

$$[V_1] = [V_a] - [V_n]$$

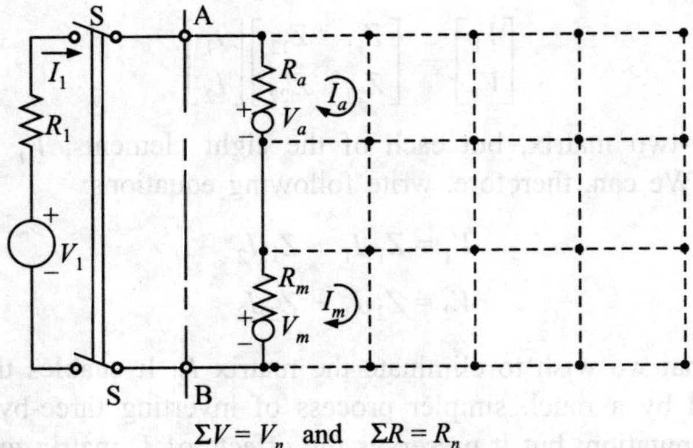

$\Sigma V = V_n$ and $\Sigma R = R_n$

FIGURE 2.1(a) Thevenin's theorem indicated (I_1 flowing when switch S is closed).

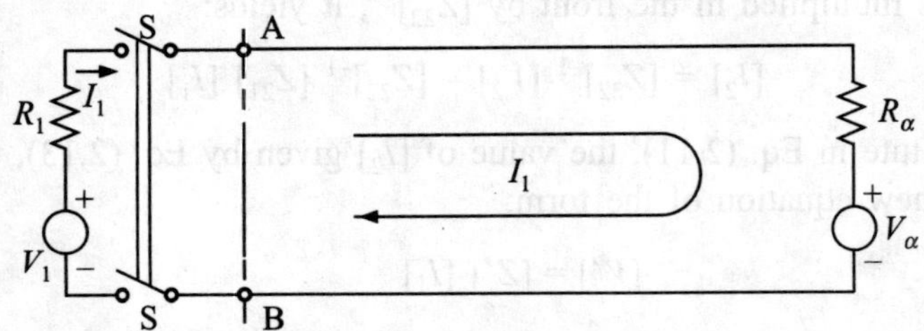

$[R_\alpha] = [R_n] - [R_{12}].[R_{22}]^{-1}.[R_{21}]$ and $[V_\alpha] = [V_n] + [R_{12}].[R_{22}]^{-1}.[V_2]$

FIGURE 2.1(b) Thevenin's theorem equivalent (I_1 flowing when switch S is closed).

Let $[V_2]$ be the matrix of the cyclic voltages in the remaining meshes. Then, after elimination of the meshes shown in Eq. (2.10) in Section 2.2.3, there is, in the notations adopted in the above Section 2.2.3

$$\begin{aligned}[V'] &= [V_1] - [R_{12}]\,[R_{22}]^{-1}\,[V_2] \\ &= [[V_a] - [V_n]] - [R_{12}][R_{22}]^{-1}[V_2] \\ &= [V_a] - [V_\alpha]\end{aligned}$$

It follows that as far as voltage sources are concerned, the network in Figure 2.1(a) and the simplified circuit shown in Figure 2.1(b), are equivalent if:

$$[V_\alpha] = [V_n] + [R_{12}]\,[R_{22}]^{-1}\,[V_2]$$

Again, let $[R_n]$ be the sum of the resistances in mesh (1), to the right of switch S of Figure 2.1(a). Then, the earlier notion of:

$$[R_{11}] = [R_a] + [R_n]$$

While $[R_{12}]$, $[R_{21}]$ and $[R_{22}]$ are the appropriate partial resistance matrices of the remaining meshes in the equivalent Eqs. (2.11) and (2.12) [when replacing Z_{12}, Z_{21} and Z_{22} by their equivalents R_{12}, R_{21} and R_{22} respectively] are,

$$[R'] = [R_{11}] - [R_{12}]\,[R_{22}]^{-1}\,[R_{21}]$$

That is:

$$[R'] = \big[[R_a] + [R_n]\big] - [R_{12}][R_{22}]^{-1}[R_{21}]$$
$$= [R_a] + [R_\alpha]$$

It follows that the two circuits shown in Figure 2.1(a) and Figure 2.1(b) are also equivalent as far as resistances are concerned. After the reduction using compound matrix concepts highlighted in Section 22.3, the current in the mesh (1) of Figure 2.1(a) will be given by:

$$[V'] = [R']\,[I_1]$$

Figure 2.1(b) clearly indicates that the same current $[I_1]$ flows through the equivalent circuit of Figure 2.1(b).

Physical interpretation of the above reduction process illustrates that the equivalence of the two depicted network circuits, is general in nature. By comparing them, it will be seen that:

(a) $[V_\alpha]$ must be the voltage that appears across the terminals A–B of Figure 2.1(a), if the switch S is opened, and

(b) $[R_\alpha]$ must be the resistance measured across the terminals A–B of Figure 2.1(a), if all the voltage sources in the right hand portion of the network are treated as shorted.

This result constitutes Thevenin's theorem, which may be stated as follows:

'Any network of resistance elements and voltage sources, if reviewed from any point in a network, may be replaced by a voltage source and an equivalent resistance in series.' Thus, if V_a and R_a denote the new branch of the network, the current in the new branch is the equivalent of the voltage in the branch, V_a, added to the voltage of the remaining part, V_α between the points of connection—A–B, before the connection is completed, divided by the new branch resistance, R_a, added to the resistance, R_α, looking into the original network from the points of connection A–B.

If we refer to Figure 2.1(a), the current flowing with the resistance R_a short-circuited is assumed to be I_0. Then, from Figure 2.1(b), we have

$$I_0 = \frac{V_a - V_\alpha}{R_\alpha}, \quad \text{and} \quad I_1 = \frac{V_a - V_\alpha}{R_a + R_\alpha}$$

But,

$$I_1 = I_0\,\frac{R_\alpha}{R_a + R_\alpha} = I_0\,\frac{1}{1 + \left(\dfrac{R_a}{R_\alpha}\right)}$$

That is, the current in branch A–B after the insertion of a new resistance R_a, is equal to the current I_0 before the insertion divided by an amount $\left(1 + \dfrac{R_a}{R_\alpha}\right)$, where, R_α is the resistance looking into the network from A–B before the addition of R_a.

This is the Norton's theorem, stated in words.

Delta-Star Transformation

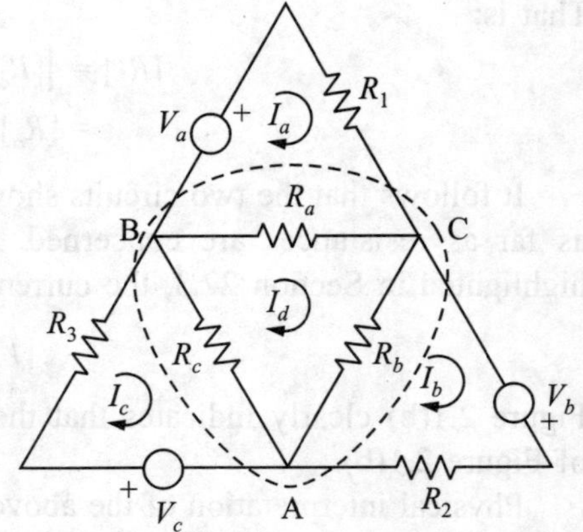

FIGURE 2.2 A four-mesh resistance network, in the form of delta-connection.

In this section, we wish to apply the compound matrix concepts to arrive at the standard relations obtained in delta-star transformation observed under three-phase circuit analysis. Let us consider the four-mesh resistance network encountered in the delta-connection mode, indicated in Figure 2.2, which is the normal form of a delta-connected system used in three-phase circuits. We ultimately, concentrate on the triangular portion marked by A-B-C and encircled by a dotted circle, in the diagram.

We apply Maxwell's rule, detailed later in Section 2.4, to write down the voltage equations of each mesh:

$$V_a = (R_a + R_1)I_a - 0.I_b - 0.I_c - R_a I_d$$
$$V_b = -0.I_a + (R_b + R_2)I_b - 0.I_c - R_b I_d$$
$$V_c = -0.I_a - 0.I_b + (R_c + R_3)I_c - 0.I_d$$
$$V_d = -R_a I_a - R_b I_b - R_c I_c + (R_a + R_b + R_c)I_d$$

We examine the above voltage equations and write down the $[V]$, $[I]$ and $[R]$ matrices. These are as follows:

$$[V] = \begin{bmatrix} V_a \\ V_b \\ V_c \\ \hdashline 0 \end{bmatrix}; \; [I] = \begin{bmatrix} I_a \\ I_b \\ I_c \\ \hdashline I_d \end{bmatrix};$$

$$[R] = \left[\begin{array}{ccc:c} (R_a + R_1) & 0 & 0 & -R_a \\ 0 & (R_b + R_2) & 0 & -R_b \\ 0 & 0 & (R_c + R_3) & -R_c \\ \hdashline -R_a & -R_b & -R_c & (R_a + R_b + R_c) \end{array}\right]$$

We arrange to eliminate the current I_d. For this, we partition the voltage matrix $[V]$ and the current matrix $[I]$, both of which are common matrices by drawing horizontal lines between the third and the fourth rows. Similarly, we partition the resistance matrix by drawing horizon and vertical dotted lines between the third and fourth rows and columns as shown. Using the compound matrix concepts, we can write the values of part-matrices as follows:

$$[V_1] = \begin{bmatrix} V_a \\ V_b \\ V_c \end{bmatrix}; \quad [V_2] = [0]; \quad [I_1] = \begin{bmatrix} I_a \\ I_b \\ I_c \end{bmatrix}; \quad [I_2] = [I_d]$$

$$[R_{11}] = \begin{bmatrix} (R_a + R_1) & 0 & 0 \\ 0 & (R_b + R_2) & 0 \\ 0 & 0 & (R_c + R_3) \end{bmatrix}; \quad [R_{12}] = \begin{bmatrix} -R_a \\ -R_b \\ -R_c \end{bmatrix}$$

$$[R_{21}] = [-R_a \quad -R_b \quad -R_c] \quad [R_{22}] = [R_s] = [R_a + R_b + R_c]$$

We apply the method of compound matrices to eliminate I_2. For this, we work out the inverse of $[R_{22}]$, first.

Inverse of $[R_{22}]$ is:

$$[R_{22}]^{-1} = \frac{1}{R_s} = \frac{1}{(R_a + R_b + R_c)}$$

And,

$$[R_{12}]\ [R_{22}]^{-1} = \frac{1}{R_s}\begin{bmatrix} -R_a \\ -R_b \\ -R_c \end{bmatrix}$$

so that,

$$[R_{12}]\ [R_{22}]^{-1}\ [R_{21}] = \frac{1}{R_s}\begin{bmatrix} -R_a \\ -R_b \\ -R_c \end{bmatrix} [-(R_a + R_b + R_c)]$$

$$= \frac{1}{R_s}\begin{bmatrix} R_a^2 & R_aR_b & R_aR_c \\ R_aR_b & R_b^2 & R_bR_c \\ R_aR_c & R_bR_c & R_c^2 \end{bmatrix}$$

Hence,

$$[R'] = [R_{11}] - [R_{12}]\ [R_{22}]^{-1}\ [R_{21}]$$

$$= \begin{bmatrix} (R_a + R_1) & 0 & 0 \\ 0 & (R_b + R_2) & 0 \\ 0 & 0 & (R_c + R_3) \end{bmatrix} - \begin{bmatrix} \frac{R_a^2}{R_s} & \frac{R_aR_b}{R_s} & \frac{R_aR_c}{R_s} \\ \frac{R_aR_b}{R_s} & \frac{R_b^2}{R_s} & \frac{R_bR_c}{R_s} \\ \frac{R_aR_c}{R_s} & \frac{R_bR_c}{R_s} & \frac{R_c^2}{R_s} \end{bmatrix}$$

or

$$[R'] = \begin{bmatrix} R_1 + \frac{R_aR_b}{R_s} + \frac{R_bR_c}{R_s} & -\frac{R_aR_b}{R_s} & -\frac{R_aR_c}{R_s} \\ -\frac{R_aR_b}{R_s} & R_2 + \frac{R_aR_b}{R_s} + \frac{R_bR_c}{R_s} & -\frac{R_bR_c}{R_s} \\ -\frac{R_aR_c}{R_s} & -\frac{R_bR_c}{R_s} & R_3 + \frac{R_aR_b}{R_s} + \frac{R_aR_c}{R_s} \end{bmatrix} \qquad ...(2.14)$$

As per compound matrix rules, we find that we can write down the modified voltage equations of the system, after elimination of the $[I_2]$ term, as follows:

$$[V'] = [R'] [I_1]$$

where,

$$[V'] = [V_1] - [R_{12}] [R_{22}]^{-1} [V_2]$$

Since, $[V_2] = 0$

$$[V'] = [V_1] = \begin{bmatrix} V_a \\ V_b \\ V_c \end{bmatrix}$$

and hence, following are the modified voltage equations:

$$\begin{bmatrix} V_a \\ V_b \\ V_c \end{bmatrix} = \begin{bmatrix} \left(R_1 + \frac{R_a R_b}{R_s} + \frac{R_b R_c}{R_s}\right) & -\left(\frac{R_a R_b}{R_s}\right) & -\left(\frac{R_a R_c}{R_s}\right) \\ -\left(\frac{R_a R_b}{R_s}\right) & \left(R_2 + \frac{R_a R_b}{R_s} + \frac{R_b R_c}{R_s}\right) & -\left(\frac{R_b R_c}{R_s}\right) \\ -\left(\frac{R_a R_c}{R_s}\right) & -\left(\frac{R_b R_c}{R_s}\right) & \left(R_3 + \frac{R_a R_b}{R_s} + \frac{R_a R_c}{R_s}\right) \end{bmatrix} \begin{bmatrix} I_a \\ I_b \\ I_c \end{bmatrix}$$

Let us now consider a three-mesh arrangement encountered in the equivalent star-connection in the region A-B-C, shown in Figure 2.3, suitably encircled by a dotted circle, and write down the voltage equations of the three constituent meshes. In writing these equations, we assume the voltages as identical to those of the three branches shown along the periphery in the diagram in both schemes of Figure 2.2 and Figure 2.3, and also assume loop currents I_a, I_b and I_c, in both arrangements. Only under these conditions, we established that the two systems would be equivalent to each other about the encircled portion, if the voltages and currents in both cases are identical.

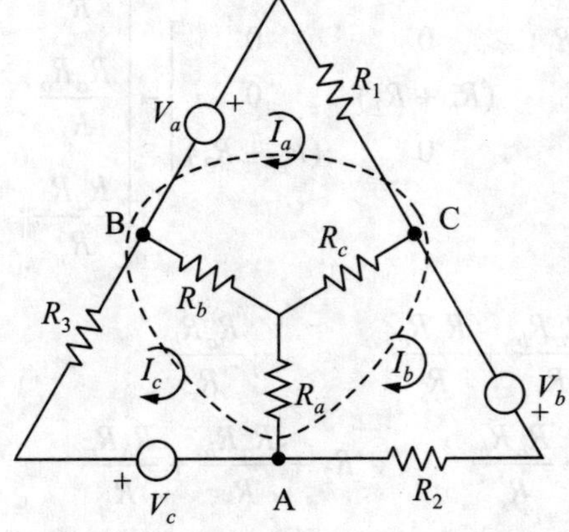

FIGURE 2.3 An equivalent three-mesh network of the circuit arrangement shown in Figure 2.2.

We write down the voltage equations of individual loops using the principles of Maxwell's rule elaborated in Section 2.4:

$$V_a = \{R_1 + (R_B + R_C)\}I_a - (R_C I_b) - (R_B I_c)$$
$$V_b = -(R_C I_a)\ \{R_2 + (R_A + R_C)\}I_b - (R_A I_c)$$
$$V_c = -(R_B I_a) - R_A I_b\ \{R_3 + (R_B + R_A)\}I_c$$

We can, easily, identify the voltage, current and resistance matrices, as follows:

$$[V]=\begin{bmatrix} V_a \\ V_b \\ V_c \end{bmatrix};\quad [I]=\begin{bmatrix} I_a \\ I_b \\ I_c \end{bmatrix};\quad [R]=\begin{bmatrix} \{R_1+(R_B+R_C)\} & -R_C & -R_B \\ -R_C & \{R_2+(R_A+R_C)\} & -R_A \\ -R_B & -R_A & \{R_3+(R_A+R_B)\} \end{bmatrix} \qquad ...(2.15)$$

Let us compare the $[R]$ matrix of Eq. (2.15) with the $[R']$ matrix of Eq. (2.14). We observe that, for equivalence of the two system in Figure 2.2 and Figure 2.3, in the encircled region:

$$R_A = \frac{R_b R_c}{R_a + R_b + R_c};\quad R_B = \frac{R_c R_a}{R_a + R_b + R_c};\quad R_C = \frac{R_a R_b}{R_a + R_b + R_c}$$

To highlight the above equivalence results, we refer to Figure 2.4. We describe the result as follows:

'The star arm of an equivalent system converging to the vertex point of a delta-connected resistance system, has value equalling the product of resistances of the original delta-arms converging towards the same vertex of the triangle, forming the original delta-connection, divided by the sum of all the constituent resistances forming the delta-connection.'

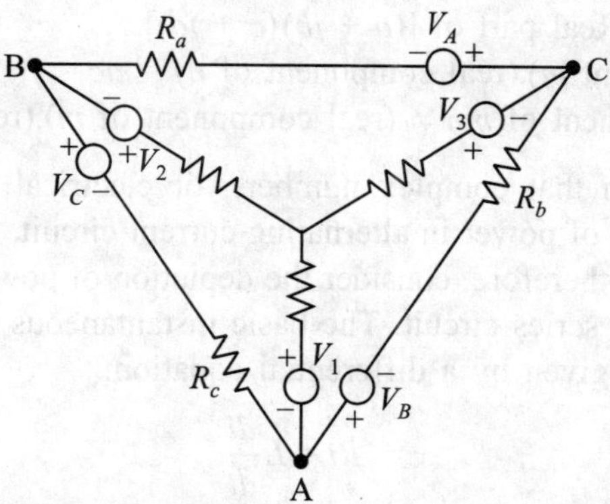

FIGURE 2.4 Delta-star transformation of voltages.

$$V_1 = \frac{V_C R_b - V_B R_c}{R_s}\quad V_2 = \frac{V_A R_c - V_C R_a}{R_s}\quad V_3 = \frac{V_B R_a - V_A R_b}{R_s}$$
$$R_s = R_a + R_b + R_c$$

Figure 2.4 shows the voltage transformation of delta to star conversion. It is left to the students to verify the above transformation.

2.2.5 Power in Alternating Current Circuits Using Phasor Depiction

A discussion on the concept of power associated with voltages and currents expressed as phasors, is desirable to overcome any confusion and difficulty in its conceptualization. The application of complexors to the solution of circuits carrying sinusoidal alternating currents is widely practiced. But complex numbers can be used with desired correct results only such that *the real part of the resulting function equals the function of the real parts.*

The only operations occurring in analysis of elementary circuits are addition (or subtraction) and differentiation of complex numbers. Let us consider two complex numbers, $m = (a + jb)$ and $n = (c + jd)$.

- When added, the real part of $(m + n)$ is $(a + c)$, or in other words:
  Real part of $(m + n)$ = Real part of m + Real part of n
- For differentiation, let the general complex number, z, expressed in polar form be $z = ke^{j\omega t}$. Then,

$$\text{Real part of } \frac{dz}{dt} = \text{Real part of } (j\omega k e^{j\omega t})$$

$$= \text{Real part of } j\omega k\,(\cos \omega t + j \sin \omega t)$$

$$= -\,\omega k \sin \omega t.$$

  and, a similar result is also obtained for integration.
- For multiplication of two complex numbers:
  Real part of mn = Real part of $[(a + jb)(c + jd)]$.
  But, (real component m).(real component of n) = ac
  Hence, real component of $mn \neq$ (real component of m).(real component of n)

It may, therefore, seem that complex numbers, or electrical phasors cannot be usefully employed for representation of power in alternating current circuit, as it involves the product of voltage and current. Let us, therefore, consider the depiction of power especially. Let us choose a simple resistive–inductive series circuit. The basic instantaneous voltage–current relationship governing such a circuit is given by a differential equation:

$$v = Ri + L\frac{di}{dt}$$

and, under steady-state condition, the current varies sinusoidally, which we can denote by $i = I_{\max}\cos \omega t$. If R and L are constants, then

$$v = RI_{\max} \cos \omega t - \omega L\, I_{\max} \sin \omega t$$

$$= ZI_{\max} \cos (\omega t + \phi)$$

$$= V_{\max} \cos (\omega t + \phi)$$

where

$$Z = \sqrt{R^2 + (\omega L)^2}$$

$$\phi = \tan^{-1} \frac{\omega L}{R}$$

$$V_{max} = ZI_{max}$$

This relationship can be conveniently written as follows:

$$V_{max} \cos (\omega t + \phi) = RI_{max} \cos \omega t - \omega L\, I_{max} \sin \omega t$$

And, by using the identities:

$$e^{j\theta} = \cos \theta + j \sin \theta$$

and

$$e^{-j\theta} = \cos \theta - j \sin \theta$$

we have the following standard relations

$$\cos \theta = \frac{1}{2}(e^{j\theta} + e^{-j\theta}) \quad \text{and} \quad \sin \theta = \frac{1}{2} j(e^{j\theta} - e^{-j\theta})$$

We arrive at the following relation by simple substitution:

$$V_{max}[e^{j(\omega t + \phi)} + e^{-j(\omega t + \phi)}] = RI_{max}[e^{j\omega t} + e^{-j\omega t}] + j\omega L\, I_{max}[e^{j\omega t} - e^{-j\omega t}]$$

Break the above equation into two parts arbitrarily as follows:

$$V_{max}e^{j(\omega t + \phi)} = (R + j\omega L)I_{max}e^{j\omega t} \qquad ...(2.16)$$

and

$$V_{max}e^{-j(\omega t + \phi)} = (R - j\omega L)I_{max}e^{-j\omega t} \qquad ...(2.17)$$

Both the part-equations satisfy the condition involved in the original equation that *the real part of the function equals the function of the real parts.*

In sinusoidal systems,

$$I_{max} = \sqrt{2}\, I_{rms} \quad \text{and} \quad V_{max} = \sqrt{2}\, V_{rms}$$

the rms values V and I are introduced in Eqs. (2.16) and (2.17) in place of V_{max} and I_{max}, and these equations reduce to:

$$Ve^{j(\omega t + \phi)} = (R + j\omega L)I\, e^{j\omega t} \qquad ...(2.18)$$

$$Ve^{-j(\omega t + \phi)} = (R - j\omega L)I\, e^{-j\omega t} \qquad ...(2.19)$$

As time progresses, both sets of phasors rotate with a uniform angular velocity, ω. The rotation is in the positive anti-clockwise direction for the first equation and in the clockwise direction for the second equation. It is conventional to accept the first equation and ignore the second equation. We can simply express Eq. (2.18), after replacing $Ve^{j(\omega t + \phi)}$ by V and $I\, e^{j\omega t}$ by I, by notionally introducing phasor concepts in which the magnitude and phasor positions of voltage and current are maintained along with their frequency. The simplified equation, therefore, assumes the form:

$$V = (R + j\omega L)I$$

We go back a little and manipulate Eqs. (2.18) and (2.19) by multiplying Eq. (2.18) by $e^{-j\omega t}$ and Eq. (2.19) by $e^{j\omega t}$, on both sides:

Equation (2.18) yields:

$$V e^{j(\omega t + \phi)} e^{-j\omega t} = (R + j\omega L)I\, e^{j\omega t} e^{-j\omega t}$$

or

$$V e^{j\phi} = (R + j\omega L)I \quad \text{...(2.20)}$$

and Eq. (2.19) results:

$$V e^{-j\phi} = (R - j\omega L)I \quad \text{...(2.21)}$$

If we critically examine the nature of Eqs. (2.20) and (2.21), we observe that the rotation of the representative phasors has disappeared, but the lagging/leading phasor relations are maintained. The significance of the new result is better appreciated by considering the Argand diagram of the two Eqs. (2.18) and (2.19) on the same diagram, shown in Figure 2.5.

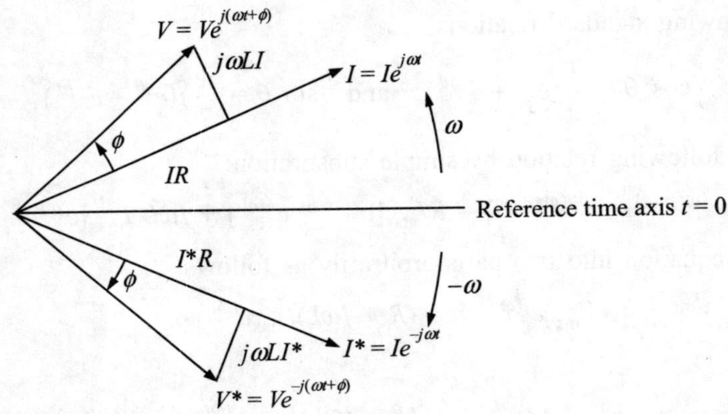

FIGURE 2.5 Argand diagram of Eqs. (2.18) and (2.19), drawn on the same diagram.

As the instantaneous power is given by (vi), we have:

$$p = vi = V_{max} I_{max} \cos(\omega t + \phi) \cos \omega t \quad \text{...(2.22)}$$
$$= 2VI \cos(\omega t + \phi) \cos \omega t$$
$$= VI [\cos(2\omega t + \phi) + \cos \phi]$$

the power is, thus, constituted of following two components:

(a) A component pulsating at double the frequency of the supply, such that, when averaged over a cycle, the net power constituted by this component is zero.
(b) A component that is independent of time, and depends on the rms values of the voltage, the current and the phase angle separating them.

The mean power over a cycle is therefore:

$$p = VI \cos \phi$$

In Eq. (2.22), the exponential form of representation of cos ϕ is introduced again and the expression becomes:

$$p = V_{max} I_{max} \left[\frac{e^{j(\omega t + \phi)} + e^{-j(\omega t + \phi)}}{2}\right]\left[\frac{e^{j\omega t} + e^{-j\omega t}}{2}\right]$$

$$= \frac{1}{2} VI \{e^{j(2\omega t + \phi)} + e^{-j\phi} + e^{j\phi} + e^{-j(2\omega t + \phi)}\} \quad ...(2.23)$$

In conventional treatment of power in phasor form, we ignore the pulsating component and consider only the mean power, $p = VI \cos \phi$. Critical examination of Eq. (2.23) reveals that:

$$p = \frac{1}{2} VI (\cdots + e^{-j\phi} + e^{j\phi} + \cdots)$$

The real parts of both $\frac{1}{2}VIe^{-j\phi}$ and $\frac{1}{2}VIe^{j\phi}$ are equal, and each of these constitutes half of the mean power. Thus, if we plot the two alternative vector representations denoted by Eqs (2.16) and (2.17) in Figure 2.6 through their Argand diagrams on the same figure, we observe that the clockwise rotating vectorial depictions are the conjugates of the corresponding counter-clockwise rotating phasors about the chosen reference axis of time and their real parts are equal, and imaginary parts are equal and opposite. In phasor/vector representation, it is a standard practice to denote conjugates of phasor/vector V and I as V^* and I^* respectively, such that:

$$V^* = Ve^{-j(\omega t + \phi)}$$

$$I^* = Ie^{-j\omega t}$$

Hence, we finally arrive at the final important relationship of power denoted in phasorical/vectorial form:

$$p = \text{Real part of } (V^* I)$$

or $$p = \text{Real part of } (VI^*)$$

Expression of Power using Matrix Algebra

We know that both $[V]$ and $[I]$ are column matrices, and hence, the first of the two matrices has to be altered into a row matrix by transposition. The result, which is a single quantity may, following the rules of matrix multiplication, be written either as:

$$p = [V_t][I] \quad \text{or} \quad p = [I_t][V]$$

For sinusoidal operation of a general circuit, power can be obtained by any one of the following four relations:

1. p = Real part of $[V_t^*][I]$
2. p = Real part of $[I_t^*][V]$
3. p = Real part of $[V_t][I^*]$
4. p = Real part of $[I_t][V^*]$

2.2.6 Transpose of a Matrix Product

Transpose of the product of two or more matrices is obtained by the product of the transposed matrices, but with the order of the multiplication reversed. Thus:

$$[[A][B]]_t = [B_t][A_t]$$

and
$$[[A][B][C]]_t = [C]_t [B]_t [A]_t$$

Let $[A]$ have a rows and b columns, and we denote the matrix $[A]$ by the symbol ${}_aA_b$. Then, for a multiplication to be possible with $[B]$, it should, necessarily, have b rows and c columns, and it may be described by the symbol ${}_bB_c$ according to the above convension, and its transverse has c rows and b columns. The product must lead to a matrix with a rows and c columns, while its transverse has c rows and a columns.

$$[A][B] = {}_aA_b\, {}_bB_c = {}_a[AB]_c$$

and
$$[[A][B]]_t = [[B]_t [A]_t] = {}_cB_b\, {}_bA_a = {}_c[BA]_a$$

We take up the following illustration to explain the point.

ILLUSTRATION 2.2 Let ${}_2[A]_3 = \begin{bmatrix} a & b & c \\ d & e & f \end{bmatrix}$ and ${}_3[B]_2 = \begin{bmatrix} g & h \\ i & j \\ k & l \end{bmatrix}$

Then,
$${}_2[A]_3\ {}_3[B]_2 = \begin{bmatrix} a & b & c \\ d & e & f \end{bmatrix} \begin{bmatrix} g & h \\ i & j \\ k & l \end{bmatrix} = \begin{bmatrix} ag+bi+ck & ah+bj+cl \\ dg+ei+fk & dh+ej+fl \end{bmatrix}$$

Also,
$${}_2[B_t]_3 \cdot {}_3[A_t]_2 = \begin{bmatrix} g & i & k \\ h & j & l \end{bmatrix} \begin{bmatrix} a & d \\ b & e \\ c & f \end{bmatrix} = \begin{bmatrix} ag+bi+ck & dg+ei+fk \\ ah+jb+cl & dh+ej+fl \end{bmatrix}$$

Hence,
$$[[A][B]]_t = \begin{bmatrix} ag+bi+ck & dg+ei+fk \\ ah+bj+cl & dh+ej+fl \end{bmatrix}$$

Hence,
$$[[A][B]]_t = [B_t][A_t]$$

2.2.7 Quadratic Forms of Matrices

Since a one-by-one matrix is a single quantity, the transpose of such a matrix is the same matrix itself. Thus:

$$[p] = [V_t][I]$$

Then,
$$[p_t] = [p] = [V_t][I] = [I_t][V_t] = [I_t][V]$$

We can reiterate that in a direct current network, the voltage and current matrices are connected by a simple relationship:

$$[V] = [R][I]$$

And hence,

$$[V]_t = [R_t][I_t]$$

But, power could be denoted by:

$$[p] = [V_t][I] = [I_t][R_t]\ [I]$$

And also

$$[p] = [I_t][V] = [I_t][R]\ [I]$$

The two values of power must be conceptually identical. This observation is supported by the fact that the resistance matrix of a d.c. circuit is symmetrical (see Illustration 2.1, that is, $[R_t] = [R]$). The quantity $[I_t]\ [R]\ [I]$ is known as a *quadratic form*, and can be expressed in the simple form as $R.I^2$. A quadratic form, given above, occurs in the expression for the torque (T) of a *commutator motor*.

$$T = \frac{\text{Power}}{\text{Angular velocity}} = \frac{p}{\omega}$$

and, ω is a scalar quantity, hence, the matrix form of torque and power are similar. Hence, we may write:

$$[T] = [I_t][G][I]$$

$$\therefore \qquad [T_t] = [T] = \big[[I_t][G][I]\big]_t = [I_t][G_t][I] \qquad \text{...(2.24)}$$

as $[I_t]_t = [I]$. But $[G]$ is not symmetrical and $[G_t] \neq [G]$.

But the quadratic forms arrived at earlier are identical. This apparent anomaly leads to the final theorem in matrix algebra:

'Any matrix, $[G]$, may be expressed as the summation of two matrices—a symmetrical matrix $[G_+]$ and an anti-symmetrical matrix $[G_-]$', because:

$$[G] = \frac{1}{2}[[G]+[G_t]] + \frac{1}{2}[[G]-[G_t]]$$

$$= [G_+] + [G_-]$$

where,

$$[G_+] = \frac{1}{2}[[G]+[G_t]] \text{ and } [G_-] = \frac{1}{2}[[G]-[G_t]]$$

We indicate the above by considering a general example.

Take a matrix $[G]$, given by

$$[G] = \begin{bmatrix} a & b \\ c & d \end{bmatrix}$$

Then,

$$[G_t] = \begin{bmatrix} a & c \\ b & d \end{bmatrix}$$

Therefore,

$$[G_+] = \frac{1}{2}[[G]+[G_t]] = \begin{bmatrix} a & \frac{1}{2}(b+c) \\ \frac{1}{2}(b+c) & d \end{bmatrix}$$

And

$$[G_-] = \frac{1}{2}[[G]-[G_t]] = \begin{bmatrix} 0 & \frac{1}{2}(b-c) \\ -\frac{1}{2}(b-c) & 0 \end{bmatrix}$$

But, earlier we had arrived at a result, see Eq. (2.24):

$$[T] = [I_t][G][I] = [I_t][G_t][I]$$

But,

$$[T] = [I_t][G][I] = [I_t][[G_+]+[G_-]][I]$$
$$= [I_t][G_+][I] + [I_t][G_-][I] \qquad \text{...(2.25)}$$

And

$$[T] = [I_t][G_t][I] = [I_t][[G_+]-[G_-]][I]$$

That is

$$[T] = [I_t][G_+][I] - [I_t][G_-][I] \qquad \text{...(2.26)}$$

A close examination of Eqs. (2.25) and (2.26) reveals that:

$$[T] = [I_t][G_+][I] \quad \text{and,} \ [I_t][G_-][I] = 0 \qquad \text{...(2.27)}$$

This leads us to the theorem, which can be stated as follows:

'The quadratic form of an anti-symmetric matrix is zero.'

In the expression for torque in Eq. (2.27), the matrix appearing twice is a column matrix, that is, a current (or voltage) matrix. Let $[G]$ have a form:

$$[G] = \begin{bmatrix} 0 & L \\ -L & 0 \end{bmatrix}, \quad \text{and} \quad [I] = \begin{bmatrix} I_1 \\ I_2 \end{bmatrix}$$

Then, torque $[T] = [I_t]\ [G]\ [I]$

$$= \begin{bmatrix} I_1 & I_2 \end{bmatrix} \begin{bmatrix} 0 & L \\ -L & 0 \end{bmatrix} \begin{bmatrix} I_1 \\ I_2 \end{bmatrix}$$

$$= [I_1 L I_2 - I_2 L I_1] = 0$$

Thus, in simplifications, the quadratic form of central matrix is amenable to its replacement by the symmetrical part. For example, if we take in a typical torque matrix, the [*G*] component of the form:

$$[G] = \begin{bmatrix} 0 & 0 & 0 \\ -M & 0 & -L \\ 0 & +L & 0 \end{bmatrix}$$

only the single term with $-M$ may be evaluated and the torque is simply:

$$T = -I_2 M I_1$$

or, it may be modified to

$$T = -M I_1 I_2$$

2.3 REVIEW OF DETERMINANTS AND THEIR EVALUATION

An *n*th order determinant is a set of n^2 quantities. Each of the n^2 term is an *element* of the determinant. These elements are arranged in ordered *n* rows and *n* columns, and are essentially square in nature.

Unlike a matrix, a determinant:

1. Must be square in nature.
2. Must have an equivalent single value other than zero.

For *first*, *second* and *third order* determinants, these values are denoted as follows:

(a) First order determinant,
Depiction: $|a|$; Value: a

(b) Second order determinant,
Depiction: $\begin{vmatrix} a & b \\ c & d \end{vmatrix}$; Value: $(ad - bc)$

(c) Third order determinant,

Depiction: $\begin{vmatrix} a & b & c \\ d & e & f \\ g & h & i \end{vmatrix}$

Value:
$$a\begin{vmatrix} e & f \\ h & i \end{vmatrix} - b\begin{vmatrix} d & f \\ g & i \end{vmatrix} + c\begin{vmatrix} d & e \\ g & h \end{vmatrix}$$

$$= a(ei - fh) - b(di - fg) + c(dh - eg)$$

or
$$-d\begin{vmatrix} b & c \\ h & i \end{vmatrix} + e\begin{vmatrix} a & c \\ g & i \end{vmatrix} - f\begin{vmatrix} a & b \\ g & h \end{vmatrix}$$

$$= -d(bi - ch) + e(ai - cg) - f(ah - bg)$$

$$\text{or} \quad +g\begin{vmatrix} b & c \\ e & f \end{vmatrix} - h\begin{vmatrix} a & c \\ d & f \end{vmatrix} + i\begin{vmatrix} a & b \\ d & e \end{vmatrix}$$

$$= g(bf - ce) - h(af - de) + i(ae - bd)$$

Thus, to find out the value of the determinant:

1. If the elements of the row and column containing a particular element of a determinant are struck off, the remaining determinant of the next lower order is called the *minor of that element.*
2. A 'minor' with appropriate sign attached to it is called a *co-factor*.
3. The *sign* attached to a co-factor in the first row and the first column is positive. If *we add the row and column positions of an element and the sum is an odd number, the sign associated with it is negative.* If the *sum is an even number, then the sign associated with the co-factor is positive.* This is indicated in the previous expansions depicted for the three-by-three determinant.
4. In estimating the value of a determinant, we can calculate by considering terms contained in any one row and their co-factors.

2.4 MAXWELL'S CYCLIC CURRENT RULE

Following are the Maxwell's cyclic current rules for writing down the voltage-current equation of circuits.

1. Assume clockwise current in each loop. These currents are hypothetical and do not have any direct bearing on the net voltage and its effective direction in a loop.
2. Observe any loop, following the chosen direction of the assumed loop current. Determine the effective potential difference around the loop by considering voltage rise based on the assumed current direction as positive and the reverse as negative.
3. In colating the resistance/inductance matrix, there has to be one positive term equal to the sum of all resistance/inductance constituting the loop; the other terms which carry currents of other loops also, have to be taken as negative.
4. Voltages form a column matrix on the left hand side of the resulting matrix equation for the network.
5. The current matrix is also a column matrix with all positive currents (which are assumed as flowing in clockwise direction).
6. Then, the equations are written down in the matrix format as follows:

$$[V] = [R][I] \quad \text{(for d.c. circuits)}$$

or

$$[I] = [R]^{-1}[V]$$

Let us illustrate this procedure with the help of a simple static d.c. circuit problem.

ILLUSTRATION 2.3 Let Figure 2.6 depict a three-mesh network selected arbitrarily.

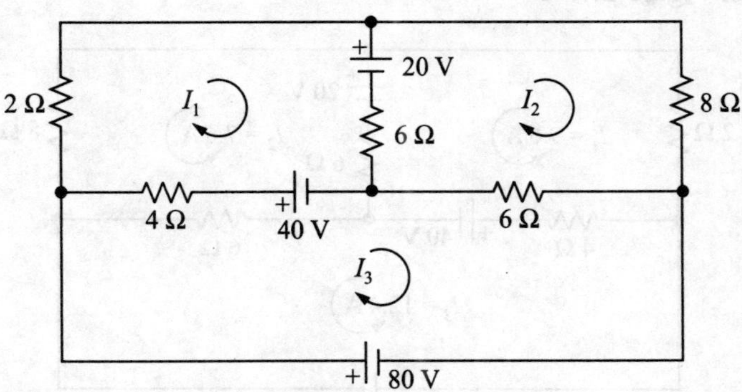

FIGURE 2.6 An arbitrarily chosen three-mesh static d.c. circuit problem.

By following the rules laid down by Maxwell, loop-1 equation is:

$$(+40 - 20) = (2 + 6 + 4)I_1 - 6I_2 - 4I_3$$

For loop-2, the *V-I* relationship is:

$$+20 = -6I_1 + (8 + 6 + 6)I_2 - 6I_3$$

and the equation for loop-3, can be similarly obtained as follows:

$$(+80 - 40) = -4I_1 - 6I_2 + (4 + 6)I_3$$

The hypothetical loop equations, therefore, assumes the form:

$$\begin{aligned} 20 &= 12I_1 - 6I_2 - 4I_3 \\ 20 &= -6I_1 + 20I_2 - 6I_3 \\ 40 &= -4I_1 - 6I_2 + 10I_3 \end{aligned}$$

which can be expressed in matrix form as follows:

$$\begin{bmatrix} V_1 = 20 \\ V_2 = 20 \\ V_3 = 40 \end{bmatrix} = \begin{bmatrix} 12 & -6 & -4 \\ -6 & 20 & -6 \\ -4 & -6 & 10 \end{bmatrix} \cdot \begin{bmatrix} I_1 \\ I_2 \\ I_3 \end{bmatrix}$$

As we have already worked out the inverse of [*R*] matrix in Eq. (2.8),

$$\therefore \qquad [I] = [R]^{-1}[V] = \frac{1}{1000}\begin{bmatrix} 164 & 84 & 116 \\ 84 & 104 & 96 \\ 116 & 96 & 204 \end{bmatrix}\begin{bmatrix} 20 \\ 20 \\ 40 \end{bmatrix}$$

or

$$[I] = \begin{bmatrix} I_1 \\ I_2 \\ I_3 \end{bmatrix} = \begin{bmatrix} 9.6 \\ 7.6 \\ 12.4 \end{bmatrix}$$

The arbitrarily chosen network of Figure 2.6, therefore, has Maxwell loop current distribution pattern as shown in Figure 2.7.

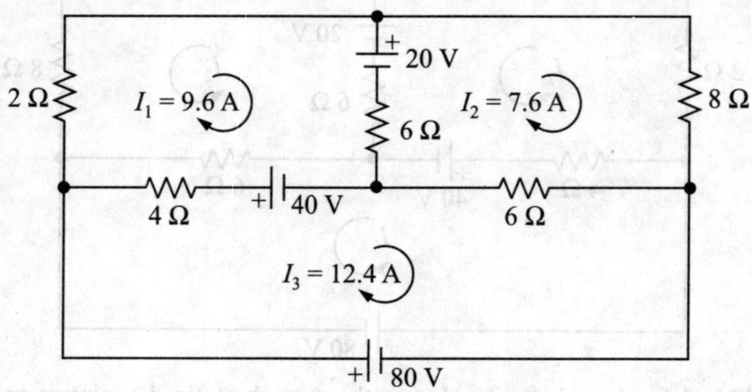

FIGURE 2.7 Solution of d.c. circuit problem chosen in Figure 2.6, by Maxwell's loop current distribution method, using matrix voltage–current equations.

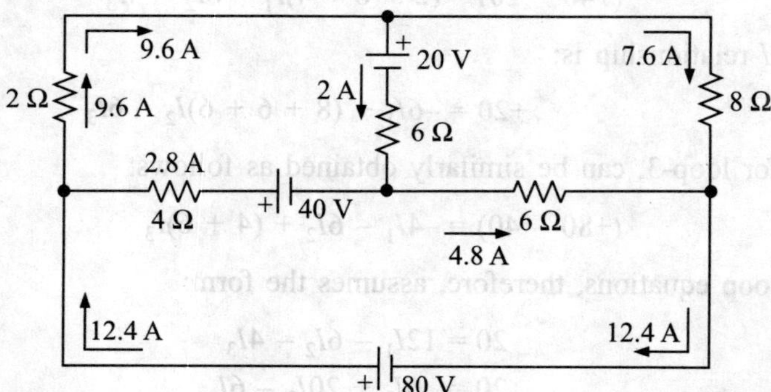

FIGURE 2.8 Estimated values of the circuit currents carried by each circuit element.

It is interesting to note that Maxwell's loop currents, as calculated, do not have any direct relevance to the actual currents carried by the circuit elements, as shown in Figure 2.8.

CHAPTER 3

Energy Storage in Magnetic Fields

3.1 INTRODUCTION

In the introductory chapter, we learnt about the mechanism of handling energy in electro-mechanical energy converters, reviewed the basic laws on which operation of these devices are based and examined the basic circuit parameters of resistance, inductance and capacitance keeping the circuit, energy and physical viewpoints as used in modelling.

In this chapter, we proceed in areas of energy storage in magnetic fields and try to polarise into configurations that we are likely to encounter with special reference to electro-mechanical energy converters, including the production of torque/force in them. In the process we shall be introduced to the concepts of *energy and co-energy* associated with magnetic circuits, with multiplicities of complications. Prior to this, we shall also briefly acquaint ourselves with different kinds of transducers on which the principles are applicable.

The basic quantities dealing with the electro-mechanical energy converters are the three important properties, to be considered when energy is involved. These are as follows:

1. Energy can be stored
2. Energy can be dissipated
3. Energy can be converted from one form to another

It is observed that an important use of the field is in storage of energy. With the force-field concept, it can be seen that it can act as a reservoir of energy which may be readily exploited by freely drawing from fields whenever required. A quantitative evaluation of energy storage properties of most commonly encountered arrangements of fields in electro-mechanical systems including machines is briefly elaborated in the following section.

While there can be a very large variety of systems in which energy can be stored they can all be included under two broad headings, for which we can enumerate a general treatment and obtain general equations. These may, then, be applied to particular cases. These are as follows:

1. Singly-excited systems
2. Multiply-excited systems

An application of the very simple general equations so derived in these considerations, serves as the system followed in this section.

3.2 SINGLY-EXCITED SYSTEMS

A single stationary coil of N turns, carrying a current i shown in Figure 3.1, produces a magnetic field around itself. The coil could be in a relay, loudspeaker or a single-phase motor. If no energy is dissipated, and since there is, at present, no mechanical movement, all the energy supplied by the source producing the current is stored in the magnetic field of the coil.

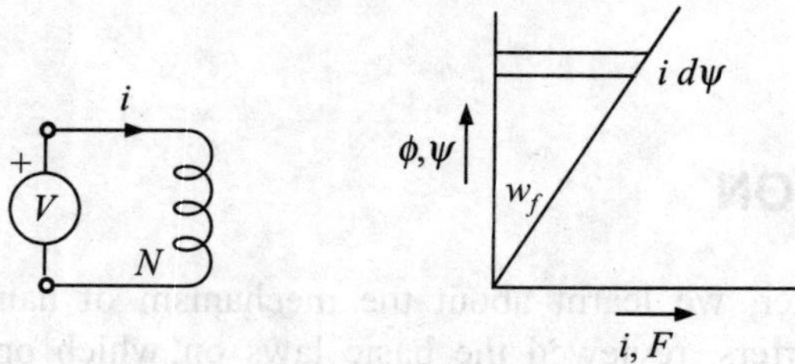

FIGURE 3.1 Singly-excited system with linear magnetization characteristic (ϕ–flux; ψ–flux-linkage; i–current and $F = Ni$ = MMF).

If the current increases to $(i + di)$, during the time interval dt, the flux augments to $(\phi + d\phi)$ and a voltage (e) is induced according to Faraday's and Lenz's law,

$$e = -N\frac{d\phi}{dt}$$

which is directed in opposition to the applied voltage. In order to maintain the applied voltage and current, the energy input from the source to the coil is as follows:

$$\begin{aligned} dW_{\text{elect}} &= e\,i\,dt \\ &= N\frac{d\phi}{dt}\,i\,dt \\ &= N\,i\,d\phi. \end{aligned}$$

The dW_{elect} could also be represented in any of the following forms:

$$dW_{\text{elect}} = e\,i\,dt = i\,d\psi = N\,i\,d\phi = F(\phi)d\phi$$

And assuming that the current changes from zero to i, while flux changes from 'zero' to ϕ, the flux-linkage changes from 'zero' to ψ and the magneto-motive force changes from $F = 0$ to $F = F(\phi)$ over the time interval $t = 0$ to T, the stored field energy:

$$W_{\text{elect}} = \int_0^{\phi} N\,i\,d\phi = \int_0^{\phi} i(N\,d\phi) = \int_0^{\psi(i)} i\,d\psi(i) = \int_0^{\phi(i)} F(\phi)\,d\phi(i) \qquad \text{...(3.1)}$$

where, $\psi(i)$ is the flux-linkage of the coil at a flux ϕ and it depends on i, and $F(\phi)$ is the mmf corresponding to flux ϕ.

Since the ratio of flux-linkage $\psi(i)$ to i is termed as the inductance of the coil, L, on the assumption that L, the coefficient of self-inductance of the coil remains constant—a condition obtained during unsaturated operation of the magnetic circuit—and leads to:

$$W_{\text{elect}} = \frac{1}{2}Li^2 \qquad \text{...(3.2)}$$

We had arrived at the same conclusion as given in Eq. (1.17). In the above expression, W_{elect} is the net electrical energy input, after copper losses in the coil have been accounted for. On the initial assumption that the coil is lossless and no movement is involved in the system, the source of voltage which forced the current into the coil, and the reaction voltage, are equal in magnitude. The product of the supplied current at a given instant and the part of the applied voltage used to overcome the induced reaction voltage, yields the instantaneous power flow into the coil system. This quantity, when integrated over the time taken by the flux to rise from the initial value of zero to the final value of ϕ, gives the total energy flow into the system. This incremental energy input dW_{elect} is, in practice, partly absorbed in the field system as enhanced field energy dW_{fld} the other part dW_{mech}, accounting for the electro-mechanical energy conversion, accounts for doing any mechanical movement. Thus, in general,

$$dW_{\text{elect}} = e\,i\,dt = dW_{\text{fld}} + dW_{\text{mech}} \qquad \text{...(3.3)}$$

In a general case, when the relationship between ϕ (or ψ) and i is not linear due to saturation in the magnetic circuit, the total energy stored can be evaluated by the following expression:

$$W_{\text{elect}} = \int_0^i L(i)\,i\,di \qquad \text{...(3.4)}$$

In Eq. (3.3), dW_{elect} is the net electrical energy input after copper losses in the coil have been accounted for, dW_{fld} is the differential energy stored in the coupling field and dW_{mech} is the net internal energy converted to mechanical form including mechanical losses. In the equation, the electrical and mechanical terms have positive values for motoring action. For generating action, both electrical and mechanical terms are indicated with negative values.

3.3 VOLTAGE AND TORQUE/FORCE EXPRESSIONS IN BASIC ELECTRICAL MACHINES

Speaking in general terminologies, the word 'electrical' in the context of modelling of electrical machines suggests the presence of current carrying coils, and the word 'machine' implies mechanical movement. A very simple form of electrical machine can, therefore, be a derivative of a solinoid-plunger type of an electro-magnet, with its magnetic core located along the axis of the exciting coil, as shown in Figure 3.2(a). The figure shows the schematic arrangement in which the magnetic core is so mounted that it is free to move along the axis of the coil horizontally, and the core is initially placed asymmetrically away from the symmetrical postion depicted in Figure 3.2(b). The core is attached to a tension spring mounted axially.

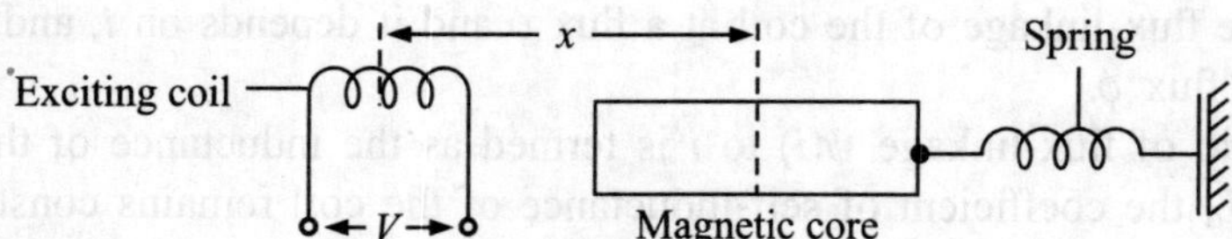

FIGURE 3.2(a) A schematic of an electro-magnet with asymetrically placed core during initial period of functioning, with voltage applied across the terminals of the exciting coil.

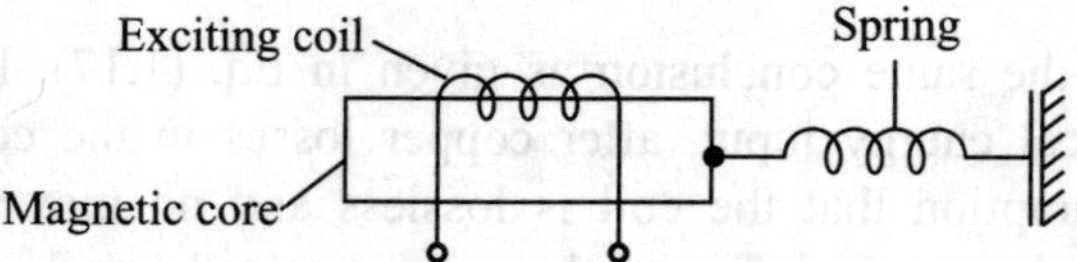

FIGURE 3.2(b) Schematic of the electro-magnet with symmetrically placed core at the final stage of functioning.

It is well-known that when a voltage V is applied across the terminals of the exciting coil, a current i flows through the coil, and a force is developed on the magnetic core along its axis, trying to force the core to occupy symmetrical position as shown in Figure 3.2(b). It is possible to write the voltage–current relationship of the system, which should have the form:

$$V = ri + \frac{d}{dt}(Li) \qquad ...(3.5)$$

where, V denotes the applied voltage across the exciting coil, r is the coil resistance and L is the coefficient of self-inductance of the coil, i being the current flowing in the coil.

In the above expression, it is important that both L and i are functions of time, while the system is initially settling down to steady-state operation.

Let us first assume that the magnetic core is not allowed to move under the action of the force developed. Under this condition the self-inductance of the exciting coil remains constant at L, till the current in the coil attains steady-state. In practice, however, the value of L does not remain constant, but varies with the magnitude of the current. But for the moment, this effect may be ignored. Since, at this stage, the core is not permitted to move under the influence of the force, there can be no change in the value of L due to the core displacement. L is therefore, treated as a constant.

The instantaneous power (p) being fed to the exciting coil assumes the following form:

$$p = Vi = ri^2 + i\frac{d}{dt}(Li) \qquad ...(3.6)$$

$$= ri^2 + L\left(i\frac{di}{dt}\right)$$

Since L is constant, we write the power equation as follows:

$$p = Vi = ri^2 + \frac{d}{dt}\left(\frac{1}{2}Li^2\right) \qquad ...(3.7)$$

In Eq. (3.7), the rate at which the electrical energy is supplied to the exciting coil, under steady-state condition, include the following:

1. The rate at which energy is dissipated in the form of (ri^2), as losses in the coil resistance.
2. Rate of energy flow given by $\frac{d}{dt}\left(\frac{1}{2}Li^2\right)$, as the energy stored in the magnetic field is increased.

Since there is no mechanical movement of the core, this equation is sufficient to describe the performance.

Let us now consider a modification in the situation so that the constraints on the core are released and the magnetic core is free to move under the influence of the force developed by the exciting coil. The core now settles in the symmetrical position as depicted in Figure 3.2(b). Let us go back to the power equation again. The first term on the right hand side continues to denote the aggregate power loss in the resistance element of the coil. Since i varies over a small range due to the nature of the magnetic circuit having sufficient air space through which the flux is developed, the current is primarily controlled by the coil resistance parameter. We can conveniently denote the loss component of energy (i^2rt) which is dissipated, as unaltered. But due to movement of the magnetic core, now the value of the magnetic permeance increases as a function of time, thus, leading to increased flux along the axis of the exciting coil, and a consequent increase in flux-linkage. Hence, L increases progressively as a function of time. The back induced emf in the exciting coil, having direct bearing on the magnitude of the resulting current i also changes as a function of time, thus, tending to reduce the value of i and, therefore, flux-linkage. This corrective effect continues during the period that the core is moving and the system is settling down. The overall effect is that both i and L manifest as functions of time. We shall, therefore, have to resort to partial derivative strategy in handling the differential term. The original voltage–current relationship is simplified to the following form:

$$V = ri + L\frac{di}{dt} + i\frac{dL}{dt} \qquad \text{...(3.8)}$$

The new expression for the instantaneous power flow in the exciting coil is, therefore, modified to the following form:

$$Vi = ri^2 + L\left(i\frac{di}{dt}\right) + i^2\frac{dL}{dt} \qquad \text{...(3.9)}$$

At a particular instant, when the value of the self-inductance of the exciting coil is L and the current is i, the value of the stored magnetic energy is $\left(\frac{1}{2}Li^2\right)$ as before, since L is now a function of time. It follow that:

$$\frac{\partial}{\partial t}\left(\frac{1}{2}Li^2\right) = Li\frac{\partial i}{\partial t} + \frac{1}{2}i^2\frac{\partial L}{\partial t} \qquad \text{...(3.10)}$$

So that power Eq. (3.7) can be rewritten as follows:

$$Vi = ri^2 + \left\{(Li)\frac{\partial i}{\partial t} + \frac{1}{2}i^2\frac{\partial L}{\partial t}\right\} + \frac{1}{2}i^2\frac{\partial L}{\partial t} \quad ...(3.11)$$

or

$$Vi = ri^2 + \frac{\partial}{\partial t}\left(\frac{1}{2}Li^2\right) + \frac{1}{2}i^2\frac{\partial L}{\partial t} \quad ...(3.12)$$

Comparison of Eqs. (3.7) and (3.12), reveals that as there was no movement of the magnetic core for Eq. (3.7), the extra term in Eq. (3.12) is to account for the power-flow into the system due to mechanical work done as a result of the movement of the magnetic core. The first three terms of Eq. (3.12) are identical with the corresponding terms of Eq. (3.7). Hence, the last term in Eq. (3.12) must denote power flow due to mechanical work done by the system. We may also note that the term associated with energy storage in the magnetic circuit has also increased by an amount of $\left(\frac{1}{2}i^2\frac{dL}{dt}\right)$. This increase is attributed to the increase in the value of L of the exciting coil, as a result of repositioning of the magnetic core.

The extra rate of energy-flow from the supply to the exciting coil during movement of the magnetic core, therefore, signifies that the extra power is flowing from the supply to accommodate:

1. The extra component of energy stored in the magnetic system-circuit due to augmentation in the value of L; and
2. The mechanical work done by the system in moving the magnetic core.

The additional energy component actually divides itself into two equal halves, so that $\left(\frac{1}{2}i^2\frac{\partial L}{\partial t}\partial t\right)$ is available for electro-mechanical energy conversion.

Let '$\bar{F}$' denote the mechanical force on the magnetic core and x be the distance between a fixed point on the armature and a fixed point on the exciting winding as shown in Figure 3.2(a). The velocity of the magnetic core is then $\left(\frac{dx}{dt}\right)$, and the mechanical power is $\bar{F}\left(\frac{dx}{dt}\right)$. Equating this to the corresponding electrical power, we have

$$\bar{F}\left(\frac{dx}{dt}\right) = \frac{1}{2}i^2\left(\frac{dL}{dt}\right)$$

$$= \frac{1}{2}i^2\left(\frac{dL}{dx}\right)\left(\frac{dx}{dt}\right) \quad ...(3.13)$$

Since, the variation of L with time arises directly from its variation with position, the required force equation is given as follows:

$$\bar{F} = \frac{1}{2}i^2\frac{dL}{dx} \quad ...(3.14)$$

As majority of electrical machines are rotating in nature, the voltage equations and power equations are applied without alteration. But now, the mechanical power into which electrical power is converted, is obtained by the product of torque T and angular velocity $\left(\frac{d\theta}{dt}\right)$.

Thus, for rotating machines, the power equation is expressed as follows:

$$T\left(\frac{d\theta}{dt}\right) = \frac{1}{2}i^2\frac{dL}{dt} = \frac{1}{2}i^2\frac{dL}{d\theta}.\left(\frac{d\theta}{dt}\right)$$

or

$$T = \frac{1}{2}i^2\left(\frac{dL}{d\theta}\right) \quad ...(3.15)$$

3.4 ELECTRO-MECHANICAL TRANSDUCERS AND THEIR ANALYSIS

The simplest and the most common electro-mechanical transducers are the loud-speakers, microphones, contactors, relays, lifting magnets, meters, small motors used in electric clocks, vacuum cleaners, refrigerators, house-hold kitchen equipment. The electric motors and generators, and transformers are more advanced and complicated versions of such transducers.

The functions of almost all these equipment are governed by Faraday's law of electro-magnetic induction and Ampere's law of force developed on current-carrying conductors. These laws are popularly known as the BLV and BLI laws.

The expressions governing their operation differ only in their performance characteristics based on their application and the amount of power handled. In all these devices, the magnetic field and the conductor length are conveniently located at right angles to each other, and during operation involves movements with respect to each at an angle, which, in majority of cases, is at right angle.

A selected group of electro-mechanical transducers are included in the following sections.

3.4.1 Electro-magnetic Pick-up

Figure 3.3(a), shows the schematic arrangement of an electro-dynamic pick-up comprising a stylus attached to a small coil, which is free to move in a magnetic field produced by a circular permanent magnet. The stylus moves in a groove and its displacement can be assumed to be Simple Harmonic Motion (SHM), denoted by an expression:

$$x = X \sin \omega t = X \sin 2\pi ft$$

where, f is the frequency of the recorded signal.

The velocity of the stylus is given by:

$$V = \frac{dx}{dt} = 2\pi fX \cos 2\pi ft$$

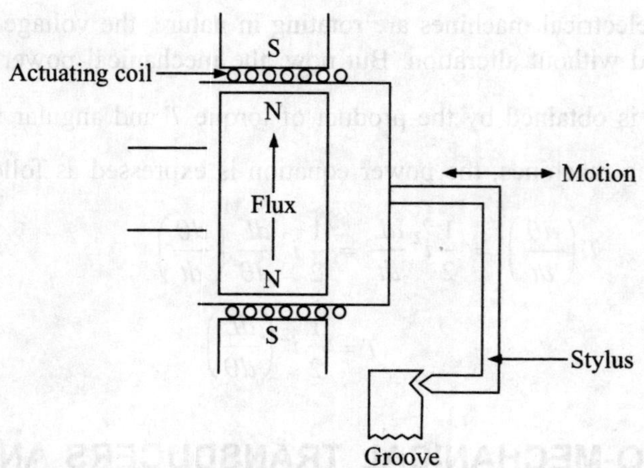

FIGURE 3.3(a) Schematic of an electro-dynamic pick-up.

and hence, the voltage picked up by the device is:

$$e(t) = BLV = 2\pi f\,XBL \cos 2\pi ft$$

where,

$L = 2\pi r.N$ = Total length of wire in the pick-up coil

in which, r = Effective radius of pick-up coil, and

N = Number of turns of the coil

The voltage developed in the pick-up coil is proportional to the frequency, and it increases with frequency. If the same gain at all frequency level is used in the pre-amplifier, the high voltage components would be very loud, while the low frequency components will be too soft or unaudible. Hence, in a practical system, an automatic reduction in gain at high frequency has to be incorporated in the accompanying amplifier.

3.4.2 Loudspeaker

An electro-dynamic loudspeaker arrangement in its schematic form is depicted in Figure 3.3(b). It consists of a voice coil, which is free to move in the field of a circular permanent magnet, and carries a time-varying current, actuated by the voltage output of the last-stage of the power-amplifier and detector.

The mechanical forces comprise force developed by the current-carrying coil in the magnetic field, and the inertial and damping forces, both oppose the driving force.

The electrical forces are composed of impressed voltage to the coil and the opposing voltages, which consist of the following:

1. Voltage generated by the movement of coil in the magnetic field; and
2. Resistance and inductance voltage drops.

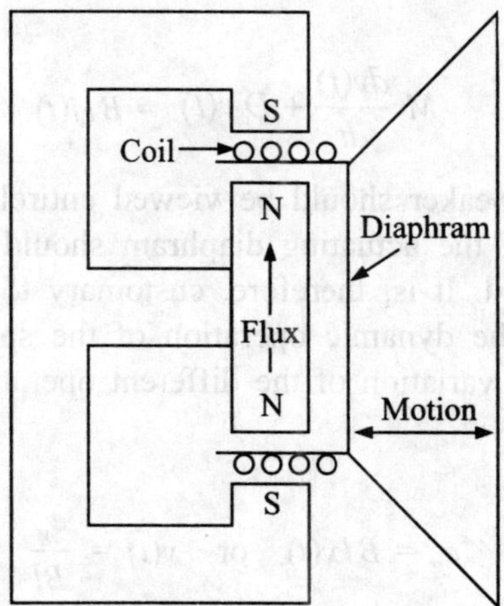

FIGURE 3.3(b) Schematic diagram of a loudspeaker.

Let:

e_g = Voltage generated in the coil due to its motion
B = Constant flus-density in the gap of magnetic structure
$e(t)$ = Voltage applied to the speaker coil
N = Number of turns of the speaker coil
r = Radius of the coil in metres
l = Length of the wire comprising the coil
R, L = Resistance and inductance of the coil
M = Mass of the moving element
D = Coefficient of damping of moving part due to air and friction
$i(t)$ = Current in speaker coil
$v(t)$ = Velocity of the coil

$v(t)$, $i(t)$ and B are at right angles to each other.

The electrical relationships consists of the following

$$\text{The driving voltage} = e(t)$$

$$\text{Opposing voltage} = Ri(t) + L\frac{di(t)}{dt} + Blv(t)$$

or

$$e(t) = Ri(t) + L\frac{di(t)}{dt} + Blv(t)$$

Similarly, the mechanical relationships consist of the following:

$$\text{Driving force, } f = B\,l\,i(t)$$

$$\text{Counter force} = M\frac{dv(t)}{dt} + Dv(t)$$

Hence,

$$M \frac{dv(t)}{dt} + D\,v(t) = B\,l\,i(t)$$

The operation of a loudspeaker should be viewed entirely in a transient state, and it is necessary that the velocity of the actuating diaphram should be faithfully replicated by the voltage in the loudspeaker coil. It is, therefore, customary to develop an electrical analogue representation for analysing the dynamic operation of the speaker, in order to subject it to experimental analysis and the variation of the different operating parameters like M, D, N, r, l and B on its functioning.

We know that,

$$e_g = B\,l\,v(t) \quad \text{or} \quad v(t) = \frac{e_g}{Bl}$$

Substituting the above value in the mechanical relationship, we get

$$\frac{M}{(Bl)^2}\frac{de_g}{dt} + \frac{D}{(Bl)^2}e_g = i(t)$$

and, the electrical force relation is:

$$Ri(t) + L\frac{di(t)}{dt} = e(t) - e_g$$

The electrical equivalent circuit, shown in Figure 3.3(c), satisfies both the above relations, if the capacitor (C) and inductance (G) shown have the following values:

$$C = \frac{M}{(Bl)^2} \quad \text{and} \quad G = \frac{D}{(Bl)^2}$$

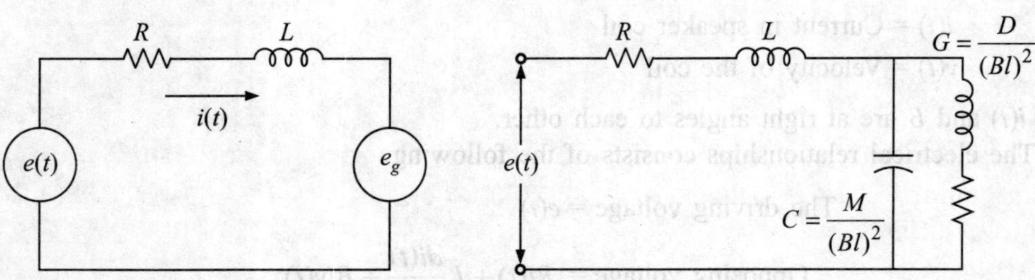

FIGURE 3.3(c) The electrical equivalent analogue.

with both connected in parallel across the voltage e_g. The current flowing in them are $C\frac{de_g}{dt}$ and Ge_g, and the sum total gives the value of $i(t)$.

The velocity of the speaker coil is indicated by the voltage across the capacitor terminals on an oscillograph, and the current in the circuit is a measure of the force on the coil.

3.4.3 Electro-magnetic Microphone

The schematic of an electro-magnetic microphone is shown in Figure 3.3(d). If comprises a light coil mounted on a movable former attached to a diaphram. The coil moves in the annular space of a round permanent magnet structure.

The system is actuated by the vibrating diaphram due to variation of pressure on the system, when spoken to in front of the mouth piece. The coil should similarly produce a voltage, faithfully reproducing the pressure variation on the diaphram. The system is in perpetual transient operation and the electrical analogue indicated in Figure 3.3(c) in connection with loudspeaker may also be used to study its performance characteristics.

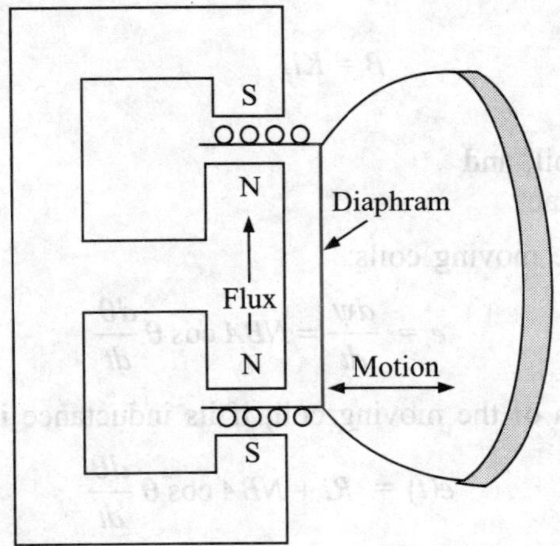

FIGURE 3.3(d) Electro-magnetic microphone.

3.4.4 Electro-dynamometer Movement

In an electro-dynamometer type of ammeter, voltmeter or wattmeter, two coils, composed of heavy conductors capable of carrying rated currents, are fixed in such a way that their planes are vertical and the flux-density produced by them is almost uniform in a small region in their intervening space and is in the horizontal direction along their axes. These two coils are excited by the circuit current. A small movable coil is provided and pivoted with its axis vertical and is free to rotate in a vertical direction. It carries a small current proportional to the applied voltage when used as a voltmeter or wattmeter, or the circuit current when used as an ammeter. The flux-linkage (ψ) of the movable coil is, shown in Figure 3.4, and has an effective value:

$$\psi = NBA \sin \theta$$

where

B = Flux-density developed by the fixed coils
A = Area of the format on which the moving coil is wound
N = Number of turns of the moving coil, and
θ = Angle between the direction of B and the plane of the moving coil.

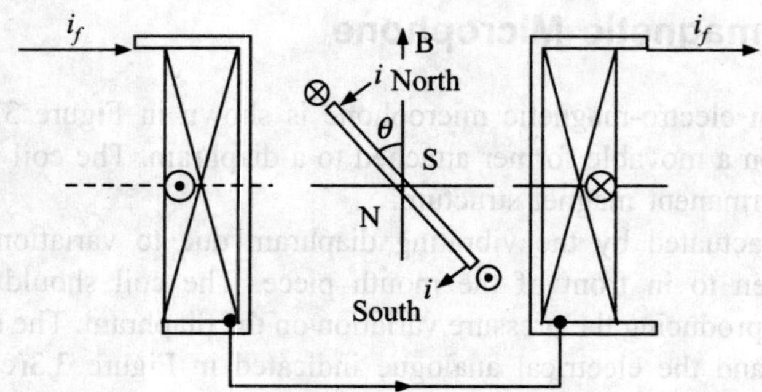

FIGURE 3.4 Electro-dynamometer movement.

Also,

$$B = Ki_f$$

where

i_f = current in fixed coil, and
K = An arbitrary constant

The voltage induced in the moving coils:

$$e = \frac{d\psi}{dt} = NBA \cos\theta \frac{d\theta}{dt}$$

and, the electrical equation of the moving coil, if its inductance is neglected, is as follows:

$$e(t) = Ri + NBA \cos\theta \frac{d\theta}{dt}$$

where, R is resistance of moving coil.

In order to quantify the torque developed by the coil current, the stored energy of the system:

$$dW_{\text{elect}} = id\psi = BANi \cos\theta \, d\theta$$

and, the torque developed:

$$T_d = \frac{dW_{\text{field}}}{d\theta} = BANi \cos\theta$$

The mechanical relationship assumes the form:

$$J\frac{d^2\theta}{dt^2} + D\frac{d\theta}{dt} + K'\theta = BANi \cos\theta$$

where

J = Polar moment of inertia of moving coil
D = Coefficient of damping, and
K' = Spring constant of the restoring spring

Now, under steady deflection, the deflecting torque equals the restraining torque provided by the spring, that is,

$$BANi \cos \theta = K'\theta$$

or

$$KANi_f \, i \cos \theta = K'\theta$$

When operating as an ammeter: Both i_f and i are in series with shunts provided across coils and $i_f \propto i$.

If the meter is used for measuring direct current, I, then

$$I^2 = \frac{K'}{KAN} = \frac{\theta}{\cos \theta}$$

The same meter can also measure alternating currents, but it requires a different calibration. In such a case:

$$i \propto i_f = \sqrt{2}I \sin \omega_0 t$$

and a time-varying torque is generated. The average value of ($i \propto i_f$) is, thus, indicated, which is denoted by:

$$I^2 = \frac{K''}{KAN} \frac{\theta}{\cos \theta}$$

where, K'' is the spring constant for restoring spring for alternating current measurements.

The meter-scale is, therefore, non-uniform.

When functioning as a voltmeter: By connecting high resistance in series with the coils—both fixed and moving—currents proportional to the voltage to be measured are passed in the coils, so that the meter reading indicates to rms value of the voltage.

When working as a wattmeter: The current in fixed coils equals the circuit current, while in the moving coil, it is proportional to the system voltage.

Let,

$$V = V_m \cos \omega_0 t$$

and

$$i_f = I_m \cos (\omega_0 t - \alpha)$$

where

V_m = Peak value of system voltage
I_m = Peak value of system current
$\omega_0 = 2\pi f$, the angular frequency of supply
α = Phase difference between V and i_f.

The current in the moving coil is:

$$i = \frac{V_m}{R} \cos \omega_0 t$$

The torque acting on the moving coil is:

$$BANi \cos \theta = KAN \, i_f \, i \cos \theta$$

$$= KAN \frac{V_m I_m}{R} \cos \theta \cos \omega_0 t \cot (\omega_0 t - \alpha)$$

And, its time-average value is given by:

$$\frac{KAN}{R} VI \cos \alpha \cos \theta$$

in which V and I are the rms voltages and currents of the ac circuit in which the wattmeter is used, and under steady-state condition:

$$T_{av} = K''\theta$$

so that,
$$VI \cos \alpha = \frac{K''R}{KAN} \frac{\theta}{\cos \theta}$$

The meter, therefore, measures the product of the rms voltage and current and the cosine of the angle between them.

3.4.5 Hall-effect Devices

The flux-density in the air-gap of a magnetic circuit can be very conveniently measured with probes working on Hall effect probes. Recent magnetometers or Gauss meters use thin wefers of semi-conducting materials like indium–antimony (In–Sb), with flux-sensitive areas as small as 1.5 mm × 3 mm having thickness of 0.1 mm.

Figure 3.5 shows a typical semi-conductor chip in a magnetic field of strength B acting along the y-direction. The opposite faces are connected to a source of voltage which results in a flow of current in the x-direction. A Lorentz force (f_L), thus, acts on electrons, whose magnitude is given by:

$$f_L = q v B$$

where

q is the charge of electrons in coulomb
v is the velocity of electrons in m/s, and
B is the flux-density in tesla.

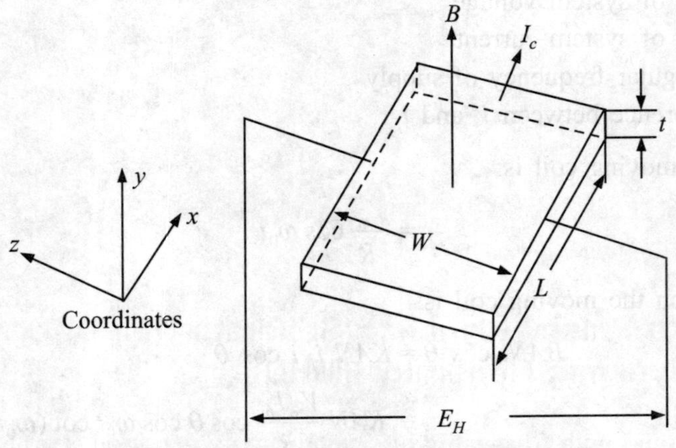

FIGURE 3.5 Schematic of a Hall probe device.

The accumulation of electrons on one side of the material results in a potential difference at right angles to both the *B*-field and the control current directions. The resulting voltage gradient can be found from the opposing force exerted on the electrons. Thus,

$$q e_H = q v B$$

or

$$e_H = v B$$

In the above expression, e_H is known as the Hall voltage gradient. The Hall voltage (E_H) may be measured by connecting a voltmeter across the width of the chip. This has the following value of:

$$E_H = w e_H = w v B$$

In terms of the control current I_c, we could write:

$$I_c = J wt = n q v w t$$

or

$$v = \frac{I_c}{n q w t}$$

where

J is the current density in the material,

n is the density of charge carries per unit volume,

t is the thickness of the chip along the *y*- (or field) direction, and

w is the width of the chip across which the voltage appears (or the *z*-direction as shown)

Therefore, the flux-density is given by:

$$B = \frac{E_H}{wv} = n q t \frac{E_H}{I_c}$$

The quantity $\left(\frac{1}{nq}\right)$ is known as Hall coefficient, R_H and its value depends on the doping material of the semi-conductor probe. Thus,

$$E_H = \frac{R_H}{t} B I_c = K B I_c$$

where, *K* is a constant of the probe material and has the dimensions of volts per tesla-amperes. Hall effect probes can be used for measuring *B* fields up to 3 tesla or more at temperature ranges over –270°C to 100°C. Since they are non-magnetic, there is no distortion of the applied magnetic field. The device has four terminals, two for the control current and two for Hall voltage output. Typical values of general-purpose probes would be a control current of 200 mA and a Hall voltage output of 500 mV at 1.0 tesla.

3.4.6 Magneto Plasma-dynamic Generator (MPD) or Magneto Hydrodynamic Generator (MHD)

If a hot gas (or plasma) with reasonably high conductivity is passed at a high velocity across a magnetic field through a duct, if generates a voltage, which can be used for power supply. The system is termed MHD or MPD generation and came in vogue in the late twentieth century. The idea was first patented in USA in August 1940, and attempts for large-scale power generation was first taken in Bosnia in 1989.

The schematic of an MHD generator is shown in its simplest form in Figure 3.6(a). A refractory duct having dimensions $L \times H \times D$ metres, carrying electrodes on two of its vertical sides has a magnetic field of B tesla in the vertical direction. The gas (or plasma) enters the duct at a velocity,

$$\overline{V} = V_1\, i + V_2\, j + V_3\, k \text{ m/s}$$

where, i, j, and k are unit vectors along the x-, y- and z-directions, as shown in Figure 3.6(b).

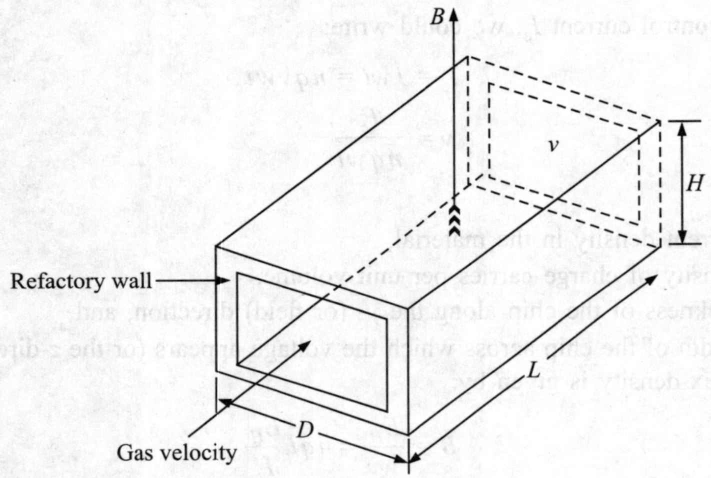

FIGURE 3.6(a) Schematic of an MHD generator.

The magnetic field is given by:

$$\overline{B} = jB \text{ tesla}$$

and, the voltage generated per unit length is given as follows:

$$\varepsilon = -\overline{V} \times \overline{B} = i\,BV_3 + j0 - kBV_1$$

FIGURE 3.6(b) Unit vector depiction assumed.

The voltage gradient $(-kBV_1)$ is termed as 'Faraday gradient', while the gradiant (iBV_3) in the x-direction is known as the 'Hall gradient. Then,

$$E_{\text{Faraday}} = \varepsilon D = BV_1 D$$

and the Hall voltage:

$$E_{\text{Hall}} = BV_3 L$$

The scheme could, therefore, be utilised both in Faraday and Hall configurations. In a Faraday generator, the side electrodes offer a short-circuit to the Hall voltages and segmenting the electrodes is desirable, leaving an electrode each along the flow of the plasma for collection of Hall voltages.

The schematics of Faraday and Hall generator configurations are depicted in Figure 3.6(c).

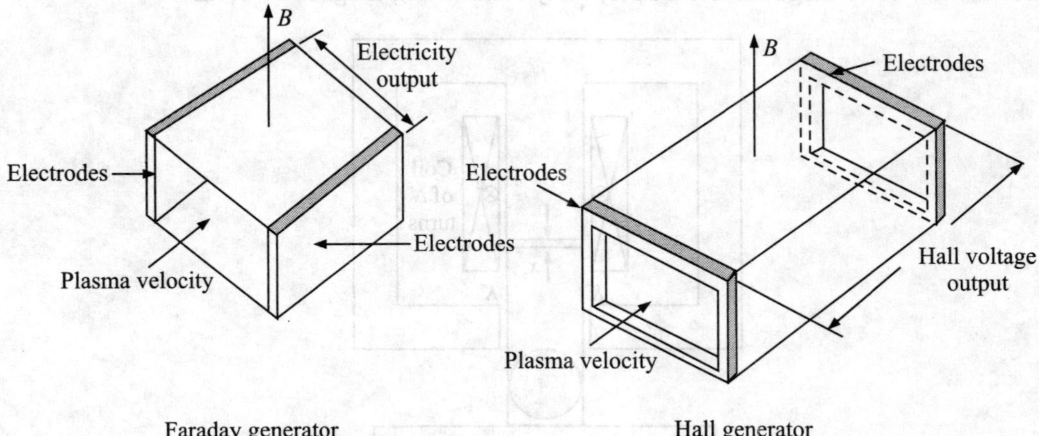

FIGURE 3.6(c) Faraday and Hall generator configurations.

Power relations for Faraday generators

Let C = loading factor = $\dfrac{\text{Voltage across load}}{\text{Voltage induced}}$

σ_g = Conductivity of gas in mho-metre

V = Volume of the duct in m^3 = LDH

A = Electrode area in m^2 = LH

and J = Current density in the electrodes in amp/m^2

Then, voltage drop across the generator = $(1 - C)\, E_{\text{Faraday}}$

and current in the generator, $I = JA$

Also, effective resistance of the duct = $\dfrac{1}{\sigma_g}\dfrac{D}{A} = R_g$

Thus,

$$I = JA = (1 - C)\,\frac{E_{\text{Faraday}}}{R_g}$$

$$= (1 - C)\, BV_1\, D\sigma_g\, \frac{A}{D}$$

or

$$J = (1 - C)\, BV_1\, \sigma_g$$

Hence, power delivered,

$$P_d = CE_{\text{Faraday}}I = \sigma_g B^2 V_1^2\, (1 - C)\, CAD$$

and, power delivered per unit volume of the duct:

$$P_0 = \sigma_g B^2 V_1^2\, (1 - C)\, C$$

3.4.7 Plunger Magnet

The schematic of a simple plunger magnet is shown in Figure 3.7. This arrangement forms the basis of all electro-magnetic contactors, relays and lifting magnet devices.

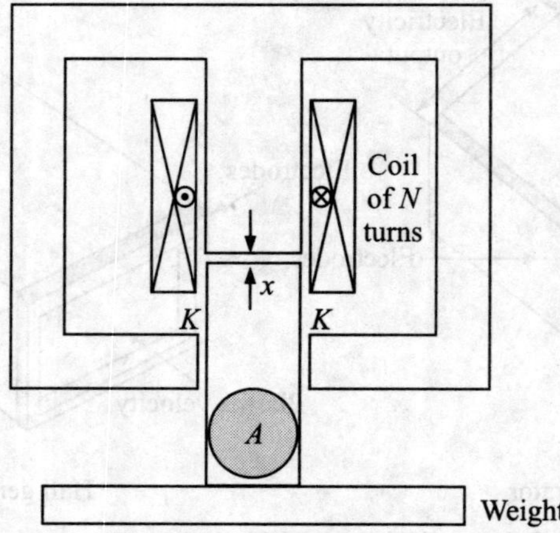

FIGURE 3.7 Schematic of a plunger magnet.

Let the existing coil in the diagram be composed of N turns and let the magnetic core be made up of materials of very high permeability. If we ignore the effect of clearance between the plunger and the yoke of the magnetic circuit at K, which permits the up and down sliding movement of the plunger, the reluctance of the gap of length x and the area A may be computed as follows:

$$\mathcal{R} = \frac{1}{\mu_0}\frac{x}{A} = \frac{1}{\mathcal{P}}$$

The flux-density in the air-gap may be expressed as follows:

$$B = \mu_0 H = \mu_0 \frac{Ni}{x}$$

Hence, the energy stored in the system around the air-gap may be expressed as:

$$W_{\text{field}} = \frac{1}{2\mu_0} B^2 Ax = \frac{1}{2}\phi_g^2 \mathcal{R}$$

$$= \frac{1}{2}(Ni)^2 P = \frac{1}{2} L(x) i^2$$

where

B = Flux density in the gap
ϕ_g = Flux in gap
$\mathcal{R}$ = Reluctance
$\mathcal{P}$ = Permeance
x = Length of air-gap
A = Area of plunger

In this expression the inductance of the exciting coil,

$$L(x) = N^2\mathcal{P}(x) = N^2\mu_0\frac{A}{x}$$

and
$$\frac{dL(x)}{dx} = -N^2\mu_0\frac{A}{x^2}$$

The reluctance of the coil $\mathcal{R}$ also changes as a function of the position of the plunger, so that

$$\frac{d\mathcal{R}}{dx} = \frac{1}{\mu_0 A}$$

Since the nature of inductance in a circuit is such that it does not permit an abrupt change in flux linked in its magnetic circuit, and hence the current flowing through it cannot alter instantaneously, we can assume two extreme cases in the operation of the actuating coil:

1. The movement of the plunger is not accompanied by a change in flux in its magnetic circuit.
2. The current in the coil remains unchanged during the plunger movement.

For the condition described in case 1, the force acting on the plunger is as follows:

$$f_{\text{mechanical}} = \frac{dW_{\text{field}}}{dx} = -\frac{1}{2}\phi_g^2\frac{d\mathcal{R}}{dx}$$

$$= -\frac{1}{2}\phi_g^2\frac{1}{\mu_0 A}\Bigg]_{\text{for } \varphi_g \text{ constant}}$$

And, for the current in the coil remaining constant during plunger movement:

$$f_{\text{mechanical}} = -\frac{1}{2}(Ni)^2\frac{\partial\mathcal{P}}{\partial x}$$

$$= -\frac{1}{2}N^2 i^2\frac{\mu_0 A}{x^2}\Bigg]_{\text{for } i \text{ constant}}$$

$$= -\frac{1}{2}\frac{\partial L}{\partial x}i^2$$

The negative sign has been introduced in the above expressions, since the force of attraction between pole-faces acts in the direction of decreasing x. If the flux is held constant, the force does not depend on the length of the air-gap, but if the current in coil is assumed to remain constant, the force on the plunger varies inversely with the square of the gap length.

3.4.8 Reluctance Motor

The principle underlying the operation of a reluctance motor is basic to many types of important rotating machines.

Figure 3.8 depicts a coil wound on a magnetic yoke. A piece of magnetic material is mounted annularly to this magnetic yoke and is free to revolve around the polar projections of the yoke. The rotating member is shaped such that the reluctance of the magnetic circuit changes in a desired manner. As the permeance of the magnetic field varies with the length of the air-gap through which flux-lines have to develop, it is evident that for the arrangement shown:

1. The permeance is maximum at $\theta = 180°$ or $\theta = 0°$, when the gap is minimum; and
2. The permeance is minimum at $\theta = 90°$ or $\theta = 270°$, for which the gap is maximum, and the flux has to traverse the longest path through the air.

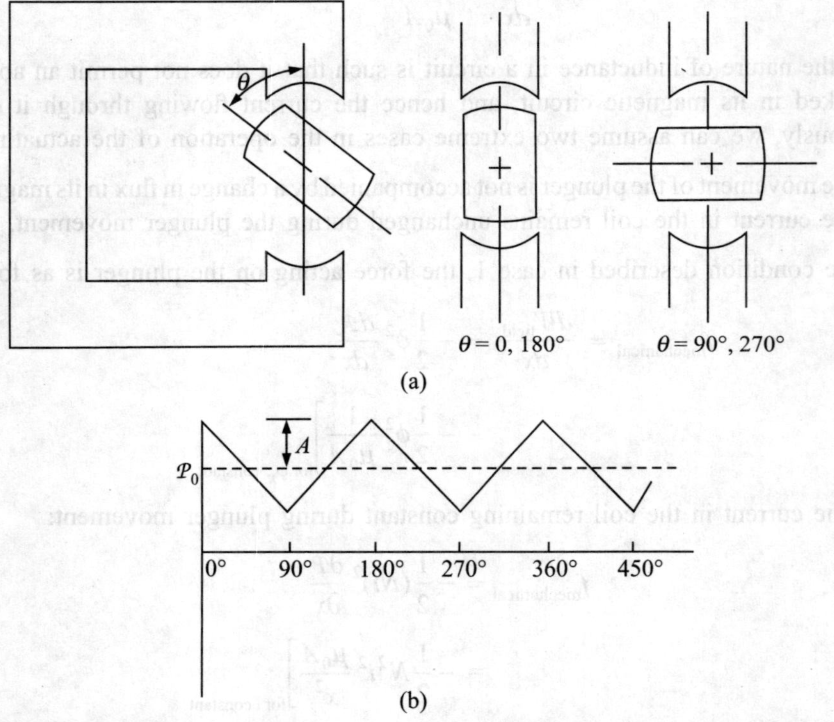

Figure 3.8 (a) Positions of rotating member at maximum and minimum permeance positions (b) Assumed gap-permeance with rotor positions.

The variation of permeance may be conveniently expressed in general by a Fourier series as follows:

$$\mathcal{P}(\theta) = \mathcal{P}_0 + \mathcal{P}_2 \cos 2\theta + \text{terms involving higher odd multiples of } 2\theta.$$

where

$\mathcal{P}$ = Permeance of air gap

$\mathcal{P}_2$ = Permeance terms of higher odd multiples appearing in Fourier series

As an extreme case, let the permeance variation be triangular with an average value of

$$\mathcal{P}_0 = \frac{1}{2}(\mathcal{P}_{max} + \mathcal{P}_{min})$$

and, the amplitude

$$A = \frac{1}{2}(\mathcal{P}_{\max} - \mathcal{P}_{\min})$$

as shown in Figure 3.8(b).

We may write-down the Fourier series having the form:

$$\mathcal{P}(\theta) = \mathcal{P}_0 + \frac{8}{\pi^2}A\left(\cos 2\theta + \frac{1}{9}\cos 6\theta + \frac{1}{25}\cos 10\theta + \dots\right)$$

The higher terms decrease very rapidly, so that only the 2θ term is important.

The permeance variation in its approximate state assume the form:

$$\mathcal{P}(\theta) = \mathcal{P}_0 + \mathcal{P}_2 \cos 2\theta$$

as we had concluded above. On this basis, we can evaluate the continuous torque (T_d).

Since,
$$W_{\text{field}} = \frac{1}{2}(Ni)^2\mathcal{P}$$

$$= \frac{1}{2}N^2 i^2 (\mathcal{P}_0 + \mathcal{P}_2 \cos 2\theta)$$

and
$$T_d = \frac{\partial W_{\text{field}}}{\partial \theta} = N^2 i^2 \mathcal{P}_2 \sin 2\theta$$

If the moving member is rotating at an angular velocity ω_0, then,

$$\theta = \omega_0 t - \delta$$

where, δ accounts for the initial angular position of the axis of the rotating member at $t = 0$. Thus,

$$T_d = -N^2\mathcal{P}_2\, i^2 \sin 2(\omega_0 t - \delta)$$

When the coil is energised by a constant voltage supply, there is no time average on continuous torque if the coil is mounted on the yoke (fixed coil), and the long-time average of $\sin 2(\omega_0 t - \delta)$ is zero.

However, if the system is fed from an ac source with sinusoidal feeding current of frequency f, then

$$i = I_m \cos \omega t = I_m \cos 2\pi f t$$

Then,
$$i^2 = \frac{1}{2}I_m^2(1 + \cos 2\omega t)$$

and
$$T_d(t) = -\frac{1}{2}N^2 I_m^2 \mathcal{P}_2[(1 + \cos 2\omega f)\sin 2(\omega_0 t - \theta)]$$

or
$$T_d(t) = -\frac{1}{2}N^2 I_m^2 \mathcal{P}_2\left[\sin 2(\omega_0 t - \delta) + \frac{1}{2}\sin 2(\omega t - \omega_0 t + \delta) - \frac{1}{2}\sin 2(\omega_0 t + \omega t - \delta)\right]$$

In this expression, all three terms within the parenthesis are varying sinusoidally with time, and the average torque seems to be zero; but when $\omega = \omega_0$, the torque expression becomes:

$$T_d(t) = -\frac{1}{2}N^2 I_m^2 \mathcal{P}_2 \left[\sin 2(\omega t - \delta) + \frac{1}{2}\sin 2(2\omega t - \delta) - \frac{1}{2}\sin 2\delta\right]$$

The long-time average of T_d is a non-zero quantity and assumes the following form

$$T_{av} = \frac{1}{4}N^2 I_m^2 \mathcal{P}_2 \sin 2\delta$$

Since, the effective value of $I = \frac{I_m}{\sqrt{2}}$, therefore, we can write:

$$T_{av} = 0.5N^2 I^2 \mathcal{P}_2 \sin 2\delta$$

The rotor, therefore, continues to rotate at only one particular speed corresponding to the frequency of current applied to the excitation coil. This called the synchronous speed (n_p) of the system.

Its value is given by:

$$n_p = \frac{2}{P}f \text{ revolution/second}$$

and P is the pole-pair.

The power developed by such a motor,

$$p = T\omega_m = T2\pi n_p = \frac{\omega}{P}\mathcal{P}_2 N^2 I^2 \sin 2\theta \text{ watt.}$$

ILLUSTRATION 3.1 An unsaturated clapper-type relay presents a linear relationship between the air-gap flux and the operating coil current a straightline characteristic that passes through a point given by $i = 10$ amp and $\phi = 5$ mWb. The operating coil has 100 turns. The arrangement and the $(i - \phi)$ characteristics are shown in Figure 3.9. Find:

(a) The energy stored in the air-gap as a function of the coil current, when the current in the coil is 10 amp.
(b) Inductance of the operating coil.

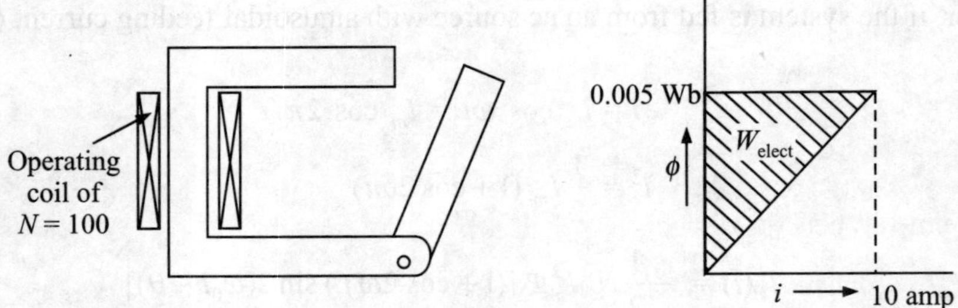

FIGURE 3.9 Clapper relay and its ϕ–i characteristic.

Solution: (a) MMF produced by the operating coil, F_{field} at an excitation of 10 amp,

$$NI = 100 \times 10 = 1000 \text{ amp-turn}$$

Also, from the given characteristic, slope of the i–ϕ characteristic is given by:

$$\frac{i}{\phi} = \frac{10}{0.005} = 2000$$

Hence, $$i = 2000\phi = 2 \times 10^3 \phi$$

Using Eq. (3.1), we have

$$W_{\text{elect}} = \int_0^{\phi} i(Nd\phi) = \int_0^{0.005} 2\times10^3 \times 10^2 \phi \, d\phi$$

$$= \int_0^{0.005} 2\times10^5 \phi \, d\phi = \frac{1}{2}\times 2\times10^5 \phi^2 \Big]_0^{0.005}$$

$$= 2.5 \times 10^{-3}$$

Alternatively, since

$$\phi = 0.5 \times 10^{-3} i$$

We could also write

$$W_{\text{elect}} = \int_0^{i} 0.5\times10^{-3} i \, di$$

At $i = 10$ amps

$$W_{\text{elect}} = 2.5 \times 10^{-3} \quad \text{or} \quad 2.5 \text{ joule}$$

Also, from Eq. (3.2), we have

$$W_{\text{elect}} = \frac{1}{2}Li^2 \quad \text{or} \quad L = \frac{2W_{\text{elect}}}{i^2}$$

(b) The effective inductance of the operating coil, L, is given by:

$$L = \frac{2\times 2.5}{10^2} = 50 \text{ mH}$$

The energy stored in the air-gap is shown by the shaded portion above the ϕ–i characteristic and the ϕ-axis in Figure 3.9.

ILLUSTRATION 3.2 Figure 3.10 shows a 400-turn coil wound on a core of silicon sheet steel of infinitely large permeability and carrying a current of 2 A. The pole faces have areas of 2.5×5.0 cm^2. When the air-gap length is 0.5 m, calculate:

(a) The flux-density and flux in the gaps
(b) The Energy stored in the magnetic field, and
(c) The Weight that it can lift.

Given the value of $\mu_0 = 4\pi \times 10^{-7}$ henry/metre

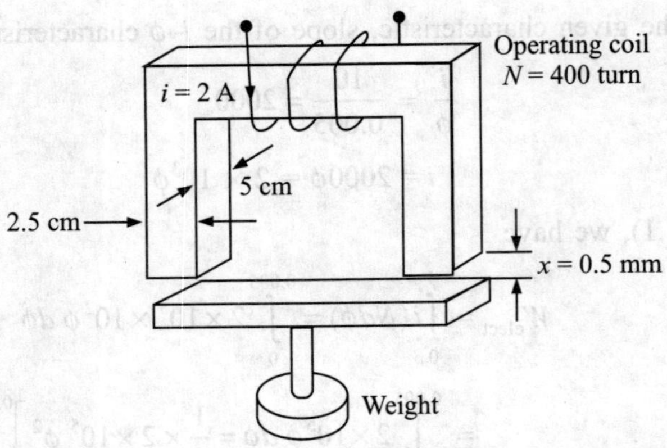

FIGURE 3.10 Electro-magnet of Illustration 3.2.

Solution:

(a) Since the permeability of steel is virtually infinite, the entire coil mmf is taken as acting across 1 mm air-gap under the two projections in the magnetic circuit together. Then,

$$H_g = \frac{Ni}{g} = \frac{400 \times 2}{10^{-3}} = 0.8 \times 10^6 \text{ A/m}$$

$$B_g = \mu_0 H_g = 1 \text{ tesla}$$

and $$\phi_g = B_g A = 1.0 \times 2.5 \times 5 \times 10^{-4} = 1.25 \text{ mWb}$$

(b) Energy stored in the gaps per unit volume

$$\frac{1}{2}\mu_0 H_g^2 = 0.4 \times 10^6 \text{ J/m}^3$$

Total volume of the air-gap (and therefore the field)

$$= 2.5 \times 5 \times 1 \times 10^{-4} \times 10^{-3}$$
$$= 1.25 \times 10^{-6} \text{ m}^3$$

Therefore, energy stored in the gap field

$$W_{\text{field}} = 0.4 \times 10^6 \times 1.25 \times 10^{-6} = 0.5 \text{ joule}$$

(c) The coil current is constant at 2 A. Therefore:

$$f_{\text{mechanical}} = 2 \times \frac{1}{2} F^2 \mu_0 \frac{A}{x^2} = \mu_0 H^2 A$$

$$= 320\pi = 1005 \text{ newton}$$

Since 1 kg wt = 9.8 newton, $f_{\text{mechanical}} = \dfrac{1005}{9.8} = 102.55$ kg

ILLUSTRATION 3.3 A dc contactor is to be made with a coil wound on the centre leg of a 3-limb core shown in Figure 3.11(a). The contractor is required to exert a pull of 9.8 newton

(1 kg) over a gap of 0.635 cm (or 0.25 inch). Find the ampere-turn required in the coil if the area of the three pole faces are 6.45, 12.9 and 6.45 sq cm respectively. Neglect mmf required for steel.

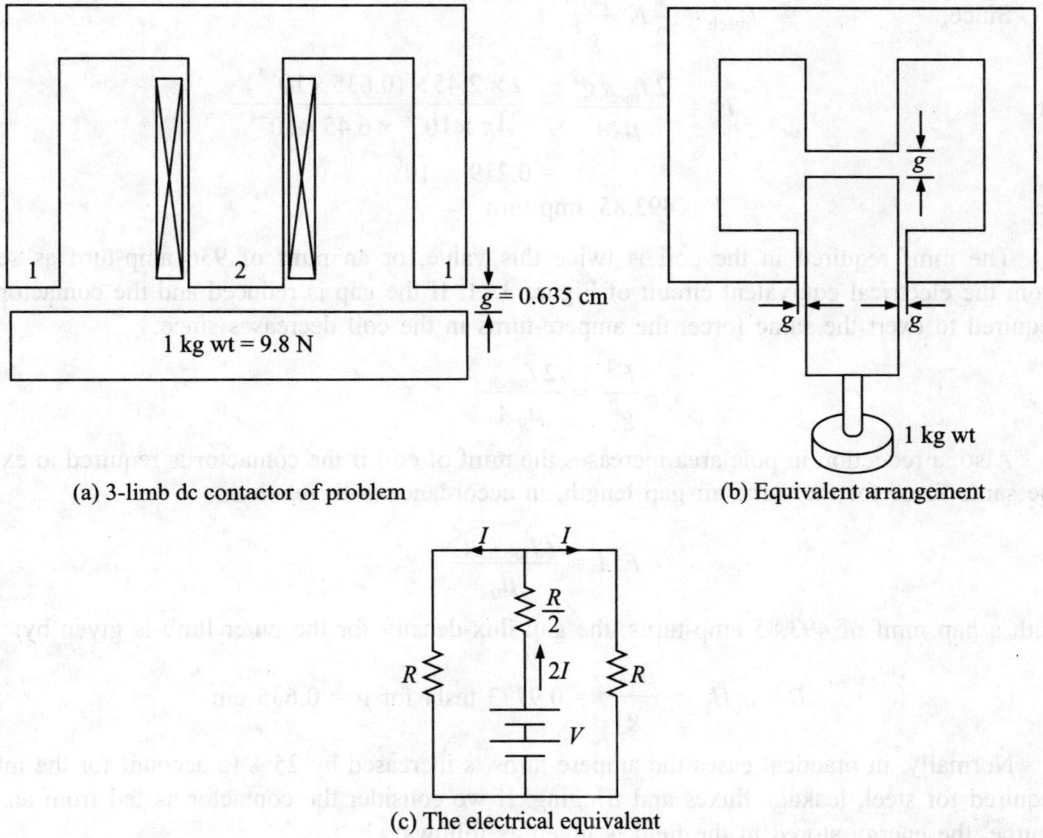

1 2 1 $g = 0.635$ cm

1 kg wt = 9.8 N

(a) 3-limb dc contactor of problem

g g g

1 kg wt

(b) Equivalent arrangement

I I $\frac{R}{2}$ $2I$ R R V

(c) The electrical equivalent

FIGURE 3.11 Schematic DC lifting magnet and electrical equivalent.

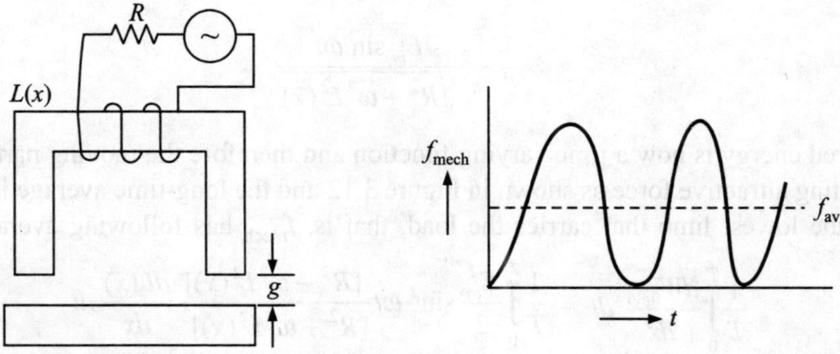

FIGURE 3.12 AC contactor showing variable force with time and average force of attraction.

Solution: In order to develop a pull of 9.8 newton, the outer limbs have to develop a pull of $\frac{9.8}{4} = 2.45$ N each, while the central limb is required to produce a force of 4.9 N.

Since,
$$f_{\text{mech}} = \frac{1}{2} F^2 \frac{\mu_0 A}{g^2}$$

or
$$F^2 = \frac{2 f_{\text{mech}} g^2}{\mu_0 A} = \frac{2 \times 2.45 \times (0.635 \times 10^{-2})^2}{4\pi \times 10^{-7} \times 6.45 \times 10^{-4}}$$
$$= 0.219 \times 10^6$$

or
$$F = 493.85 \text{ amp-turn}$$

The mmf required in the coil is twice this value, or an mmf of 936 amp-turn as seen from the electrical equivalent circuit of Figure 3.11. If the gap is reduced and the contactor is required to exert the same force, the ampere-turns in the coil decreases since:

$$\frac{F^2}{g^2} = \frac{2 f_{\text{mech}}}{\mu_0 A}$$

Also, a reduction in pole area increases the mmf of coil if the contactor is required to exert the same force for the same air-gap length, in accordance with relation:

$$F^2 A = \frac{2 f_{\text{mech}}\, g^2}{\mu_0}$$

with a gap mmf of 493.85 amp-turns, the gap flux-density for the outer limb is given by:

$$B = \mu_0 H_g = \frac{\mu_0 F}{g} = 0.9773 \text{ tesla for } g = 0.635 \text{ cm}$$

Normally, in practical cases the ampere-turns is increased by 25% to account for the mmf required for steel, leakage fluxes and fringing. If we consider the contactor as fed from an ac source, the energy stored in the field is given as follows:

$$W_f = \frac{1}{2} L(x) i^2$$

and
$$i = \frac{E_m \sin \omega t}{\sqrt{R^2 + \omega^2 L^2(x)}}$$

The stored energy is now a time varying function and therefore the moving part is subjected to an undulating attractive force as shown in Figure 3.12 and the long-time average is responsible for pulling the lowest limb that carries the load, that is, f_{mech} has following average value:

$$\frac{1}{T}\int_0^T \frac{dW_{\text{field}}}{dx}\, dt = \frac{1}{T}\int_0^T \frac{E_m^2}{2} \sin^2 \omega t \frac{[R^2 - \omega^2 L^2(x)]}{[R^2 + \omega^2 L^2(x)]^2} \frac{dL(x)}{dx}\, dt$$
$$= \frac{1}{2} E^2 \frac{[R^2 - \omega^2 L^2(x)]}{[R^2 + \omega^2 L^2(x)]^2} \frac{dL(x)}{dx}$$

The inductance of the coil is given by:

$$L(x) = N^2\mathcal{P}(x) = N^2\mu_0 \frac{A}{2x}$$

and
$$\frac{dL(x)}{dx} = -\frac{1}{2}N^2\mu_0 \frac{A}{x^2}$$

If the coil resistance is neglected,

$$\text{average of } f_m \approx \frac{E^2}{\omega^2}\frac{1}{N^2\mu_0 A}$$

Taking up the problem with alternating voltage applied to the coil, if f_{mech} average is 9.8 N, E = 220 volt, ω = 314 and $A = 6.45 \times 10^{-4}$ m^2, then number of turns required in the coil is:

$$N^2 = \frac{E^2}{\omega^2}\frac{1}{\mu_0 A\, f_{\text{mech}}\,(\text{av})}$$

$$= \frac{(220)^2}{(314)^2}\frac{1}{4\pi\times 10^{-7}\times 6.45\times 10^{-4}\times 2.45}$$

$$= 247{,}201.1 \times 10^{10}$$

or
$$N = 157{,}226.3055 \text{ turns}$$

Its inductance, when the gap is 0.635×10^{-2} m is as follows:

$$L = N^2\mu_0 \frac{A}{2x}$$

$$= \frac{247{,}201.1\times 10^{15}\times 4\pi\times 10^{-7}\times 6.45\times 10^{-4}}{2\times 63.5\times 10^{-4}}$$

$$= 157.77 \text{ henry}$$

and the coil has a reactance,

$$X_L = 314 \times 157.77$$

$$= 49.540 \times 10^3 \text{ ohm}$$

The rms value of current drawn from the supply is

$$\frac{220}{49.540\times 10^3} = 4.441 \text{ mA}$$

The force is independent of the gap spacing when excited by a constant voltage source. But the current required to exert the force depends on the coil reactance, and hence, on the gap spacing. In order to prevent chatter caused by the double frequency oscillatory component of force, shading rings are normally provided on the pole-faces. The current induced in these short-circuited copper rings gives rise to interaction with high-frequency flux, thus, creating a steadier force of attraction and a quieter operation.

3.5 MACHINES WITH TWO OR MORE EXCITING WINDINGS

We would, now, like to extend our consideration to rotating electrical machines with two or more exciting windings mounted on the stationary and rotating member of the machine frame.

The schematic of a two-winding rotating machine in its simplest form is depicted in Figure 3.13. The angle separating the axis of the fixed and the moving coil mounted on the rotor and the stationary coil on the stator is denoted by θ_r.

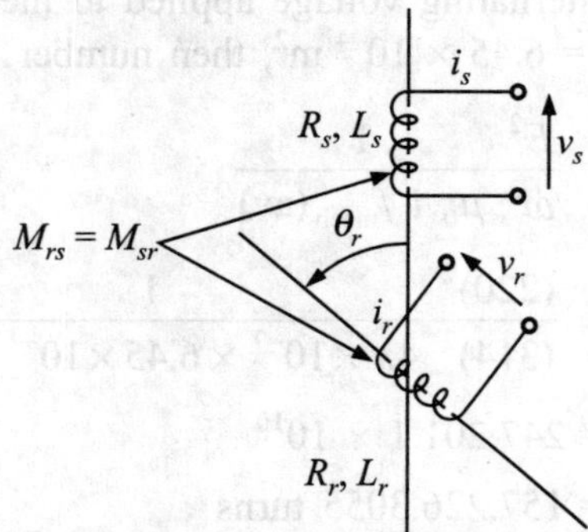

FIGURE 3.13 Schematic of a two-winding electrical machine.

In addition to the self-inductance of the coils, L_s and L_r are the inductance of stator and rotor coils respectively, we have to take into account the coefficient of mutual inductance M_{sr} and M_{rs} of the stator for exciting current in the rotating coil and that of the rotor coil for excitation current in the stationary coil respectively. Hence, the system is governed by the following voltage equations:

$$V_s = R_s i_s + \frac{d}{dt}(L_s i_s) + \frac{d}{dt}(M_{sr} i_r)$$

and
$$V_r = R_r i_r + \frac{d}{dt}(L_r i_r) + \frac{d}{dt}(M_{rs} i_s)$$

We can derive the torque equation following similar lines as the single winding case considered earlier. The above equations can be expanded by applying partial differentiation technique, as both inductance coefficients and currents are functions of time. Thus:

$$V_s = R_s i_s + L_s \frac{\partial i_s}{\partial t} + i_s \frac{\partial L_s}{\partial t} + M_{sr} \frac{\partial i_r}{\partial t} + i_r \frac{\partial M_{sr}}{\partial t}$$

and
$$V_r = R_r i_r + L_r \frac{\partial i_r}{\partial t} + i_r \frac{\partial L_r}{\partial t} + M_{rs} \frac{\partial i_s}{\partial t} + i_s \frac{\partial M_{rs}}{\partial t}$$

If we multiply, the former equation by i_s and the latter equation by i_r and add the two together, the result gives the total instantaneous power. Thus:

$$v_s i_s + v_r i_r = (R_s i_s^2 + R_r i_r^2) + L_s i_s \frac{\partial i_s}{\partial t} + i_s^2 \frac{\partial L_s}{\partial t} + L_r i_r \frac{\partial i_r}{\partial t} + i_r^2 \frac{\partial L_r}{\partial t}$$
$$+ M_{sr} \frac{\partial i_r}{\partial t} + i_r i_s \frac{\partial M_{sr}}{\partial t} + M_{rs} i_r \frac{\partial i_s}{\partial t} + i_s i_r \frac{\partial M_{rs}}{\partial t}$$

It has been shown through tests that coefficients of mutual inductance M_{sr} and M_{rs} are always equal under equal conditions of saturation of the common magnetic circuit, and hence, $M_{sr} = M_{rs} = M$. The above equation can be rewritten as follows:

$$v_s i_s + v_r i_r = (R_s i_s^2 + R_r i_r^2) + L_s i_s \frac{\partial i_s}{\partial t} + i_s^2 \frac{\partial L_s}{\partial t} + L_r i_r \frac{\partial i_r}{\partial t} + i_r^2 \frac{\partial L_r}{\partial t}$$
$$+ M \frac{\partial i_s}{\partial t} + M \frac{\partial i_r}{\partial t} + 2 i_s i_r \frac{\partial M}{\partial t}$$

If we consider the special case where the rotating member carrying the winding with suffix r as stationary, the values of i_s, i_r and M are constant. Hence, $\frac{\partial L_s}{dt} = \frac{\partial L_r}{dt} = \frac{\partial M}{dt}$ are all zeros. Then, the above relationship reduces to:

$$v_s i_s + v_r i_r = R_s i_s^2 + R_r i_r^2 + L_s i_s \frac{\partial i_s}{\partial t} + L_r i_r \frac{\partial i_r}{\partial t} + M i_s \frac{\partial i_r}{\partial t} + M i_r \frac{\partial i_s}{\partial t}$$
$$= R_s i_s^2 + R_r i_r^2 + \frac{\partial}{\partial t}\left(\frac{1}{2} L_s i_s^2 + \frac{1}{2} L_r i_r^2 + M i_s i_r\right)$$

The quantity within the brackets is the stored magnetic energy. When the rotating machine is moving, the value of the derivative assumes an amount given by:

$$\frac{\partial}{\partial t}\left(\frac{1}{2} L_s i_s^2 + \frac{1}{2} L_r i_r^2 + M i_s i_r\right) = L_s i_s \frac{\partial i_s}{\partial t} + \frac{1}{2} i_s^2 \frac{\partial L_s}{\partial t} + L_r i_r \frac{\partial i_r}{\partial t} + \frac{1}{2} i_r^2 \frac{\partial i_r}{\partial t} + M i_s \frac{\partial i_r}{\partial t}$$
$$+ M i_r \frac{\partial i_s}{\partial t} + i_r i_s \frac{\partial M}{\partial t}$$

The original power-flow equation can, now, be written as follows:

$$v_s i_s + v_r i_r = (R_s i_s^2 + R_r i_r^2) + \frac{\partial}{\partial t}\left(\frac{1}{2} L_s i_s^2 + \frac{1}{2} L_r i_r^2 + M i_s i_r\right)$$
$$+ \left(\frac{1}{2} i_s^2 \frac{\partial L_s}{\partial t} + \frac{1}{2} i_r^2 \frac{\partial L_r}{\partial t} + i_s i_r \frac{\partial M}{\partial t}\right) \quad \text{...(3.16)}$$

The terms, considered in groups in Eq. (3.16), have the same significance as in case of a single winding. The torque (T) may be obtained by replacing t by θ in the last group, so that:

$$T = \frac{1}{2} i_s^2 \frac{\partial L_s}{\partial \theta_r} + \frac{1}{2} i_r^2 \frac{\partial L_r}{\partial \theta_r} + i_s i_r \frac{\partial M}{\partial \theta_r} \quad \text{...(3.17)}$$

The performance of different types of machines may be analysed from the two fundamental relationships given by Eqs. (3.16) and (3.17)—the voltage and torque equations—after suitably modifying them by taking into account the actual winding arrangement. In order to show that Eqs. (3.16) and (3.17) are general, we can extend the scope to cover a machine with three windings. These windings may be denoted by using suffices 1, 2 and 3. Thus, the voltage equations become:

$$V_1 = R_1 i_1 + \frac{\partial}{\partial t}(L_1 i_1) + \frac{\partial}{\partial t}(M_{12} i_2) + \frac{\partial}{\partial t}(M_{13} i_3)$$

$$V_2 = R_2 i_2 + \frac{\partial}{\partial t}(L_2 i_2) + \frac{\partial}{\partial t}(M_{21} i_1) + \frac{\partial}{\partial t}(M_{23} i_3)$$

$$V_3 = R_3 i_3 + \frac{\partial}{\partial t}(L_3 i_3) + \frac{\partial}{\partial t}(M_{31} i_1) + \frac{\partial}{\partial t}(M_{32} i_3)$$

and finally the appropriate torque equation is given by:

$$T = \frac{1}{2} i_1^2 \frac{\partial L_1}{\partial \theta_r} + \frac{1}{2} i_2^2 \frac{\partial L_2}{\partial \theta_r} + \frac{1}{2} i_3^2 \frac{\partial L_3}{\partial \theta_r} + i_1 i_2 \frac{\partial M_{12}}{\partial \theta_r} + i_1 i_3 \frac{\partial M_{13}}{\partial \theta_r} + i_2 i_3 \frac{\partial M_{23}}{\partial \theta_r}$$

3.5.1 Representation of Matrix Form of Basic Voltage and Torque Equations

The basic voltage and torque equations for a two-winding electrical machine may be written down in matrix form as follows:

$$\begin{bmatrix} V_s \\ V_r \end{bmatrix} = \begin{bmatrix} (R_s + pL_s) & pM \\ pM & (R_r + pL_r) \end{bmatrix} \begin{bmatrix} i_s \\ i_r \end{bmatrix}$$

In which p replaces $\left(\frac{d}{dt}\right)$ operator. We appreciate that the operator p appears before the product (Li), both of which are functions of time.

We rearrange the above equations in the matrix from by breaking the impedance matrix into three components as follows:

$$V_s = R_s i_s + L_s p i_s + i_s p L_s + M p i_r + i_r p M$$

and

$$V_r = R_r i_r + L_r p i_r + L_i p L_r + M p i_s + i_s p M$$

Written in matrix form, the above set of equation may be denoted by:

$$\begin{bmatrix} V_s \\ V_r \end{bmatrix} = \left\{ \begin{bmatrix} R_s & 0 \\ 0 & R_r \end{bmatrix} + \begin{bmatrix} L_s & M \\ M & L_r \end{bmatrix} p + \omega_r \left(\frac{\partial}{\partial \theta_r} \right) \begin{bmatrix} L_s & M \\ M & L_r \end{bmatrix} \right\} \begin{bmatrix} i_s \\ i_r \end{bmatrix} \quad \text{...(3.18)}$$

In the last impedance matrix component, we substituted $\frac{\partial}{\partial t} = p = \frac{\partial}{\partial \theta_r} \cdot \frac{\partial \theta_r}{\partial t}$ and replaced $\frac{\partial \theta_r}{\partial t}$ by ω_r, the angular velocity of the rotor for attributing special property to this component. In general, we affix the first component as the resistance matrix associated with resistance loss, the second component with mutual and self-inductance voltages to denote the transformer-induced voltages in the machine and the third component as the speed-induced voltages, to be described as $[G]$ matrix component coupled with ω_r, so that:

$$[G] = \frac{\partial}{\partial \theta_r}[L]$$

where,

$$[L] = \begin{bmatrix} L_s & M \\ M & L_r \end{bmatrix}$$

Thus, the voltage equations reduce to a matrix relation of the form:

$$[V] = \{[R] + [L]p + \omega_r[G]\}[i]$$

where, each of the resistance and inductance matrices are two-by-two in form.

The physical significance of rotational voltages can be understood better if the coefficients of the $[G]$ matrix are expressed in terms of permeances.

Thus:

$$M = M_{sr} = M_{rs} = N_s N_r \Lambda_{sr},\ L_s = N_s^2 \Lambda_s \text{ and } L_r = N_r^2 \Lambda_r$$

where, Λ_{sr}, Λ_s and Λ_r are the gap permeances for the mutual flux, and the stator and rotor self-flux respectively. Hence:

$$[G] = \begin{bmatrix} N_s^2 \frac{\partial \Lambda_s}{\partial \theta_r} & N_s N_r \frac{\partial \Lambda_{sr}}{\partial \theta_r} \\ N_s N_r \frac{\partial \Lambda_{sr}}{\partial \theta_r} & N_r^2 \frac{\partial \Lambda_r}{\partial \theta_r} \end{bmatrix}$$

Hence, $[G]$ indicates that rotational voltage terms are due to variation of permeance during rotation. The stator coil of a cylindrical rotor machine finds constant air-gap permeance, that is, $\frac{\partial \Lambda_s}{\partial \theta_r} = 0$. The rotor coil of a smooth stator machine also sees a constant gap permeance, or,

$$\frac{\partial \Lambda_r}{\partial \theta_r} = 0$$

But for all stator–rotor configurations, the effective air gap between the stator and rotor coils alters with rotation, and Λ_{sr} is a function of θ_r.

Then substituting the mutual and self-inductance values appropriately in Eq. (3.18), we get the stator and rotor voltage equations as shown below:

Case 1: Smooth air-gap stator and rotor:

L_s = constant; L_r = constant

$M_{sr} = M = M_m \cos \theta_r$

$$\therefore \begin{bmatrix} V_s \\ V_r \end{bmatrix} = \left\{ \begin{bmatrix} R_s & 0 \\ 0 & R_r \end{bmatrix} + \begin{bmatrix} L_s & M_r \cos\theta_r \\ M_m \cos\theta_r & L_r \end{bmatrix} p \right.$$

$$\left. + \omega_r \begin{bmatrix} 0 & -M_m \sin\theta_r \\ -M_m \sin\theta_r & 0 \end{bmatrix} \right\} \begin{bmatrix} i_s \\ i_r \end{bmatrix}$$

Case 2: One degree of saliency—either the stator or the rotor is smooth or cylindrical.
For salient-pole stator and cylindrical rotor:
L_s = constant, $L_r = L_{r1} + L_{r2} \cos 2\theta_r$
$M_{sr} = M = M_{rs} = M_m \cos \theta_r$

$$\begin{bmatrix} V_s \\ V_r \end{bmatrix} = \left\{ \begin{bmatrix} R_s & 0 \\ 0 & R_r \end{bmatrix} + \begin{bmatrix} L_s & M_m \cos\theta_r \\ M_m \cos\theta_r & (L_{r1} + L_{r2} \cos 2\theta r \end{bmatrix} p \right.$$

$$\left. + \omega_r \begin{bmatrix} 0 & -M_m \sin\theta_r \\ -M_m \sin\theta_r & -(2 \cdot L_{r2} \sin 2\theta) \end{bmatrix} \right\} \begin{bmatrix} i_s \\ i_r \end{bmatrix}$$

The expression for the machine with a salient-pole rotor and a smooth stator can be obtained by interchanging *s* and *r* suffices in the above equations. Thus:

$$\begin{bmatrix} V_s \\ V_r \end{bmatrix} = \left\{ \begin{bmatrix} R_s & 0 \\ 0 & R_r \end{bmatrix} + \begin{bmatrix} (L_{s1} + L_{s2} \cos\theta_r) & M_m \cos\theta_r \\ M_m \cos\theta_r & L_r \end{bmatrix} p \right.$$

$$\left. + \omega_r \begin{bmatrix} -(2L_{s2} \sin 2\theta_r) & -M_m \sin\theta_r \\ -M_m \sin\theta_r & 0 \end{bmatrix} \right\} \begin{bmatrix} i_s \\ i_r \end{bmatrix}$$

Case 3: Two degrees of saliency: Such a case is very rare, and has not been included.

Torque expressions: Torque expressions may be written down, following expression in the form of a matrix product:

$$T_e = \frac{1}{2} \left[\left\{ i_s \frac{\partial L_s}{\partial \theta_r} + i_r \frac{\partial M}{\partial \theta_r} \right\} \cdot \left\{ i_s \frac{\partial M}{\partial \theta_r} + i_r \frac{\partial L_r}{\partial \theta_r} \right\} \right] \cdot \begin{bmatrix} i_s \\ i_r \end{bmatrix}$$

However, the first matrix itself may be expressed as a product of two matrices, thus leading to torque equations:

$$T_e = \frac{1}{2} [i_s \, i_r] \begin{bmatrix} \dfrac{\partial L_s}{\partial \theta_r} & \dfrac{\partial M}{\partial \theta_r} \\ \dfrac{\partial M}{\partial \theta_r} & \dfrac{\partial L_r}{\partial \theta_r} \end{bmatrix} \begin{bmatrix} i_s \\ i_r \end{bmatrix}$$

And, it is possible to take out the operator part $\left(\dfrac{\partial}{\partial \theta_r} \right)$ outside the second matrix so that the above equation may be simplified in the following representative form:

$$[T_e] = \frac{1}{2}[i_t]\left(\frac{\partial}{\partial\theta_r}\right)[L][i] \quad ...(3.19)$$

where,

$$[L] = \begin{bmatrix} L_s & M \\ M & L_r \end{bmatrix}$$

We can, therefore, observe that the expression for torque is in quadratic form.

The voltage and torque expressions are general ones and can be applied to all machines subject to limitations already discussed.

In Eq. (3.19) for torque, one very important word of caution must be emphasised. In the estimation of T_e—the electro-magnetic torque. We had taken a 2-pole model, for which the physical angle θ_r and the electrical angle have the same value. Since, in electrical machines of multipolar configurations θ_r is indicated in electrical radians, T_e must be associated with the pole-pairs also. Hence,

$$T_e = (\text{Pole pairs})\frac{\partial W_{\text{mech}}}{\partial\theta_r}$$

In the present text, we adopted 2-pole structures consistently and avoided the repeated use of the multiplier—pole pair—in the expressions. However, at the end of the Section 3.5 it is added in the form of a special mention. It should be included in the expression for torque suitably.

$$T_e = (\text{Pole pairs})\left[\frac{1}{2}i_s^2\frac{\partial L_s}{\partial\theta_r} + \frac{1}{2}i_r^2\frac{\partial L_r}{\partial\theta_r} + i_s i_r\frac{\partial M}{\partial\theta_r}\right]$$

and in the matrix form,

$$T_e = \frac{1}{2}(\text{Pole pairs})[i_s \quad i_r]\begin{bmatrix} \dfrac{\partial L_s}{\partial\theta_r} & \dfrac{\partial M}{\partial\theta_r} \\ \dfrac{\partial M}{\partial\theta_r} & \dfrac{\partial L_r}{\partial\theta_r} \end{bmatrix}\begin{bmatrix} i_s \\ i_r \end{bmatrix}$$

which in simplified form yields:

$$T_e = \frac{\text{Pole pair}}{2}[i_t]\left[\frac{\partial L}{\partial\theta_r}\right][i]$$

or

$$T_e = \frac{\text{Pole pair}}{2}[i_t][G][i] \quad ...(3.20)$$

3.6 ENERGY AND CO-ENERGY ASSOCIATED WITH MAGNETIC CIRCUITS

Let us refer to Eq. (3.1) again. We should observe that in W_{fld}, the current $i(\psi)$ is a function of the flux-linkage of the circuit, and the effective mmf, $F(\phi)$ in this expression is a function of ϕ in the magnetic circuit. This functional relation between them depends on the geometry of the coil, the magnetic circuit configuration and the magnetic properties of the core material. When the magnetic circuit contains magnetic material, the functional relationships become non-linear, as depicted in Figure 3.14(a) and Figure 3.14(b). Hence, computation of stored field energy, *which involves the area enclosed above the* $[i(\psi) - \psi]$ *characteristic and is bounded by the final value of* ψ_0, *that is the area marked* $0\psi10$ *in Figure 3.14(a) or the area* $0\phi20$ *in the MMF* $[F(\phi) - \phi]$ *characteristic shown in Figure 3.14(b) with horizontal shadings* requires point-by-point aggregation of elemental areas depicted in the two figures. We must remember that the field energy has been explicitly expressed in terms of flux-linkage, ψ, or flux, ϕ, as the variable of integration.

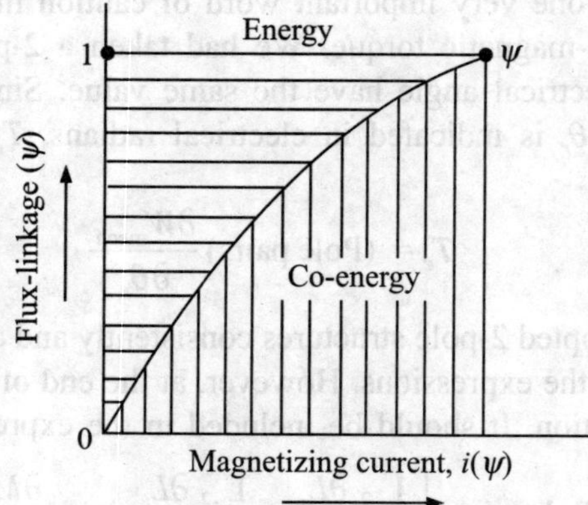

FIGURE 3.14(a) Magnetization characteristic of core material with saturation plot between flux-linkage ψ and magnetizing current, $i(\psi)$.

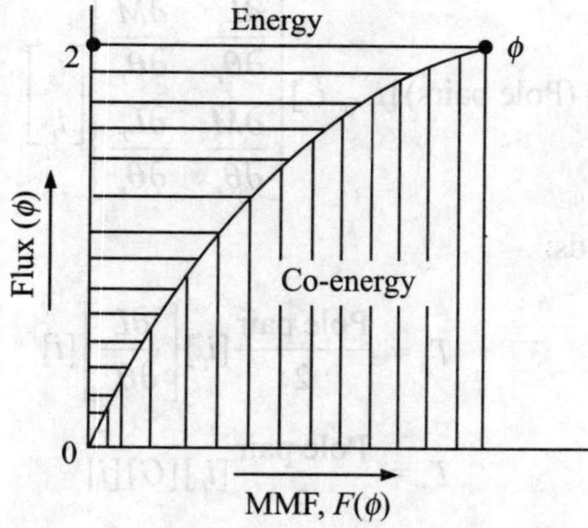

FIGURE 3.14(b) Magnetization characteristic of core material with saturation plot between MMF, $F(\phi)$ and flux ϕ.

The quantity,

$$W_{\text{fld}} = \int_{0}^{i(\psi)} \psi i(\psi) di(\psi) = \int_{0}^{F(\phi)} \phi F(\phi) dF(\phi) \qquad ...(3.21)$$

where, $i(\psi)$ denotes the magnetizing current and $F(\phi)$ denotes the mmf.

If the magnetic circuit is assumed to operate under unsaturated conditions, the magnetizing characteristic assumes a straightline, for which:

$$W_{\text{fld}} = \frac{1}{2}\psi i = \frac{1}{2}F\phi = \frac{1}{2}Li^2 \qquad \text{...(3.22)}$$

The areas depicted by vertical-line shadings are areas bounded by magnetizing current $i(\psi)$-axis in Figure 3.14(a) and $F(\phi)$-axis in Figure 3.14(b), and are explicitly expressed in terms of $i(\psi)$or $F(\phi)$, as the integration variable. They also have the principal dimensions of energy. However, this does not lead itself readily to any physical quantity associated directly with energy in magnetic circuit, that is the field. This is called *co-energy*. It is often useful in interpreting the energy distribution scenario in electro-magnetical systems with co-energy.

3.7 MECHANISM OF TORQUE PRODUCTION IN ROTARY TYPE ELECTRO-MECHANICAL ENERGY CONVERTORS

In majority of electro-mechanical energy converters, discussed in the later chapters of this book, energy conversion is achieved through production of torque on a rotating member using principles elaborated under the aegis of Ampere's law.

In this section we propose to discuss how using the concept of magnetic flux production by current-carrying electric circuits, production of rotational torque can be visualized. At first, we shall attempt to understand physically the mechanism of torque production and their distinguishing features. This will be followed by analysis to evolve suitable torque expressions of general nature in terms of operational parameters obtaining these devices.

3.7.1 A Qualitative Analysis

In order to appreciate the general mechanisms of torque production in rotating type energy converters, let us assume a hypothetical structure for a basic electro-mechanical device comprising a normal magnetic circuit of a single-phase, two winding transformer of the shell-type comprising three limbs as shown in Figure 3.15.

For representation, it will be assumed that the primary winding is split into two equal parts and wound around two outer limbs, while the secondary electrical circuit is housed around the inner central limb. The magnetic circuit provides a common flux-path through the central limb, providing a common flux shared between the two half electrical circuits working as the primary and the single central electric circuit acting as the secondary. The magnetic circuit provides a common flux-path for effective magnetic coupling between these windings. Furthermore, it will be assumed that the central limb structure is modified by introducing thin air-gaps at the extremes of it in the magnetic core, housing the secondary. Let it further be symmetrically supported by a shaft mounted on bearings to enable smooth rotation under the influence of torque produced due to current flowing in the primary components. The final arrangement of the schematic is shown in Figure 3.15.

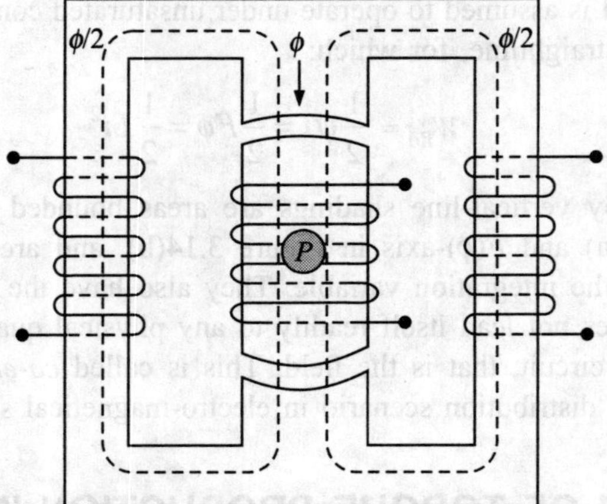

FIGURE 3.15 Schematic arrangement of electric circuits and magnetic circuits, with air-gap separation, making secondary member to rotate around its pivot P, under influence of torque of a shell-type transformer configuration.

3.7.2 Singly-excited System—Production of Reluctance Torque

To start with, let us assume that the two primary windings on the outer limbs are fed with direct current such that the direction of flux-distribution is as shown in the core. The schematic can be considered to have been folded into a two-limb structure as shown in Figure 3.16, and the two primary halves are combined to form the primary with double the number of turns but carrying the earlier value of current. Let the movable portion of the core housing the secondary be displaced from its symmetrical position by an angle θ_r as shown.

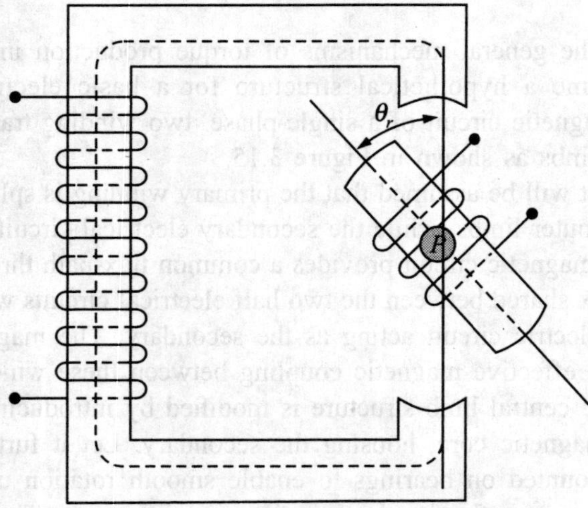

FIGURE 3.16 Modified schematic arrangement with the secondary limb disturbed from its symmetrical position by rotating by angle θ_r.

The current produces an mmf around the magnetic circuit and develops a flux ϕ, in the core as shown in Figure 3.15. We can consider an unsaturated magnetic circuit due to the presence of air-gaps, and majority of the mmf appears as effective across the air-gaps. We may liken the situation with a condition in which the portion of the core housing the open-circuited secondary has induced poles of opposite polarities to those appearing in the air-gap on the primary side. For the direction of flux shown, the upper polar projection on the primary is a North Pole, corresponding to which the upper side of the secondary has an induced South Pole and vice-versa for the lower projection interface. There will, therefore, be a natural tendency for the part of core housing the secondary to be attracted to the poles formed on the primary side of the core. This natural tendency of the core to realign in case of disturbance from its original symmetrical position, is because the magnetic materials always tend to orient themselves along paths of least reluctance, and the flux lines behave as elastic bands with least length of flux paths. If now, we try to disturb the alignment, we shall have to apply an external torque on the shaft to do work because the magnetic circuit responds by producing a counter torque tending to realign the members. A careful examination of the nature of counter torque produced in the arrangement shows that with an increase in misalignment from the natural equilibrium position, the magnitude of the counter torque initially rises and attains a peak value when the misalignment angle is approximately 45°. Thereafter, it will start decreasing and eventually become zero in the region of 90° misalignment, thus, attaining a state of unstable equilibrium. Any further increase in misalignment now produces an aiding torque that will tend to rotate the misalignment part such that it stabilizes in the new position of 180° misalignment. One way of explaining the phenomenon is to indicate that the lines of force behave like elastic bands tied between opposite polarities and have a natural tendency to shrink to their minimum length as mentioned earlier. Any misalignment will stretch them, thereby, causing to develop a counter force so directed as to effect restoration of the alignment. The component of this force, which is tangential to the circle of rotation during misalignment, appears as the cause for the restoring torque on the system. Another way of viewing the physical phenomenon obtaining the situation is that the magnitude of the flux crossing the air-gaps is reduced as the members are turned out of alignment due to effective increase in magnetic reluctance offered by the magnetic circuit as a result of increasing of gap length. It is pointed out that under unsaturated conditions, maintained in the metallic parts of the magnetic circuit, approximately the whole of the mmf is used up for developing flux across the air-gaps. The energy stored in the magnetic field under these conditions is contained in the air-gaps, and in the position of misalignment, the energy shored attains a maximum value for a given excitation current in the primary. As the torque developed during misalignment is due, exclusively, to there being a position of minimum reluctance, which a rotating member always tries to attain, the torque so produced is called the *reluctance torque*.

Flux in the magnetic circuit with proper alignment between portions of the core housing the primary and the secondary electrical circuits, is the maximum for a given value of primary exciting current. From energy storage viewpoint, therefore, maximum amount of energy is stored in the magnetic circuit under aligned condition, and self-inductance of the coils has their maximum values for this position.

When a misalignment is imposed on the system, the coil inductance, which is a function of the position of the member, changes, thus changing the amount of energy it can keep stored. The balance of the stored energy is, therefore, given out by the system in the form of counter reluctance. Reluctance torque can, therefore, be regarded as caused by an excited winding only if its self-inductance is a function of the nature and relative position of the members. It is important to note that the nature of air-gap presented by the non-flux producing member actually decides how the self-inductance of the flux-producing coil changes with position. Accordingly, if the secondary transformer limb is replaced by a rotor with smooth air-gap in the arrangement shown in Figure 3.16, no reluctance torque is developed on the rotor even by delibrately misalign it from the primary flux axis.

The modified shell-type mechanical configuration detailed in Figure 3.15, with three limbs described at the beginning of Section 3.7, was further modified by an equivalent two limb structure depicted in Figure 3.16 and described in the beginning of Section 3.7.2, for a particular reason. Although both systems are equivalent, visualization of the system in Figure 3.16 is simpler. But it suffers from the lack of symmetry in structure—the member housing the primary (or exciting) electrical circuit is not symmetrically disposed with respect to the unexcited secondary electrical circuit. This lack of symmetry makes visualization to transition in electrical circuits using uniform but distributed conductor arrangement—a common feature in practical electro-mechanical devices—difficult. Because the use of uniformly distributed circuit arrangement pre-supposes the requirement of symmetrical magnetic circuit configuration, it is easier to transit from the three-limb structure of Figure 3.15 to conceptualize the modification in the mechanical arrangement. We can now see that the mechanical arrangement in electro-mechanical devices can be one of the following four types. But this classification is based on the nature of the air-gap between the exciting member or primary and the secondary.

If the air-gap surface between the stationary member housing the exciting winding and the rotating member is smooth, that is none of the circuit configurations is mounted with projections, it is referred to as having *non-salient pole structure*. When one of the members—either the stationary member or the movable member—is mounted with projections, thus resulting in non-uniformity in air-gap length along its periphery, the solid structure is said to have salient structure with projections attributed to the concerned member. Thus, it may be called *stator salient, smooth rotor* when the air-gap is non-uniform at the stator or *smooth stator, salient rotor*. When both members contribute to non-uniformity in air-gap, it is called *doubly salient machine*.

Hence, the mechanical arrangement can lead to any one of the following configurations:

- *Non-salient stator and rotor* with smooth air-gap
- *Stator salient, rotor smooth* structure
- *Stator smooth, rotor salient* structure
- *Doubly salient* with non-uniform air gap around the entire periphery

The arrangements are shown in Figure 3.17.

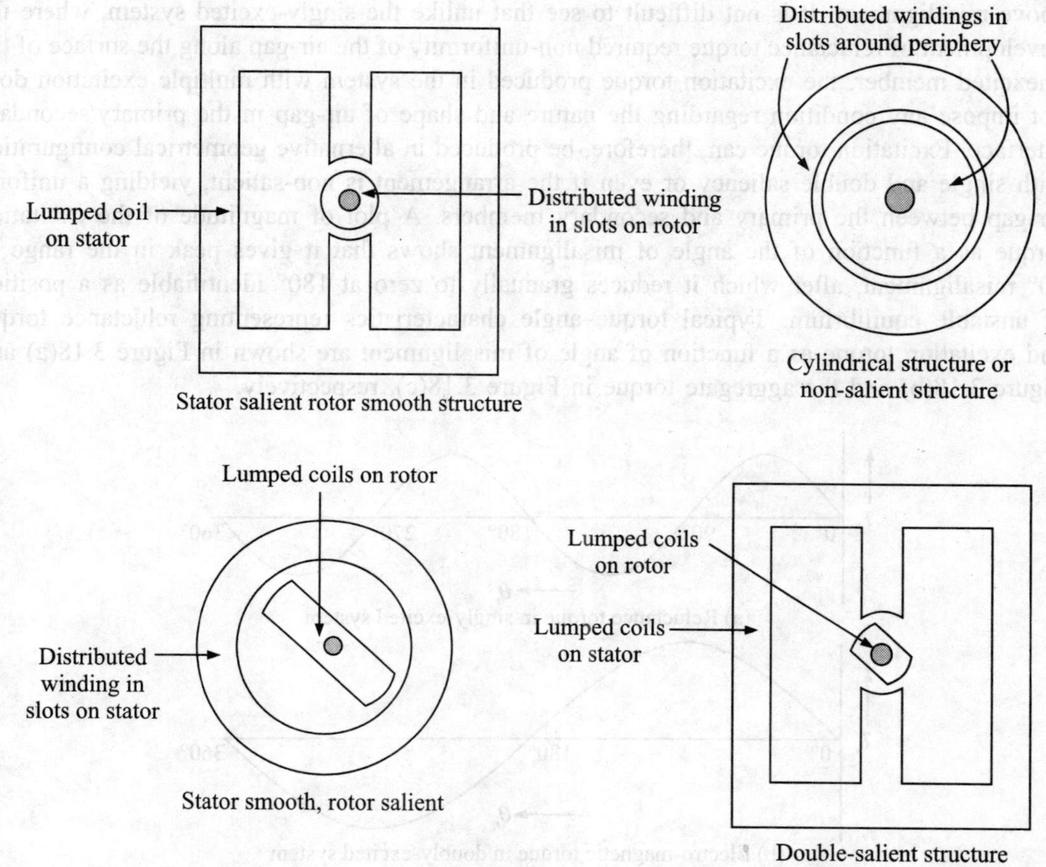

FIGURE 3.17 Alternative mechanical structures of electro-mechanical rotating devices.

3.7.3 Systems with Multiple Excitation—Production of Excitation Torque

Now, let us consider that both, primary and secondary windings of the arrangement, shown in Figure 3.15 are excited from direct current sources. Both members produce independent fluxes along their axes, their effective directions being dependent upon the nature of the winding layout and current input. A little consideration shows that the primary and secondary cores strive to attain a position of alignment such that opposite polarities face each other on the primary and secondary surfaces. At this normal position of stable equilibrium, the net flux in the magnetic circuit is contributed jointly by the primary and secondary members, and the flux across the air-gap maximizes. Any attempt to disorient the part of the core hosing the secondary member from this position distorts the air-gap flux distribution such that a tangential component of flux-distribution now appears, while the net air-gap flux reduces in magnitude. This tangential component introduces a restoring torque directed such that the secondary displays an urge for realignment. This means that work has to be done on the rotating member to maintain the above misalignment. It is not difficult to see that unlike the singly-excited system, where the development of reluctance torque required non-uniformity of the air-gap along the surface of the unexcited member, the excitation torque produced in the system with multiple excitation does not impose any condition regarding the nature and shape of air-gap in the primary/secondary interface. Excitation torque can, therefore, be produced in alternative geometrical configuration with single and double saliency or even if the arrangement is non-salient, yielding a uniform air-gap between the primary and secondary members. A plot of magnitude of the excitation torque as a function of the angle of misalignment shows that it gives peak in the range of 90° misalignment, after which it reduces gradually to zero at 180° identifiable as a position of unstable equilibrium. Typical torque–angle characteristics representing reluctance torque and excitation torque as a function of angle of misalignment are shown in Figure 3.18(a) and Figure 3.18(b) and the aggregate torque in Figure 3.18(c), respectively.

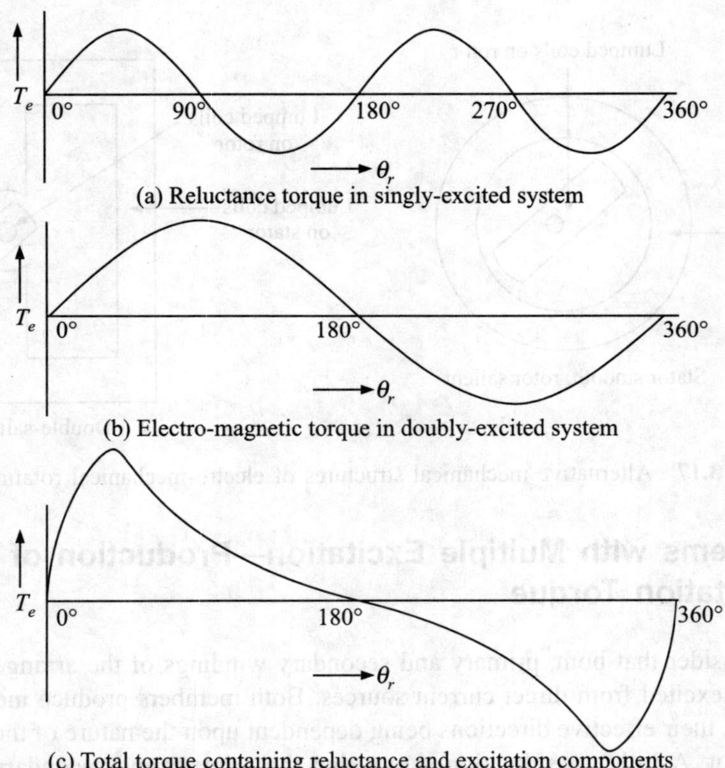

FIGURE 3.18 Typical torque-angle characteristic, derived on qualitative basis.

Use of the phenomenon of excitation torque production is more prevalent in conventional rotating machines with multiple excitation systems. Such a torque can be viewed as resulting from interaction of rotor current and the air-gap flux. However, it becomes necessary that the concerned flux and current maintain a correct relative position both in the space and time domain in order to yield a useful torque. It is not hard to see that the necessary and sufficient condition for torque production for dc excitation to be that the air-gap flux built by the primary current and that formed by the secondary should be at space-quadrature but in time-phase.

3.7.4 Estimation of Stored Energy and Developed Torque

This section is devoted to deriving expressions for energy stored in the magnetic fields associated with the two magnetically coupled circuits. The approach resembles the procedure adopted earlier in respect of energy associated with a single coil. The strategy envisages that we consider one coil at a time and estimate the energy supplied to it when the current through it is raised from zero to the final value. For simplicity, resistance of electrical circuits is neglected, leading to a condition that, at all times, the applied terminal voltage equals the induced emf in the coil produced due to change of flux linked by it, while current through it changes from zero to the final value over an appropriate time. Further, the magnetic circuit is assumed to remain unsaturated for this analysis. This idealization permits the values of self and mutual-inductances presented by each circuit to remain independent of the magnitude of circuit current and the principle of superposition can be applied in estimating the total energy-flow into the system.

Let us suppose that, at first, current in the primary circuit is raised from zero to i_p over a time of $t = \tau_p$. At this time, it is considered that the secondary circuit does not carry any current, and therefore, does not contribute to flux production in the common magnetic circuit. At the end of the period, τ_p, the energy input (W_p) to the system is given by:

$$W_p = \frac{1}{2} L_p i_p^2 \qquad ...(3.23)$$

Now, let the current in the secondary circuit be raised from zero to the final value of i_s, keeping the primary current constant at i_p. From similar considerations, the energy input to the secondary circuit is given by $\left(\frac{1}{2} L_s i_s^2\right)$. However, while this process of building up of secondary current was going on, the common magnetic circuit shared by primary and secondary members caused the flux to change along the axis of the primary circuit as well, thereby including an emf in it. An equal voltage must, therefore, have been applied to counter the effect of the additional induced emf, in order to maintain the current constant at i_p during the said time. The direction of the energy flow in this case is decided by the sign of the additional voltage, which in turn, is decided by the direction of flow of current i_s in the secondary member. If we assume that the mutual inductance between the primary and secondary coil is denoted by M_{ps}. The value of instantaneous power flow into the primary winding due to change in current is given by:

$$\pm i_p \frac{d}{dt}(M_{ps} i_s) \qquad ...(3.24)$$

If the system is assumed to be conservative in nature, that is, there is no loss of energy in the system, the total stored energy (W_{fld}) in the magnetic system is given by:

$$W_{\text{fld}} = \frac{1}{2} L_p i_p^2 + \frac{1}{2} L_s i_s^2 \pm M_{ps} i_p i_s \qquad ...(3.25)$$

With the above expression in hand, we are now ready to obtain the value of reluctance torque for a singly excited system discussed earlier. For quantitative estimation of torque, we shall assume that the maximum value of self-inductance of the exciting coil is L_p corresponding to a perfect alignment between primary and secondary projections of the limb and its value changes as a known function of the angle of misalignment, θ_r. Furthermore, the non-excited member (in this case the secondary) presents non-uniformity in air-gap around the periphery. This was depicted in Figures 3.13 and 3.14(a) and (b), and repeated in Figure 3.18 [(a), (b) and (c)]. It may be necessary to elaborate the strategy used for the derivation of the expression of the instantaneous torque as a function of misalignment angle, θ_r.

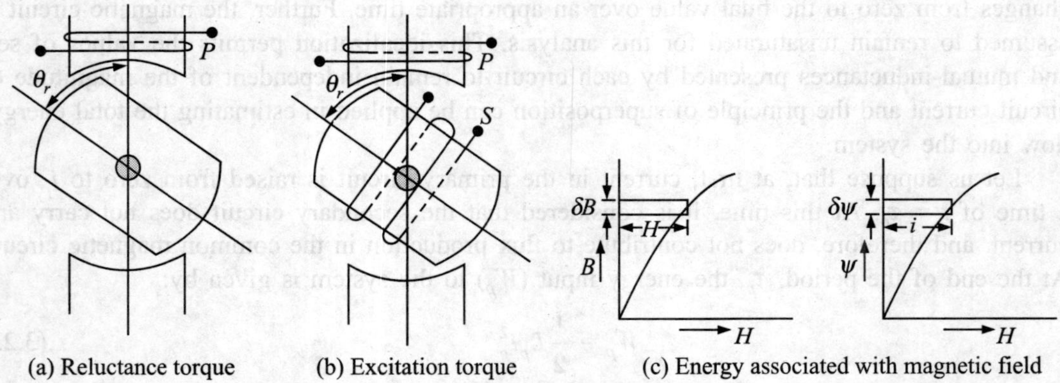

FIGURE 3.19 Estimation of torque in electro-mechanical system.

We apply the *principle of virtual work*, which envisages that we allow a small involvement to occur while the current in the circuit is kept constant and evaluate the change in energy level in the system due to the work done, under the principle of conservation of energy. With reference to Figure 3.19(a), let there be a small movement of the secondary member in the positive direction. As shown in the figure, θ_r decreases and in the process, a small amount of mechanical work will be done due to the movement, while at the same time, the flux in the magnetic circuit tends to increase slightly, thereby, implying that more energy will be stored in the magnetic field. Increase in flux causes an increase in flux-linkage. This, in turn, induces a voltage in the coil tending to oppose the change in flux-level and consequently the circuit current. In order to maintain a constant current through the exciting coil circuit, a small amount of applied voltage is necessary to counter the effect of the induced voltage. As before, the product of coil current and the applied voltage constitutes the input power to the system at any moment, while the instantaneous power when aggregated over the time interval during the moment gives the total energy input. This energy partly accounts for the purported increase in the magnetic field energy, the remaining part is representing the mechanical work done. Assuming a conservative system as before, the elemental electrical energy input, δW_{elect}, is given by:

$$\delta W_{\text{elect}} = i_p (\text{rate of change of flux-linkage})\, \delta t$$

where, δt denotes the time duration of the change.

Thus,
$$\delta W_{\text{elect}} = i_p \frac{d\psi_p}{dt} \delta t = i_p\, d\psi_p \qquad ...(3.26)$$
$$= i_p d(L_p i_p) = i_p^2 dL_p$$

since, i_p is assumed to remain constant during the change. But the energy already stored in the field is $\frac{1}{2} i_p^2 L_p$. The part of the additional energy that goes into storage in the form of field energy is $\frac{1}{2} i_p^2 dL_p$.

Hence, the energy used up to perform mechanical work on the system, is as follows:

$$dW_{\text{elect}} = T_{\text{mech}}\, \delta\theta_r = \frac{1}{2} i_p^2\, \delta L_p \qquad ...(3.27)$$

Hence,

$$T_{\text{mech}} = \frac{1}{2} i_p^2 \frac{\partial L_p}{\partial \theta_r} \qquad ...(3.28)$$

This point was made in a general way at the end of Section 3.3, in the explanation after Eq. (3.12).

By using a similar logic, it is also possible for us to develop an idealized torque expression for excitation torque for system with multiple excitation. For this consideration, let us set aside the possibility of production of reluctance torque at this stage. This amounts to assuming that both the primary and secondary members present smooth surfaces at the air-gap and both i_p and i_s remain constant and unaffected for different values of misalignment angle θ_r. It should be noted that even with a smooth air-gap, the mutual inductance M_{ps}, between the primary and secondary continues to be a function of the misalignment angle, θ_r.

In order to determine an expression for the *excitation torque*, we take recourse to the *principle of virtual work* again, assuming that the principle of *energy conservation* is applied and the system is *conservative*. We further, imagine that the secondary member moves through a small angle, $\delta\theta_r$ in the direction of realignment of the primary and secondary flux axes, while maintaining the primary current i_p and the secondary current i_s constant at their original values. The small angular movement causes mechanical work to be done on the system to the extent of ($T_{\text{mech}}\, \delta\theta_r$) as a result. As the angle of misalignment decreases, the mutual flux tends to increase, leading to an inflow of energy from both primary and secondary input systems, and there is an augmentation of field energy storage in the magnetic circuit. From earlier considerations, it may be seen that the additional electrical energy input (dW_{elect}) during the movement, expressed in differential form is as follows:

$$dW_{\text{elect}} = i_p \times \text{(rate of change of flux-linkage of primary)} \times dt$$
$$+ i_s \times \text{(rate of change of flux-linkage at the secondary)} \times dt$$
$$= i_p\, d\psi_s + i_s\, d\psi_p$$
$$= i_p(\pm\, dM_{ps}\, i_s) + i_s(\pm\, dM_{sp}\, i_p)$$

and since, in case of properly coupled magnetic circuits,

$$M_{ps} = M_{sp} = M$$

$$\therefore \qquad dW_{\text{elect}} = \pm\, 2 i_p\, i_s\, dM$$

The first component of dW_{elect} is contributed by the secondary member, and the second part is due to the contribution of the primary. From earlier considerations and arguments leading to Eq. (3.25), we already know that the energy originally stored in the mutual field is ($\pm\, i_p i_s M$), and by extension of the same logic leading to Eq. (3.28), we can infer that the change of energy stored in the mutual field during this small movement through $\delta\theta_r$ amounts to ($\pm\, i_p i_s dM$). The balance of net energy input during movement, ($\pm\, i_p \cdot i_s \cdot dM$) is available for doing mechanical work, through the production of *excitation torque* (T_{exc}). As has already been shown that

$$\text{Mechanical work done, } dW_{\text{mech}} = T_{\text{exc}}\, d\theta_r$$
$$= \pm\, i_p i_s\, dM$$

and

$$T_{\text{exc}} = \pm\, i_p\, i_s \frac{dM}{d\theta_r} \qquad \text{...(3.29)}$$

In view of the significance of the expression for torque developed in Eq. (3.29), it will be prudent to represent the findings in a consolidated manner.

There are only two alternative methodologies by which torque is produced in rotating type electro-mechanical energy converters. These include:

(i) Production of reluctance torque occurs when the self-inductance of the excited windings becomes a function of the misalignment angle between the axes of primary and secondary member constituting a device. In other words, the reluctance torque can be produced, if at least one of its components—primary or secondary—has projected construction. Production of reluctance torque in a singly excited system can occur only when the component member (which does not house the exciting winding circuit) has a projected configuration. Reluctance torque is produced in rotating type energy converters with double-salient structure. The reluctance torque (T_{rel}) is given by:

$$T_{rel} = \frac{1}{2} i_p^2 \frac{dL_p}{d\theta_r} + \frac{1}{2} i_s^2 \frac{dL_s}{d\theta_r}$$

for a doubly-salient structure, where i_p (and i_s) are currents in exciting coils and L_p (and L_s) are the self-inductances, while θ_r is the positional angle depicting misalignment between the primary and secondary flux axis.

(ii) Production of excitation torque occurs in systems with multiple excitation only, when magnetic fluxes are developed both by primary and secondary members. The shape of the air-gap presented by the primary or the secondary magnetic circuits, i.e., their projections along the air-gap, does not play any part in the production of this type of torque, because, at least the mutual inductance between the primary and secondary exciting coils is always a function of the relative angular displacement between the axes of the primary and secondary fluxes.
Excitation torque has a value of:

$$T_{\text{exc}} = \pm i_p i_s \frac{dM}{d\theta_r}$$

where, i_p and i_s denote primary and secondary currents and M denotes the mutual inductance between the primary and the secondary, θ_r being the misalignment angle, as before.

(iii) Both reluctance torque and excitation torque can contribute to total torque production in electro-mechanical systems, provided that the basic considerations elaborated in para (i) and para (ii) are satisfied. For a double salient structure, where non-uniform air-gap is provided to its counter-part member—due to projections in the magnetic circuit of the exciting coils, when both primary and secondary circuits are excited—an expression for total torque (T_d) can be obtained by the principle of superposition, and it is given as follows:

$$T_d = T_{\text{rel}} + T_{\text{exc}} = \frac{1}{2} i_p^2 \frac{dL_p}{d\theta_r} + \frac{1}{2} i_s^2 \frac{dL_s}{d\theta_r} \pm i_s i_r \frac{dM}{d\theta_r} \quad \text{...(3.30)}$$

The first two terms in the expression denote the contribution due to reluctance torque, and the third term accounts for the excitation torque developed by interaction of the primary and secondary circuits.

(iv) Inspection of the various torque expressions indicates that, generally speaking, constant current situations and unsaturated condition of operation, half the electrical energy flow goes to manifest itself as mechanical work done by (or on) an electro-mechanical system, the other half-being stored in the magnetic field. A general expression for torque is, therefore, of the form:

$$T_d = \frac{\partial W_{\text{fld}}\,(i_p, i_s, \theta_r)}{\partial \theta_r} \quad \text{...(3.31)}$$

In general, without mentioning that a similar expression derived above, can also be used for estimating forces developed in linear devices, we can calculate that with the small modification the misalignment angle, θ_r is replaced by a distance variable x. Hence, the general expression for production of forces, F_d in linear devices given as follows:

$$F_d = \frac{\partial W_{\text{fld}}\,(i_p, i_s, x)}{\partial x} \quad \text{...(3.32)}$$

It is important for us to remember that the above expressions have been derived for instantaneous torque/force only with a given position, and hence these can be liken to expressions for torque/force under instantaneous *static conditions* qualified by relative positions of stationary and movable members. Further, these results were obtained under idealized assumptions of non-saturation of magnetic circuits, which point to situation where the total energy input divided itself equally—*half of the total incremental energy as field energy due to altered magnetic situation in the core and air-gap and the other half being available for energy conversion, manifesting itself in the form of useful torque or forces, as the case may be.*

Figure 3.20 shows the energy flow diagram under this idealized condition and demarkates components stored as magnetic energy and mechanical work done.

For the sake of completeness, let us go back, once again, to the general situation when flux-linkage changes from an initial value of ψ to a final value of $(\psi + \delta\psi)$ over a time interval of δt, while the secondary moves through an angle of $\delta\theta_r$. Treating the current through the circuit to be a variable such that it varies from an initial value of zero to a final value of $i = I_1$, as ψ varies from zero to ψ or $(\psi + \delta\psi)$ as the case may be, the total incremental energy flow is as follows:

$$\delta W_{\text{elect}} = e\, i\, \delta t = i\frac{\delta\psi}{\delta t}\delta t = i\,\delta\psi \qquad ...(3.33)$$

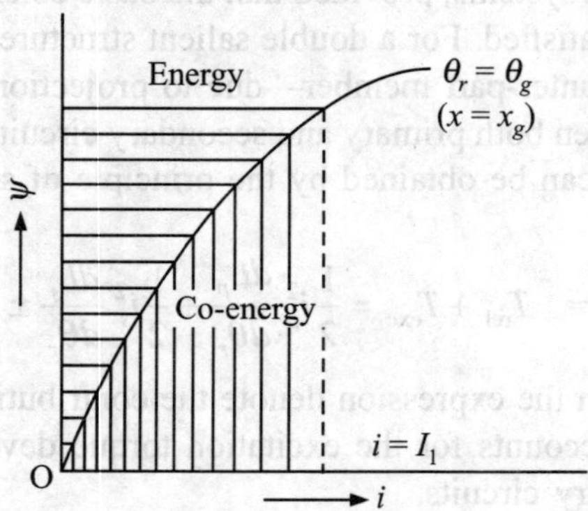

FIGURE 3.20(a) i–ψ characteristics of magnetic circuit, for $\theta_r = \theta_g$ (or $x = x_g$).

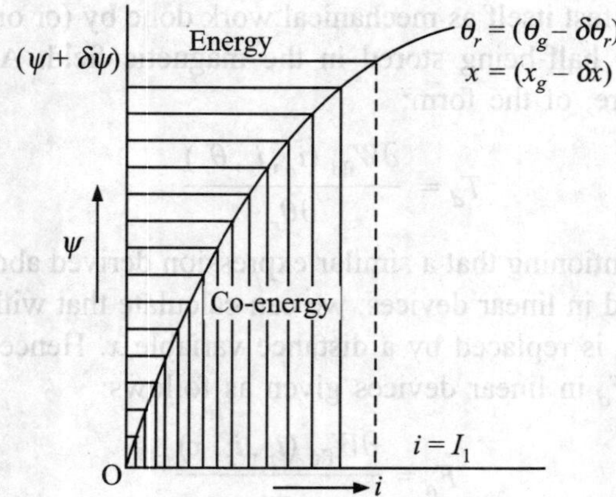

FIGURE 3.20(b) i–ψ characteristics of magnetic circuit, for $\theta_r = (\theta_g - \delta\theta_r)$ or $x = (x_g - \delta x)$.

During the same time, δt, the value of mechanical work done by the system, δW_{mech}, is given as follows:

$$\delta W_{\text{mech}} = T_d \delta\theta_r \qquad \text{...(3.34)}$$

and the incremental change in magnetic energy (δW_{fld}) during this displacement, $\delta\theta_r$, is as follows:

$$\delta W_{\text{fld}} = \int_0^{(\psi+\delta\psi)} i\, d\psi[(\theta_g - \delta\theta_r), i] - \int_0^{\psi} i\, d\psi[\theta_g, i] \qquad \text{...(3.35)}$$

These are pictorially represented by Figure 3.20(a) and Figure 3.20(b), and the diagram shown in Figure 3.20(c). By superposing horizontal hatchings for energy and vertical hatchings for co-energy in Figures 3.20(a) and (b) and double hatching for incremental field energy as well as dot shading for incremental mechanical work done, Figure 3.20(c) depicts these component energies.

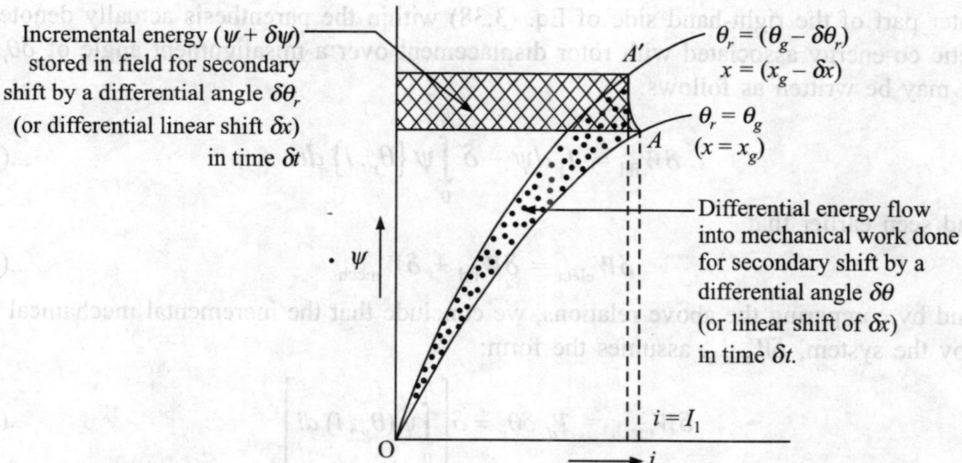

FIGURE 3.20(c) Figure obtained by superposing Figures 3.20(a) and (b).

In Eq. (3.35), we may note that the first term of integration gives the magnetic field energy of the system in which the secondary axis is positioned at $\theta_r = (\theta_g - \delta\theta_r)$, the integration being carried out with respect to ψ over a limit of zero to $(\psi + \delta\psi)$, with θ_r maintained at $(\theta_g - \delta\theta_r)$. The solution is depicted in Figure 3.20(b). Likewise, the second term of the integration represents the situation indicated in Figure 3.20(a), for which the integration is to be performed with respect to ψ over the limits of zero and ψ, with θ_r fixed at θ_g. It may also be noted that the rectangular area bounded by the sides ψ and I in Figure 3.20(a) and sides $(\psi + \delta\psi)$, and I in Figure 3.20(b), stand for the total energy flow into the system during the respectively indicated operating conditions. Each integration is executed *in parts*, as follows:

$$\int_0^{(\psi+\delta\psi)} i\, d\psi = i\left[\psi\right]_0^{\psi+\delta\psi} - \int_0^{I_1} \psi[\theta_g - \delta\theta_r, i]\, di$$

$$= I(\psi + \delta\psi) - \int_0^{I_1} \psi[\theta_g - \delta\theta_r, i]\, di \qquad \text{...(3.36)}$$

Similarly, the second term of Eq. (3.35), becomes:

$$\int_0^{\psi} i\, d\psi = I\psi - \int_0^{I_1} \psi[\theta_g, i]\, di \qquad ...(3.37)$$

We shall come back to the physical significance of the area marked OA′A in Figure 3.20(c), again in Section 3.7.5. It is interesting to note that the second terms of the integrations in Eqs. (3.36) and (3.37) denote magnetic co-energy of the respective states of the operations. The above statements may be confirmed with reference to Figure 3.20(a) and Figure 3.20(b).

Thus, by substitution:

$$\delta W_{\text{fld}} = I\, d\psi - \left[\int_0^{I_1} \{\psi(\theta_g - \delta\theta_r)\, i\}\, di - \int_0^{I_1} \{\psi(\theta_g), i\}\, di\right] \qquad ...(3.38)$$

The later part of the right-hand side of Eq. (3.38) within the parenthesis actually denotes the magnetic co-energy associated with rotor displacement over a misalignment angle of $\delta\theta_r$, and hence, may be written as follows:

$$\delta W_{\text{fld}} = I_1\, d\psi - \delta \int_0^{I_1} \psi\{\theta_r, i\}\, di \qquad ...(3.39)$$

We had seen earlier that

$$\delta W_{\text{elect}} = \delta W_{\text{fld}} + \delta W_{\text{mech}} \qquad ...(3.40)$$

And by comparing the above relations, we conclude that the incremental mechanical work done by the system, δW_{mech}, assumes the form:

$$\delta W_{\text{mech}} = T_d\, \delta\theta_r = \delta\left[\int_0^{I_1} \psi(\theta_g, i)\, di\right] \qquad ...(3.41)$$

Also, the general expression of the torque developed, T_d, has the form:

$$T_d = \delta \int_0^{I_1} \frac{\psi(\theta_g, i)\, di}{\delta\theta_r}$$

That is,

$$T_d = \left.\frac{\delta\,(\text{Magnetic co-energy})}{\delta\theta_r}\right]_{\text{at constant } I} \qquad ...(3.42)$$

By similarity, general expression for force developed, F_d, is given by replacing θ_r by x, as follows:

$$F_d = \frac{\delta \int_0^{I_1} \psi\,(x_g, i)\, di}{\delta x_g} = \frac{\delta\,(\text{Magnetic co-energy})}{\delta x_g} \qquad ...(3.43)$$

at constant value of I_1.

These expressions are fundamental in the estimation of torque or force, as the case may be, when dealing with saturation effects of devices. Also, for a multiple excited system with n

electrical systems contributing to flux production, the general expressions for torque and force modify respectively to:

$$T_d = \frac{\sum_{i=I_1}^{i=I_n} \delta\psi(\theta_g i)\, di}{\delta\theta_r}; \quad F_d = \frac{\sum_{i=I_1}^{i=I_n} \delta\psi(x_g i)\, di}{\delta x_g}$$

We can also treat the above analysis by revisiting the energy distributional diagram in Figure 1.1) for an electro-mechanical system. The descriptive energy distribution pattern is elaborated in Figure 3.21.

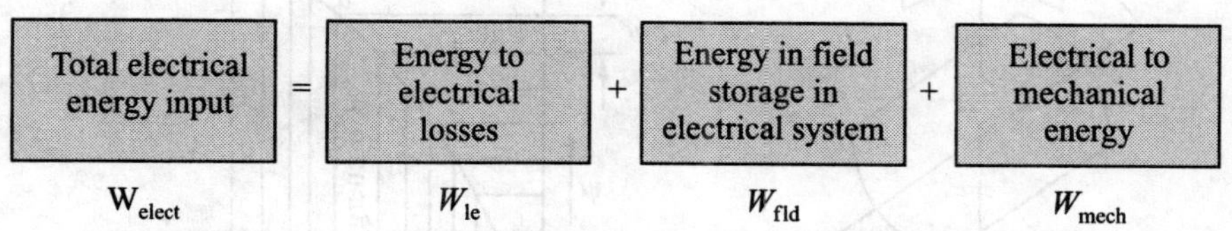

FIGURE 3.21 Energy distribution diagram of Figure 1.1 revisited.

Hence, $$W_{\text{elect}} = W_{\text{le}} + W_{\text{fld}} + W_{\text{mech}} \qquad \text{...(3.44)}$$

Electrical and mechanical power do not make themselves available to direct exploitation but are lost as friction and windage losses (W_{lm}), and some may remain stored in the system, (W_{sm}) in rotating inertia,

$$W_{\text{mech}} = W_{\text{mt}} + W_{\text{lm}} + W_{\text{sm}}$$

Hence,

$$W_{\text{elect}} = (W_{\text{le}} + W_{\text{lm}}) + (W_{\text{fld}} + W_{\text{sm}}) + W_{\text{mt}}$$

and similarly, for generator operation

$$W_{\text{mech}} = (W_{\text{le}} + W_{\text{lm}}) + (W_{\text{fld}} + W_{\text{sm}}) + W_{\text{elect}}$$

We can, therefore, combine these two equations by a single relation:

$$W_{\text{mech}} + W_{\text{elect}} = (W_{\text{le}} + W_{\text{lm}}) + (W_{\text{fld}} + W_{\text{sm}})$$

In this expression, for the electrical to mechanical energy conversion (motoring action) W_{elect} is positive and W_{mt} is negative. For mechanical to electrical energy conversion, (generating operation) W_{elect} is negative and W_{mt} (i.e., W_{mech}) is positive.

3.7.5 Generalized Equation for Torque

It may not be realistic to assume that current flow remains constant during flow of energy in the device. For the sake of completeness, let us go back once again to the general situation when the flux linkage increases from an initial value of ψ to a value of $(\psi + \delta\psi)$, over a time of δt, while the secondary moves through a small angle of $\delta\theta_r$. Treating the current through the circuit to be a variable, and assuming that it varies from an initial value of zero to a final value of $i = I_1$, as the flux-linkage ψ varies from zero to ψ or to $(\psi + \delta\psi)$, as the case may be, the initial incremental energy flow, can be worked out by considering critically the energy distribution diagram in some detail. For this purpose, we shall choose a singly excited electro-mechanical system for two alternative static positions of misalignment between the secondary and primary magnetic axis corresponding to angle of misalignment, $\theta_r = \theta_{r1}$, and $\theta_r = \theta_{r2}$. These are depicted in Figure 3.22(a) and (b) and Figure 3.23(a) and (b) respectively, by similar diagrams as in Figure 3.23(a) and (b) respectively.

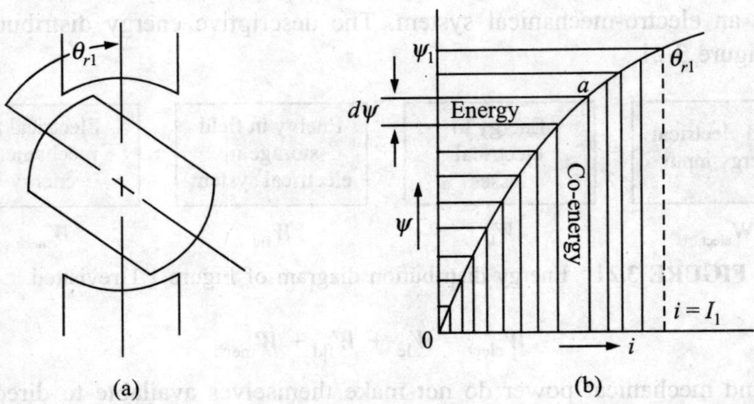

FIGURE 3.22 For static secondary member position making θ_{r1} with primary flux axis [Figure 3.22(a)] and the corresponding energy distribution position showing total electrical energy = Area of the rectangle $(0\psi_1\theta_{r1}I_10)$.
Total field energy stored in magnetic circuit = Horizontally hatched area $(0\psi_1\theta_{r1}a0)$ [Figure 3.11(b)]

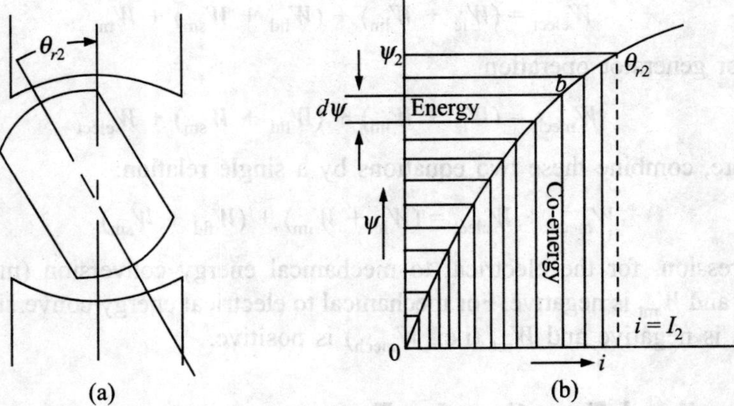

FIGURE 3.23 For static secondary member position making θ_{r2} with primary flux-axis 3.23(a), and the corresponding energy distribution position showing total electrical energy = Area of the rectangle $(0\psi_2\theta_{r2}I_20)$.
Total field energy stored in magnetic circuit = Horizontally hatched area $(0\psi_2\theta_{r2}b0)$ [Figure 3.23(b)].

It can be noted from Figure 3.22(a) and (b) and Figure 3.23(a) and (b), that the two diagrams have $\theta_{r1} > \theta_{r2}$. The corresponding i–ψ-diagrams for the two cases are presented in

Figure 3.22(b) and Figure 3.23(b). Further, non-linearity in the characteristics indicates that saturation in magnetic circuit is not neglected as in earlier considerations. This leads to values of self-inductances of the current carrying coil in the primary, not to be independent of i. We had seen in Section 3.5 in Eq. (3.16) that the energy stored in the magnetic field has the form $(i\,d\psi)$, integrated over current limits of $i = 0$, $i = I_1$ for Figure 3.22(b) and the $(i–\psi)$ curve for the static position of misalignment θ_{r1}. The same conclusions can also be arrived at from Figure 3.23(b), for the position misalignment θ_{r2}. We have shown these positions with horizontal hatchings.

Thus, the field energy stored for θ_{r1} misalignent is indicated by $(0\psi_1\theta_{r1}a0)$ in Figure 3.22(b). The field stored for θ_{r2} misalignment is similarly indicated by the area $(0\psi_2\theta_{r2}b0)$ in Figure 3.23(b).

We also note that the total electrical energy input, W_{elect} for the above static operations are denoted by the rectangular areas:

W_{elect} for θ_{r1} misalignment is the area $(0\psi_1\theta_{r1}I_10)$ in Figure 3.22(b), and W_{elect} for θ_{r2} misalignment is the area $(0\psi_2\theta_{r2}I_20)$ in Figure 3.23(b).

Figure 3.22(b) and 3.23(b) also include areas denoted by $(0a\theta_{r1}I_10)$ for $\theta_r = \theta_{r1}$ and area $(0b\theta_{r2}I_20)$ for $\theta_r = \theta_{r2}$. We had indicated that these areas, in Figures 3.14(a) and (b) and its general expression has form $(\psi\,di)$, or the area bounded by the parts located below the current–flux-linkage characteristic, bound by current I_1 (or I_2). These portions are shown with vertical hatchings. We shall superpose Figure 3.22(b) and Figure 3.23(b) and further consider two more $(i–\psi)$ characteristics for secondary misalignment angles $\theta_r = \theta_g$ and $\theta_r = \theta_g - \delta\theta_r$. Here, $\delta\theta_r$ is taken as a small angular displacement, so that $\delta\theta_r$ tends to reduce the misalignment angle, which makes

$$\theta_g - \delta\theta_r < \theta_g$$

The superposed diagram is shown in Figure 3.24.

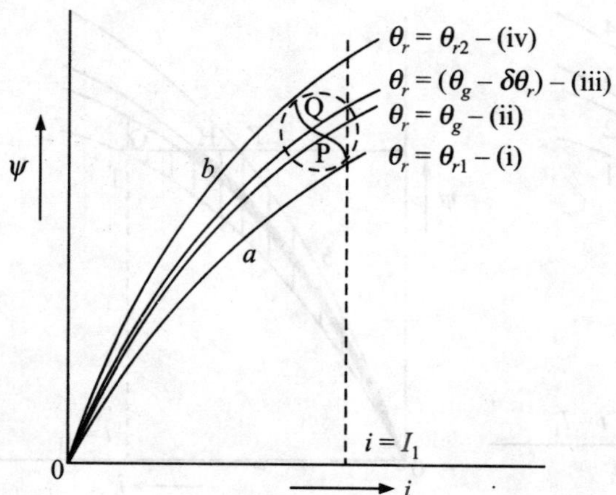

FIGURE 3.24 $(i–\psi)$ characteristics for different static secondary member positions with misalignment angles (i) $\theta_r = \theta_{r1}$, (ii) $\theta_r = \theta_g$, (iii) $\theta_r = (\theta_g - \delta\theta_r)$ and (iv) $\theta_r = \theta_{r2}$ $[\theta_{r1} > \theta_g > (\theta_g - \delta\theta_r) > \theta_{r2}]$.

3.7.6 Dynamic Energisation

Before considering the general transient operation of the energy converter arrangement, it is interesting to study the alternative types of controlled movement from misalignment positions $\theta_r = \theta_{r1}$ to $\theta_r = \theta_{r2}$ in Figure 3.24.

Under Condition 1—very slow movement

If the secondary member is constrained to move very slowly, the current i is sensibly constant as $\left(\dfrac{d\psi}{dt}\right)$ is very small. The flux-linkage, then, increases from ψ_1 to ψ_2 at an assumed constant current $i = I_1$.

The useful electric current is given by:

$$\int_{\psi_1}^{\psi_2} i\, d\psi = i \int_{\psi_1}^{\psi_2} d\psi = I_1(\psi_2 - \psi_1)$$

Figure 3.25(a) shows that this amounts to the field energy changes, and the mechanical work done, W_{mech}, is the difference between the input energy and the change in field energy. This is depicted by the area (0*a*14*b*0). The average electro-magnetic torque (T_e) during the so called, movement is given by:

$$T_e \text{ (average)} = \frac{\text{Mechanical work done}}{\text{Angular movement}}$$

$$= \frac{\text{Area } (0a14b0)}{\theta_{r2} - \theta_{r1}}$$

The above area is shown as hatched and blackened in Figure 3.25(a)

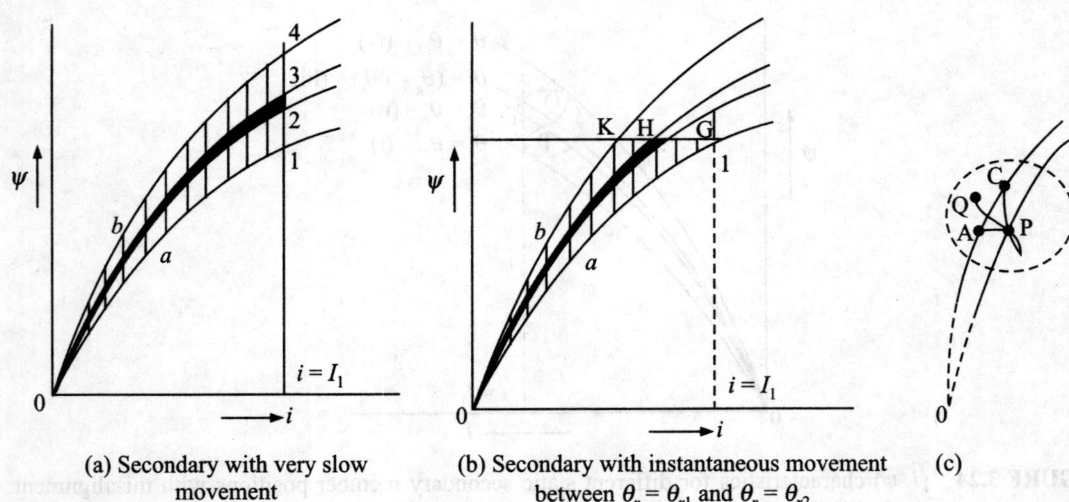

(a) Secondary with very slow movement

(b) Secondary with instantaneous movement between $\theta_r = \theta_{r1}$ and $\theta_r = \theta_{r2}$

(c)

FIGURE 3.25 Details of the circled part of Figure 3.24 showing energy to mechanical work for the differential angular movement of secondary member by $\delta\theta_r$.

Under Condition 2—instantaneous movement

Under this condition, the secondary is moved instantaneously from position corresponding to misalignment angle $\theta_r = \theta_{r1}$ to $\theta_r = \theta_{r2}$. Realistically, if the secondary movement is not instantaneous but is very fast, a small change of flux in the primary field induces a high voltage in the exciting primary winding. This voltage causes the current to change so as to oppose the flux change. In practice, the resulting flux-change is very small and can be neglected. *The principle of assuming no change in flux-linkage is known as the theorem of constant flux-linkage.* The current in the coil reduces from $i = I_1$ to $i = I_2$ at constant flux-linkage. This is depicted in Figure 3.25(b), with the secondary held in position corresponding to $\theta_r = \theta_{r2}$. The current and flux-linkage have to increase gradually to the final steady-state values, $i = I_2$ and $\psi = \psi_2$. During this time, the field stored energy is increased by an amount equal to the electrical input.

No energy is taken from the supply during the secondary movement under condition 2, as $\int i d\psi = 0$. Mechanical energy is obtained from the change in field energy. This is shown in Figure 3.25(b).

The average torque, during condition 2, movement can be written as follows:

$$T_e\ (\text{average}) = \frac{\text{Mechanical work done}}{(\text{Angular movement})}$$

$$= \frac{\text{Area } (0a1GHKb0)}{\theta_{r1} - \theta_{r2}}$$

The area denoting the mechanical work done—$(0a1GHKb0)$ is shown as hatched and darkened area in Figure 3.25(b). For a small angular movement of $\delta\theta_r$ from a stationary initial position corresponding to $\theta_r = \theta_g$ to $\theta_r = (\theta_g - \delta\theta_r)$, the area is also indicated in Figure 3.25(a) and (b) by darkened portions. We are, now ready to write the expressions for torque T_e by inspection as follows:

(i) For condition 1—very slow movement:

$$T_e = \frac{dW_{fld}}{d\theta_r} = \lim_{\delta\theta_r \to 0}\left[\frac{\text{Area (0230)}}{\delta\theta_r}\right] = \frac{\partial W_{fld}}{\partial\theta_r} \qquad \text{...(3.45)}$$

(ii) For condition 2—very fast (or instantaneous) movement:

$$T_e = -\frac{dW_{fld}}{d\theta_r} = -\lim_{\delta\theta_r \to 0}\left[\frac{\text{Area (0GH0)}}{\delta\theta_r}\right] = -\frac{\partial W_{fld}}{\partial\theta_r} \qquad \text{...(3.46)}$$

Equation (3.46) *is shown as negative as the field energy decreases with movement.*

In the above expressions W_{fld} is a function of θ_r and i. In Figure 3.25(a) and (b), the rate of change of W_{fld} with respect to θ_r occurs at constant value of $i = I_1$ and constant of $\psi = \psi_1$, respectively. Since a differential with respect to one variable, with the other variable, considered as constant, is customarily denoted by partial derivative, $\left(\frac{\partial}{\partial\theta_r}\right)$:

For Condition 1:
$$T_e = \left.\frac{\partial W_{\text{fld}}}{\partial \theta_r}\right]_{\text{at constant current}}$$

and similarly, for Condition 2:
$$T_e = \left.\frac{\partial W_{\text{fld}}}{\partial \theta_r}\right]_{\text{at constant flux-linkage}}$$

Under Condition 3—under transient movement

This is the most plausible condition of operation, and both current and flux-linkage have time to change during the movement. A typical movement from $\theta_r = \theta_{r1}$ to $\theta_r = \theta_{r2}$ including possible overshooting on reaching position with angle of non-alignment, $\theta_r = \theta_{r2}$, was shown in Figure 3.24. The figure also indicated the small movement from $\theta_r = \theta_g$ to $\theta_r = (\theta_g - \delta\theta_r)$, within the encircled part on the figure. This part has been shown blown up in Figure 3.25(c), and is of special interest in this section.

It will be clearly evident that the energy converted to medicinal work lies between the values obtained by moving the secondary very slowly and instantaneously. For a small movement indicated by linear displacement depicted by PQ of the flux-linkage of the secondary member we see from Figure 3.25(c) (in combination with Figure 3.24) that the energy to mechanical work is denoted by the area (OPQO) with controlled movement at P, the energy to mechanical work is denoted by (OPAO) at constant flux-linkage and (OPCO) at constant current $i = I_1$. Hence, for the actual movement over $\delta\theta_r$ from the initial position of $\theta_r = \theta_g$ to $\theta_r = (\theta_g - \delta\theta_r)$:

$$T_e = \left.\frac{\partial W_{\text{fld}}}{\partial \theta_r}\right]_{\text{at constant current}}$$

or
$$T_e = -\left.\frac{\partial W_{\text{fld}}}{\partial \theta_r}\right]_{\text{at constant flux-linkage}}$$

which is same as:

$$\lim_{\delta\theta_r \to 0}\left[\frac{\text{Area(OPQO)} + \text{Area (CPQC)}}{\delta\theta_r}\right]_{\text{at constant } I_1} \quad \text{...(3.47)}$$

And for controlled movement at constant flux-linkage over the above values of θ_r, it is as follows:

$$T_e = \lim_{\delta\theta_r \to 0}\left[\frac{\text{Area(OPQO)} - \text{Area (APQA)}}{\delta\theta_r}\right] \quad \text{...(3.48)}$$

But in the limit of ($\delta\theta_r \to 0$), both the areas (CPQC) and (APQA) vanish, and the final torque expression takes the form:

$$T_e = \frac{\partial W_{\text{fld}}}{\partial \theta_r} = \frac{\text{Change in energy stored in field}}{\partial \theta_r} \quad \text{...(3.49)}$$

Thus, the torque can be calculated at any angle of misalignment between the flux axes, from a knowledge of the change of field energy or field co-energy under controlled movement at constant current i or at constant flux-linkage ψ.

We should note that since the above expressions have been developed for 2-pole configuration, the final value of torque has also to be multiplied by pole-pairs.

The mechanical force developed in electro-mechanical devices, always tends to act in a direction to decrease the stored field energy at constant flux, or to increase the stored energy and co-energy at constant, mmf, which amounts to resulting in decrease in reluctance of the interfacing magnetic circuit or an increase in inductance of the concerned electrical circuits.

3.7.7 Dynamic Equations and Nature of Voltages Induced in the System

In order to extend the equations to cover dynamic condition, we cannot assume constancy of current magnitudes any further. Besides, the total torque comprises inertia torque of the movable member in addition to torque due to mechanical load. The dynamic equation of the system, therefore, has the form:

$$J\frac{d^2\theta_r}{dt^2}+T_{\text{mech}}=\frac{1}{2}(i_p)^2\frac{dL_p}{dt}+\frac{1}{2}(i_s)^2\frac{dL_s}{dt}\pm i_p i_s\frac{dM}{dt} \qquad \text{...(3.50)}$$

Here, J signifies the angular mass moment of inertia presented by the mobile member.

We shall use the above relation to study the nature of voltage in respect of the primary and secondary electric circuits under general condition of operation. For this purpose, we shall, once again, revert back to assumption to conservative system and express voltages in terms of currents and circuit impedances—both self and mutual—which are functions of time.

Since for a conservative system, the applied voltage at all times balances the induced emf, we may write that:

$$v_p=\frac{d}{dt}[L_p i_p\pm M i_s]$$

$$=L_p\frac{\partial i_p}{\partial t}+i_p\frac{\partial L_p}{\partial t}\pm M\frac{\partial i_s}{\partial t}+i_s\frac{\partial M}{\partial t}$$

which can be rearranged as follows:

$$v_p=\left\{L_p\frac{\partial i_p}{\partial t}\pm M\frac{\partial i_s}{\partial t}\right\}+\left\{i_p\frac{\partial L_p}{\partial \theta_r}\pm i_s\frac{\partial M}{\partial \theta_r}\right\}\frac{\partial \theta_r}{\partial t} \qquad \text{...(3.51)}$$

Similarly, it can be shown that:

$$v_s=\left\{\pm M\frac{\partial i_p}{\partial t}+L_s\frac{\partial i_s}{\partial t}\right\}+\left\{\pm i_p\frac{\partial M}{\partial \theta_r}+i_s\frac{\partial L_s}{\partial \theta_r}\right\}\frac{\partial \theta_r}{\partial t} \qquad \text{...(3.52)}$$

The expressions for v_p and v_s may be subjected to elaborate interpretations. A close examination reveals that these voltages can be seen as composed of two distinct components—the first exclusively depending on the time rate of change of currents and the second derived by the time rate of change of misalignment angle or the relative speed of rotation of the movable member with respect to its stationary counterpart. The first part, therefore, is of the nature of voltages induced in static systems, where the voltages are induced on the basis of ratio of change of flux linked by individual circuits like in transformers. This part of voltage induction is termed as *transformer induced voltages*. The other component is distinctly associated with angular velocity of the rotating member with respect to the stationary member. Hence, this component of voltage is identified as *speed induced voltages*. Significance of these two component voltages can be appreciated immediately. When the individual circuit voltages are multiplied by respective values of circuit currents, the result yields instantaneous power flow. The first component obtained from the product of transformer induced voltage and current yields power input towards storage of the magnetic energy in field systems, while the second component resulting from the product of speed induced voltage and current denotes power component related to electro-mechanical energy conversion, through production of mechanical torque. This information proves to be of special significance later in the analysis of special geometry electro-mechanical devices.

Physical interpretation of transformer-induced and speed-induced voltages and their interrelationship in the space and time domain, may offer an interesting study in the light of the above observation, especially in relation to their specific role in the process of energy conversion. Let us consider a simple model consisting of spatially distributed single-phase stator winding housed in uniformly distributed slots of an annular structure, and fed from an alternating current source of frequency f_s. In the air-gap of this stator arrangement is concentrically mounted a rotating member with uniformly distributed copper bars along the rotor periphery. This model has close resemblance to the actual arrangement obtained in ideal single-phase induction machines. Let us, further, assume that the rotor is rotating at some angular speed of ω_r radian per second, while the synchronous speed of the system corresponding to the input supply frequency of f_s is given by ω_s. Two distinct phenomena occur simultaneously in the rotor conductors. Thus:

1. The alternating stator flux, linking the rotor bars, produces in them instantaneous induced voltages by *transformer action*. The nature of this induced voltage is such that it always tends to oppose the changing stator-flux. Frequency of this transformer induced voltage equals the frequency f_s of the excitation voltage. If the rotor induced voltage were permitted to cause a current flow in its bars, it will lead to development of rotor flux in exact space and time opposition to the stator alternating flux.
2. Rotor conductors also cut across the same stator flux distribution due to their movement at an angular speed of ω_r. The resulting component of induced voltage in rotor conductors is exclusively due to *speed action*, its magnitude depending on the instantaneous value of the flux-density at the instant of flux-cutting, and the speed at which the flux is cut, in accordance with Faraday's *BLV* law.

Let us adopt a general convention that all quantities related with performance through transformer action, are represented with a suffix T, while those associated with speed action have a suffix S. A schematic of the model is shown in Figure 3.26.

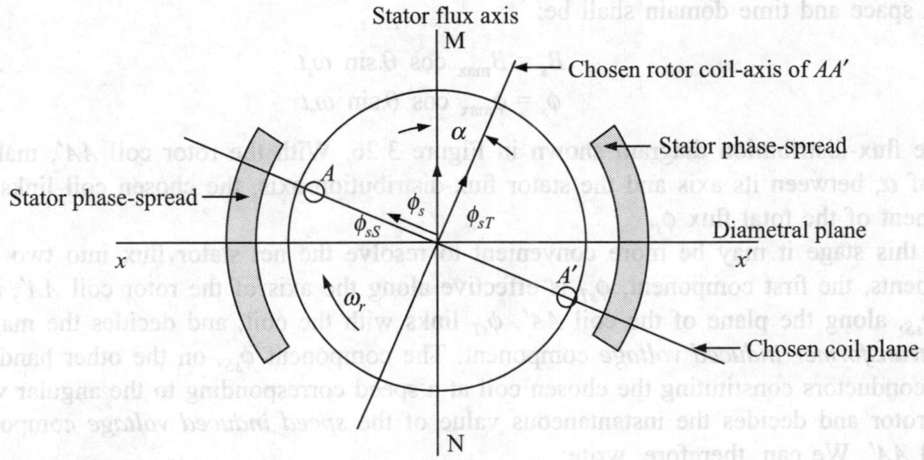

FIGURE 3.26 Flux distribution diagram for the stator mounted phase winding with a general rotor coil AA'.

The stator flux-axis, MN, is assumed to be vertical. We shall consider a general rotor turn corresponding to two rotor bars AA' situated diametrically opposite to each other. The axis of this rotor turn is at a space angle of α with respect to the stator flux axis, as shown. Clockwise direction of rotation along which the rotor is assumed to rotate, is taken as positive direction of rotation in the present conventions.

Assume that the total flux produced by the exciting stator coil is denoted by ϕ_s. The value of ϕ_s is a function of both space and time, since the winding itself is distributed in space and carries a time-varying current, i, through it. If the total excitation flux ϕ_s produced by the maximum value of excitating current, ${}_{\max}I_s$, in the stator winding is denoted by $\phi_{\max}$, then on the assumption that the flux is distributed in the air-gap such that the flux-density ($B_{s\theta}$) at any point θ is given by:

$$B_s = B_{s\theta} \text{ where, } B_{s\theta} = B_{\max} \text{ along axis MN at } i = {}_{\max}I_s$$

$$B_{s\theta} = B_{\max} \cos \theta$$

$$\phi_s = \phi_{\max} \cos \theta$$

The above expression indicates the nature of flux-density at different points on the periphery, on the assumption of co-sinusoidal flux distribution in space, when, $\phi_s = \phi_{\max}$ at $\theta = 0$. However, ϕ_s varies with the instantaneous magnitude of i_s, which may be assumed to vary in time in accordance with the relationship:

$$i_s = {}_{\max}I_s \sin \omega_s t$$

Assume that the instantaneous magnitude of excitation current is zero at $t = 0$, in the above expression. Also $\omega_s = 2\pi f_s$, represents the angular velocity corresponding to the frequency of the exciting current. The value of flux-density B_s at a spatial angular position θ (say $\theta = \alpha$), from the axis MN of the exciting stator winding shall then have the following form:

$$B_s = B_{s\theta}$$

And in space and time domain shall be:

$$B_s = B_{\max} \cos \theta . \sin \omega_s t \quad \text{...(3.53)}$$

$$\phi_s = \phi_{\max} \cos \theta . \sin \omega_s t \quad \text{...(3.54)}$$

The flux-distribution diagram shown in Figure 3.26. With the rotor coil *AA'*, making an angle of α, between its axis and the stator flux-distribution axis, the chosen coil links with a component of the total flux ϕ_s.

At this stage it may be more convenient to resolve the net stator flux into two distinct components, the first component, ϕ_{sT} is effective along the axis of the rotor coil *AA'*, and the other ϕ_{sS}, along the plane of the coil *AA'*. ϕ_{sT} links with the coil, and decides the magnitude of the *transformer induced voltage* component. The component ϕ_{sS}, on the other hand, is cut by the conductors constituting the chosen coil at a speed corresponding to the angular velocity of the rotor and decides the instantaneous value of the *speed induced voltage* component in the coil *AA'*. We can, therefore, write:

$$\phi_{sT} = \phi_{\max} \cos \alpha \cos \omega_s t \quad \text{...(3.55)}$$

and

$$\phi_{sS} = \phi_{\max} \sin \alpha \cos \omega_s t \quad \text{...(3.56)}$$

First, let us determine the magnitude of transformer voltage in the rotor bars. Based on the consideration that the chosen rotor coil *AA'* (with electrical continuity envisaged through short-circuited ends as obtained in rotors of single-phase induction machines) form a single turn coil, and link with flux ϕ_{sT}. e_{T1}, the transformer induced voltage in the coil *AA'* of the rotor,

$$e_T = \frac{d}{dt}\phi_{sT} = \phi_{\max} \cos \alpha \frac{d}{dt}(\cos \omega_s t)$$

or

$$e_T = -\,\omega_s\, \phi_{\max} \cos \alpha \sin \omega_s t \quad \text{...(3.57)}$$

Determination of an expression for the instantaneous value of speed-induced voltage in the same pair of rotor bars corresponding to the chosen instantaneous position, needs a slightly modified strategy as detailed below:

1. It is assumed that the magnitude ϕ_{sS} is unvarying at the particular instant chosen, so that ϕ_{sS} is momentarily independent of time. The same is also considered for B_{sS}.
2. However, this component of flux ϕ_{sS} (like its counterpart flux component ϕ_{sT}) is also distributed in space sinusoidally.
3. For generality, consider that the distributed winding on the stationary member produced a 2*p*-pole flux-distribution in the air-gap. Figure 3.15 depicts the arrangement for a number of poles, $2p = 2$. Over a complete revolution, a rotor conductor cuts across a total flux of $(2p\,\phi_{sS})$ units, at an average flux-density of B_{av}, and produces in the chosen single-turn full-pitch, coil *AA'*, an average value of speed induced voltage E_s, given by:

$$E_s = 2 B_{av} L V \text{ volts}$$

where, L is the length of each rotor bar cutting fluxes at a linear speed of V. Also:

$$B_{av} = \frac{\text{Total flux cut by a conductor over one revolution}}{\text{Total surface area over which the flux is distributed}}$$

$$= \frac{2p\phi_{sS}}{\pi DL}$$

where, D denotes the diameter of rotor. As the conductor rotates, it finds that the flux-distribution is sinusoidal along the circumference of the rotor/stator air-gap. In other words, the general pattern of flux-distribution (B_x) at a point x as seen by the rotating conductors, is given as follows:

$$B_x = B_{max} \sin \theta$$

where, θ is the space angle measured between the plane passing through x. For the coil AA', we can observe that $\theta = \alpha$.

Thus, for such a system with sinusoidal flux-density distribution along the rotor circumference:

$$\frac{B_{av}}{B_{max}} = \frac{2}{\pi}$$

where, $$B_{max} = \frac{p\phi_{max}}{DL}$$

This relationship helps us to conclude that ***the assumption of a uniform flux-density distribution, B_{max} along the diametral plane perpendicular to the axis of the exciting coil is the same thing as considering that the nature of flux-density distribution along the circumferential surface of the rotor is sinusoidal.***

Hence,

$$E_S = 2 L V B_{av}$$

and, with $V = \pi D n_r$, where, $n_r = \dfrac{\omega_r}{2p \cdot \pi}$ in revolution/second,

$$E_S = 2L\left(\frac{\pi D\omega_r}{2p\pi}\right)\left(\frac{2p\phi_{sS}}{\pi DL}\right) = \frac{2\phi_{sS}}{\pi}\omega_r \qquad \text{...(3.58)}$$

E_s in this expression stands for the average speed-induced voltage, which varies at an angular frequency of ω_r. The maximum value of this voltage is given by:

$$_{max}E_S = \phi_{sS}\, \omega_r$$

and the general expression for the speed-induced voltage E_S can now be written down by substitution of the value of ϕ_{sS}, as follows:

$$E_S = \phi_{max}\, \omega_r \sin \theta \cos \omega_s t \qquad \text{...(3.59)}$$

A critical comparison of general expressions of E_T and E_S, in Eqs. (3.57) and (3.59) reveal that the transformer-induced voltage component has a magnitude proportional to ω_s, that alternates at a frequency of f_s, while the speed-induced voltage component has a magnitude proportional to ω_r, pulsating at a frequency of f_r.

The frequency of pulsation of the total voltage in rotor conductors is a complex function of f_s and f_r. When the rotor is rotated at an angular frequency of ω_s, both these components have the same value of frequency, but space and time-phase relationships differ by 90°.

When dealing with standard machines, it is usually sufficient to consider any one of these effects while analysing performance under steady-state conditions. However, any analysis under instantaneous conditions requires consideration of both transformer induced and speed-induced effects. Hence, these concepts assume special importance in model-based generalised analysis of electro-mechanical devices.

3.7.8 Effect of Saturation of Magnetic Circuits

In this section, we discuss the effect of saturation of magnetic circuit on force/torque production and energy conversion in electro-mechanical devices.

We know that it is a standard practice to assume, for purposes of simplifications in analysis, that magnetic circuits are considered unsaturated over the entire range of normal operation of such devices. In reality, however, almost all electro-mechanical devices inexorably operate with varying degrees of magnetic saturation as this results in more effective utilization of materials and improvement of operational efficiency in the burgain. In fact, it is a common practice in certain electro-mechanical devices like hysteresis motors, stepper motors, etc. to deliberately utilize saturation effects and use magnetic materials with rectangular hysteresis loops of large specific hysteresis loss.

Magnetic saturation leads to non-linear behaviour, and devices working under its influence do not lend themselves to general analysis using standard analytical techniques. A case-by-case study with graphical aid is often the only recourse available for such analysis, that too with high degree of approximation.

Although non-linearity does not directly encroach upon the force/torque equations discussed in the earlier sections of this chapter. We should not over look that saturation often may have an important indirect effect on the force/torque produced by a specified coil current, since the air-gap flux and mmf will be determined by the magnetic characteristics of the entire magnetic circuit. The force/torque depends primarily on energy changes at the variable air-gap configuration, and is independent of the source of the field. The force/torque equations in terms of conditions at the air-gap, therefore apply equally well when the field is produced by permanent magnets.

OBJECTIVE TYPE QUESTIONS

3.1 Force on a current-carrying conductor placed in a magnetic field of flux-density B is given by:

(a) BLV (b) iBL

(c) None of (a) and (b) is true (d) Both (a) and (b) are true

3.2 The stored energy W_{field} in a magnetic field may be expressed as:

(a) $\frac{1}{2}$ flux^2 × reluctance (b) $\frac{1}{2}$ mmf^2 × permeance

(c) $\frac{1}{2}$ inductance × current^2 (d) All the above

3.3 Which of the following statements is correct?

(a) Energy can be transformed from one from to another
(b) Energy can be stored
(c) Energy can be dissipated
(d) All the above

3.4 An electro-dynamometer movement can be utilized in:

(a) Voltmeter configuration
(b) Ammeter configuration
(c) Wattmeter configuration
(d) All the above

3.5 A magnetic circuit shows a linear B–H or ϕ–i relation, with ψ = 1.2 Weber-turns at i = 5 amperes. The energy stored in the field is:

(a) 6 joules (b) 0.24 joules
(c) 3 joules (d) 3.6 joules

3.6 Two coils with self-inductance of 1 H and 2 H, have a mutal-inductance of 1 H. The coupling factor is:

(a) $\frac{1}{2}$ (b) $\frac{1}{\sqrt{2}}$

(c) $\sqrt{\frac{1}{4}}$ (d) 1

3.7 If the coils in Problem 6 were to carry currents of 2 A and $\sqrt{2}$ A respectively, the total energy stored in the magnetic field is:

(a) $2(2+\sqrt{2})$ joule (b) $2(1+\sqrt{2})$ joule
(c) 6 joule (d) 3 joule

3.8 The permeance of a magnetic circuit is given by:

(a) reciprocal of inductance (b) $\mu_0 \mu_r \frac{l}{A}$

(c) $\mu_0 \mu_r \frac{A}{l}$ (d) $\frac{1}{\mu_0 \mu_r} \frac{l}{A}$

where l is the length and A the cross-sectional area of the magnetic circuit.

3.9 In magneto-hydrodynamic generator shown in Figure 3.6(a), with unit vector depiction shown in Figure 3.6(b), and with effective plasma velocities denoted by v_1, v_2, v_3 respectively along the i, j and k unit vectors, that is $v = v_1 i + v_2 j + v_3 k$ m/s, the voltage developed is (in the Faraday mode):

(a) $-v_2iB - v_1jB + v_3kB$ (b) $v_1iB + v_2jB + v_3kB$

(c) $v_2iB - v_1jB + 0kB$ (d) $v_2i_B + v_1jB + 0kB$

where Bk is the duct flux-density.

3.10 If saturation in the magnetic field is taken into account, the stored energy is always:

(a) greater than the magnetic co-energy

(b) smaller than the magnetic co-energy

(c) equal to the magnetic co-energy

(d) can not be concluded

Answers

3.1 (b)	**3.2** (d)	**3.3** (d)	**3.4** (d)	**3.5** (c)	**3.6** (b)
3.7 (a)	**3.8** (c)	**3.9** (c)	**3.10** (b)		

NUMERICAL PROBLEMS

3.1 A doughnut shaped toroidal iron ring has inner and outer diameters of 10 cm and 12 cm, and is made up of iron of circular cross-section of 1 cm diameter. If the relative permeability is 2000 at the flux-density of 1.0 tesla in the core and the coil has 200 turns wound on the core, evaluate:

(i) the flux; (ii) the coil current, (iii) coil inductance (iv) energy stored in the coil and (v) energy stored per unit volume of iron in joules/m^2

[**Ans.** (i) 0.785×10^{-4} Wb; (ii) 0.69 A; (iii) 22.85 mH
(iv) 5.435×10^{-3} joules (v) 199 J/m^3]

3.2 If a radial air-gap is cut in the magnetic core of Problem 1, all other data remaining unchanged, estimate (i) to (v) quantities. Neglect the fringing effect in the magnetic circuit. [**Ans.** (i) 0.785×10^{-4} Wb; (ii) 20.59 A; (iii) 0.762 mH
(iv) 161.6×10^{-3} joules (156.2×10^{-3} in air-gap and 5.4×10^{-3} in iron)
(v) 199 J/m^3 in iron and 39.8×10^4 J/m^3 in air-gap]

3.3 Please refer to Figure 3.10 Example 3.2. If the lifting coil has 800 turn carrying a constant 2 amp current in it, obtain: (i) the flux; (ii) energy stored in coil; (iii) energy stored in unit volume of air-gap; (iv) lifting power of the magnetic.

[**Ans.** (i) $B_g = 2$ T, $\phi_g = 2.5$ m Wb (ii) $E = 1.6 \times 10^6$ J/m^3
$W_{\text{field}} = 2$ joule; (iii) $f_{\text{mech}} = 4000$ N]

3.4 An MHD generator shown in Figure 3.6, has a plasma velocity of $v = 300i + 0j + 500k$ m/s and the magnetic flux-density is $B = 2j$ tesla. Estimate: (i) Faraday voltage (ii) Hall voltage, developed by the generator, if the duct dimensions are $0.5i + 0.5j + 2k$ metres.

[**Ans.** (i) $E_F = Bv_1D = 500$ V; $E_H = B \cdot V_3 \cdot L = 1200$ V]

3.5 A Hall-effect device develops 500 mV at a flux-density of 1 Tesla, with a control current of 200 mA. Estimate the Hall coefficient if its thickness is 10^{-4} m, and specify its units.

[**Ans.** $\frac{1}{400}$ Vm/TA (or Ohm-metre/tesla]

3.6 (i) A conductor moves at right angles in a magnetic field of 0.3 tesla at 6 m/s. The voltage developed across the ends of the conductor is measured as 0.54 volt. Find the length of the conductor.

(ii) A conductor 0.5 m long is moved at 50 m/s at an angle of 30° to the direction of the lines of force having density of 0.5 tesla. Determine the voltage induced between the ends of the conductor.

Derive the relation used from first principles. [**Ans.** (i) 0.3 m; (ii) 6.25 volt.]

3.7 Figure 3.27 shows a cylindrical electromagnet. The reluctance of the iron may be neglected, as may the radial space between the plunder and the yoke. The coil has 500 turns, the plunger area is 20 cm^2, and the air-gap length, g, is at the limits of travel of the plunger, 0.5 cm and 1.0 cm respectively. Leakage, fringing and friction may be neglected and μ_0 may be taken as $4\pi \times 10^{-7}$ H/m. Derive expressions for:

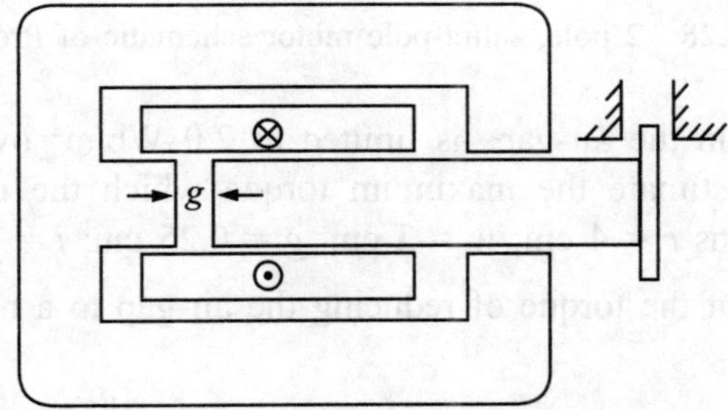

FIGURE 3.27 Cylindrical electromagnet of Problem 3.7.

(i) the coil inductance
(ii) the energy stored in the gap in terms of gap length g, and coil current I. Hence, for a coil current of 6.0 ampere calculate.
(iii) the external force required just to keep the plunger in the maximum air-gap position, and state, with suitable reasons, whether the condition is stable.
If the plunger is moved slowly from the maximum to the minimum gap position, find out:
(iv) the work done on it, and explain briefly the energy changes in the system.

[**Ans.** (i) $L = \mu_0 \frac{500}{g}$ henry (ii) $W_g = \mu_0 \left(\frac{250}{g}\right) I^2$ joule

(iii) 113 newton, unstable (iv) 1.13 joule]

3.8 Figure 3.28 shows a 2-pole salient-pole rotor, which is mounted on bearings inside a salient-pole stator. Both members have an axial length of l. Ignoring leakage and fringing, that is, assuming that the magnetic flux passes radially across the air-gaps, and assuming that the current in the coils, which are connected in series, is maintained constant, express the torque produced by the device in terms of the magnetic flux density, B existing in the air-gap, and of the dimensions given.

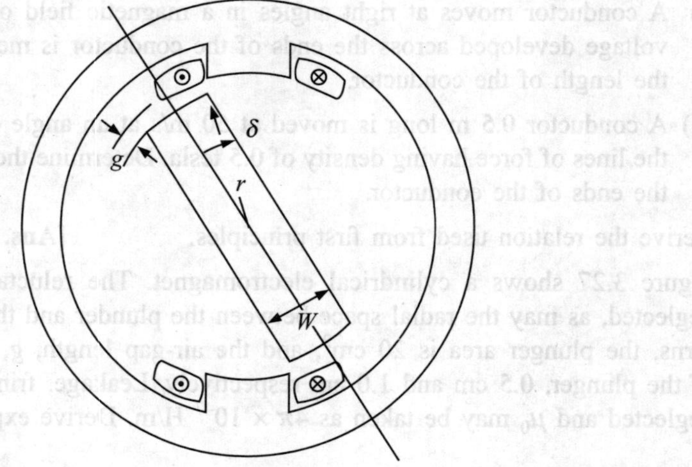

FIGURE 3.28 2 pole, salint-pole motor schematic of Problem 3.8.

If the flux-density in the air-gaps is limited to 2.0 Wb/m^2 owing to saturation of the magnetic circuit, estimate the maximum torque which the device can exert for the following dimensions r = 4 cm; w = 3 cm, g = 0.25 cm; l = 3 cm.

Explain the effect of the torque of reducing the air-gap to a negligibly small value.

[Ans. $T = \left(\dfrac{B^2}{\mu_0}\right) r\, l\, g$; 9:55 Nm]

CHAPTER 4

Layout of Electrical Circuits Used in Rotating Electro-mechanical Devices

4.1 INTRODUCTION

In the previous chapter we learnt about energy storage in electro-magnetic fields—both singly and multiply-excited systems—and studied about the behaviour of magnetic circuits which always tend to prefer their positioning such that the inductance offered by constituent circuits maximizes, that is, the path of the magnetic flux has the minimum reluctance. This causes alignment forces/torques to be developed between the fixed and movable parts of the magnetic circuits in longitudinal configurations or torque developed in rotary systems. We also studied the concept of energy and co-energy associated with magnetic circuits and production of torque in rotary-type energy converters, and thus the rudiments of electro-mechanical energy conversion using the concept of co-energy.

In this chapter, we intend to study about the electrical circuits used in rotating machines in terms of their layouts and classification when exposed to magnetic field distribution produced by flow of current in the conductors, and, therefore, the nature of mmf distribution along the air-gap surface. We shall, also, learn about the effectiveness of distribution and coil-spanning, by comparing with the most ideal effective distribution, leading to induced voltages in distributed-type of windings.

4.2 ELECTRICAL WINDINGS AND THEIR CLASSIFICATION

The term *winding,* used in connection with electrical machines, signifies a scheme of interconnection of spatially distributed conductors providing a continuous electrical circuit such that it is capable of developing a preferred and pre-designed magnetic field distribution along the air-gap surface, when an electric current flows through the electrical circuit thus formed.

In the study of electrical machines, we often use terms like *conductors*, *turns*, *coil-sides*, *coil*, *windings*, etc. Sometimes it becomes important to appreciate the subtle distinctions between and implications of the usage of these terms. We must also understand the mechanism and basis of forming a coil and the inter-connection of coils to develop a winding. Conductors are basic to all electric circuit formations. In relation to a winding, however, a reference to conductor is associated with those portions of every turn of a coil which actively participate in electro-magnetic induction process. The portion of the conductor partaking in this process is identified as its *active length*. A *turn* of a *coil* is constituted of two conductors separated in space and suitably inter-connected so as to link flux developed along its axis. The *axis* of a turn is the planet perpendicular to the plane which contains the conductors constituting the turns in question, located along the line of longitudinal symmetry of the said conductors. A coil may or may not contain several turns, but *each turn of a multi-turn coil* is identical in shape and is placed identically with reference to other turns of the coil, so that all of them have a common axis. A coil is, in effect, an independent component of an electrical circuit, and several coils are suitably inter-connected to form a *winding*.

Just as a turn has two constituent conductors a coil has two sides, called *coil-sides*, which are physically separated in space. Figure 4.1 clarifies pictorially the terms detailed above.

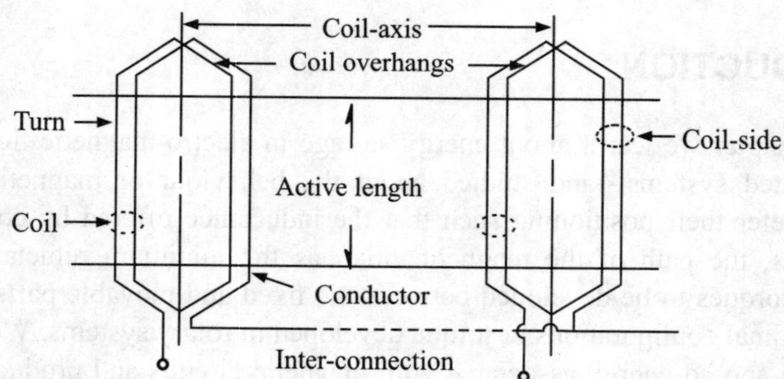

FIGURE 4.1 A schematic showing the general winding arrangement in electrical machines.

We shall, now, look into the basis for selecting two conductors to form a turn and the mechanism of desirable inter-connection between them. If a conductor is assumed to be located in a sinusoidally distributed field system, and if it is rotated at a constant angular velocity with respect to the field, the emf induced in it can be idealized to be sinusoidal in time, and is position-dependent. Two conductors mounted on the rotating system have similar pattern of emf induction, but these emfs have *phase displacements* determined by their effective spatial angle of separation, as the maxima of the sinusoidal voltage induced in them does not occur at the same instant. If, however, we select two conductors which are located 180° apart in space, for a 2-pole field distribution, the emfs induced in them is in time-opposition, for all times to come, as shown in Figure 4.2. θ_r is the angle that the plane of the coil makes with neutral magnetic axis—NM and ω_r is the rotor speed in radians/sec for 2-pole configuration.

These two conductors, *A* and *A′*, if *connected back-to-back*, that is, in series-opposition, always yield a net voltage of twice the voltage induced in each conductor at all times. Such an arrangement leads to the best utilization of a turn, and has the additional advantage of not having to cross-over from one side to the other to effect a series connection.

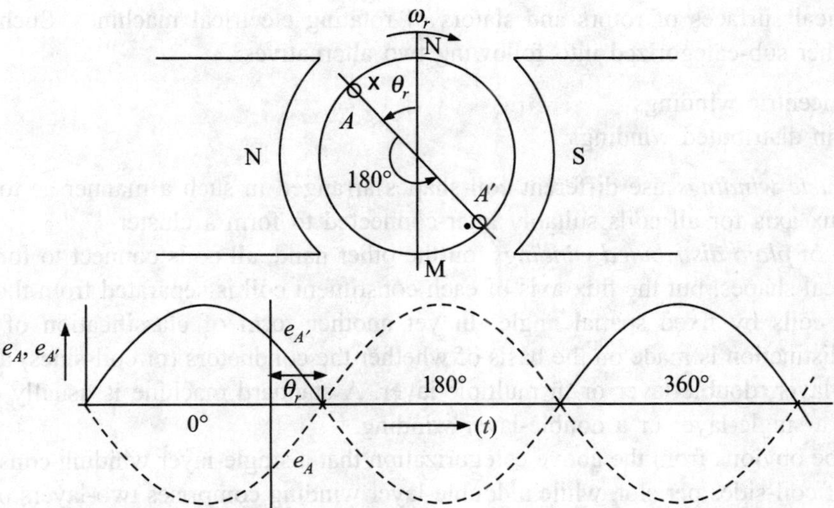

FIGURE 4.2 Nature of voltage induced in conductors related in magnetic field distribution.

Unless otherwise specifically required by a designer, formation of a turn is always made by choosing, as far as possible, conductors effectively separated by 180° (electrical degrees) apart in space, and by connecting them back-to-back.

A multi-turn coil is formed by using a number of identical turns and by connecting each turn in series with the others. Hence, each turn of a multi-turn coil configuration consists of identically-shaped turns so placed as to have a common magnetic axis.

Before extending our discussion any further, it may be prudent to examine the different types of winding that one is likely to encounter in practice in the study of electrical machines. As stated earlier, a winding comprises several coils distribution in space, and is suitably inter-connected to form a continuous electrical circuit. A number of considerations which include geometry of the device on which the winding is to be housed, effective utilization of space desirable shapes of mmf distribution, space available for placement of conductors, mechanical forces associated with operation, etc., decide the type of winding that is to be chosen for machines. This calls for *classification* of windings based on their shapes, inter-connection used and overall get-up of windings. It is customary to group windings into closed-type of windings and open-type of windings depending on whether the final winding is an *end-less type*, without any specific beginning or end (as encountered in commutator-type armature windings), or an *open-ended type*, with clearly identifiable beginning and end available for effective initialization (such a winding is invariably used in ac systems only).

A second configuration is based on the shape of the coils used in the formation of windings and the relative positions of the axis of effectiveness of each coil. Under this category, we have *concentrated* or *distributed-type* of windings. The concentrated-type uses multiple coils of identical shape connected in *series-addition* for a single cluster such that all coils forming a cluster share a common flux-axis in space. This type of winding is connected in field-coils of machines and in transformers. By contrast, a distributed-type of winding is obtained along the cylindrical surfaces of rotors and stators of rotating electrical machines. Such a winding can be further sub-categorized into following two alternatives:

1. Concentric windings
2. Plain distributed windings.

Concentric windings use different coil-shapes arranged in such a manner as to result in a common flux axis for all coils suitably inter-connected to form a cluster.

In case of *plain distributed windings*, on the other hand, all coils connect to form a cluster have identical shapes, but the flux axis of each constituent coil is separated from those of other constituent coils by fixed spatial angle. In yet another form of classification of distributed windings, distinction is made on the basis of whether the conductors (or coil-sides) are arranged in a single-layer, double-layer or in multiple-layer. A standard machine is usually constructed with either a single-layer or a double-layer winding.

It will be obvious from the above categorization that a single-layer winding consists of only one-layer of coil-sides per slot, while a double-layer winding comprises two-layers of coil-sides in each slot. Only in case of special machines we may sometimes come across multiple-layer of windings, and such windings are kept out of our discussion in the present deliberations.

Figure 4.3 presents the schematics of different categories of windings discussed in this section.

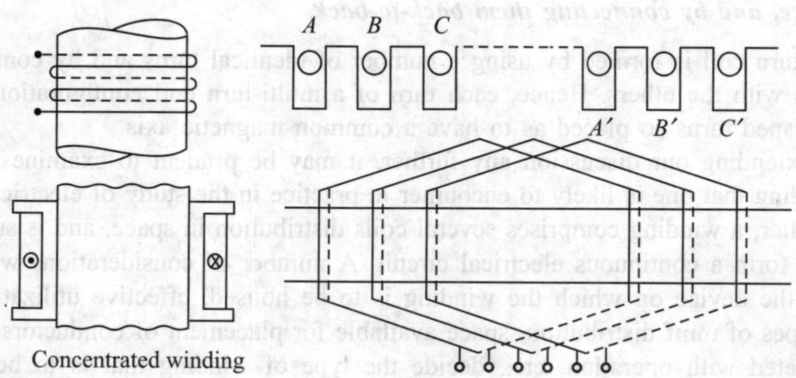

Concentrated winding

Double-layer plain distributed winding

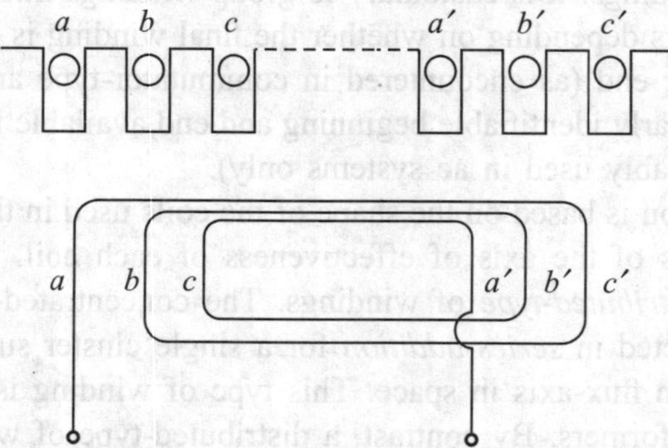

Single-layer concentric winding (distributed)

FIGURE 4.3 Different categories of windings discussed in this section.

4.3 EFFECT OF COIL DISTRIBUTION

We shall, now, examine the question of *effective utilization of copper* associated with windings formed by connecting specially distributed coils in series, in terms of their capability in producing mmfs and emfs.

If we take a *concentrated winding* and if it is subjected to a changing flux-linkage (or flux-cutting), such that the net rms value of emf induced in a turn is ε, then the rms value of the emf induced in each m-turn coil of the winding is given as $(m\varepsilon)$, and the net emf available from the output terminals of such a winding formed of M coils in series is $(Mm\varepsilon)$ volts.

Let us examine how a *distributed winding* having M series connected coils, each coil developing a voltage of $(m\varepsilon)$ under similar conditions of flux interaction will behave. Let the axes of two successive coils be separated in space by an electrical angle of θ. Since the coils are identical, each comprising m turns, the emf induced in each coil has identical nature and pattern, but there exists a phase displacement between each coil-voltage profile. In order to deal with such situation, it is customary to use phasorical depiction of induced emfs for each coil. Thus, when winding is formed by connecting M coils in series, the net voltage available for the winding constitutes a *phasorical sum* of all the coil voltages in the cluster. The net rms voltage, E, appearing across the terminals of the winding equals the length of the closing side of the *voltage polygon* made up of M sides constituting the coil-emf phasors as shown in Figure 4.4.

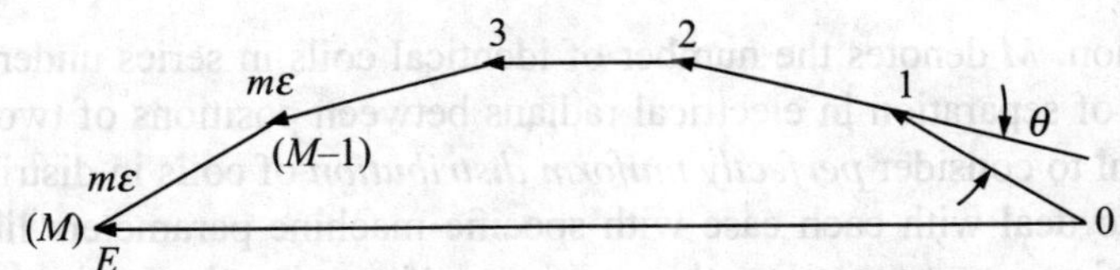

FIGURE 4.4 Voltage polygon indicating M phasors added up phasorically at an angle of θ between successive phasors, to result in a total emf of E.

When expressed in complexor form:

$$E = m\varepsilon[e^{j0} + e^{j\theta} + e^{j2\theta} + \dots + e^{j(M-1)\theta}]$$

The above M-term series is in geometric progression such that:

$$E = m\varepsilon S$$

where, S is the summation of the G.P. series, and

$$S = \frac{a(r^n - 1)}{(r - 1)}$$

in which, $a = e^{j0} = 1$; $r = e^{j\theta}$; and $n = M$

$$\therefore \qquad E = m\varepsilon \frac{[e^{jM\theta} - 1]}{[e^{j\theta} - 1]}$$

As we are only interested to get the magnitude,

$$|E| = m\varepsilon \frac{\sqrt{(\cos M\theta - 1)^2 + \sin^2 M\theta}}{\sqrt{(\cos\theta - 1)^2 + \sin^2\theta}}$$

or

$$|E| = m\varepsilon \left[\frac{\sin \dfrac{M\theta}{2}}{\sin \dfrac{\theta}{2}} \right] \qquad \text{...(4.1)}$$

We had already concluded earlier that if M coils in question were arranged to form a *concentrated winding*, which affords the most effective utilization of copper, the net emf obtained would have a value of $(Mm\varepsilon)$.

The ratio of these two values would tell us about the effectiveness of a distributed winding in comparison to its concentrated counterpart. This ratio, K_d, variously known as distribution factor, spread factor or layer factor of a distributed winding, is exclusively contributed as a result of winding distribution. Its value is given by:

$$K_d = \frac{\sin \dfrac{M\theta}{2}}{M \sin \dfrac{\theta}{2}} \qquad \text{...(4.2)}$$

In this expression, M denotes the number of identical coils in series under each pole, and θ is the spatial angle of separation in electrical radians between positions of two successive slots.

It is often useful to consider *perfectly uniform distribution* of coils in distributed coil system to avoid the need to deal with each case with specific machine parameters like the number of coils per pole. It is important to realize that *perfect uniform distribution* is a purely theoretical concept. It emphasises that *the total copper volume under each pole used to form the winding is kept constant along with the spread of each phase under each pole, while at the same time, the total number of coils M under each pole tends to be infinitely large and θ tends to zero.*

Under this hypothetical condition, the idealized winding is likened with a *current sheet*. In physical terms, it means that the winding consists of a very large number of coils of smaller number of turns, each coil-side being separated from its immediate neighbour by a very small space angle, but the overall phase-distribution is unchanged. The idealised situation is depicted in Figure 4.5.

In mathematical terms the new value of K_d, the distribution factor of a perfectly uniformly distributed winding, is given as follows:

$$K_d = \lim_{\substack{M \to \infty \\ \theta \to 0}} \frac{\sin \dfrac{M\theta}{2}}{M \sin \dfrac{\theta}{2}} \qquad \text{...(4.3)}$$

ψ is the phase-spread and (Mm) will indicate that the copper volume remains constant.

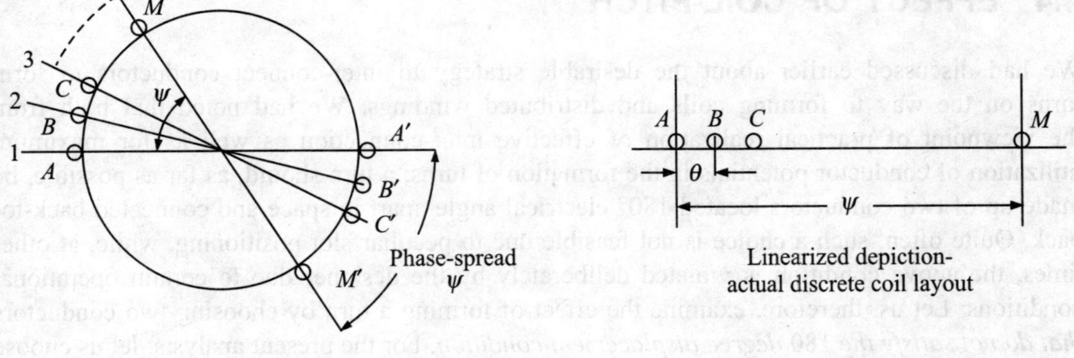

$AA', BB' ..., MM'$ —Dentoes coils

1, 2, 3..., M —Indicates successive coil-numbers.

ψ —Represents phase-spread stands for slot-angle,

θ —i.e., angular separation between adjacent coils

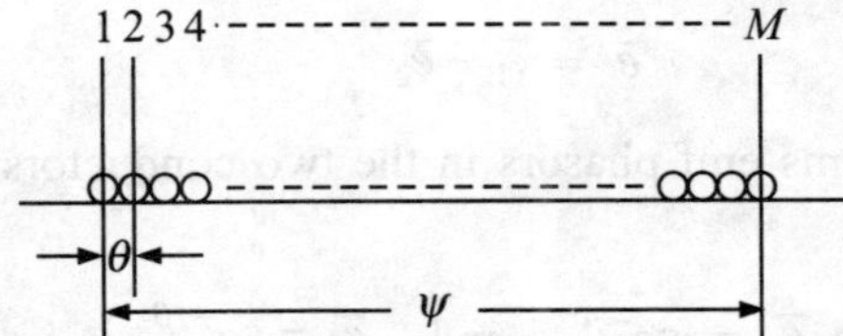

Principle of 'current sheet' diagramatically depicted.

ψ - Phase spread, constant

$M \longrightarrow \infty$

$\theta \longrightarrow 0$

FIGURE 4.5 A distributed winding idealized as a current sheet.

It may be seen from scrutiny of the later part of Figure 4.5 that ψ is related to θ, in accordance with the following relation:

$$\psi = (M - 1)\theta \quad \text{or} \quad M = \frac{\psi + \theta}{\theta}$$

By substituting the value of M from the above relationship, it is possible to reduce the complexity of limit under two variables into limit with one variable. Thus:

$$K_d = \lim_{\theta \to 0} \frac{\sin\left(\dfrac{\psi + \theta}{2}\right)}{\left(\dfrac{\psi + \theta}{2}\right)\sin\dfrac{\theta}{2}} = \frac{\sin\dfrac{\psi}{2}}{\left(\dfrac{\psi}{2}\right)} \qquad \text{...(4.4)}$$

This expression is commonly used to consider the effect of distribution of windings under idealised conditions.

4.4 EFFECT OF COIL-PITCH

We had discussed earlier about the desirable strategy to inter-connect conductors to form turns on the way to forming coils and distributed windings. We had noted that both from the viewpoint of practical realization of effective inter-connection as well as for maximum utilization of conductor potentials in the formation of turns, a turn should, as far as possible, be made up of two conductors located 180° electrical angle apart in space and connected back-to-back. Quite often, such a choice is not feasible due to peculiar slot positioning, while, at other times, the above condition is violated deliberately by the designer due to certain operational conditions. Let us, therefore, examine the effect of forming a turn by choosing two conductors *that do not satisfy the* 180 *degree displacement condition*. For the present analysis, let us choose two conductors which are spaced β electrical degrees apart in space ($\beta \neq 180°$ electrical angle) and connect them back-to-back.

The emfs induced in the two conductors are out of phase by $\beta°$ electrical angle, when exposed to a common flux distribution, and the voltage ($\overline{e}$) obtainable from the formed turn is given by:

$$\overline{e} = \overline{e}_1 - \overline{e}_2$$

where, $\overline{e}_1$ and $\overline{e}_2$ denote the rms emf phasors in the two conductors. Complexor representation of these emfs are as follows:

$$\overline{e}_1 = \varepsilon e^{-j0} \quad \text{and} \quad \overline{e}_2 = \varepsilon e^{-j\beta}$$

on the assumption that the arbitrarily chosen time reference axis coincides with $\overline{e}_1$. Figure 4.6(a), shows a linearized depiction of the formed turn. Two alternative possibilities are shown. In the first case, the return (or second) conductor, used to provide the return-path for the turn in association with the first conductor, is located at a space angle β behind the first, where $\beta < 180°$, that is, *the pitch of the turn* falls short of 180° by an angle α, such that $\alpha + \beta = 180°$. The second case consider, $\beta > 180°$, and $|\beta - \alpha| = 180°$. In the first case, the turn so produced is named as *a short-pitched turn*, while in the second case, it is called an *over-pitched turn*.

The resulting voltage available across the output terminals of the turn, in both cases, I and II, can be found to be:

$$|e| = 2\varepsilon \sin \frac{\beta}{2} = 2\varepsilon \cos \frac{\alpha}{2}$$

against the ideal value of (2ε), for a full-pitched turn, for which, $\beta = 180°$ and $\alpha = 0°$.

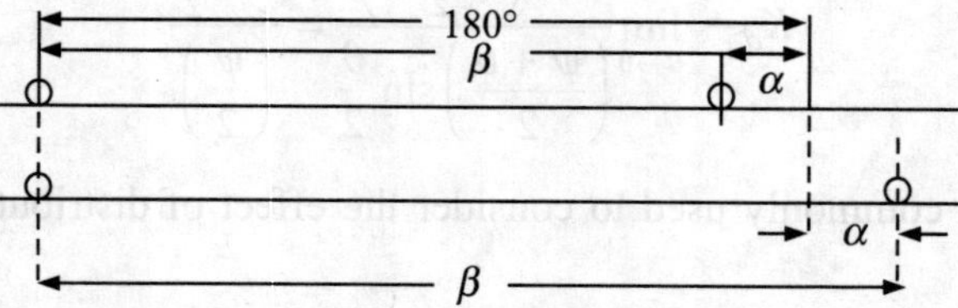

FIGURE 4.6(a) Relative position of two conductors chosen to form a turn.

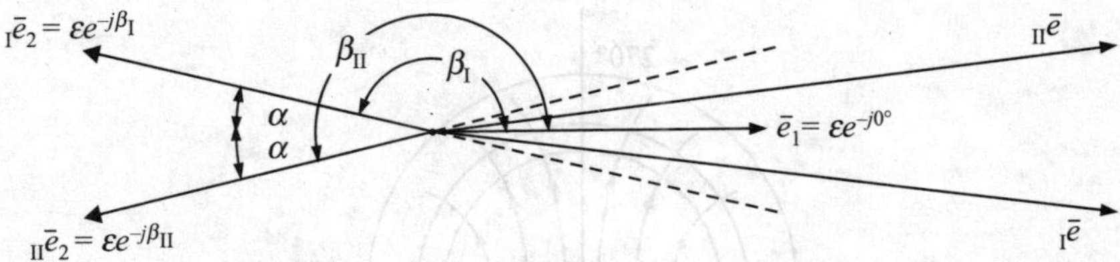

FIGURE 4.6(b) Phasor representation of the two conductor voltages separated in space by β electrical radians [Case I denotes $\beta < 180°$, and Case II represents $\beta > 180°$].

The effect of short-pitching or over-pitching in a formation of a turn is to introduce a factor, called variously as pitch factor, K_p, or breadth factor, K_b, or coil-span factor, K_c. The said factor is, therefore, represented as a ratio of phasor sum of conductor voltage in a turn and the algebraic sum of conductor voltages in a turn, i.e.,

$$K_p(\text{or } K_b \text{ or } K_c) = \frac{2\varepsilon \sin\dfrac{\beta}{2}}{2\varepsilon} = \sin\left(\frac{\beta}{2}\right) = \cos\left(\frac{\alpha}{2}\right) \qquad ...(4.5)$$

It should be pointed out that, although both relations are equally valid for determining the value of K_b, the former is safer to use when dealing with estimation of values for K_b for the harmonics, to obviate committing error in the determination of the effective value of α in such individual cases.

The effect of *distribution* and *coil-spanning* is to be considered together in order to describe the behavioural property of distributed windings. To do this, a new factor, called the *winding factor*, K_w, is introduced, where,

$$K_w = K_p K_d \qquad ...(4.6)$$

The factor, K_w, determines the effectiveness of a distributed winding system, vis-à-vis its concentrated counterpart.

4.5 ELECTRICAL WINDINGS AND THEIR CLASSIFICATION

Whenever an electric current flows through the conductors, a magnetic flux is produced. The magnitude and distribution of the current and the nature of the surrounding medium (see Section 1.1) decide the nature of flux distribution. The current produces a magnetizing force, H, and its coil in two diametrically opposite slots. We can see the flux pattern, which would be produced when the conductors of the coil carry I ampere of current. For the direction of the currents, shown in dots and crosses, the inner member sees a north pole over the upper half circumference and a south pole over its lower half circumference. The other member also has induced on it similar polarities of opposite nature. We wish to calculate the mmf and flux-distribution patterns produced by the arrangement using the unfurled diagrams of Figure 4.7(b). We can construct the mmf diagram easily, by taking recourse to the following strategy. We consider that a person, sensitive to mmf produced by current-carrying conductors, is taking a walk along the circumference in the air-gap space from the left to

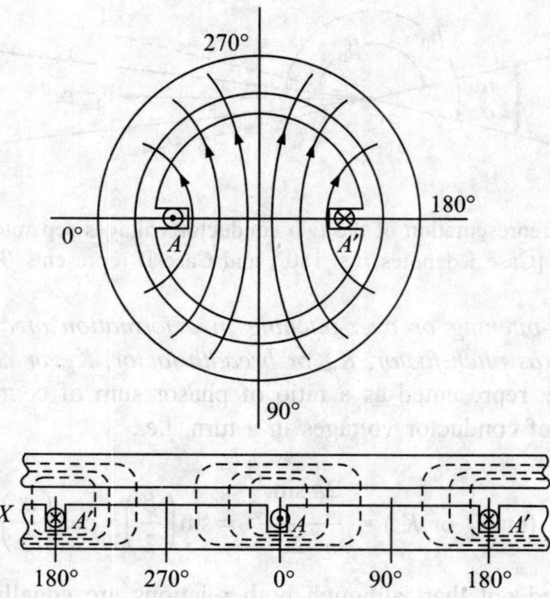

FIGURE 4.7(a) Flux-pattern produced by single-coil near parallel air-gap.

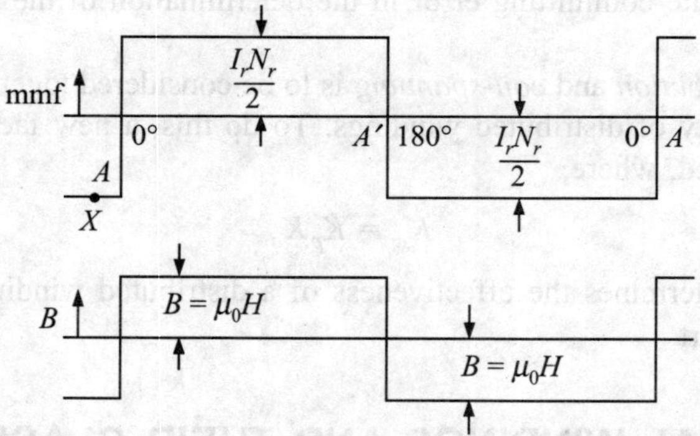

FIGURE 4.7(b) MMF pattern produced by a single-coil of N_r turns, carrying I_r amperes in its conductors and the flux-density pattern, for the chosen model with smooth air-gap.

the right, starting at an arbitrarily chosen point, marked as *X*, both on Figure 4.7(a) and Figure 4.7(b). Whenever he crosses the centre line of a slot (that is, the centre line passing through a current-carrying coil-side), he goes up (or down) a step depending on whether the instantaneous current flowing in the coil-side is a 'dot' or a 'cross' (a positive or a negative), the height of the step being equal to $(N_r i_r)$ units. Since, on going round the circumference of the circle once he ends up at the point *X* once again, from where he had started in the first place and also since the current in the various coil-sides he had crossed during the traverse is such that for all the coils taken together, the number of dot currents (positives) always equals the number of cross currents (negatives), he would naturally end up at the same height of *X*—the point from where he had started his walk.

The mean line of the diagram, which is the undulating path of his traverse, can now be drawn after tracing the contour of his traverse around the periphery. This is shown in Figure 4.7(b). The resulting diagram depicts the mmf distribution line along the circumference of the air-gap. Such an mmf diagram can be readily drawn for any conductor distribution of a winding. All that we require is a knowledge of the distribution of conductors of the winding and the values of instantaneous current-flow in them.

If we assume, as earlier, that saturation of the magnetic material part of the circuit is negligible and we know the air-gap geometry—its length—then the flux-density at each point of the air-gap can be readily estimated, and geometry of the magnetic path—its length: the peripheral, rasial and axial lengths: the length of the air-gap separating the stator and the rotor members—then the flux-density along the air-gap surface can be readily estimated. We can also quantify the inductance of the winding produced by the flux-linkage. This will be the magnetizing inductance, X_m.

We can extend the discussion on Figure 4.7(a) and Figure 4.7(b), by considering mmfs contributed by two identical *m* turn coils housed in two adjacent slot-pairs of a uniformly slotted structure, with parallel air-gap, by using the above strategy by simple superposition. The result is shown in Figure 4.8.

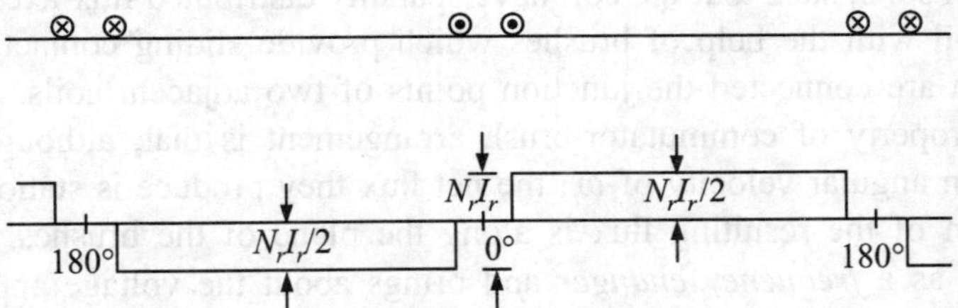

FIGURE 4.8 MMF distribution pattern (or flux-density pattern) produced by two identical N_r turn coils with conductors carrying current of I_r through them and with uniform air-gap of length *g*.

If L_m denotes the flux-linkage per unit ampere of the winding due to that flux component which crosses the air-gap and hence links with any secondary winding on the other member, this is the mutual flux, that is the coupling flux. L_m is then the *mutual inductance* between the two windings. There is also a *leakage inductance* component, L_l, not considered so far, which is caused by flux that does not cross the air-gap and interact with the winding on the other member. L_l is usually small in magnitude when compared with L_m if the magnetic circuit remains unsaturated. The sum of L_m and L_l components comprises the '*total inductance*' of the winding, L. Thus:

$$L = L_m + L_l \qquad ...(4.7)$$

This is also known as the *magnetizing inductance* of a winding.

Computation of the value of the *magnetizing inductance*, L can be done from the knowledge of the total flux produced by it per unit ampere carried by its winding conductors, or the air-gap flux-density distribution pattern that it produces per unit ampere along with the physical dimension and parameters of the machine. Alternatively, the total energy stored in its magnetic field produced by the concerned winding for unit current may be computed, which is numerically one-half of the magnetizing inductance.

For complicated conductor distribution pattern, the calculation by stored-energy method is found to be more straightforward and is usually recommended.

The mutual inductance with another winding, usually situated on the other side of the air-gap, can be estimated by inspection of the superposed diagrams showing the conductor layout of the second winding and the flux-density distribution at points around the air-gaps produced by the first winding. The flux-distribution produced by distributed windings is approximately close to sinusoidal ones.

Before proceeding any further, we should consider the resulting ampere-conductor distribution, spatial mmf distribution pattern and resulting flux-distribution configuration of distributed windings. For this to be done, we must review alternative possibilities of distributed-type windings and their nature of ampere–conductor distribution, suitable for their specific applications.

We encounter three alternative distributed-type windings in electrical machines. These are:

Commutator-type of continuously distributed windings: Such a winding arrangement uses identical coil-shapes. This type of winding is invariably placed on the rotor member and is constituted of double-layer endless type—one cannot identify its starting and finishing points—and are arranged in slots, such that the coil have spatially distributed flux axes. Contact is made with individual coil with the help of brushes which provide sliding contact with commutator segments to which are connected the junction points of two adjacent coils.

The unique property of commutator-brush arrangement is that, although the coils move with the rotor at an angular velocity of ω_r, the net flux they produce is stationary in space and the spatial position of the resulting flux is along the plane of the brushes. The commutator, therefore, behaves as a *frequency changer* and brings about the voltage appearing across the brushes, as decided by the relative speed of the brush assembly vis-a-vis the speed of the excitation field (this point is elaborated and indicated in detail in Section 5.1.4 later) from the frequency of voltage generated in the rotor conductors due to their speed of revolution with respect to the exciting field.

Poly-phase distributed windings: Such a winding also uses identical coil-shapes housed on the stationary member or the rotor, but has different identifiable clusters of inter-connected coils under each pole, producing phase winding, with clearly identifiable beginnings and endings of phase clusters. When mounted on the rotor member, contact is made to the beginnings and endings (if they are not already connected to form poly-phase windings with a minimum number of contact points) by stationary brushes with slip-ring arrangements. For a typical 3-phase winding on the rotor, the number of slip rings required is three, when the other ends of each phase are suitably connected to form delta or star connection. When such windings are housed on the stator, rigid connections are directly used with beginning and ending of poly-phase windings.

Single-layer distributed windings of the short-circuited cage type: Typically used as secondary winding of squirrel-cage induction motors and as damper winding of synchronous machines, they constitute evenly distributed short-circuited windings. They are made of stout, short-circuited bar, and are laid in one layer, uniformly spread along the air-gap surface around the periphery.

We shall consider, in detail, the commutator winding and the phase winding. The latter can be found both on the stator and on the rotor members. We shall consider the other member to present a smooth surface at the air-gap interface when considering the mmf and flux-distribution pattern presented by the distributed-type windings of these types.

4.5.1 Commutator Winding

The armature–conductor distribution of a commutator winding is detailed in Figure 4.9, over its entire periphery. As before, we follow the procedure already laid down earlier in this chapter for tracing the resulting mmf diagram. Let us consider that N_r turns for each coil of the rotor configuration carry a current of i_r ampere, but the other side across the air-gap is smooth with uniform air-gap around the periphery. If the effect of slotting is ignored and an mmf sensitive observer takes a walk along the circumference from left to right of the unfurled diagram of Figure 4.9, the undulating movement of the person closely resembles stepped movement denoted by continuous 'rise' from zero to 90° axis, followed by continuous loss in height between 90° and 270° axes, and similarly a continuous rise from 270° to 360° axis.

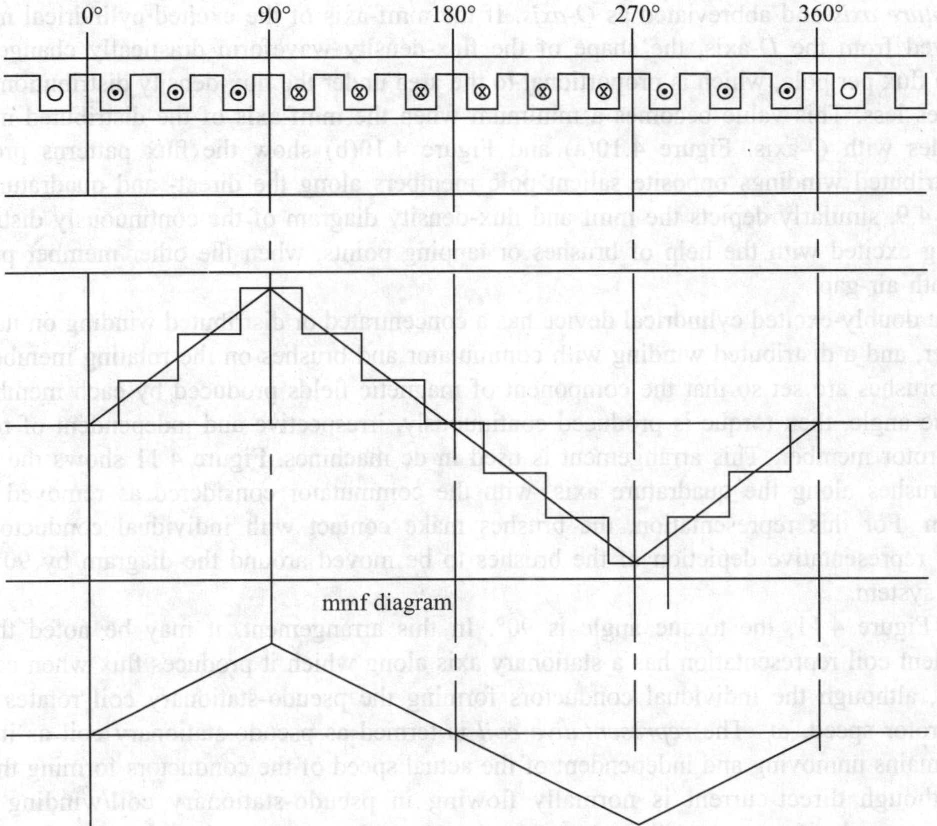

FIGURE 4.9 MMF and flux-density distrition diagram of an excited smoothly distributed ampere–conductor layout, with the adjacent member presenting a smooth air-gap.

We can easily eliminate the steps by considering uniform sloping, as very little flux passes down the slots and the flux passing the teeth is constant.

When a large number of slots carry the same current, it is often worth simplifying the mmf diagram by assuming that the total current for all the slots is uniformly spread over the portion of the periphery covered by slots, and the conductor is approximated by treating it as a *current sheet*. The resulting mmf diagram then assumes the slope of an *undulating ramp instead of steps* as shown in the diagram. The mmf diagram then has the shape of two isosceles triangles, and this simplified mmf pattern is quite adequate.

It is worth noting that both mmf and flux-density configurations will remain unaffected if the stator and rotor members are interchanged in their arrangement. The flux-density for the salient-cylinder configuration will remain almost the same if the rotor with smoothly distributed winding is excited instead of salient poles, provided that the axes of the two members remain aligned, that is, the centre line of the mmf distribution of the distributed winding coincides with that of the salient-poles of the excitation windings.

The centre lines of the salient-poles are called the *direct axis* and abbreviated as *D-axis*. This is the line of symmetry of the salient-pole flux distribution when the member is excited. The centre line of the inter-polar space at 90° electrical angle with the *D*-axis, is termed as *quadrature axis* and abbreviated as *Q-axis*. If the mmf-axis of the excited cylindrical member is moved from the *D*-axis, the shape of the flux-density waveform drastically changes. The air-gap flux per pole, which is proportional to the area under the flux-density distribution curve, becomes less. This value becomes a minimum when the mmf-axis of the distributed member coincides with *Q*-axis. Figure 4.10(a) and Figure 4.10(b) show the flux patterns produced by distributed windings opposite salient pole members along the direct- and quadrature-axis. Figure 4.9, similarly depicts the mmf and flux-density diagram of the continuously distributed winding excited with the help of brushes or tapping points, when the other member presents a smooth air-gap.

If a doubly-excited cylindrical device has a concentrated or distributed winding on its stator member, and a distributed winding with commutator and brushes on the rotating member, and if the brushes are set so that the component of magnetic fields produced by each member has a torque angle, then torque is produced continuously, irrespective and independent of rotation of the rotor member. This arrangement is used in dc machines. Figure 4.11 shows the device with brushes along the quadrature axis, with the commutator considered as removed in the diagram. For this representation, the brushes make contact with individual conductors and affords representative depiction of the brushes to be moved around the diagram by 90° for a 2-pole system.

In Figure 4.11, the torque angle is 90°. In this arrangement, it may be noted that the equivalent coil representation has a stationary axis along which it produces flux when carrying current, although the individual conductors forming the pseudo-stationary coil rotates at the actual rotor speed, ω_r. The *representative coil* is termed as pseudo-stationary coil as its flux-axis remains unmoving and independent of the actual speed of the conductors forming the coil.

Although direct-current is normally flowing in pseudo-stationary coil/winding of dc machines, producing constant flux-pattern stationary with respect to the other member (stator, in this case), the individual coils of the actual winding carry pulsating currents, because the connection to every coil is reversed twice by the commutator brush arrangement, over every

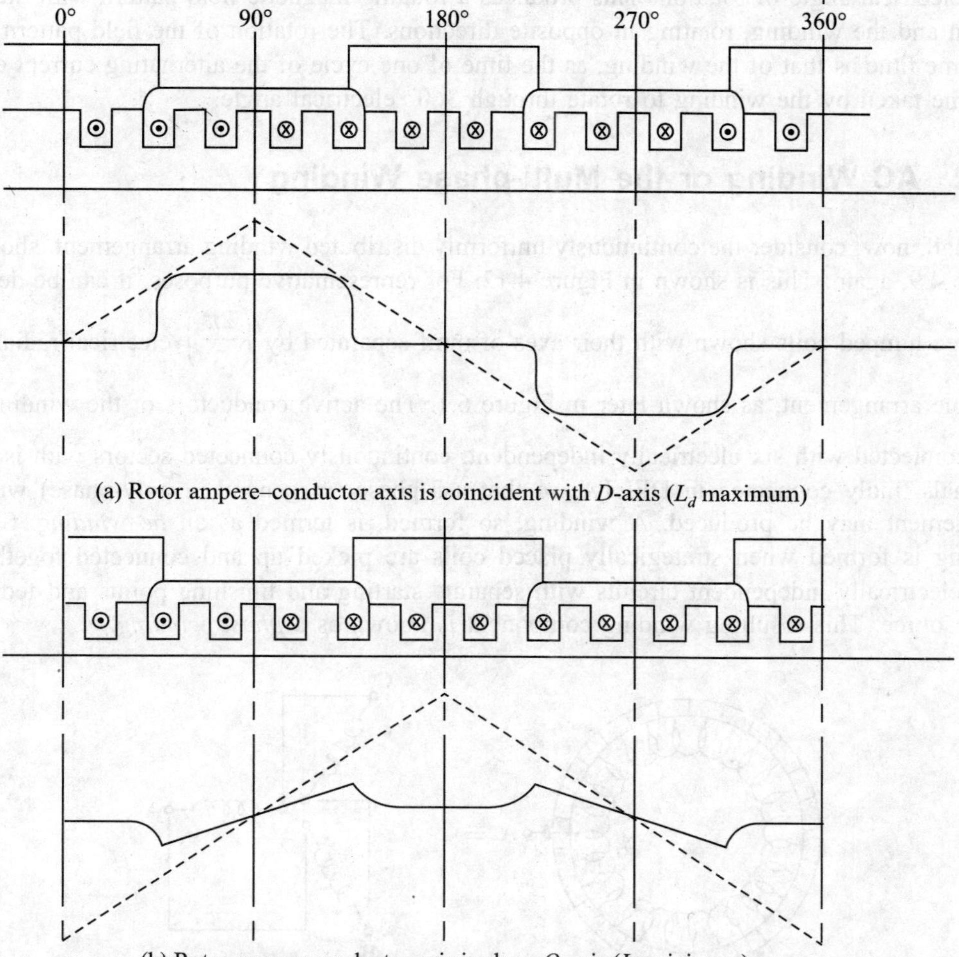

(a) Rotor ampere–conductor axis is coincident with D-axis (L_d maximum)

(b) Rotor ampere-conductor axis is along Q-axis (L_q minimum)

FIGURE 4.10 Flux patterns produced by distributed winding opposite salient-pole centred on (a) D-axis and (b) Q-axis.

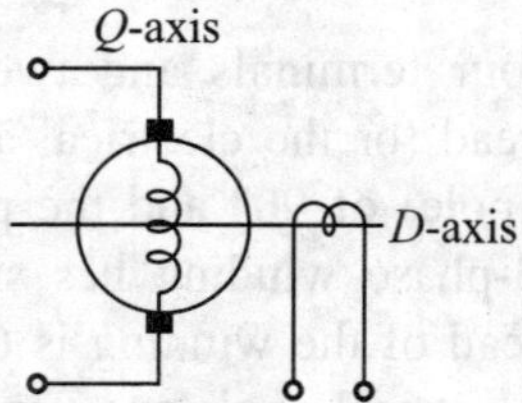

FIGURE 4.11 Schematic depiction of a dc machine with pseudo-stationary coil representation of the armature.

360° electrical angle of rotation. This produces a rotating magnetic field pattern with the field pattern and the winding, rotating in opposite directions. The rotation of the field pattern takes the same time as that of the winding, as the time of one cycle of the alternating current equals the time taken by the winding to rotate through 360° electrical angle.

4.5.2 AC Winding or the Multi-phase Winding

We shall, now, consider the continuously uniformly distributed winding arrangement, shown in Figure 4.9, again. This is shown in Figure 4.12. For representative purposes, it can be denoted by three lumped coils shown with their axes of mmf separated by $\left(\frac{2\pi}{3}\right)$ electrical radians for a 2-pole arrangement, as shown later in Figure 6.1. The active conductors of the winding can be reconnected with six electrically independent, continuously connected sectors with isolated terminals findly connected in DELTA, so that a 3-phase (in general, a poly-phase) winding arrangement may be produced. A winding, so formed, is termed as an *ac winding*. Such a winding is formed when strategically placed coils are picked up and connected together to form electrically independent circuits with separate starting and finishing points and fed from an ac source. This resulting winding component is known as a *phase winding*.

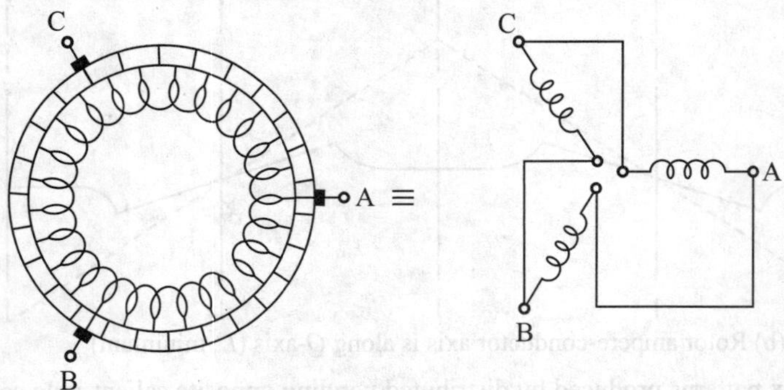

FIGURE 4.12 Strategic representation of a 3-phase commutator winding and its star-equivalent layout with phases A-B-C connected with commutator-brush interface.

Thus a 2-phase winding has four terminals and two electrically independent circuits. Each phase winding has a phase spread (or the electrical angle over which the conductors lie continuously distributed under each pole) of 90° and the phase centre-lines are 90° electrical angle apart in space. Similarly, a 3-phase winding has six terminals with three electrically independent circuits. Each phase-spread of the winding is 60° electrical angle under each pole and the phase centre lines are 120° electrical angle apart in space or 120° electrical angle over every two poles, and the phase centre lines are 120° electrical angle apart in space. The former is identified as 60°-phase spread, 3-phase windings and the latter is called the 120°-phase spread, 3-phase windings.

When fed from a balanced 2-phase or 3-phase supply, a poly-phase winding produces an approximately constant mmf and executes one revolution about the conductor distribution during one cycle of alternating current supply. The arrangement for a 3-phase supply is depicted in Figure 4.12. We assume that the currents in the conductors of the winding arrangement are in the form of a *current sheet*, so that the need to show the effect of slotting on the mmf wave is no-longer necessary, and the resulting mmf waveform can be smoothened. We consider two instances:

1. When *a*-phase is fed with positive-maximum current, at which instant the current in both *b*- and *c*-phases are both minus one-half maximum, as shown in the representative phasor diagram, and
2. When the instantaneous current in phase *b* is zero, while current in *a*-phase is positive 0.732 times maximum, while that in *c*-phase is negative 0.732 times maximum as shown in the representative phasor diagram.

Comparing the mmf diagrams in (a) and (b) of Figure 4.13, we note that the centre-line of the resulting mmf waveform in (b) is 30° behind that in (a), which is one-twelfth of a cycle

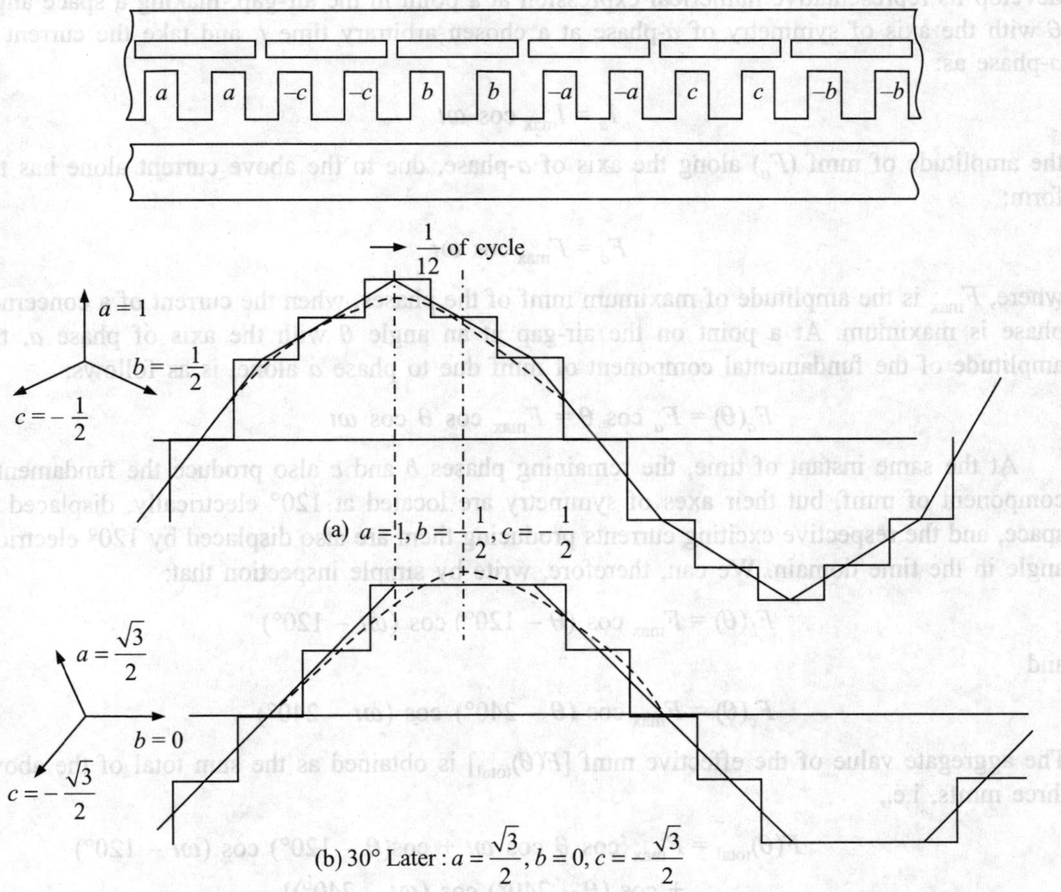

FIGURE 4.13 MMF and flux-density distribution diagram with excited 3-phase distributed winding at two extreme cases: (a) when current in *a*-phase attains positive maximum value and (b) $\frac{1}{12}$th of a cycle later.

apart. The mmf diagrams are plotted following the method already elaborated earlier, when plotting the mmf diagrams of continuously distributed windings fed with direct current supply (see Figure 4.9 and Figure 4.10). The mmf pattern developed by a 3-phase balanced winding, supplied with 3-phase balanced currents, changes its shape under two extreme conditions, which occur in every one-twelfth of one cycle of operation. A close scrutiny of these two extreme cases shows that if we analyse the resulting mmf waveform applying Fourier series, we shall find very close similarity at the different operating conditions. The higher order harmonics have little influence on the behaviour, while the fundamental component remains constant. We can also see how a uniformly rotating constant mmf-pattern can be produced by a 3-phase winding, when fed from a balanced 3-phase source. For this to be done, we assume each of the three phases as a space sinusoidal mmf pattern. Each of these mmf-patterns varies sinusoidally in amplitude when fed with sinusoidal currents, but as there is a space and time difference existing among the individual phase excitations, a travelling wave of constant amplitude is developed.

If we consider only the fundamental component of mmf distribution of each phase and develop its representative numerical expression at a point in the air-gap, making a space angle θ with the axis of symmetry of a-phase at a chosen arbitrary time t, and take the current in a-phase as:

$$I_a = I_{\max} \cos \omega t$$

the amplitude of mmf (F_a) along the axis of a-phase, due to the above current alone has the form:

$$F_a = F_{\max} \cos \omega t$$

where, $F_{\max}$ is the amplitude of maximum mmf of the phases, when the current of a concerned phase is maximum. At a point on the air-gap at an angle θ with the axis of phase a, the amplitude of the fundamental component of mmf due to phase a alone, is as follows:

$$F_a(\theta) = F_a \cos \theta = F_{\max} \cos \theta \cos \omega t$$

At the same instant of time, the remaining phases b and c also produce the fundamental component of mmf, but their axes of symmetry are located at 120° electrically, displaced in space, and the respective exciting currents producing them are also displaced by 120° electrical angle in the time domain. We can, therefore, write by simple inspection that:

$$F_b(\theta) = F_{\max} \cos (\theta - 120°) \cos (\omega t - 120°)$$

and

$$F_c(\theta) = F_{\max} \cos (\theta - 240°) \cos (\omega t - 240°)$$

The aggregate value of the effective mmf [$F(\theta)_{\text{total}}$] is obtained as the sum total of the above three mmfs, i.e.,

$$\begin{aligned} F(\theta)_{\text{total}} &= F_{\max}\{\cos \theta \cos \omega t + \cos(\theta - 120°) \cos (\omega t - 120°) \\ &\quad + \cos (\theta - 240°) \cos (\omega t - 240°)\} \\ &= \frac{3}{2} F_{\max} \cos(\theta - \omega t) \qquad \text{...(4.8)} \end{aligned}$$

Thus, simultaneous excitation of three phases with ac currents produces a space-distributed mmf wave of constant amplitude of one-and-half times the maximum amplitude of mmf produced by each phase, and it rotates at a constant angular velocity. The mmf pattern coincides with the positive peak value of each phase when it is carrying the positive maximum value of current through it.

If the 3-phase winding is rotated at the same angular velocity but in opposite direction, then the resulting mmf is at stand-still with respect to the stationary member, and the condition with continuously distributed commutator winding with stationary excited field winding is attained. Hence, a balance of mmf can be achieved by providing direct current excitation to the outer winding and a steady torque production occurs. But, for this condition to be met, the rotor is to be provided with slip-rings, and the rotor is required to be rotated at a speed decided by the frequency of the exciting currents in the rotor. This is termed as the *synchronous speed* of the machine.

4.6 MATHEMATICAL REPRESENTATION ELECTRICAL MACHINE PARAMETERS

We would, now, describe in mathematical terms, the physical principles of action of most of the conventional rotating machines, and develop a basic theory, in order to estimate the performance of all such machines. We would like to introduce possible simplifying assumptions in order to make the analytical method suitable for mathematical operations without compromising on the authenticity of the analysis:

1. Saturation of the magnetic circuit is neglected. This assumption permits superposition of magnetic fields. Further, self- and mutual-inductance of all machine windings become independent of the value of the winding current.
2. Air-gap mmfs and fluxes will be representable by the fundamental components of their spatial distribution. These distributions are symmetrical about the magnetic-axis of the winding developing them.
3. Slotting effect is ignored. Distributed windings are replaced by equivalent *current sheets*.
4. Commutation process is taken as ideal. The current reversal in commutators is instantaneous.
5. There is no hysteresis and eddy current losses in magnetic circuits.

With the above idealizing assumptions, it is possible to develop linear equations including voltage equations and torque equations for machines—called general machine equations—involving coefficients of self- and mutual-inductances of machine windings. We shall, now, develop expressions of self- and mutual-inductances.

We had already discussed the physical arrangements of practical machine windings in Section 4.5. As observed there we have two basic winding types—those which are distributed in slots around a magnetic surface and those which are lumped or concentrated and wound around projected magnetic protrusions. Expressions, in the following sections, for self- and mutual-inductances, first derived for non-sinusoidal air-gap flux-density distribution are subsequently modified to conform to the simplifying assumption of a fundamental air-gap flux-density distribution. At the end, we describe a distributed winding to be represented in one basic form.

4.6.1 Concentrated Stator Coil—Self- and Mutual-inductances

A simple 2-pole machine with a cylindrical rotor of diameter d and length l, and a salient-pole stator is shown in Figure 4.14(a). Such a depiction is analogous to the representation used in Figure 4.2 earlier. Concentrated salient pole winding of N_s turns, supplied with a current i_s provides air-gap excitation. The rotor coil of N_r concentrated turns is short-chorded with an angular pitch of 2β.

As shown in the diagram, the D-axis and Q-axis which are in space quadrature, represent the main axes of symmetry of the machine. The rotor coil occupies an arbitrary position such that the coil-axis makes an angle of θ_r electrical radian with D-axis.

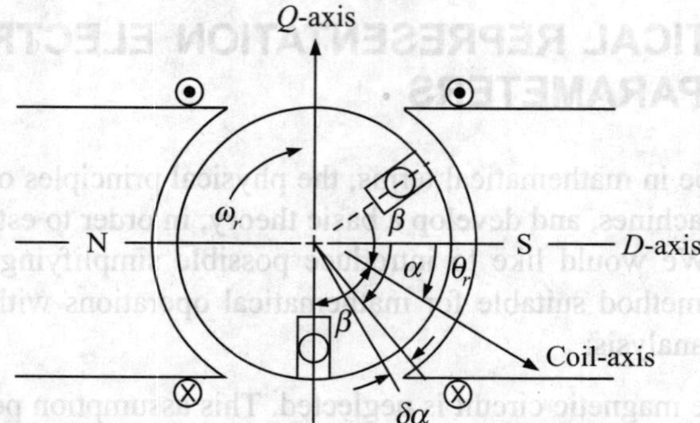

FIGURE 4.14(a) General machines schematic with concentrated short-chorded rotor coil.

Mutual-inductance

The unfurled diagram of Figure 4.14(b) shows the air-gap distribution of mmf $F(\alpha)$ and the flux-density $B(\alpha)$ of the above arrangement, corresponding to D-axis taken as reference. If we take $B(\alpha)$ to have a rectangular distribution is space, of amplitude $B_{\max}$ and having an angular width of $(\pi - 2\gamma)$, then $B(\alpha)$ can be described by a general Fourier series expression:

$$B(\alpha) = \frac{4}{\pi} B_{\max} \sum_{n=1}^{n=\infty} \left[\frac{(-1)^{(n+1)}}{(2n-1)} \cos(2n-1)\gamma \cos(2n-1)\alpha \right] \qquad \text{...(4.9)}$$

We may note that harmonic crest values:

$$B_{(2n-1)} = \frac{4}{\pi} B_{\max} \left[\frac{(-1)^{(n+1)}}{(2n-1)} \cos(2n-1)\gamma \right] \qquad \text{...(4.10)}$$

reduce rapidly with ascending order of n, thus justifying the assumption of considering only sinusoidal fundamental component for consideration. Hence, assuming that the stator mmf is employed entirely in driving flux across the air-gap:

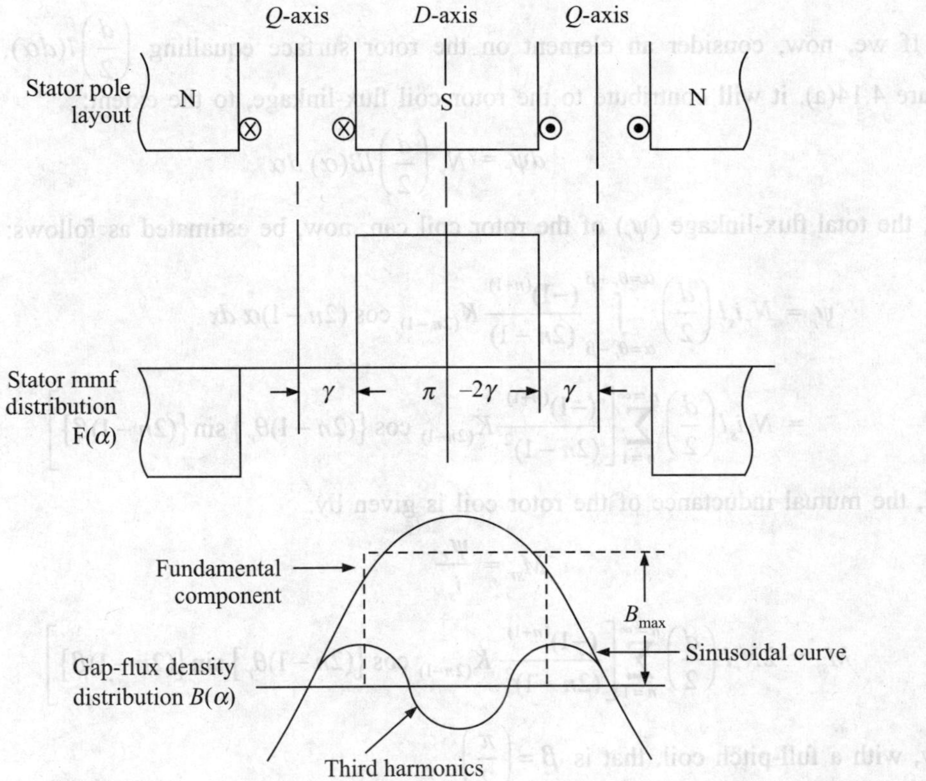

FIGURE 4.14(b) Unfurled diagram of Figure 4.14(a) showing air-gap distribution of mmf and flux-density for stator excitation only.

$$B_{\max} = \mu_0 H = \frac{\mu_0}{g} N_s i_s \qquad ...(4.11)$$

where, g is the length of the air-gap, the radial width of the air-gap being $(\pi - 2\gamma)$ as shown in Figure 4.14(a) and Figure 4.14(b).

If we substitute Eq. (4.11) in Eq. (4.9), we shall obtain:

$$B(\alpha) = \frac{4\mu_0}{\pi g} N_s i_s \sum_{n=1}^{n=\infty} \left[\frac{(-1)^{(n+1)}}{(2n-1)} \cos(2n-1)\gamma \cos(2n-1)\alpha \right]$$

which can be rewritten as:

$$B(\alpha) = i_s \sum_{n=1}^{n=\infty} \left[\frac{(-1)^{(n+1)}}{(2n-1)} K_{(2n-1)} \cos(2n-1)\alpha \right] \qquad ...(4.12)$$

where, the constant term:

$$K_{(2n-1)} = \frac{4\mu_0 N_s}{\pi g} \cos(2n-1)\gamma \qquad ...(4.13)$$

If we, now, consider an element on the rotor surface equalling $\left(\frac{d}{2}\right) l(d\alpha)$, shown in Figure 4.14(a), it will contribute to the rotor coil flux-linkage, to the extent:

$$d\psi_r = N_r \left(\frac{d}{2}\right) lB(\alpha)\, d\alpha$$

and, the total flux-linkage (ψ_r) of the rotor coil can, now, be estimated as follows:

$$\psi_r = N_r i_s l \left(\frac{d}{2}\right) \int_{\alpha=\theta_r-\beta}^{\alpha=\theta_r+\beta} \frac{(-1)^{(n+1)}}{(2n-1)} K_{(2n-1)} \cos(2n-1)\alpha\, dx$$

$$= N_r i_s l \left(\frac{d}{2}\right) \sum_{n=1}^{n=\infty} \left[\frac{(-1)^{(n+1)}}{(2n-1)^2} K_{(2n-1)} \cos\{(2n-1)\theta_r\} \sin\{(2n-1)\beta\} \right] \quad ...(4.14)$$

And, the mutual inductance of the rotor coil is given by:

$$M_{sr} = \frac{\psi_r}{i_s}$$

or $$M_{sr} = 2N_r l \left(\frac{d}{2}\right) \sum_{n=1}^{n=\infty} \left[\frac{(-1)^{(n+1)}}{(2n-1)^2} K_{(2n-1)} \cos\{(2n-1)\theta_r\} \sin\{(2n-1)\beta\} \right] \quad ...(4.15)$$

Now, with a full-pitch coil, that is $\beta = \left(\frac{\pi}{2}\right)$,

$$M_{sr} = N_r ld \left[K_1 \cos\theta_r + \frac{K_3}{3^2} \cos 3\theta_r + \frac{K_5}{5^2} \cos 5\theta_r + \cdots \right] \quad ...(4.16)$$

K_1, K_3, K_5 etc. are the constants specified in Eq. (4.13) corresponding to fundamental, third harmonic, fifth harmonic components, etc. and, the fundamental flux-density distribution corresponding to $n = 1$, is as follows:

$$M_{sr} = N_r l d K_1 \sin\beta \cos\theta_r$$
$$= M_{\max} \sin\beta \cdot \cos\theta_r \quad ...(4.17)$$

With a full-pitch rotor coil (that is, $\sin\beta = 1$), and a fundamental flux-density distribution:

$$M_{sr} = N_r l d K_1 \cos\theta_r = M_{\max} \cos\theta_r \quad ...(4.18)$$

Self-inductance

For the machine schematic of Figure 4.13(a), the mmf on the rotor coil acts on an air-gap permeance which varies in a very complicated manner, as the position of the rotor coil is changed with rotation. As a result, the air-gap flux-density distribution constantly changes with magnitude and symmetry with respect to *D*-axis as well as *Q*-axis. At these continuous changes, it is almost impossible to evolve a mathematical expression for the air-gap flux-density distribution pattern and, adopt the usual strategy of estimating the rotor coil flux-linkage by integration of the flux-density distribution pattern.

We shall discuss, here, an alternative methodology, evolved with permissible approximation. This method makes the flux-density pattern possible to alter with only one-degree of freedom, and also makes it possible to adopt integration of its distribution curve.

Consider a rotor coil as full-pitched, and the rectangular mmf distribution yielded by it, when excited with acceptable precision, with the space distribution of the fundamental mmf, F, having a crest value of F_r. If we resolve the crest value into two components along the D- and Q-axis, so that the resolved values are F_{rd}, and F_{rq}, of the two space-stationary sinusoidally distributed mmf of F_d and F_q respectively, these mmfs—F_d and F_q—along paths of different permeance and produce space-stationary, non-sinusoidal gap flux-density distribution, symmetrically disposed about their respective axes. These can be denoted fairy accurately by their fundamental components of spatial distribution, B_d and B_q, having their crest values of B_{rd} and B_{rq}.

The machine has, now, two effective fictitious air-gap lengths g_d and g_q, corresponding to D-axis and Q-axis with D-axis permeance greater than that of Q-axis, or in other words, $g_q > g_d$.

We can resolve F_r along D-axis and Q-axis and obtain:

$$\left.\begin{aligned} F_{rd} &= F_r \cos\theta_r \\ F_{rq} &= F_r \cos\left(\frac{\pi}{2} + \theta_r\right) = -F_r \sin\theta_r \end{aligned}\right\} \quad ...(4.19)$$

and, take the D-axis as reference.

The mmf component acting along a point displaced by an angle α, leads to:

$$\left.\begin{aligned} F_{\alpha d} &= F_{rd} \cos\alpha = F_r \cos\theta_r \cos\alpha \\ F_{\alpha q} &= -F_{rq} \cos\left(\frac{\pi}{2} + \alpha\right) = F_r \sin\theta_r \sin\alpha \end{aligned}\right\} \quad ...(4.20)$$

The permeances of an element, having an area of $\left[\left(\frac{d}{2}\right) l \, d\alpha\right]$ are:

$$\left.\begin{aligned} d\Lambda_d &= \frac{\mu_0 \left(\frac{d}{2}\right) l \, d\alpha}{g_d} \\ d\Lambda_d &= \frac{\mu_0 \left(\frac{d}{2}\right) l \, d\alpha}{g_q} \end{aligned}\right\} \quad ...(4.21)$$

Hence, the component flux-linkage with the rotor coil:

$$\begin{aligned} d\psi_r &= N_r[(d\Lambda_d)F_{\alpha d} + (d\Lambda_q)F_{\alpha q}] \\ &= N_r \mu_0 \left(\frac{d}{2}\right) l \, d\alpha \left[\frac{F_{\alpha d}}{g_d} + \frac{F_{\alpha q}}{g_q}\right] \end{aligned} \quad ...(4.22)$$

But, the permeance for a full-pitched coil-span:

$$\Lambda_d = \frac{\mu_0\left(\frac{d}{2}\right)l\pi}{g_d}; \text{ and } \Lambda_q = \frac{\mu_0\left(\frac{d}{2}\right)l\pi}{g_q} \quad ...(4.23)$$

So that,

$$d\psi_r = \frac{N_r}{\pi} d\alpha\,[\Lambda_d F_{\alpha d} + \Lambda_q F_{\alpha q}] \quad ...(4.24)$$

This leads to substitution of Eq. (4.19) to Eq. (4.24), and integration to obtain the total rotor coil flux-linkage:

$$\psi_r = \frac{F_r N_r}{\pi} \int_{\alpha=\left(\theta_r - \frac{\pi}{2}\right)}^{\alpha=\left(\theta_r + \frac{\pi}{2}\right)} [\Lambda_d (\cos\theta_r \sin\alpha) + \Lambda_q (\cos\theta_r \sin\alpha)]\, d\alpha \quad ...(4.25)$$

or $$\psi_r = \frac{2F_r N_r}{\pi}[\Lambda_d \cos^2\theta_r + \Lambda_q \sin^2\theta_r]$$

which is the same as saying:

$$\psi_r = \frac{F_r N_r}{\pi}[(\Lambda_d + \Lambda_q) + (\Lambda_d - \Lambda_q)\cos 2\theta_r] \quad ...(4.26)$$

But, the rectangular mmf distribution has a maximum value of:

$$F_{\max} = \frac{N_r \cdot i_r}{2}$$

A Fourier series expression for the rotor mmf distribution can be written, with $\gamma = 0$ and α replaced by $(\alpha - \theta_r)$, that is, considering full-pitched rotor coils and knowing that the mmf distribution is symmetrically disposed about the axis of the rotor coil, as:

$$F(\alpha - \theta_r) = \frac{4}{\pi} F_{\max} \sum_{n=1}^{n=\infty} \frac{(-1)^{n+1}}{(2n-1)} \cos(2n-1)(\alpha - \theta_r) \quad ...(4.27)$$

The crest value of the fundamental, obtained when $\alpha = \theta_r$ is given by:

$$F_r = \frac{4}{\pi} F_{\max} = \frac{2}{\pi} N_r i_r \quad ...(4.28)$$

Hence, combining expressions (4.26) and (4.28), we shall get:

$$\psi_r = \frac{2N_r^2}{\pi^2} i_r [(\Lambda_d + \Lambda_q) + (\Lambda_d - \Lambda_q)\cos 2\theta_r] \quad ...(4.29)$$

Thus, the rotor self-inductance is:

$$L_r = \frac{\psi_r}{i_r} = [L_{r1} + L_{r2}\cos 2\theta_r] \quad ...(4.30)$$

where

$$L_{r1} = \frac{2}{\pi^2} N_r^2 (\Lambda_d + \Lambda_q) \text{ and } L_{r2} = \frac{2}{\pi^2} N_r^2 (\Lambda_d - \Lambda_q) \qquad ...(4.31)$$

In other words, the self-inductance of a concentrated rotor coil, rotating in the annular space of a salient-pole structure has a form indicated by a second-harmonic term superposed on a constant value.

4.6.2 Self- and Mutual-inductance of a Distributed Rotor Winding

A uniformly distributed 2-pole rotor winding with Z-series conductors per electrical radian, is depicted in Figure 4.15(a) and (b).

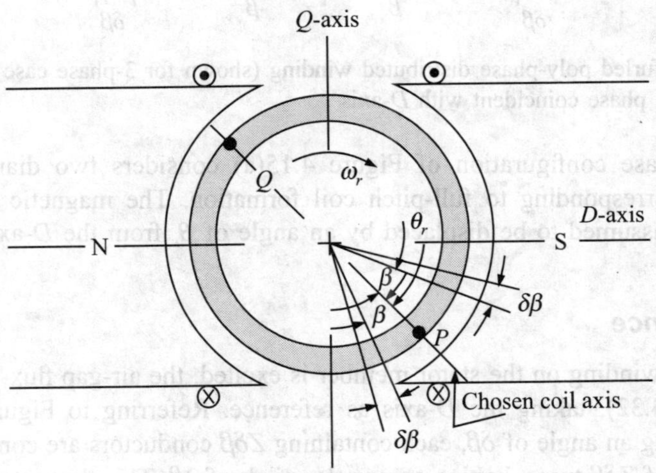

(a) A uniformly distributed single-phase winding on rotor

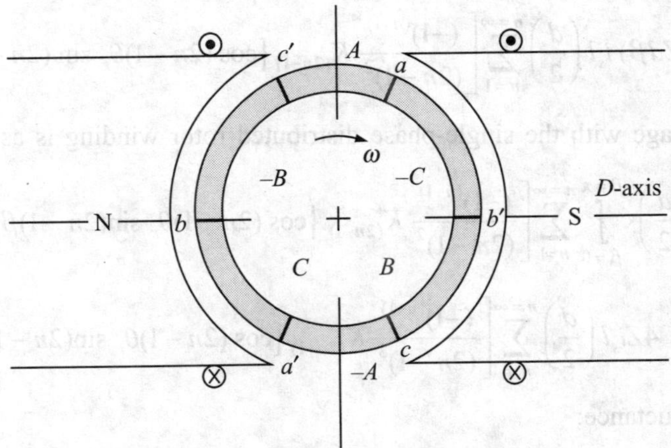

(b) Uniformly distributed 3-phase winding on rotor

FIGURE 4.15 A uniformly distributed rotor winding, with Z-series conductors per electrical radians and with salient-pole stator structure.

The part (a) of Figure 4.15(a) considers the uniformly distributed 2-pole rotor winding of single-phase type, while part (b) indicates the poly-phase version of same rotor winding. The unfurled diagram of Figure 4.15(b) has been shown in Figure 4.16, with the axis of A phase belt coinciding with *D*-axis of the excitation field system.

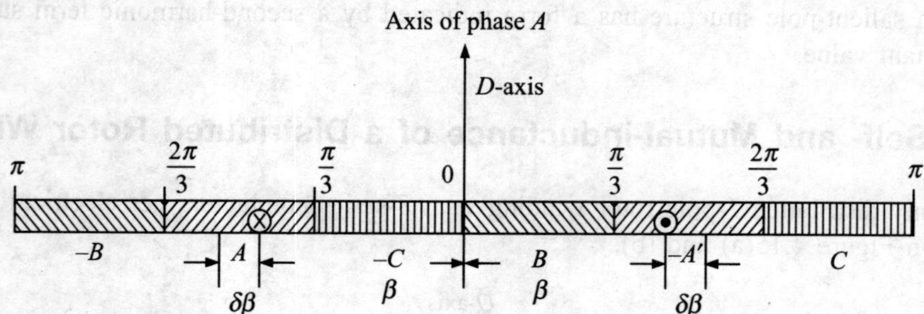

FIGURE 4.16 Unfurled poly-phase distributed winding (shown for 3-phase case) with the axis of the *AA*′ phase coincident with *D*-axis.

The single-phase configuration of Figure 4.15(a) considers two diametrically opposite tappings points corresponding to full-pitch coil formation. The magnetic axis of the single-phase winding is assumed to be displaced by an angle of θ_r from the *D*-axis of the stationary stator field.

Mutual-inductance

If the salient-pole winding on the stator member is excited, the air-gap flux-density distribution is given by Eq. (4.32), taking the *D*-axis as reference. Referring to Figure 4.14(a), the two elements subtending an angle of $\delta\beta$, each containing $Z\delta\beta$ conductors are considered to form the concentrated coil of $Z\delta\beta$ turns, having an angular pitch of 2β. The elementary coil flux-linkage can, now be written down directly, by inspection from Eq. (4.14) by substituting $(Zd\beta)$ for N_r.

$$d\psi_r = 2(Zd\beta)\, i_s l \left(\frac{d}{2}\right) \sum_{n=1}^{n=\infty} \left[\frac{(-1)^{(n+1)}}{(2n-1)^2} K_{(2n-1)} \{\cos (2n-1)\theta_r \sin (2n-1)\beta\} \right] \quad ...(4.32)$$

The total flux-linkage with the single-phase distributed rotor winding is as follows:

$$\psi_r = 2Zi_s l \left(\frac{d}{2}\right) \cdot \int_{\beta=0}^{\beta=\pi} \sum_{n=1}^{n=\infty} \left[\frac{(-1)^{(n+1)}}{(2n-1)^2} K_{(2n-1)} \{\cos (2n-1)\theta_r \sin(2n-1)\beta\} \right] d\beta \quad ...(4.33)$$

or

$$\psi_r = 4Zi_s l \left(\frac{d}{2}\right) \sum_{n=1}^{n=\infty} \left[\frac{(-1)^{(n+1)}}{(2n-1)^3} K_{(2n-1)} \{\cos (2n-1)\theta_r \sin(2n-1)\beta\} \right] \quad ...(4.34)$$

Hence, mutual inductance:

$$M_{sr} = \frac{\psi_r}{i_s} = 2Zld \sum_{n=1}^{n=\infty} \frac{(-1)^{(n+1)}}{(2n-1)^3} K_{(2n-1)} \cos (2n-1)\theta_r \quad ...(4.35)$$

for the full-pitched coil $\left(\text{with } \beta = \dfrac{\pi}{2}\right)$, and

$$M_{sr} = 2Zld\left[K_1 \cos\theta_r - \frac{K_3}{(3)^3}\cos 3\theta_r + \frac{K_5}{(5)^3}\cos 5\theta_r + \cdots\right] \quad ...(4.36)$$

Comparison of Eqs. (4.16) and (4.36) reveals that for identical flux-density distribution, the amplitudes of the higher order mutual inductance harmonics decrease more rapidly for a distributed winding than in case of a concentrated winding. Then, with an air-gap flux-density distribution denoted by a space fundamental:

$$M_{sr} = 2ZldK_1 \cos\theta_r = M_{\max}\cos\theta_r \quad ...(4.37)$$

we are, now, ready to take up poly-phase windings on the rotor. For this, we refer to Figure 4.15(b) and Figure 4.16. We redraw the unfurl diagram of Figure 4.16 in Figure 4.17, and clearly indicate the three tapping points $\underline{A}$, $\underline{B}$ and $\underline{C}$ on the diagram of the rotor winding identified as starting points of the three phases.

To develop an expression for mutual inductance of one of the three phases, say phase AA', with D-axis taken as reference, we shall use Eq. (4.33) with a change of limits of integration. For phase AA', we observe that the new limits of integration are $\beta = \left(\dfrac{2\pi}{3}\right)$ and $\beta = \left(\dfrac{\pi}{3}\right)$, by observation of the unfurled diagrams of Figures 4.16 and 4.17. The total flux-linkage with phase AA' is as follows:

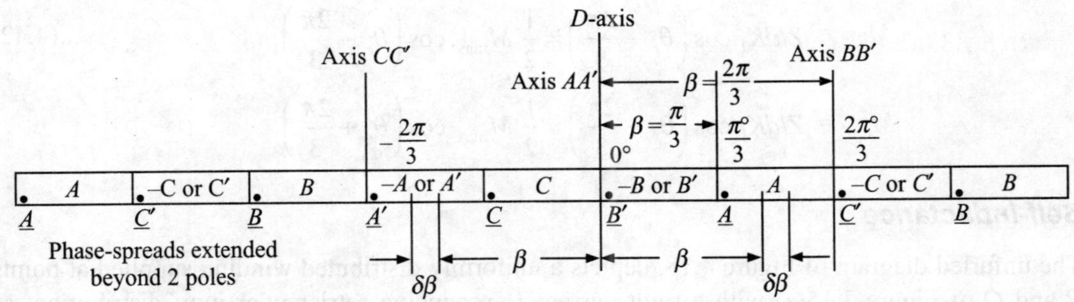

FIGURE 4.17 Uniformly distributed 3-phase winding with tapping points.

$$\psi_{A\text{-}A'} = 2Zi_s l\left(\frac{d}{2}\right)\int_{\beta=\frac{\pi}{3}}^{\beta=\frac{2\pi}{3}}\left[\frac{(-1)^{(n+1)}}{(2n-1)^2}K_{(2n-1)}\cos(2n-1)\,\theta_r \sin(2n-1)\beta\right]d\beta$$

And,

$$M_{AA'} = \frac{\psi_{AA'}}{i_s} = 4Zl\left(\frac{d}{2}\right)\sum_{n=1}^{n=\infty}\left\{\frac{(-1)^{(n+1)}}{(2n-1)^3}K_{(2n-1)}\cos(2n-1)\theta_r \cdot\left[\cos(2n-1)\frac{\pi}{3} - \cos(2n-1)\frac{2\pi}{3}\right]\right\} \quad ...(4.38)$$

By expansion, we shall get:

$$M_{AA'} = Zld\left(K_1 \cos\theta_r + 2\frac{K_3}{(3)^3}\cos 3\theta_r + \frac{K_5}{(5)^3}\cos 5\theta_r + \cdots\right) \quad ...(4.39)$$

With axis of phase BB' at $\left(\theta_r - \frac{2\pi}{3}\right)$:

$$M_{BB'} = Zld\left[K_1 \cos\left(\theta_r - \frac{2\pi}{3}\right) + 2\frac{K_3}{(3)^3}\cos 3\left(\theta_r - \frac{2\pi}{3}\right) + \frac{K_5}{(5)^3}\cos 5\left(\theta_r - \frac{2\pi}{3}\right) + \cdots\right] \quad ...(4.40)$$

and, similarly with the axis of phase CC' at $\left(\theta_r + \frac{2\pi}{3}\right)$:

$$M_{CC'} = Zld\left[K_1 \cos\left(\theta_r + \frac{2\pi}{3}\right) + 2\frac{K_3}{(3)^3}\cos 3\left(\theta_r + \frac{2\pi}{3}\right) + \frac{K_5}{(5)^3}\cos 5\left(\theta_r + \frac{2\pi}{3}\right) + \cdots\right] \quad ...(4.41)$$

If we assume that the stator gap flux-density distribution is describable with adequate accuracy by its space distributed fundamental, with M_{max}, defined by Eq. (4.37) then:

$$M_{AA'} = ZldK_1 \cos\theta_r = \frac{1}{2} M_{max} \cos\theta_r$$

$$M_{BB'} = ZldK_1 \cos\left(\theta_r - \frac{2\pi}{3}\right) = \frac{1}{2} M_{max} \cos\left(\theta_r - \frac{2\pi}{3}\right) \quad ...(4.42)$$

$$M_{CC'} = ZldK_1 \cos\left(\theta_r + \frac{2\pi}{3}\right) = \frac{1}{2} M_{max} \cos\left(\theta_r + \frac{2\pi}{3}\right)$$

Self-inductance

The unfurled diagram of Figure 4.18, depicts a uniformly distributed winding supplied at points P and Q of Figure 3.15(a) with a rotor current i_r, producing a triangular mmf distribution, as had been discussed earlier in connection with Figure 4.9.

Enclosed in the space bounded by dotted lines on the figure is a total current $(Z\pi i_r)$. The maximum value of the triangular mmf (F_{max}) distribution is given by:

$$F_{max} = \frac{Z\pi i_r}{2} \quad ...(4.43)$$

With the winding axis taken as reference (that is, at $\beta = 0$), the mmf distribution pattern can be described by a Fourier series expression:

$$F(\beta) = \frac{8}{\pi^2} F_{max}\left[\cos\beta + \frac{1}{(3)^2}\cos 3\beta + \frac{1}{(5)^2}\cos 5\beta + ...\right] \quad ...(4.44)$$

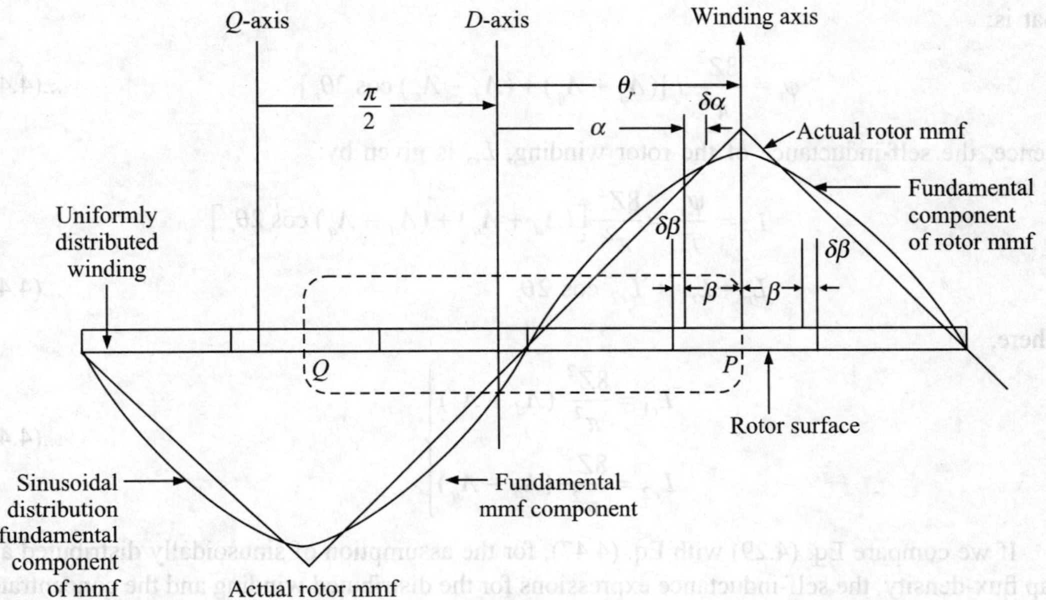

FIGURE 4.18 MMF distribution pattern for a uniformly distributed winding excited by single-phase supply in rotor and fed at diametrically opposite points on a 2-pole arrangement shown in Figure 4.15(a).

If we assume that the triangular mmf distribution can be adequately represented by its fundamental component

$$F(\beta) = \frac{8}{\pi^2} F_{\max} \cos\beta = \frac{8}{\pi^2} F_{\max} \text{ at } \beta = 0$$

with the fundamental crest value

$$F_r = \frac{8}{\pi^2} F_{\max} = \frac{4}{\pi} Z i_r \qquad ...(4.45)$$

As discussed earlier, F_r can, now, be resolved into D-axis and Q-axis components by the technique evolved in Section 4.6.1 for the self-inductance of a concentrated winding.

The flux-linkage for the elementary coil of $Zd\beta$ turns having an angular pitch of 2β can be found using expression (4.25) with a change of the limits of integration.

$$d\psi_r = \frac{Z(d\beta)}{\pi} \frac{4Z}{\pi} i_r \int_{\alpha = \theta_r - \beta}^{\alpha = \theta_r + \beta} [\Lambda_d \cos\theta_r \cos\alpha + \Lambda_q \sin\theta_r \sin\alpha]\, d\alpha$$

$$= \frac{8Z^2}{\pi^2} (d\beta)\, i_r [\Lambda_d \cos^2\theta_r + \Lambda_q \sin^2\theta_r] \sin\beta \qquad ...(4.46)$$

Then, the total flux-linkage with the rotor distributed winding:

$$\psi_r = \frac{8Z^2}{\pi^2} i_r (\Lambda_d \cos^2\theta_r + \Lambda_q \sin^2\theta_r] \int_{\beta=0}^{\beta=\pi} \sin\beta\, d\beta$$

that is:

$$\psi_r = \frac{8Z^2}{\pi^2} i_r[(\Lambda_d + \Lambda_q) + (\Lambda_d - \Lambda_q)\cos 2\theta_r] \qquad ...(4.47)$$

Hence, the self-inductance of the rotor winding, L_r, is given by:

$$L_r = \frac{\psi_r}{i_r} = \frac{8Z^2}{\pi^2}\left[(\Lambda_d + \Lambda_q) + (\Lambda_d - \Lambda_q)\cos 2\theta_r\right]$$

$$L_r = L_{r1} + L_{r2}\cos 2\theta_r \qquad ...(4.48)$$

where,

$$\left.\begin{aligned} L_{r1} &= \frac{8Z^2}{\pi^2}(\Lambda_d + \Lambda_q) \\ L_{r2} &= \frac{8Z^2}{\pi^2}(\Lambda_d - \Lambda_q) \end{aligned}\right\} \qquad ...(4.49)$$

If we compare Eq. (4.29) with Eq. (4.47), for the assumption of sinusoidally distributed air-gap flux-density, the self-inductance expressions for the distributed winding and the concentrated winding are of similar forms.

4.7 CONCENTRATED AND DISTRIBUTED WINDING EQUIVALENCE

When we compare the derived expressions of self- and mutual-inductances of distributed and concentrated windings, with mmf distribution taken as sinusoidal, we observe that for uniformly distributed windings:

$M_{sr} = 2ZldK_1 \cos \theta_r$, as seen in Eq. (4.37)

and, $L_r = \dfrac{8Z^2}{\pi^2}[(\Lambda_d + \Lambda_q) + (\Lambda_d - \Lambda_q)\cos 2\theta_r]$, as in Eq. (4.48), as against:

$M_{sr} = N_r ldK_1 \cos \theta_r$, as observed in Eq. (4.18)

and, $L_r = \dfrac{2N_r^2}{\pi^2}[(\Lambda_d + \Lambda_q) + (\Lambda_d - \Lambda_q)\cos 2\theta_r]$ as obtained in Eq. (4.31).

By equating the expressions for self- and mutual-inductances separately, it is obvious that a concentrated and a distributed winding are magnetically similar and equivalent, when:

$$N_r = 2Z \qquad ...(4.50)$$

With the above relationship, it is established that a distributed winding can be represented by a lumped coil under the convention, so that the current and mmf are both effective along the coil-axis in the same direction as shown in the equivalence diagram in Figure 4.19.

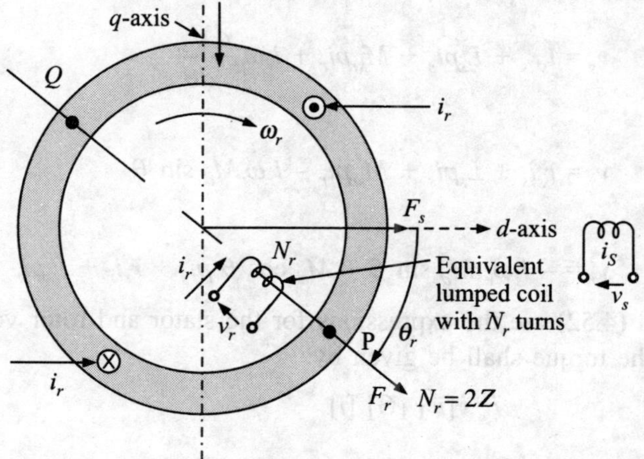

FIGURE 4.19 Diagrammatic depiction of equivalent concentrated and distributed windings.

ILLUSTRATION 4.1 A rotating transducer has a salient-pole stator and a rotating coil connected to slip-rings as shown in Figure 4.20. The mutual coupling varies as:

$$M_d \cos \theta$$

where, θ is the angle shown.

(a) Derive from first principles expressions for the instantaneous torque and the rotor and stator voltages.

(b) How is the torque expression modified when the rotating winding is connected to a commutator and the current enters through a fixed brush pair inclined at an angle α to the salient pole axis?

(c) Discuss the physical concepts of torque production for the two types of rotor connection.

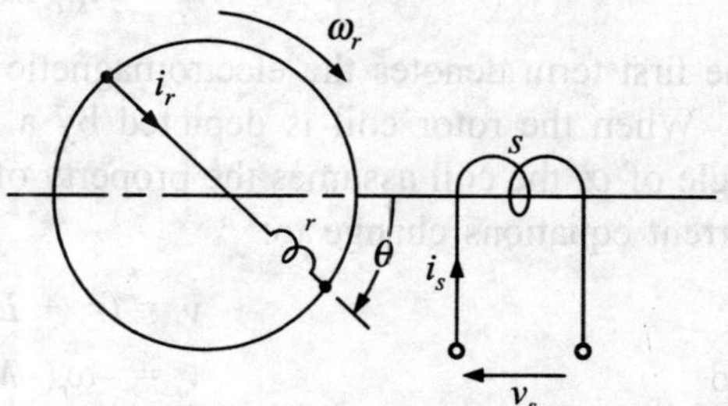

FIGURE 4.20 A rotating transducer with a salient pole stator and a rotating coil connected to slip rings.
Given: $M_{sr} = M_{rs} = M_d \cos \theta$
$L_r = L_{r1} + L_{r2} \cos 2\theta$
(Stator is salient pole)

Solution: Voltage equation for the s-winding is given by:

$$v_s = r_s i_s + \frac{d}{dt}(L_s i_s) + \frac{d}{dt}(M_{sr} i_r)$$

$$= r_s i_s + L_s \frac{d}{dt}(i_s) + M_{sr}\frac{d}{dt}(i_r) + \left\{\omega_r \frac{d}{d\theta}(L_s) + \omega_r \frac{d}{d\theta}(M_{sr})\right\} i_r$$

Since no mention about properties of L_s has been made in the problem, we assume L_s to be independent of θ and rewrite v_s as:

$$v_s = i_s r_s + L_s p i_s + M_{sr} p i_r + i_r \omega_r \frac{dM_{sr}}{d\theta}$$

which leads to:

$$v_s = r_s i_s + L_s p i_s + M_{sr} p i_r - i_r \omega_r M_d \sin\theta \qquad \text{...(4.51)}$$

Similarly,

$$v_r = -i_s \omega_r M_d \sin\theta + M_d \cos\theta\, p i_s + r_r i_r + L_r p i_r \qquad \text{...(4.52)}$$

Equations (4.51) and (4.52) are the expressions for the stator and rotor voltages.

Expression for the torque shall be given by:

$$T = [i^*]\,[G]\,[i]$$

$\because\quad [i] = \begin{bmatrix} i_s \\ i_r \end{bmatrix}$, therefore, $\quad [i^*] = [i_s \;\; i_r]$

and

$$[G] = \begin{bmatrix} 0 & -M_d \sin\theta \\ -M_d \sin\theta & -2L_{r2}\sin 2\theta \end{bmatrix} \qquad \text{...(4.53)}$$

$$\therefore \qquad T = -i_s i_r M_d \sin\theta - i_s i_r M_r \sin 2\theta \qquad \text{...(4.54)}$$

The first term denotes the electromagnetic torque, while the second term is *reluctance*.

When the rotor coil is depicted by a commutator winding with a brush axis fixed at an angle of α, the coil assumes the property of a pseudo-stationary coil. Accordingly, the voltage–current equations change to:

$$v_s = (r_s + L_s p) i_s + M_d \cos\alpha\, p i_r \qquad \text{...(4.55)}$$

and

$$v_r = -\omega_r(-M_d \sin\alpha) i_s + (r_r + L_r p) i_r \qquad \text{...(4.56)}$$

and the resulting torque is given by:

$$T = M_d \sin\alpha\, i_r i_s \text{ units} \qquad \text{...(4.57)}$$

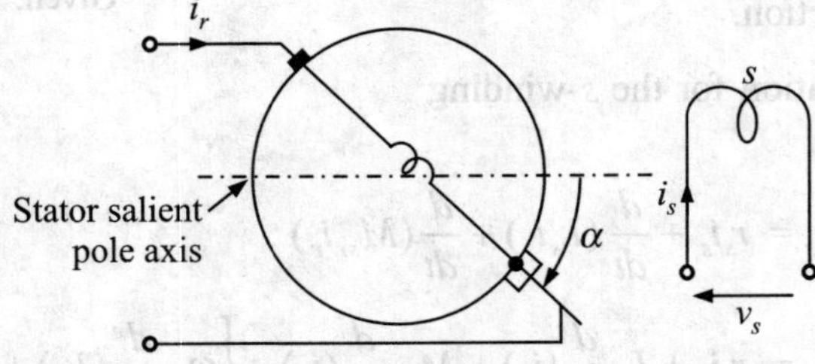

FIGURE 4.21 Rotating winding is connected to a commutator [Problem 4.1(b)].

ILLUSTRATION 4.2

(a) Explain how a rotating winding may be represented as an equivalent pseudo-stationary coil along the brush axis and describe the special properties of this type of coil.

(b) Write the general voltage and torque equations for the 2-pole commutator machine shown in Figure 4.22. Explain how the rotational inductance coefficients $[G]$ may be obtained in terms of the L and M inductance coefficients.

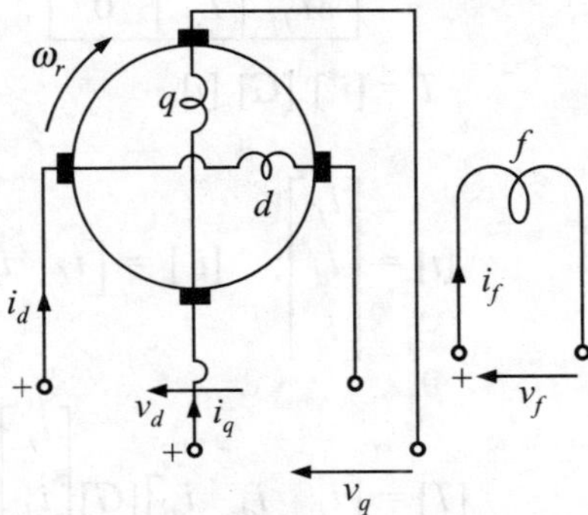

FIGURE 4.22 A 2-pole commutator winding arrangement mounted on the rotor member and denoted by its equivalent with pseudo-stationary coils.

Solution: The voltage equations are:

$$v_f = (r_f + L_f p) i_f + M_{df} p\, i_d \quad \text{...(4.58)}$$

$$v_d = M_{fd}\, p i_f + (r_d + L_d p) i_d + \frac{d}{d\theta}(M_{qd})(i_q)$$

$$= M_{fd}\, p i_f + (r_d + L_d p) i_d - M_{qd}\, \omega_r\, i_q \quad \text{...(4.59)}$$

$$v_q = i_f \frac{d}{d\theta} M_{fq} + i_d \frac{d}{d\theta} M_{dq} + (r_q + L_q p) i_q \quad \text{...(4.60)}$$

$$\Rightarrow \begin{bmatrix} v_f \\ v_d \\ v_q \end{bmatrix} = \begin{bmatrix} r_f + L_f p & M_{df} p & 0 \\ M_{fd} p & r_d + L_d p & -\omega_r M_{qd} \\ \omega_r M_{fq} & \omega_r M_{dq} & r_q + L_q p \end{bmatrix} \begin{bmatrix} i_f \\ i_d \\ i_q \end{bmatrix} \quad \text{...(4.61)}$$

In order to relate M_{fd}, M_{df}, M_{fq}, M_{qf}, M_{dq} and M_{qd} to normal self and mutual-inductance parameters of circuits, we note the following points:

- The length of the air-gap along the d- and q-axis may be different (for salient-pole structure for the stator construction of commutator machines).
- All the mutual-inductances may be assumed to be reversible nature, that is, $M_{df} = M_{fd}$, etc. and
- $G_{fq} = M_{fq}$, the notional mutual-inductance coefficient between f and q, if q is considered as shifted to share a common magnetic axis with f, without changing the magnetic property along the axis of the q-coil.

Similarly, $G_{fd} = M_{fd}$

and, $G_{dq} = L_q$ (since d and q are identical coils)

- Thus, [G] matrix of the machine assumes the form shown in Eq. (4.62)

$$[G] = \begin{bmatrix} 0 & 0 & 0 \\ 0 & 0 & -L_q \\ M_{fq} & L_d & 0 \end{bmatrix} \quad ...(4.62)$$

$$T = [i^*]\,[G]\,[i]$$

And, as

$$[i] = \begin{bmatrix} i_f \\ i_d \\ i_q \end{bmatrix}; \quad [i^*] = \begin{bmatrix} i_f & i_d & i_q \end{bmatrix}$$

$$\therefore \quad [T] = \begin{bmatrix} i_f & i_d & i_q \end{bmatrix} [G] \begin{bmatrix} i_f \\ i_d \\ i_q \end{bmatrix}$$

$$T_e = M_{fq}\, i_f i_q\, M_{fq} + i_{dq}\,(L_d - L_q) \quad ...(4.63)$$

ILLUSTRATION 4.3 A 4-pole commutator winding has 576 uniformly distributed conductors. How many turns has the equivalent concentrated coil, if the winding is

(a) lap-wound

(b) wave wound.

Solution: The machine has 576 conductors or $\left(\dfrac{576}{2}\right)$, that is, 288 single-turn coils.

(a) In the simplex lap winding, the brushes divide the total turns into parallel paths = Number of poles

$\therefore$ Gross number of turns per parallel to path

$$= \frac{288}{4} = 72 \text{ turns}$$

But the effectiveness of distributed coils in commutator windings (brushes are located effectively 180° or π radians apart) is $\dfrac{2}{\pi}$.

$\therefore$ Number of turns effective per paralllel path

$$= \frac{2}{\pi} \times 72 = 45.8$$

Now, for a total current of i in the armature, the current per parallel path is $\left(\dfrac{1}{4}\right)$, and the total ampere-turn of armature $= 4 \times \left(\dfrac{1}{4}\right) \times 45.8i = 45.8i$

Hence, the number of turns in the representative equivalent concentrated coil denoting the armature is 45.8 or 46 turns.

(b) For the simplex wave winding, number of parallel path = 2 (independent of the pole-number).

Hence, N, the number of turns of the representative coil

$$= \frac{2}{\pi} \times \frac{288}{2} = 91.6 \simeq 92 \text{ turns}$$

ILLUSTRATION 4.4 A 2-pole commutator machine has a sinusoidal space-distributed stator flux with a maximum density of 1.5 weber per metre^2. The effective length and radius of the rotor are 0.8 metre and 0.25 metre respectively.

(a) Calculate from first principles of flux-linkage of a full-pitch rotor coil with its axis inclined at an angle α to the stator pole axis.

(b) Hence, calculate the maximum flux-linkage (per parallel path) of the complete rotor winding, if there are 200 conductors.

Solution: For Figure 4.23 shown, the angle θ (space angle) is measured with reference chosen along *d*-axis. The main flux-density occurs along the *d*-axis and the general expression for flux-density distribution along the air-gap has the form:

$$B = B_{\max} \cos \theta$$

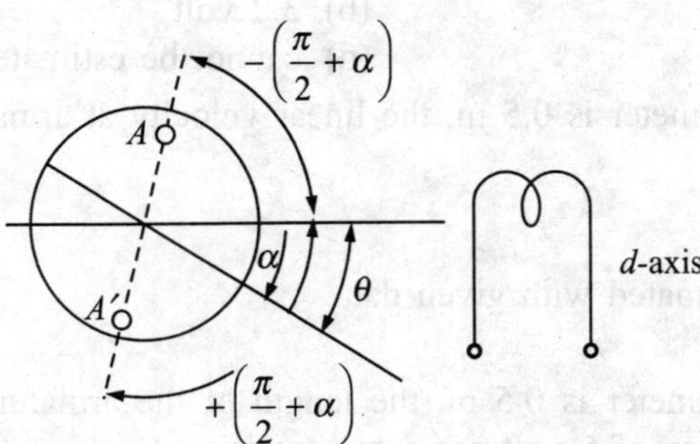

FIGURE 4.23 Schematic of 2-pole commutator machine.

(a) For the given position of the full-pitch rotor coil AA', the flux linked by the coil:

$$\psi_{AA'} = \int BLR \, d\theta$$

over the area embraced by the coil between spatial positions of A and A', that is,

between limits of $-\left(\frac{\pi}{2}+\alpha\right) \le \theta \le +\left(\frac{\pi}{2}+\alpha\right)$

$$\therefore \qquad \psi_{AA'} = 1.5 \times 0.8 \times 0.25\, N \left\{ \sin\left(\frac{\pi}{2}+\alpha\right) - \sin\left(-\frac{\pi}{2}+\alpha\right) \right\}$$

$$= 0.6 \text{ N} \cos \alpha \text{ Wb-turns}$$

where, N is the number of turns (effective) of the rotor commutator winding.

(b) With 200 conductors and 2-poles, effective number of turns per parallel path

$$= \frac{2}{\pi} \times 100 \times \frac{1}{2} = \frac{100}{\pi} \text{ turns}$$

Total flux linked, if the axis of the winding is at an angle α

$$= \frac{100}{\pi} \times 0.6 \cos \alpha \text{ Wb-turns}$$

and flux-linkage is maximized when the axis of the winding and that of the flux, are co-axial, that is, $\alpha = 0$.

Hence, the maximum flux-linkage of the rotor winding

$$= \frac{60}{\pi} = 19.1 \text{ Wb-turns}$$

OBJECTIVE TYPE QUESTIONS

4.1 The flux in the air-gap of a 2-pole machine is a rectangle with a total flux of 10 m-Wb per pole. The armature rpm is at 1200 rpm. The armature coil has 4 turns. The emf generated in a coil is:

(a) 8 volt (b) 3.2 volt
(c) 5 volt (d) cannot be estimated with given data

4.2 If the armature diameter is 0.5 m, the linear velocity at armature periphery is:

(a) 10π m/s
(b) 20π m/s
(c) Cannot be estimated with given data
(d) 40π m/s

4.3 If the armature diameter is 0.5 m, the length of the armature is also 0.5 m, the flux-density, for a voltage of 5 volt per 4 turn coil is:

(a) 0.0255 tesla (b) 0.5 tesla
(c) 1.0 tesla (d) cannot be estimated with given data

A 3-phase ac generator has 4 poles, the flux is sinusoidally distributed with B_{max} = 0.8 tesla at the pole centre. Stator diameter is 1 m, stator length is also 1 m, rotor speed is 1500 rpm. Number of conductors per phase = 160, full-pitch coils are used having 4 slots per pole per phase.

For the above machine determine the following:

4.4 The area per pole is:

(a) 5 m^2 (b) 3.927 m^2
(c) 5π m^2 (d) None of the above

4.5 Flux per pole is:

(a) π Wbs (b) 2 Wbs

(c) 100 m Wbs (d) 5 Wbs

4.6 Frequency of the generated voltage is:

(a) 50 Hz (b) 60 Hz

(c) 25 Hz (d) 83.3 Hz

4.7 The average flux-density under a pole is:

(a) 0.8 tesla (b) 0.5093 tesla

(c) 1.0 tesla (d) 0.4 tesla

4.8 Peripheral velocity of the conductors in m/s is:

(a) 100 (b) 50π

(c) 25π (d) 12.5π

4.9 The coil-span factor (or breadth factor) of the windings is:

(a) 1 (b) cos 60°

(c) 0.866 (d) None of the above

4.10 The distribution factor is:

(a) 1 (b) 0.256

(c) sin 15°/4 sin 7.5° (d) None of the above

4.11 Generated rms value of voltage per phase is:

(a) 25.13 kV (b) 8.0 kV

(c) 17.025 kV (d) 11.0 kV

Answers

4.1 (b)	**4.2** (b)	**4.3** (d)	**4.4** (b)	**4.5** (b)	**4.6** (a)
4.7 (b)	**4.8** (c)	**4.9** (a)	**4.10** (b)	**4.11** (c)	

NUMERICAL PROBLEMS

4.1 A dc machine has the following details: 6-poles, 600 conductors, $2a = 2$; $\phi_p = 33$ mWb current external to the machine, $I_a = 1000$ A. Estimate at 600 revolutions per minute the:

(a) developed torque

(b) generated voltage

(c) power developed

when the machine is operating at steady-state.

[**Ans.** (a) 9450 Nm; (b) 594 volt (c) $9450 \times 20\pi$ W]

4.2 The magnetic loading of a 3-ph alternator is $P\phi_p = 5\pi/3$ weber and the electric loading $ZI_m = 8 \times 10^5$ ampere-conductors. Estimate:

P – Total number of poles

Z – Total number of armature conductors:

(a) The maximum value of the developed torques.

(b) The maximum synchronous watt developed at synchronous speed, if the machine has 12 poles. [**Ans.** (a) $T_{dm} = 5 \times 10^5$ Nm (b) 26.15 MW]

4.3 The flux-density acting on conductors in a 3-phase non-salient pole, 2-pole machine with 4-turn coils, having active length 2 m, is rectangular over the pole-pitch with constant amplitude of 1 tesla. The coil is full-pitch. The frequency is 50 Hz, and the rotor diameter is 1 m.

(a) Obtain the peak and rms values of induced voltage in the coil due to fundamental only.

(b) If the coil is chorded by 20° corresponding to $\left(\frac{8}{9}\right)$ of full-pitch, estimate the peak and rms values of voltage induced in a coil for fundamental component.

(c) The phase winding has $q = 4$ slots/pole/phase, a coil-span of $\left(\frac{8}{9}\right)$ full-pitch. Calculate the value of K_w for fundamental. Also the rms value of the induced voltage per phase.

[**Ans.** (a) 3200 volt, 2262 volt, (b) 3150 volt, 2230 volt (c) 0.9406, 8500 volt]

CHAPTER 5

Voltage and Torque Equations for Rotating Electro-mechanical Devices

5.1 INTRODUCTION

In the previous chapter, we learnt about some important terminologies associated with common windings and also about classification of windings employed in electro-mechanical energy converters. We also noted that distribution of windings along the air-gap periphery, in general, renders more effective utilization of available space. It leads to attainment of near-sinusoidal flux-density distribution in the air-gap. It, therefore, controls space and time harmonics. The price for this harmonic control and more effective space utilization is less than using lumped concentric windings in terms of the effectiveness of the winding. We noted that the effect of distribution introduced *distribution factor* and the effect of chording coils led to *coil-span factor*, both of which have values of less than unity. These two factors together are called the winding factor of distributed windings, which is a measure of effectiveness of distributed windings. Its value poses a challenge to winding designers to make designs with a minimum loss of utilization by finding suitable trade-offs with other considerations in design. We also studied about the mmf and flux-density distribution in space when current was supplied to the windings. We evolved strategies for estimating the value of energy stored in the magnetic field. We also developed methods of estimation of self- and mutual-inductances of different configurations of stator and rotor windings of alternative layouts like concentrated, continuously distributed phase and commutator windings. Finally, we arrived at a relationship for denoting distributed winding by their equivalent concentrated-coil counter parts.

In the present chapter, we look at and derive expressions to formulate voltage induction in windings, and physically evolve their nature and significance. Finally, we wish to develop the concept of **primitive machine** and identify methods of developing different types of particular machines in practical use from them, including the development of relationships leading to expressions of torque production in them.

5.2 EMFS INDUCED IN DISTRIBUTED WINDINGS

The process of voltage induction in a coil rotating at a uniform angular velocity in a stationary field system, is identical with that in relation to induction of voltage in a stationary coil subject to the influence of a magnetic field rotating with respect to the coil a uniform speed. In the latter case, if the field has a fixed magnitude, it is possible to view the mechanism of voltage induction due to rate of change of flux linked by the coil along its stationary axis over one revolution of the flux system (*transformer-induced voltage*) or the rate of flux-cutting by the moving coil with respect to the stationary flux (*speed-induced voltage*).

In estimating the voltage induced in an electrical winding system through interaction with a flux of constant magnitude, *it suffices to consider any one of the above voltage induction processes but not both. Aggregation of transformer-induced and speed-induced voltages in a winding becomes necessary only in special case, where the amplitude of the flux varies as a function of time while, there is also a relative motion between the flux and the winding* (For example, in case of a single phase induction machine).

We shall, now undertake to derive expressions for the quantum of voltage induction due to interaction of time-invariant flux, with a distributed-type electrical winding, with a relative motion between them. We shall use both the approaches, and show that the result is the same in either case.

5.2.1 Induced emf in a Phase Winding Considering Effect of Transformer-induced Voltage

We shall consider a perfectly uniformly distributed winding with M coils per pole and having a phase-spread of ψ. The above winding is exposed to a flux system, having a flux ϕ weber per pole with relative angular velocity of ω_r radians per second between them. ($\omega_r = 2\pi f_r$, where f_r is the angular frequency of motion).

When viewed with respect to a chosen reference axis fixed to a coil of the winding, each coil effectively experiences an alternating flux along the axis of symmetry of the winding, such that the amplitude of the flux alternates at a frequency of f_r over magnitudes of ($\pm\ \phi_m$). The situation is indicated in Figure 5.1. Two coils AA' and BB' have been especially chosen in the diagram as indicated. To start with, we shall assume that the coils making up the winding are *full-pitched.* Further, we shall choose coil AA' such that it shares a common axis with the axis of symmetry of the winding, which is also the considered axis of the field system at the chosen instant.

If each coil comprises m turns, the magnitude of the emf ($e_{AA'}$) induced in the coil AA' is as follows:

$$e_{AA'} = -\frac{d}{dt}(m\phi) \qquad ...(5.1)$$

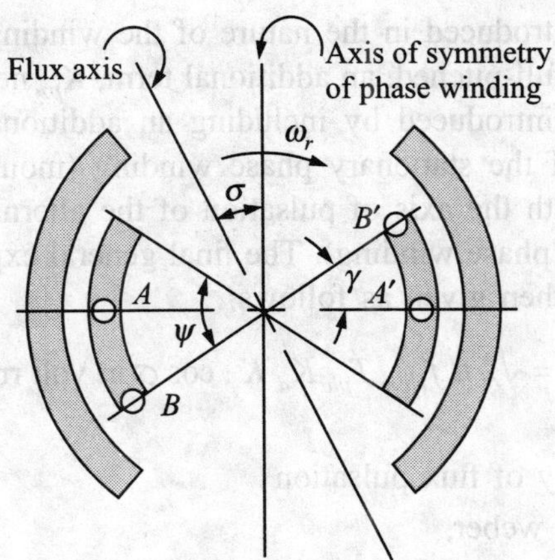

FIGURE 5.1 Phase winding and flux-distribution schematic for a 2-pole system—transformer-induced voltage in phase winding.

since, $\phi = \phi_m \sin \omega_r t$, $e_{AA'} = -m\ \omega_r\ \phi_m \cos \omega_r t$ which has an rms value of

$$E_{AA'} = \frac{1}{\sqrt{2}} m\omega_r\ \phi_m = \sqrt{2}\pi f_r m \phi_m \text{ volt} \qquad \text{...(5.2)}$$

Now, choosing the general coil BB', displaced from AA' by an angle γ, we can write by inspection, that:

$$E_{BB'} = \sqrt{2}\pi f_r m\phi_m \cos\gamma \text{ volt} \qquad \text{...(5.3)}$$

If we consider all M coils of the winding in series, the rms value of the induced emf in the winding is given as follows:

$$E_T = \sqrt{2}\pi f_r m\phi_m \sum_{M=1,2,\ldots M} \cos\gamma$$

However, for a perfectly uniformly distributed winding, by definition indicated in Section 4.3, $M \to \infty$, and

$$E_T = \sqrt{2}\pi f_r\ (mM)\,\phi_m \frac{1}{\psi}\int_{-\frac{\psi}{2}}^{\frac{\psi}{2}} \cos\gamma\ d\gamma$$

$$= \sqrt{2}\,\pi f_r\ (mM)\,\phi_m \frac{\sin\left(\frac{\psi}{2}\right)}{\left(\frac{\psi}{2}\right)} \qquad \text{...(5.4)}$$

we can readily identify the term mM as the *total number of turns in series*, T_{ph}, constituting the phase winding. Similarly, the term $\dfrac{\sin\left(\frac{\psi}{2}\right)}{\left(\frac{\psi}{2}\right)}$ denotes the *distribution factor*, K_d.

If further generality is introduced in the nature of the winding by assuming that the coils forming the winding are not full-pitched, an additional term, K_p, needs to be added in Eq. (5.4). A further generality can be introduced by including an additional factor (cos σ), to signify that the axis of symmetry of the stationary phase winding (mounted on the stator) makes a general spatial angle of σ with the axis of pulsation of the alternating field (in case it is also stationary with respect to the phase winding). The final general expression for the induced emf per phase, in rms values, is then given as follows:

$$E_T = \sqrt{2}\pi f_r \phi_m T_{ph} K_d K_p \cos\sigma \text{ in volt rms} \quad ...(5.5)$$

where

f_r is the angular frequency of flux pulsation

ϕ_m is the flux per pole in weber,

$T_{ph} = mM$, is the total turns in series per phase

K_p is the pitch factor

$K_d = \dfrac{\sin\left(\dfrac{\psi}{2}\right)}{\dfrac{\psi}{2}}$ is the distribution factor

ψ is the phase-spread of the winding and

σ is the spatial angular displacement between the axis of symmetry of the phase winding and the flux axis.

5.2.2 Induced emf in a Phase Winding Using the Effect of Speed-induced Voltage

Now, we shall consider a $2p$-pole magnetic field system, in which a phase winding, having M coils, is rotated at a uniform speed of n_r revolutions per second, thus producing an angular velocity of $\omega_r = 2\pi n_r$ radians per second. As in the earlier case section 5.2.1, we shall consider that each coil of the phase winding consists of m-turns and to begin with, we assume that each turn is full-pitched. Additionally, let the flux-distribution be such that the resulting flux-density along the diametral plane XX' is uniform and is equal to B_m, as shown in Figure 5.2.

Take a general coil BB' mounted on the rotor as shown. In a $2p$-pole field system, a conductor mounted on the rotor cuts across $2p\phi_m$ units of flux for each revolution around the periphery. This total flux is distributed over the entire periphery having a surface area of πDl, where, D is the diameter of the rotor and l is the axial length perpendicular to the plane of the paper. Hence, the value of B_{av} is given by:

$$B_{av} = \frac{2\pi\phi_m}{\pi Dl} \quad ...(5.6)$$

along the surface of the rotor member.

Although the flux-density at any point on the diametral plane, XX', is taken as uniform, a single conductor is seen as instantaneously moving at a tangent with the rotor periphery

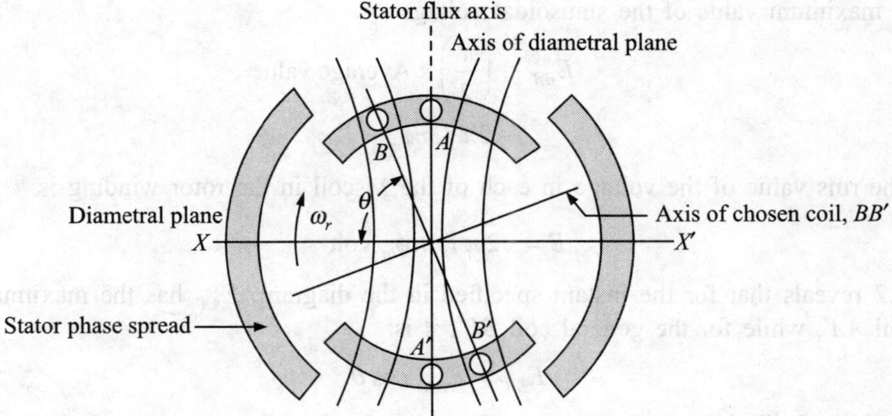

FIGURE 5.2 Speed-induced voltage in distributed phase winding due to rotation of armature in a magnetic field of constant magnitude.

at any given time. A single conductor, therefore, cuts across sinusoidal component of flux-density at any point during rotation. This point was established in Section 3.7.6 with reference to Figure 3.18 earlier. In general terms, B, the flux-density cut across by a single conductor, is thus related to B_m according to the following relation:

$$B = B_m \sin \theta$$

θ being the spatial angle between the diametral plane and the plane of the two conductors forming a coil, as indicated for the general coil BB'. For this case,

$$\frac{B_{\text{av}}}{B_m} = \frac{2}{\pi} \qquad \text{...(5.7)}$$

But, total flux across XX' is always $(p\phi_m)$ distributed uniformly over an area of (Dl). Hence:

$$B_m = \frac{p\phi_m}{Dl}$$

The expression for the average voltage in the general coil $E_{BB'}$ is given as follows:

$$E_{BB'} = 2B_{\text{av}} l v m \qquad \text{...(5.8)}$$

provided that the coil BB' has m turns and the coil is full-pitched. Here, we may substitute:

$$v = \pi D n_r \quad \text{and} \quad B_{\text{av}} = \frac{2p\phi_m}{\pi Dl} \qquad \text{...(5.9)}$$

Hence,

$$E_{BB'} \text{ (average value)} = 2\frac{2p\phi_m}{\pi Dl} l(\pi D n_r)m$$

$$= 2(2pn_r)\phi_m m$$

$$= 4 f_r m \phi_m \qquad \text{...(5.10)}$$

since, $f_r = pn_r$

And, the maximum value of the sinusoidal voltage

$$E_{BB'} = \left(\frac{\pi}{2}\right) \times \text{Average value}$$

$$= 2\pi f_r m\phi_m$$

Hence, the rms value of the voltage in each of the M coil in the rotor winding is:

$$E = \sqrt{2}\pi f_r m\phi_m \text{ volt} \qquad \text{...(5.11)}$$

Figure 5.2 reveals that for the instant specified in the diagram, $E_{AA'}$ has the maximum value in the coil AA', while for the general coil BB', it is:

$$E_{BB'} = E_{AA'} \cos\theta \qquad \text{...(5.12)}$$

Consideration of all M coils in series in the rotor, leads to aggregation of all the induced voltages. This is, the speed-induced voltage of the rotor winding, E_S. Thus,

$$E_S = \sqrt{2}\pi f_r m\phi_m \sum_{M=1,2,\ldots,M} \cos\theta \qquad \text{...(5.13)}$$

which, for $M \to \infty$ and when while θ is limited to the total phase-spread of ψ, assumes a value,

$$E_S = \sqrt{2}\,\pi\, f_r(mM)\frac{1}{\psi}\int_{-\frac{\psi}{2}}^{\frac{\psi}{2}} \cos\theta\, d\theta$$

as was done in case of E_T, the transformer induced voltage in Eq. (5.4). This leads to the following equation:

$$E_S = \sqrt{2}\pi f_r\, \phi_m\, T_{ph}\, K_d \qquad \text{...(5.14)}$$

As before, if the coils are not full-pitched, a further generalization can be incorporated in the expression, so that it assumes a general form,

$$E_S = \sqrt{2}\,\pi\, f_r\, \phi_m\, T_{ph}\, K_d\, K_p \text{ volt (rms)} \qquad \text{...(5.15)}$$

Comparison of Eq. (5.5) derived for E_T (rms value) and Eq. (5.15) for E_S (rms value) shows that they are identical in magnitude. Any one consideration can, therefore develop the emf equation per phase for an ac machine with distributed-type armature windings.

5.2.3 Induced emf in a Commutator-type Distributed Winding, Using the Effect of Transformer-induced Voltage

We had observed in Section 4.2 that a *commutator-type of distributed winding* is physically different from its phase-type counterpart. The former is a *closed winding* arrangement, with the beginning of the winding connected to the end, such that it gives the impression of an *endless* winding, after it is completed. A *phase winding*, on the other hand, has distinct beginning and end, and hence it is also described as an *open-type winding*. A commutator winding invariably uses a commutator/brushes interface to establish contact between stationary brushes and the moving conductors, and at different instants of time, different conductors come in direct contact with brush probes. Placement of brushes on the commutator connected to the winding invariably leads to the formation of parallel paths. Also the symmetry associated with placement of brush ensures that between two adjacent brushes such a winding has an effective phase-spread of π electrical radians. Thus, for a $2p$-pole commutator winding, brushes divide the armature conductors into $2a$ parallel paths ($2a = 2p$, for simplex lap winding and $2a = 2$, for simplex wave windings). On the other hand, phase-type windings are either permanently connected with supply at fixed selected points (when such windings are located on the stationary member), or to slip-ring/brush mechanism. Further, inter-connection between different phase windings constituting a multi-phase system are often required to be made in order to achieve compatibility between the total terminals of a winding and the supply terminals. Figure 5.3 presents the schematic of a commutator-type winding arrangement with brushes, for a $2p$-pole configuration. To start with, the winding is assumed to be subjected to a pulsating magnetic field, with the axis of pulsation coinciding with the plane of the brushes. Further, the conductors constituting the winding, and also the magnetic field system, are assumed to be stationary in space. It is seen that each parallel path subtends a spread of angle π and contains $\left(\dfrac{T}{2a}\right)$ turns, where, T is the total number of turns per phase. Modification of the expression for transformer-induced voltage in phase type windings by appropriate substitution yields.

$$E_T = \sqrt{2}\,\pi\, f\, \phi_m \left(\frac{T}{2a}\right) K_p K_d \cos\sigma \text{ volt (rms)}$$

It may, however, be noted that for this type of winding, $\psi = \pi$, so that, $K_d = \left(\dfrac{2}{\pi}\right)$. Hence, the general expression for E_T in a commutator winding has the following form

$$E_T = \sqrt{2} f \phi_m \left(\frac{T}{a}\right) K_p \cos\sigma \text{ volt (rms)} \qquad \text{...(5.16)}$$

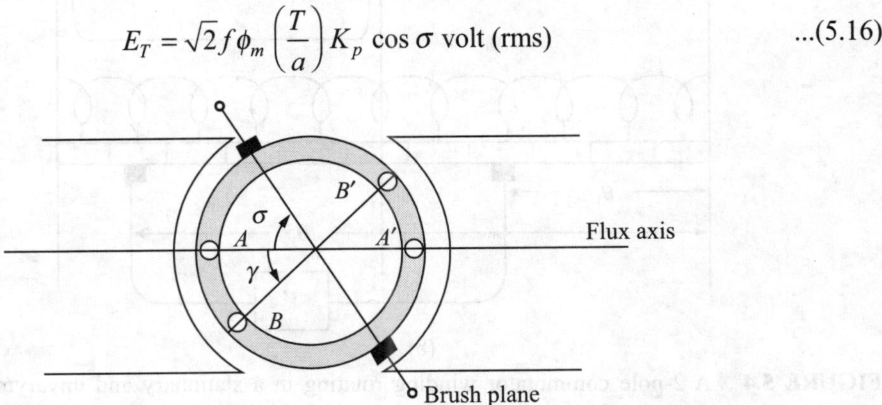

FIGURE 5.3 Schematic of a commutator-type winding with brushes for a 2-pole configuration.

5.2.4 Induced emf in a Commutator-type Distributed Winding using Effect of Speed-induced Voltage

As before, we shall consider a $2p$-pole commutator winding having a total number of turns, T, and rotating in an unvarying $2p$-pole distributed field system at a uniform speed of n_r revolutions per second. For generality, we shall assume that the brushes are stationary and are placed at electrical angles θ_1 and θ_2 respectively from the plane of the neutral magnetic axis (that is the plane perpendicular to the flux-axis) for each pole-pair. The arrangement is shown with respect to a 2-pole schematic in Figure 5.4. Also, shown in Figure 4.5(b), is the corresponding linearized (or unfurled) version of the scheme, complete with specially distributed commutators and brushes.

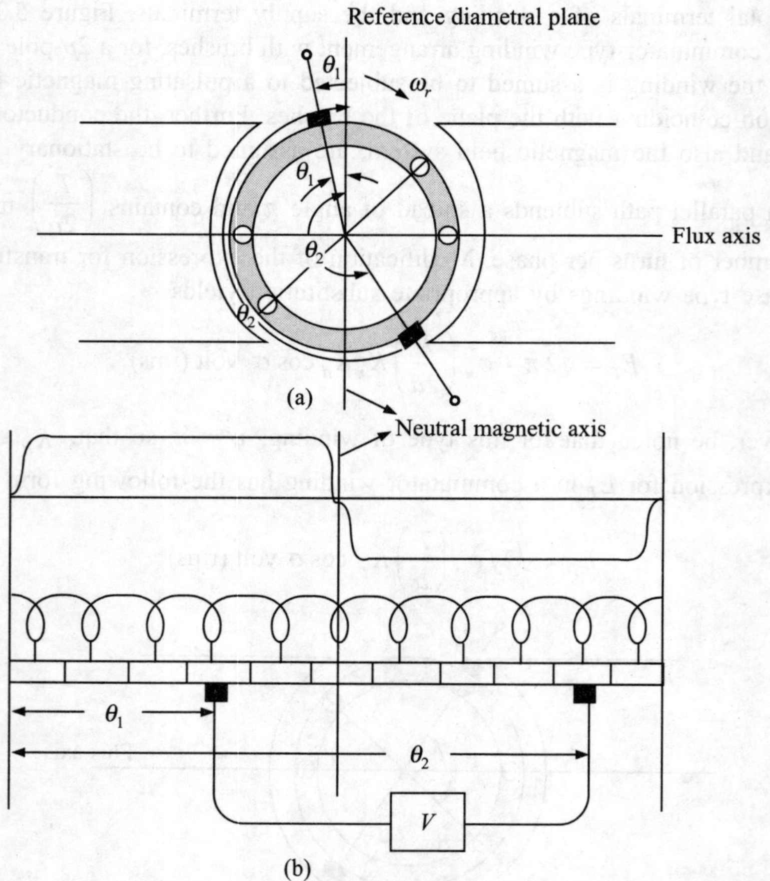

FIGURE 5.4 A 2-pole commutator winding rotating in a stationary and unvarying field distribution: (a) A 2-pole commutator winding rotating in 2-pole stationary and unvarying flux system, and (b) Linearized (or unfurled) schematic of 5.4(a) together with the flux-density distribution curve.

If a voltmeter is connected across the brushes as shown in Figure 5.4(b), it will record a voltage equating emfs of all coils between the brushes, that is,

$$e_{\text{brush}} = \sum_{\theta=\theta_1}^{\theta=\theta_2} e_{\text{coil}} \qquad \text{...(5.17)}$$

It can be seen that although the emf of a specific coil is a function of its instantaneous spatial position relative to the axis of the stationary flux, the emf of a coil in a given fixed position in space always remains the same. This is obvious because the coils themselves keep on changing due to rotation of the member housing the coils with respect to stationary brushes. Further, the net emf appearing across the brushes always remains constant (that is at zero frequency and constant amplitude). For a given position, the emf induced in a coil will be given as follows:

$$e_{\text{coil}} = 2B\,l\,v\,m \text{ volt}$$

where, B represents the value of the flux-density cut across by the coil in the spatial position in question, l is the effective length of the conductors constituting a coil, v is the velocity of flux-cutting and m is the number of turns making up a coil. B is, in fact, given by the ordinate of the flux density distribution curve corresponding, to the position of a coil-side in question as shown in Figure 5.4(b). If there are M coils between the brushes, there are M such ordinates to be considered, and the resulting emf between brushes as measured by the voltmeter v has the following value:

$$E_{\text{brush}} = 2\{l\,v\,(m\,M)\,(\text{Average height of ordinates of } B \text{ between } \theta_1 \text{ and } \theta_2)\}$$

The flux-density distribution curve, in general, may take a repetitive shape as indicated in Figure 5.4(b). But at this stage, if we consider the fundamental component of the waveform to simplify the determination of the average value of the flux-density ordinates between the limits of θ_1 and θ_2, this leads to the following equation:

$$E_{\text{brush}} = 2\,l\,v\,mM \frac{1}{\theta_2 - \theta_1} \int_{\theta=\theta_1}^{\theta=\theta_2} B_m \sin\theta \, d\theta \qquad \text{...(5.18)}$$

B_m and v in the above relation are known terminologies from the earlier analysis in connection with speed-induced voltage in phase windings. Their values can be re-recorded as follows:

$$B_m = \frac{p\phi_m}{Dl} \quad \text{and} \quad v = \pi D n_r$$

Again, if the winding has a total of C coils on the rotor, arranged in $2a$ parallel paths due to commutator and brush positioning, then $\left(\frac{C}{2a}\right)$ coils are effectively in series per parallel path, each path subtending a spread angle of π, as discussed earlier, so that,

$$\frac{M}{\left(\frac{C}{2a}\right)} = \frac{\theta_2 - \theta_1}{\pi}$$

Substitution of these values in Eq. (5.18) yields:

$$E_{\text{brush}} = \left(m\frac{C}{a}\right)(pn_r)\,\phi_m\,(\cos\theta_2 - \cos\theta_1)$$

$$= \left(\frac{T}{a}\right) f_r\phi_m\,(\cos\theta_2 - \cos\theta_1) \qquad \text{...(5.19)}$$

We shall, now, refer to the new brush-position diagram in Figure 5.5. This reveals the new general feature, where the axis of symmetry of the stationary brush-pair and the stationary flux-axis do not occupy the same spatial position. The angle of separation δ between these two axes, and the angle of separation, ν, between the brushes, given by, $\nu = \theta_2 - \theta_1$, can now be employed to represent the values of θ_1 and θ_2, as follows:

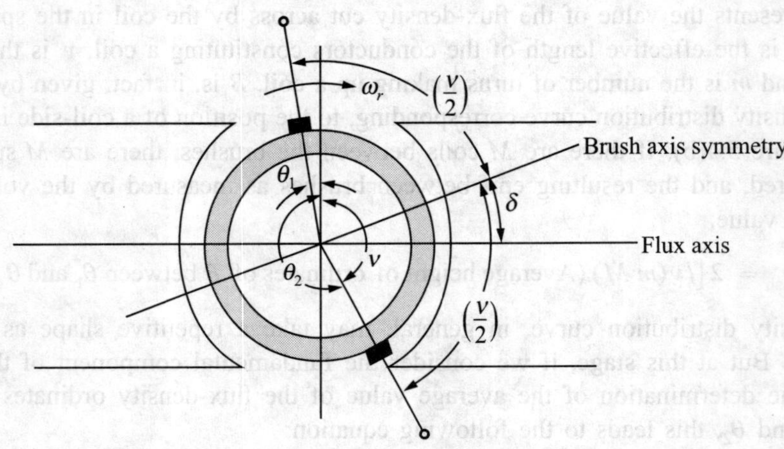

FIGURE 5.5 Brush position diagram [Figure 5.4(b)], revisited with additional details.

By inspection of Figure 5.5, we shall observe that

$$\theta_1 = 90° + \delta - \left(\frac{\nu}{2}\right) \qquad \text{...(5.20)}$$

and

$$90° + \theta_2 + \left(\frac{\nu}{2}\right) = (360° + \delta)$$

which can be easily shown in trigonometric equivalent as follows:

$$\theta_2 = 90° + \delta + \left(\frac{\nu}{2}\right) \qquad \text{...(5.21)}$$

Putting the above values of θ_1 and θ_2, and simplifying expression (5.19), the new expression for E_{brush}, is given as follows:

$$E_{\text{brush}} = 2\left(\frac{T}{a}\right) f_r\phi_m \sin\left(\frac{\nu}{2}\right)\cos\delta \qquad \text{...(5.22)}$$

A critical scrutiny of Eq. (5.22) should lead to the following observations and conclusions:

1. *The frequency of the emf induced in the conductors of the winding depends on the relative angular velocity with which the conductors cut across fluxes, and is given by* f_r.
2. *The magnitude of the output voltage available across brushes is a function of* f_r.
3. *The magnitude and nature of* E_{brush} *are governed by the value of* δ, *the angular separation between the brush axis and the flux axis. If both brush and flux are stationary in the space frame,* δ *has a fixed value. For a given system, all things remaining unchanged, the value of* E_{brush} *maximizes when* $\delta = 0$, *that is, when the spatial positions of the brush axis and the flux axis coincide, the value of the brush voltage being zero, for* ($\delta = 90°$ *electrical*).
4. *The angle of brush separation,* ν *likewise influences the value of* E_{brush} *through a factor of* $\left(\sin\dfrac{\nu}{2}\right)$. *When* $\nu = 180°$ *electrical, that is, when the brushes are so placed, that they are at the extremities of the parallel paths for a particular winding, maximum value of* E_{brush} *results.*
5. E_{brush}, *the value of emf across the brushes, has a constant amplitude when* ν *and* δ *are constant. The voltage is, therefore, at zero frequency for the condition considered although the frequency of the emf induced in the individual conductors of the windings is* f_r.

It is interesting to examine what happens if the flux is rotated with respect to the stationary brushes at a speed of n_s revolutions per second. Since the rotor is rotating at a speed of n_r rps with respect to the stator frames, the relative speed of flux-cutting by the conductors is changed to a new value of $(n_s \pm n_r)$ rps, the negative sign being considered when both rotate in the same direction, while the positive sign is taken when the two rotations are in opposite directions. The resulting frequency of the emf-induced in the conductors of the winding is given by $[p\,(n_s \mp n_r)]$ or by $(f_s \mp f_r)$, in addition to affecting their magnitude. But this is not all. In such a case, δ does not remain constant as the spatial position of the flux axis changes continuously. The potential difference appearing across the brushes also changes in magnitude in accordance with the changing value of δ. This voltage, which is a function of $\cos\delta$, is co-sinusoidal when plotted in time domain with reference to the time-axis, such that it is maximum in amplitude when $\delta = 0$, thus:

$$\text{Max}\, E_{\text{brush}} = 2\left(\frac{T}{a}\right)(f_s \mp f_r)\,\phi_m \sin\frac{\nu}{2} \qquad \text{...(5.23)}$$

Therefore, the rms value of potential difference between the brushes under this condition is given as follows:

$$E_{\text{brush}} = \sqrt{2}\left(\frac{T}{a}\right)(f_s \mp f_r)\,\phi_m \sin\left(\frac{\nu}{2}\right) \qquad \text{...(5.24)}$$

We may note that the frequency of the above voltage is decided by the angular frequency associated with the changing of δ. But $\delta = \omega t$ and $\omega = 2\pi f_s$, when time is measured from the instant $t = 0$, and when $\delta = 0$. Hence, frequency of E_{brush}, *appearing across brushes is* f_s.

It must be mentioned, even at the cost of repetition that frequency of the induced emf in the conductors of the winding is $(f_s \mp f_r)$, *but frequency of the voltage appearing across the*

brushes is f_s, when the flux-axis of the stator flux has an angular frequency of f_s. The frequency of the conductor-voltage depends on the relative speed of the flux with respect to conductors, while the frequency of the brush-voltage depends on the relative speed of the flux with respect to the brushes. This change in frequency is exclusively due to the use of the commutator brush arrangement. Thus, the commutator acts as a frequency changer in electrical machines.

5.2.5 Induced Voltages Represented as Functions of Inductance Voltages

In Section 3.7.5 we had seen that for a conservative system, the applied voltage at all times was required to balance the induced emfs. Hence, we can express voltages in terms of currents and circuit elements—both self- and mutual. Thus, writing down the equations for voltage induction, the total flux-linkage of a coil comprises the sum of self- and mutual flux-linkages. This results

Stator flux-linkage

$$\psi_s = L_s i_s + M_{sr}\, i_r \qquad \text{...(5.25)}$$

Rotor flux-linkage

$$\psi_r = L_r\, i_r + M_{sr}\, i_s \qquad \text{...(5.26)}$$

But, the instantaneous applied coil voltages are as follows:

$$\left.\begin{aligned} v_s &= i_s R_s + \frac{d}{dt}(\psi_s) \\ v_r &= i_r R_r + \frac{d}{dt}(\psi_r) \end{aligned}\right\} \qquad \text{...(5.27)}$$

If, for neatness, $\left(\dfrac{d}{dt}\right)$ is replaced by an operator p without compromising on its nature, we shall have

$$\left.\begin{aligned} v_s &= i_s R_s + p(L_s i_s + M_{sr} i_r) \\ v_r &= i_r R_r + p(M_{sr} i_s + L_r i_r) \end{aligned}\right\} \qquad \text{...(5.28)}$$

Since, the inductance terms are functions of θ_r and t, we have to expand the above equations to the following forms:

$$v_s = i_s R_s + L_s p i_s + i_s p L_s + M_{sr} p i_r + i_r p M_{sr}$$

and

$$v_r = i_r R_r + L_r p i_r + i_r p L_r + M_{sr} p i_s + i_s p M_{sr}$$

which can be further generalized and rearranged in the following forms:

$$\left.\begin{aligned} v_s &= i_s R_s + L_s p i_s + M_{sr} p i_r + \omega_r i_s \frac{dL_s}{d\theta_r} + \omega_r i_r \frac{dM_{sr}}{d\theta_r} \\ \text{and} \quad v_r &= i_r R_r + L_r p i_r + M_{sr} p i_s + \omega_r i_r \frac{dL_r}{d\theta_r} + \omega_r i_s \frac{dM_{sr}}{d\theta_r} \end{aligned}\right\} \qquad \text{...(5.29)}$$

It is very convenient to denote these pairs of equations jointly in a matrix-form, as follows:

$$\begin{bmatrix} v_s \\ v_r \end{bmatrix} = \left\{ \begin{bmatrix} R_s & 0 \\ 0 & R_r \end{bmatrix} + p \begin{bmatrix} L_s & M_{sr} \\ M_{sr} & L_r \end{bmatrix} + w_r \left(\frac{d}{d\theta_r} \right) \begin{bmatrix} L_s & M_{sr} \\ M_{sr} & L_r \end{bmatrix} \right\} \begin{bmatrix} i_s \\ i_r \end{bmatrix} \qquad ...(5.30)$$

The induced voltage terms in the above matrix equations are of two different types:

1. Those which are functions of time derivatives of currents and self- and mutual inductance voltages.
2. Those which are angular derivatives of inductance and are rotational voltages.

If, for representation purposes, we denote the matrix forms as follows:

$$\begin{bmatrix} R_s & 0 \\ 0 & R_r \end{bmatrix} = [R] \qquad ...(5.31)$$

$$\begin{bmatrix} L_s & M_{sr} \\ M_{sr} & L_r \end{bmatrix} = [L] \qquad ...(5.32)$$

and

$$\left(\frac{d}{d\theta_r} \right) \begin{bmatrix} L_s & M_{sr} \\ M_{sr} & L_r \end{bmatrix} = [G] \qquad ...(5.33)$$

and also

$$\begin{bmatrix} v_s \\ v_r \end{bmatrix} = [v]; \quad \begin{bmatrix} i_s \\ i_r \end{bmatrix} = [i] \qquad ...(5.34)$$

we can write Eq. (5.30) in its simple form:

$$[v] = \{[R] + p[i] + \omega_r[G]\}\,[i] \qquad ...(5.35)$$

But L_s, L_r and M_{sr} can be expressed in terms of the permeances as seen by the stator, rotor and the mutual flux, and $[G]$ takes the following general form:

$$[G] = \begin{bmatrix} N_s^2 \dfrac{d\Lambda_s}{d\theta_r} & N_s N_r \dfrac{d\Lambda_{sr}}{d\theta_r} \\ N_s N_r \dfrac{d\Lambda_{sr}}{d\theta_r} & N_r^2 \dfrac{d\Lambda_r}{d\theta_r} \end{bmatrix} \qquad ...(5.36)$$

and, we can denote

$$\left.\begin{aligned} L_s &= N_s^2 \Lambda_s \\ L_r &= N_r^2 \Lambda_r \\ M_{sr} &= N_s N_r \Lambda_{sr} \end{aligned}\right\} \qquad ...(5.37)$$

It is seen that the rotational voltage terms are the result of permeance variation during rotation of the rotor. The stator coils of a cylindrical rotor machine see a constant air gap/permeance, or $\dfrac{d\Lambda_s}{d\theta_r} = 0$. Similarly, the rotor coils rotating in the annular space of a smooth stator also see a constant air-gap permeance, that is, $\dfrac{d\Lambda_r}{d\theta_r} = 0$. But, for all stator and rotor configurations, the effective area of the air-gap between the stator and the rotor coil changes when rotated, and, therefore, Λ_{sr} is always a function of θ_r.

If we substitute the value of M_{sr} obtained from Section 4.5 in Eqs. (5.32), (5.33) and (5.34), we shall get the stator and rotor voltage equations.

Then, for zero-saliency, that is, with a uniform air-gap between the stator and rotor member (as in case of induction machines and non-salient pole synchronous machines):

L_s is of constant value, unvarying with θ_r

L_r is of constant value, unvarying with θ_r

$M_{sr} = M_m \cos \theta_r$ and

$$[V] = \left\{ [R] + p \begin{bmatrix} L_s & M_m \cos\theta_r \\ M_m \cos\theta_r & L_r \end{bmatrix} + \omega_r \begin{bmatrix} 0 & -M_m \sin\theta_r \\ -M_m \sin\theta_r & 0 \end{bmatrix} \right\} [i] \quad \text{...(5.38)}$$

And, for order of saliency of one, (that is, one of the members presents a smooth surface like in a dc machine or a salient-pole synchronous machine) say a cylindrical rotor and a salient-pole stator

L_s = constant in value, unvarying with θ_r

$L_r = L_{r1} + L_{r2} \cos 2\theta_r$

$M_{sr} = M_m \cos \theta_r$ and

$$[V] = \left\{ [R] + p \begin{bmatrix} L_s & M_m \cos\theta_r \\ M_m \cos\theta_r & L_{r1} + L_{r2} \cos 2\theta_r \end{bmatrix} + \omega_r \begin{bmatrix} 0 & -M_m \sin\theta_r \\ -M_m \sin\theta_r & -2L_{r2} \sin 2\theta_r \end{bmatrix} \right\} [i] \quad \text{...(5.39)}$$

Similarly, for machines with salient-pole rotor and smooth stator, we use Eq. (5.39), and interchange *s* and *r* suffices, as follows:

$$L_s = L_{s1} + L_{s2} \cos 2\theta_r$$

L_r is constant in value, unvarying with θ_r and

$$M_{sr} = M_m \cos \theta_r$$

so that

$$[V] = \left\{ [R] + p \begin{bmatrix} L_{s1} + L_{s2} \cos 2\theta_r & M_m \cos\theta_r \\ M_m \cos\theta_r & L_r \end{bmatrix} + \omega_r \begin{bmatrix} -2L_{s2} \sin 2\theta_r & -M_m \sin\theta_r \\ -M_m \sin\theta_r & 0 \end{bmatrix} \right\} [i] \quad \text{...(5.40)}$$

We shall not consider the double saliency cases, as such machine geometries are rare.

5.2.6 An Alternative Approach to Quantification of Torque of the Electro-magnetic Kind

Principle of energy conversion revisited

We shall devote this section to the study of the production of torque in electro-mechanical devices, as a part of the energy conversion process. We shall start with a basic device having a single coil P_sQ_s on the stator and a single coil, P_rQ_r on the rotor, both are full-pitched with magnetic linkage between them. A linearized version of the system is presented in Figure 5.6(a) with instantaneous position of the axis of the stator coil, spatially co-axial to that of the rotor. We shall assume that the stator coil is energized, so that a constant current flows in the coil-sides and results in a flux-distribution along the stator–rotor interface as shown. On consideration, that the stator and the rotor members present equi-potential surfaces on both sides of the air-gap and the magnetic permeance of the material constituting the stator and the rotor is infinitely large, a uniformly distributed field is developed in the air-gap. This is indicated in Figure 5.6(b). Since the principle of electro-mechanical energy conversion concerns itself with movement of the rotor coil with respect to the flux produced in the stator, we concentrate on the reaction of the rotor coil under the influence of the uniform stator field thus produced.

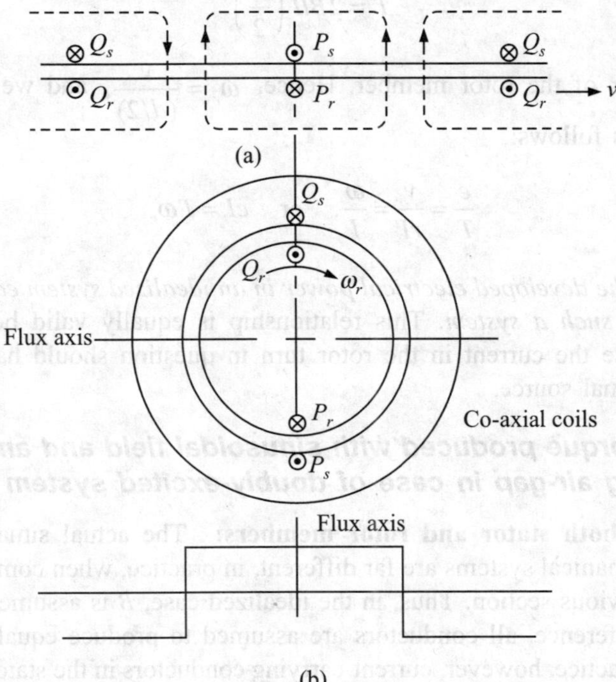

FIGURE 5.6 Linearized-version of a typical double-excited electro-mechanical system around air-gap: (a) Co-axial coils P_sQ_s and P_rQ_r placed on stator and rotor and (b) Field distribution developed by equi-potential surfaces in the air-gap.

Now, for the assumed reference at time $t = 0$, the total flux linked by the rotor coil is zero. Also, after the lapse of a short time-interval of δt, let the rotor coil move by a peripheral distance of δx at a constant velocity of v. If the flux-density distribution in space is constant and uniform as shown for the case under consideration, the movement leads to a change in the flux-linkage by the rotor coil $P_r\theta_r$ by an amount given by $\delta\phi = 2B\delta A$, where B stands for the flux-density and δA is the differential area through which the field penetrates. In this case, $\delta A = l\,dx$ and $\delta x = v\,\delta t$. Thus

$$\delta\phi = 2Blv\delta t$$

which is the same thing as syaing that:

$$|e| = \frac{\delta\phi}{\delta t} = 2Blv \text{ for each turn of the rotor coil.}$$

Also a constant current, I, flowing through the turns of the rotor-coil, P_rQ_r, interacts with B to produce a force on each conductor comprising the coil-sides, thus

$$F_{\text{conductor}} = BlI$$

and a torque per turn,

$$T = 2BlI\left(\frac{d}{2}\right)$$

d being the diameter of the rotor member. Hence, $\omega_r = \dfrac{v}{(d/2)}$, and we obtain the basic of energy conversion as follows:

$$\frac{e}{T} = \frac{v}{rI} = \frac{\omega}{I}, \quad \text{or} \quad eI = T\omega_r$$

which signifies that *the developed electrical power in an idealized system equals the mechanical power developed in such a system*. This relationship is equally valid both for generator or motor action, because the current in the rotor turn in question should have been induced or forced from an external source.

Expression for torque produced with sinusoidal field and ampere–conductor distribution along air-gap in case of doubly-excited system

With excitation in both stator and rotor members: The actual situation encountered in practical electro-mechanical systems are far different, in practice, when compared with idealized conditions of the previous section. Thus, in the idealized case, B is assumed to have a constant magnitude and by inference, all conductors are assumed to produce equal torque in the same direction. In actual practice, however, current-carrying conductors in the stator and rotor members of a generally doubly-excited system jointly contribute to the production of magnetic field in the air-gap. This *combined magnetic field* reacts with the ampere–conductor distribution along the rotor surface to produce desired energy conversion, through development of torque. It is more realistic to assume a sinusoidal magnetic field distribution both in stator and rotor members with a space and time difference between these two fields, because in majority of the devices, the windings mounted on the stator and the rotor are intentionally designed to yield, as nearly as possible, such spatial distributions. The resulting field distribution along the air-gap leads to development of a sinusoidal flux-density distribution under idealized conditions. Besides the armature winding is also distributed along the periphery of the rotor to yield, as far as possible, a sinusoidal ampere–conductor distribution in space.

If M_s and M_r respectively denote the instantaneous space distribution of mmfs in the stator and the rotor, as function of space angle θ with an arbitrary space reference frame, then:

$$M_s = M_{s\text{Max}} \cos\theta \quad \text{and} \quad M_r = M_{r\text{Max}} \cos(\theta - \alpha) \qquad ...(5.41)$$

where, α denotes the spatial angular divergence between the two mmfs, then

$$M_{sr} = M_s + M_r = M_{s\text{Max}} \cos\theta + M_{r\text{Max}} \cos(\theta - \alpha) \quad ...(5.42)$$

It is often convenient to express this in the convention complexor form as follows:

$$M_s = M_{s\text{Max}}\, e^{j\theta} \quad \text{and} \quad M_r = M_{r\text{Max}}\, e^{j(\theta-\alpha)} \quad ...(5.43)$$

so that, M_{sr} assumes the following value:

$$M_{sr} = \text{Real part of } [M_{s\text{Max}} + M_{r\text{Max}}\, e^{-j\alpha}]\, e^{j\theta} \quad ...(5.44)$$

M_{sr} can be expressed in normal sinusoidal form as follows:

$$M_{sr} = M_{sr\text{Max}} \cos(\theta - \alpha) \quad ...(5.45)$$

indicating that the resulting field distribution is also sinusoidal in nature, and is displaced from and lags behind M_s by an angle δ (which is less than α). The distribution is indicated in Figure 5.7(a), along with the corresponding pictorial representation in phasor form in Figure 5.7(b). From Figure 5.7(b), it is readily seen that the amplitude $M_{sr\text{Max}}$ is given by the following relation

$$M_{sr\text{Max}} = \sqrt{(M_{s\text{Max}})^2 + (M_{r\text{Max}})^2 + 2M_{s\text{Max}} M_{r\text{Max}} \cos\alpha} \quad ...(5.46)$$

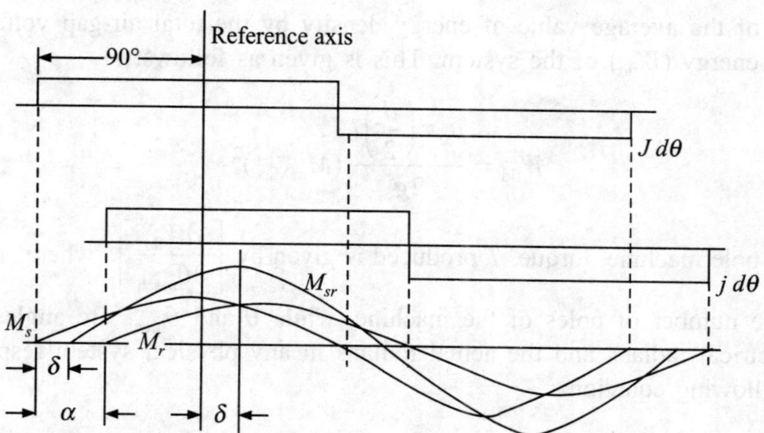

FIGURE 5.7(a) Pictorial depiction of M_s, M_r and M_{sr} in terms of their relative positions.

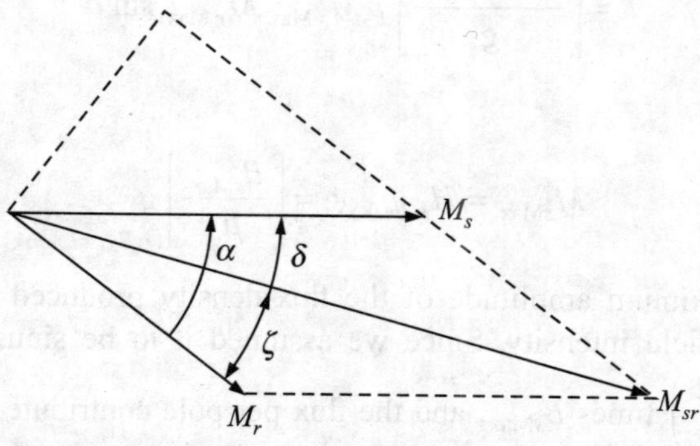

FIGURE 5.7(b) Depiction by phasor diagram.

Ampere's law gives a relationship between the mmfs and the magnetic field intensity. In a simplified form, it is expressed as

$$Hg = M_{sr}$$

for typical electro-mechanical devices under assumptions that stator and rotor cores present unsaturated equi-potential surfaces on both sides of the air-gap with very high magnetic permeability when compared to that of the air-gap. In the forgoing relation, H stands for the magnetic field intensity and g is the effective length of the air-gap along the flux-path in the combined magnetic circuit of the device. Hence,

$$M_{sr\text{Max}} = H_{\text{Peak}}\, g \qquad \text{...(5.47)}$$

and, as the magnetic field energy stored in the air-gap is given by $\left[\frac{\mu_0}{2}\right] H^2$, where, $\mu_0 = 4\pi \times 10^{-7}$ is the permeability of air or vacuum in MKS units, the average value of the square of a sine wave of maximum amplitude A is given as follows:

$$\left(\frac{\mu_0}{2}\right)(H_{\text{av}})^2 = \left(\frac{\mu_0}{2}\right)\left[\frac{H_{\text{peak}}^2}{2}\right] = \left(\frac{\mu_0}{4}\right)\left[\frac{M_{sr\text{Max}}}{g}\right]^2 \qquad \text{...(5.48)}$$

Multiplication of the average value of energy density by the total air-gap volume yields the total magnetic energy (W_{fld}) of the system. This is given as follows:

$$W_{\text{fld}} = \frac{\mu_0 \pi \left(\frac{d}{2}\right) l}{2g} (M_{sr\text{Max}})^2 \qquad \text{...(5.49)}$$

Now, for a $2p$-pole machine, torque, T produced is given by $\left[\frac{\partial W_{\text{fld}}}{\partial \theta_{\text{mech}}}\right]$, where, $\theta_{\text{mech}} = \frac{\theta}{p}$ and $2p$ signifies the number of poles of the machine, while θ and θ_{mech} are angles measured in equivalent electrical radians and the actual radians in any physical system respectively. This leads to the following equation:

$$T = \left[\frac{\mu_0 \pi \left(\frac{d}{2}\right) l}{g}\right] p M_{s\,\text{Max}}\, M_{r\,\text{Max}} \sin\delta \qquad \text{...(5.50)}$$

But, in this expression,

$$M_{s\text{Max}} = H_{s(\text{Peak})}\, g = \left[\frac{B_{s\text{Max}}}{\mu_0}\right] g \qquad \text{...(5.51)}$$

where, $B_{s\text{Max}}$ is the maximum amplitude of the flux-density produced due to the peak value of the stator magnetic field intensity. Since we assumed it to be sinusoidally distributed, its average value, B_{av} is $\left(\frac{2}{\pi}\right)$ times $B_{s\text{Max}}$, and the flux per pole contributed by the stator member is given as follows

$$\phi_s = \left(\frac{2}{\pi}\right)\left[\frac{\pi\left(\frac{d}{2}\right)l}{p}\right]B_{s\text{Max}} = \left[\frac{dl}{p}\right]B_{s\text{Max}} \qquad \text{...(5.52)}$$

Hence, $B_{s\text{Max}} = \dfrac{p\phi_s}{dl}$, and $M_{s\text{Max}} = \left[\dfrac{pg}{\mu_0 dl}\right]\phi_s$ ...(5.53)

The modified expression for T is, therefore, as follows:

$$T = \left(\frac{\pi}{2}\right)p^2\,\phi_s\,M_{r\text{Max}}\sin\alpha \qquad \text{...(5.54)}$$

Figure 5.7(b) reveals that:

$$M_{s\text{Max}}\sin\alpha = M_{sr\text{Max}}\sin\zeta$$

where, $\zeta = (\alpha - \delta)$. Hence, taking the combined contribution of the ampere–conductor distribution in the stator and the rotor, in production of the net air-gap mmf, the equation for torque produced is given as follows:

$$T = \left(\frac{\pi}{2}\right)(p)^2\,\phi_{sr}\,M_{r\text{Max}}\sin(\alpha - \delta)\text{ in Nm} \qquad \text{...(5.55)}$$

A critical scrutiny of Eq. (5.55) clearly indicates that three conditions must be met for production of torque in doubly-excited system, when both stator and rotor members carry excitations in their electrical circuits. These are the following:

1. ***There must be a magnetic field distribution effective along the air-gap surface.***
2. ***This field distribution must be able to interact with an mmf distribution produced in the rotor member by virtue of appropriate ampere–conductor distribution effective along the air-gap surface. The ampere–conductor distribution leads to the development of rotor mmf, and consequently the rotor flux-density along the air-gap surface; but the differential current-length element l di, contributing to the production of ampere–conductor distribution (J, where, di = Jdθ) has an orientation in quadrature with the resulting magnetic field intensity, and the flux-density it develops.***
3. ***Finally, there must exist a space displacement between the air-gap flux-distribution and the rotor ampere–conductor distribution.***

Expression for torque produced in multiply-excited systems, with excitation provided in one member only, the other being shorted

The case of induction torque: In the study of electro-mechanical devices, it is not uncommon to encounter devices (and instruments) in which torque is produced by induction due to interactive effect of two or more fluxes originated in one member only–either in the stator or in the rotor. In the following section, we shall consider the interactive effect of two sets of fluxes, time and space, displaced from one another, with a conducting drum strategically mounted so as to react under the influence of these fluxes.

Figure 5.8(a) shows the schematic of a system with two coils mounted on polar projections spaced 90° apart on the stator and a drum made up of conducting material, mounted co-axially with respect to the stator member. The coils serve as exciting windings for the device and are fed from alternating current sources producing alternating flux along respective axis of projection. We identify these projections as *R* and *L* as shown in Figure 5.8(a). Since we assumed that the exciting coils on the projections are fed from sources which have time displacement between them, the respective currents *i* and i_L flowing through them exhibit time-displacements between themselves. In the phasor diagram, shown in Figure 5.8(b), i_L is indicated as lagging behind i_R by a time angle θ.

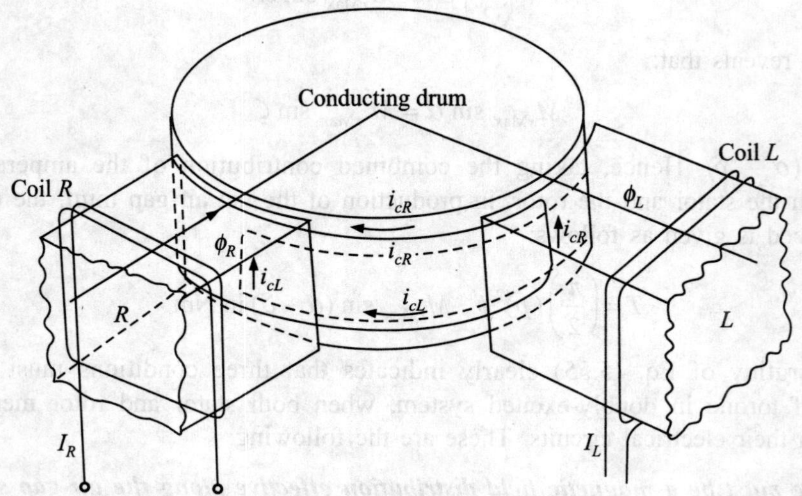

FIGURE 5.8(a) Schematic of a system with two coils mounted on polar projections spaced 90° apart on stator and a drum mounted co-axially with respect to the stationary member.

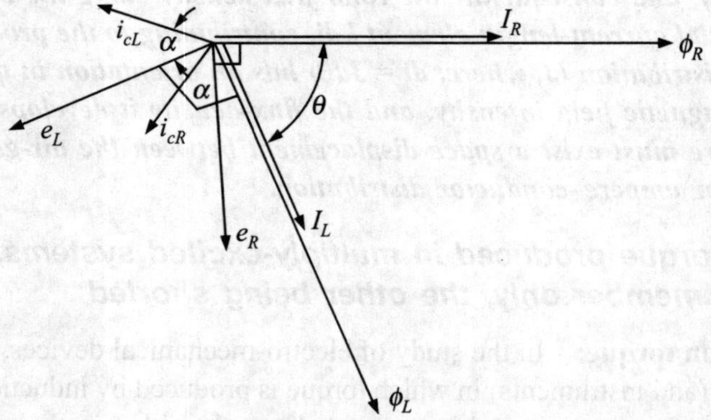

FIGURE 5.8(b) Phasor diagram of the system.

The conducting drum individually sees each alternating flux ϕ_R and ϕ_L produced by respective alternating currents, i_R and i_L. In the phasor diagram, these fluxes have been shown is time phase with respective currents.

Each flux produces voltages on the rotor. These voltages e_R and e_L have space-phase, but time-quadrature relationship with respective fluxes as indicated. Since the drum is continuous and conducting, each induced emf produces circulating currents i_{cR} and i_{cL} respectively. The paths of these circulating currents can be assumed as indicated. Further, such circulating currents lag in time domain behind respective induced emfs, depending on the inductances offered to currents by the nature of the circulating current paths. A review of Figure 5.8(a) shows that i_{cL} will interact with ϕ_R and i_{cR} with ϕ_L, to produce forces and the resulting torque is due to the difference between the interactive forces contributed by $\phi_L i_{cR}$ and $\phi_R i_{cL}$. With reference to the phasor diagram in Figure 5.8(b), the mean value of the net torque is given as follows:

$$T = K\,[\phi_L\, i_{cR} \cos(90° + \alpha - \theta) - \phi_R\, i_{cL} \cos(90° + \alpha + \theta)] \qquad \text{...(5.56)}$$

where, K is a constant accounting for the radius of the drum and magnetic properties of the arrangement governing the flux current interaction along with the physical dimensions of the system. The above expression may be further simplified by substituting the following:

$$i_{cR} = \frac{e_R}{Z_C};\ i_{cL} = \frac{e_L}{Z_C};\ e_R \propto f\phi_R \quad \text{and} \quad e_L \propto f\phi_L \qquad \text{...(5.57)}$$

This yields the new form:

$$T = K'\left[\frac{f\phi_R\phi_L}{Z_C}\right][\sin(\alpha + \theta) - \sin(\alpha - \theta)]$$

$$= \left[\frac{2K'\phi_R\phi_L}{Z_C}\right]\cos\alpha \sin\theta \qquad \text{...(5.58)}$$

Z_C signifies impedence offered by drum to circulating currents i_{cR} and i_{cL}.

In the above expression K' is a new constant. Since ϕ_R and ϕ_L are proportional to i_R and i_L respectively, T has a final form:

$$T \propto f\left[\frac{i_R i_L}{Z_C}\right]\cos\alpha \sin\theta \qquad \text{...(5.59)}$$

A review of the above expression reveals that:

1. ***T is increased when θ, the time displacement between i_R and i_L, is increased, and it is the highest when $\theta = 90°$.***
2. ***There is a need for at least two pace displaced fluxes ϕ_R and ϕ_L for producing a torque. It was assumed in the analysis that ϕ_R and ϕ_L were displaced by 90° in space (that was how i_{cL} came under the influence of ϕ_R and i_{cR} was under the influence of ϕ_L). Ideally, the space-displacement should be 90° for best results.***

By inference, induction torque can be produced only when at least two sets of alternating fluxes, which are time and space displaced from one another, interact jointly on a system. This torque maximizes when both space and time displacements are 90°. Of these, the space displacements are of paramount importance.

ILLUSTRATION 5.1 A 2-pole commutator machine has two windings on the stator *d*-axis and a brush pair on the *q*-axis. Write down the general equations for the coil-voltage and the machine torque. Hence, show that a cumulatively compounded dc motor has a steady-state armature current,

$$i_a = \frac{V}{(r_a + r_s) + \omega_r M_{as}}\left[1 - \frac{\omega_r M_{af}}{r_f}\right]$$

suffices a, f and s refer to the armature, shunt and series field respectively, V is the supply voltage and ω_r is the angular velocity.

Solution: The arrangement is shown in Figure 5.9.

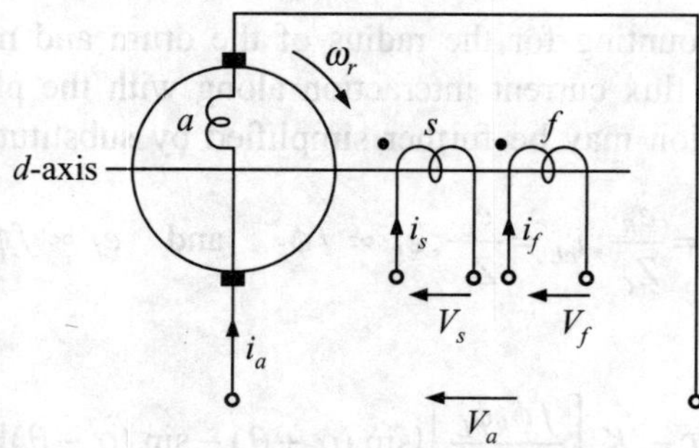

FIGURE 5.9 Schematic of a 2-pole commutator machine with two windings on the stator *d*-axis and a brush pair along *q*-axis.

The general voltage–current equations, governing the operation of such a configuration are as follows:

$$\begin{bmatrix} V_f \\ V_s \\ V_a \end{bmatrix} = \begin{bmatrix} (r_f + L_f p) & M_{fs} p & 0 \\ M_{sf} p & (r_s + L_s p) & 0 \\ \omega_r M_{fa} & \omega_r M_{sa} & r_a + L_a p \end{bmatrix} \begin{bmatrix} i_f \\ i_s \\ i_a \end{bmatrix}$$

Introducing inter-connection relationships and reducing circuit parameters, we have to note the following boundary conditions:

- For cumulatively compounded motors,

$$i_s = +i_a$$
$$v = v_a + v_s \text{ and } v = v_f$$

- Assuming reciprocity in magnetic circuits,

$$M_{sf} = M_{fs}$$

Hence,

$$\begin{bmatrix} V_f \\ V_s \\ V_a \end{bmatrix} = \begin{bmatrix} (r_f + L_f p) & M_{fs} p & 0 \\ M_{sf} p & (r_s + L_s p) & 0 \\ \omega_r M_{fa} & \omega_r M_{sa} & (r_a + L_a p) \end{bmatrix} \begin{bmatrix} i_f \\ i_a \\ i_a \end{bmatrix}$$

or

$$\begin{bmatrix} v_f \\ v \end{bmatrix} = \begin{bmatrix} (r_f + L_f p) & M_{fs} p \\ M_{fs} p + \omega_r M_{fa} & (r_s + L_s p) + (r_a + L_a p) + \omega_r M_{sa} \end{bmatrix} \begin{bmatrix} i_f \\ i_a \end{bmatrix}$$

For steady-state operation, all terms associated with p vanish.

$$\therefore \quad \begin{bmatrix} V \\ V \end{bmatrix} = \begin{bmatrix} r_f & 0 \\ \omega_r M_{fa} & (r_a + r_s) + \omega_r M_{sa} \end{bmatrix} \begin{bmatrix} I_f \\ I_a \end{bmatrix} \qquad ...(5.60)$$

From the first part of Eq. (5.60)

$$i_f = \frac{V}{r_f}$$

which enables us to write, using the second part of Eq. (5.60)

$$V = \omega_r \, M_{fa} \frac{V}{r_f} + \{(r_a + r_s) + \omega_r M_{sa}\} \, i_a$$

or

$$V\left\{1 - \frac{\omega_r M_{fa}}{r_f}\right\} = \{(r_a + r_c) + \omega_r \, M_{sa}\} i_a$$

Hence,

$$i_a = \frac{V}{(r_a + r_s) + \omega_r M_{sa}} \left\{1 - \frac{\omega_r M_{fa}}{r_f}\right\}$$

Thus proved.

ILLUSTRATION 5.2 A 2-pole series dc machine is connected to a steady supply voltage V. Express the voltage and torque equations for the machine in general terms. Plot the steady-state speed and torque curves against motoring current in the range of 1.0 to 10.0 ampere. The supply voltage is 500 volt, the mutual inductance between the armature and the series field is 10 henry, and resistance of the armature and series field is 1.0 ohm.

Solution: The primitive representation is shown in Figure 5.10, from which the general equations are indicated as follows:

$$\begin{bmatrix} v_s \\ v_a \end{bmatrix} = \begin{bmatrix} (r_s + L_s p) & 0 \\ \omega_r M_{sa} & (r_a + L_a p) \end{bmatrix} \begin{bmatrix} i_s \\ i_a \end{bmatrix}$$

For a series machine under motoring mode,

$$i_s = i_a$$

and

$$v = v_s + v_a$$

Hence,

$$v = \{(r_s + L_s p) + (r_a + L_a p) + \omega_r M_{sa}\} i_a \qquad ...(5.61)$$

and torque T is given by:

$$T = (\text{Pole pairs}) \; M_{sa} i_a^2 \text{ Nm} \qquad ...(5.62)$$

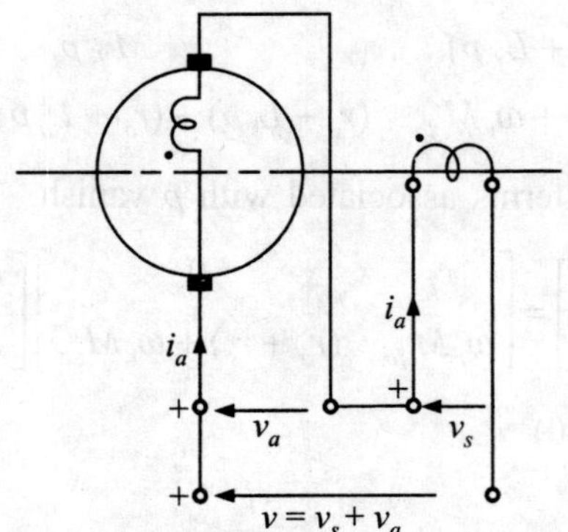

FIGURE 5.10 'Primitive' representation of a 2-pole dc series machine working as a motor.

From Eq. (5.61), under steady state condition, we have

$$i_a = \frac{v}{(r_a + r_s) + \omega_r M_{sa}}$$

which may be substituted in Eq. (5.62) to yield:

$$T = \frac{v^2 M_{sa}}{[(r_a + r_s) + \omega_r \; M_{sa}]^2}$$

or

$$\omega_r^2 M_{sa}^2 + 2\omega_r M_{sa}(r_a + r_s) + (r_a + r_s)^2 = \frac{v^2 M_{sa}}{T}$$

$$\Rightarrow \quad \omega_r^2 + 2\omega_r \frac{(r_a + r_s)}{M_{sa}} + \left\{ \frac{(r_a + r_s)^2}{M_{sa}^2} - \left[\frac{V^2}{M_{sa}} \right] \frac{1}{T} \right\} = 0$$

The roots of the above equation can be found out by the following formula:

$$\omega_r = \frac{-b \pm \sqrt{b^2 - 4ac}}{2a}$$

where,

$$b = \frac{2(r_a + r_c)}{M_{sa}}$$

and

$$c = \frac{(r_a + r_s)^2}{M_{sa}^2} - \left(\frac{V^2}{M_{sa}} \right) \frac{1}{T}$$

or

$$\omega_r = \frac{\dfrac{-2(r_a + r_s)}{M_{sa}} \pm \sqrt{4\left\{ \dfrac{r_a + r_s}{M_{sa}} \right\}^2 - 4\left(\dfrac{r_a + r_s}{M_{sa}} \right)^2 + \dfrac{4V^2}{M_{sa}T}}}{2}$$

$$= -\left(\frac{r_a + r_s}{M_{sa}} \right) \pm \frac{V}{\sqrt{M_{sa}T}}$$

or ω_r is of the form:

$$\omega_r = -A \pm \frac{V}{B} \times \frac{1}{\sqrt{T}}$$

where,
$$A = \frac{r_a + r_s}{M_{sa}} \quad \text{and} \quad B = \sqrt{M_{sa}}$$

Since armature and series field resistances are small compared to mutual inductance between armature and field, ω_r is proportional to $\frac{1}{\sqrt{T}}$, when the applied voltage is constant.

ILLUSTRATION 5.3 A 4-pole, separately excited dc generator driven at 2000 rpm, has the following parameters:

R_f = 100 ohm; L_f = 60 henry; M_{af} = 0.16 henry; R_a = 0.1 ohm; L_a = 0.5 mH.

Suffices a and f refer to the armature and field respectively.

(a) Determine the output voltage/load current characteristic for a field current of 2 amp. and a load current in the range of zero to 200 amp.

(b) Sketch the rise of armature voltage 200 volt is suddenly applied to the field winding with the armature on open-circuit.

(c) Repeat (b) with the armature initially short-circuited.

Solution: The schematic of the machine is shown in Figure 5.11.

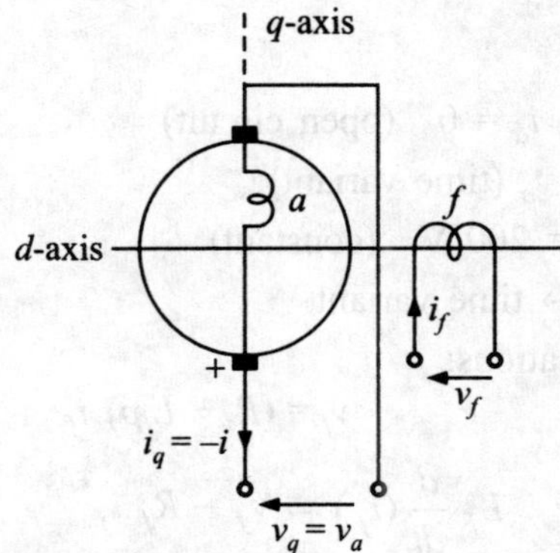

FIGURE 5.11 Schematic of a separately excited dc generator in representative form.

The basic equations for the given configuration are:

$$\begin{bmatrix} v_f \\ v_q \end{bmatrix} = \begin{bmatrix} (r_f + L_f p) & 0 \\ \omega_r M_{af} & (r_a + L_{ap}) \end{bmatrix} \begin{bmatrix} i_f \\ i_q \end{bmatrix}$$

(a) The given terminal condition yields,

$$p = 2, \quad v_q = v_a \quad \text{and} \quad i_q = -i_a$$

and, for steady-state condition, all p terms vanish, i.e.,

$$\begin{bmatrix} V_f \\ V_a \end{bmatrix} = \begin{bmatrix} r_f & 0 \\ \omega_r M_{af} & r_a \end{bmatrix} \begin{bmatrix} I_f \\ -I_a \end{bmatrix}$$

or
$$V_a = \omega_r M_{af} I_f - I_a r_a$$

Now,
$$\omega_r = \frac{2\pi}{60} N_r \left(\frac{2p}{2}\right) = \frac{2\pi p}{60} \times 2000$$
$$= 209.3\ p \text{ rad/sec}$$

And with $M_{af} = 0.16$ henry, $I_f = 2$ amp and $2p = 4$,

$$V_a = 133.95 - 0.1\ I_a$$

(b) *Solving the problem with differential equational approach:*

Step function input voltage $V_{t=0}$ 200 volt, has been applied across the field terminals. Hence, the boundary conditions are:

At $t < 0^-$:

$$\left.\begin{array}{l} \text{Armature current, } i_a = 0 \\ \text{Open circuit voltage, } V_a = 0 \\ \text{Field voltage (open circuited), } V_f = 0 \\ \text{field current, } i_f = 0 \end{array}\right\}$$

Also, at $t \geq 0^+$:

Armature current, $i_a = 0$ (open circuit)
Terminal voltage v_a (time variant)
Field voltage, $v_f = 200$ V (constant)
Field current, $i_f \rightarrow$ time variant
From voltage equations:

$$v_f = (R_f + L_f p)\ i_f$$

or
$$L_f \frac{d}{dt}(i_f) = v_f - R_f\, i_f$$

The solution of this first order linear differentiation equation is of the general form:

$$i_f = e^{-\left(\frac{R_f}{L_f}\right)t} + \frac{V_f}{R_f}$$

where, K is the constant of integration.
Now, at $t = 0$, $i_f = 0$

or
$$0 = K + \frac{v_f}{R_f} \quad \text{or} \quad K = -\frac{V_f}{R_f}$$

Hence, $$i_f = \left(1 - e^{-\frac{t}{\tau_f}}\right)\frac{V_f}{R_f}$$

where, $\tau_f = \dfrac{L_f}{R_f}$, the field time constant.

Also, $\dfrac{V_f}{R_f} = \dfrac{200}{100} = 2$ amp and $\tau_f = \dfrac{60}{100} = 0.6$ sec.

Hence, $$i_f = 2(1 - e^{-\frac{t}{0.6}})\ \text{amp}$$

The open circuit armature voltage, for $i_a = 0$ is as follows:

$$\begin{aligned} v_a(t) &= \omega_r M_{af} i_f(t) + (R_a + L_a p)\, i_a \\ &= \omega_r\, M_{af} i_f(t) \\ &= \omega_r\, M_{af} \cdot 2(1 - e^{-\frac{t}{0.6}})\ \text{volt} \end{aligned}$$

Since $$N = 2000\ \text{rpm}$$

$$\omega_m = \frac{2\pi \times 2000}{60} = 209.3\ \text{rad/sec}$$

$\therefore$ $$\omega_r = (\text{pole pair})\ \omega_m = 2 \times 209.3 = 418.6\ \text{rad/sec}$$

and $$V_a(t) = 418.6 \times 0.16 \times 2(1 - e^{-\frac{t}{0.6}})$$

Hence, $$V_a(t) = 133.95\,(1 - e^{-\left(\frac{t}{0.6}\right)})\ \text{volt}$$

Solving the problem by Laplace transform approach:

Using the same equations in differential form and introducing boundary conditions:

$$\begin{bmatrix} V_f ⌐ \\ 0 \end{bmatrix} = \begin{bmatrix} (R_f + L_f p) & 0 \\ \omega_r M_{af} & (R_a + L_a p) \end{bmatrix} \begin{bmatrix} i_f(t) \\ i_a(t) \end{bmatrix}$$

Under transformation to *s*-plane:

$$\begin{bmatrix} V_f/s \\ 0 \end{bmatrix} = \begin{bmatrix} (R_f + L_f s) & 0 \\ \omega_r M_{af} & R_a + L_a s \end{bmatrix} \begin{bmatrix} i_f(s) \\ i_a(a) \end{bmatrix} - \begin{bmatrix} L_f & 0 \\ 0 & L_a \end{bmatrix} \begin{bmatrix} i_f \\ i_a \end{bmatrix}$$

But at $t \geq 0$,

$$v_f = V_f = 200\ ⌐\text{volt}$$

$i_f = 0$; $V_a = 0$; $i_a = 0$

(Inductive current) (No excitation) (Open circuit)

$$\therefore \quad \begin{bmatrix} V_f/s \\ V_a(s) \end{bmatrix} = \begin{bmatrix} R_f + L_s s & 0 \\ \omega_r M_{af} & R_a + L_a s \end{bmatrix} \begin{bmatrix} i_f(s) \\ i_a(s) \end{bmatrix}$$

$$\tilde{v}_a(s) = \omega_r M_{af} i_f = \frac{\omega_r M_{af} v_f / R_f}{s(1+\tau_1 s)}$$

$$= \frac{\omega_r M_{af} V_f}{R_f} \left[\frac{1}{\tau_f s \left(\dfrac{1}{\tau_1} + s \right)} \right]$$

Obtaining the roots by partial fraction:

$$\frac{1}{\tau_f s \left(\dfrac{1}{\tau_f} + s \right)} = \frac{K_1}{\tau_f \left(\dfrac{1}{\tau_f} + s \right)} + \frac{K_2}{\tau_f s}$$

where,

$$K_1 = \left. \frac{1}{\tau_f \left(\dfrac{1}{\tau_f} \right)} \right]_{\text{at } s=0} = 1 \quad \text{and} \quad K_2 = \left. \frac{1}{\tau_f (s)} \right]_{\text{at } s = -\frac{1}{\tau_f}} = -1$$

By inverse transform:

$$v(t) = \frac{\omega_r M_{af} V_f}{R_f} \left[1 - e^{-\frac{t}{\tau_f}} \right]$$

or $\qquad v(t) = 133.95 \left[1 - e^{\frac{t}{0.6}} \right]$ volt

(c) From the basic voltage equation:

$$v_a(t) = \frac{v_f \omega_r M_{af}}{(R_f + L_f p)} + (R_a + L_a p)\, i_a$$

$$\therefore \quad i_a(t) = \frac{1}{(R_a + L_a p)} \left[v_a(t) - \frac{v_f \omega_r M_{af}}{(R_f + L_f p)} \right]$$

And, when armature is short-circuited [that is $V_a(t) = 0$]:

$$i_a(t) = -\frac{V_f \omega_r M_{af}}{(R_a + L_a p)(R_f + L_f p)}$$

But, $i_a(t)$ itself is negative by convention as it is generator current.

Our boundary conditions are as follows:

at $t \geq 0$,
$$V_f(t) = V\lrcorner = 200 \text{ volt}$$
$$V_a(t) = 0 \quad \text{(shorted armature)}$$
$$i_f \to 0 \quad \text{(time variant)}$$
$$i_a \to 0, \quad \text{(time variant)}$$

Now, by Laplace transformation:

$$\begin{bmatrix} V_f/s \\ 0 \end{bmatrix} = \begin{bmatrix} R_f + L_f s & 0 \\ \omega_r M_{af} & R_a + L_a s \end{bmatrix} \begin{bmatrix} \tilde{i}_f(s) \\ \tilde{i}_a(s) \end{bmatrix} - \begin{bmatrix} L_f & \\ & L_a \end{bmatrix} \begin{bmatrix} i_f = 0 \\ i_a = 0 \end{bmatrix}$$

By solving:

$$\tilde{i}_f(s) = \frac{V_f}{s(R_f + L_f s)}$$

and
$$\tilde{i}_a(s) = \frac{1}{(R_a + L_a s)} \times \frac{\omega_r M_{af} V_f}{s(R_f + L_f s)}$$

$$= \frac{\omega_r M_{af} V_f}{R_f R_a} \left[\frac{1}{s(1 + \tau_a s)(1 + \tau_f s)} \right]$$

where,
$$\tau_a = \frac{L_a}{R_a} \quad \text{and} \quad \tau_f = \frac{L_f}{R_f}$$

Applying partial fractions, we have

$$\frac{K_1}{s} + \frac{K_2}{(1 + \tau_f s)} + \frac{K_3}{(1 + \tau_a s)} = \frac{1}{s(1 + \tau_f s)(1 + \tau_a s)}$$

$$K_1 = \left. \frac{1}{\tau_f \tau_s \left(s + \frac{1}{\tau_f} \right) \left(s + \frac{1}{\tau_a} \right)} \right]_{\text{at } s=0} = 1$$

$$K_2 = \left. \frac{1}{\tau_a s \left(s + \frac{1}{\tau_a} \right)} \right]_{s = \left(-\frac{1}{\tau_f} \right)} = -\tau_f \left[\frac{\tau_f}{\tau_f - \tau_a} \right]$$

and
$$K_3 = \left. \frac{1}{\tau_f s \left(s + \frac{1}{\tau_f} \right)} \right]_{s = -\frac{1}{\tau_a}} = -\tau_a \left[\frac{\tau_a}{\tau_a - \tau_f} \right]$$

Applying inverse transformation:

$$i_a(t) = \frac{\omega_r M_{af} V_f}{R_a R_f}\left[\frac{\tau_a \tau_f}{\tau_a \tau_f}\left\{1 - \frac{\tau_f}{\tau_f - \tau_a} e^{-\frac{t}{\tau_f}} - \frac{\tau_a}{\tau_a - \tau_f} e^{-\frac{t}{\tau_a}}\right\}\right]$$

or $$i_a(t) = \frac{\omega_r M_{af} V_f}{R_a R_f}\left\{1 - \frac{\tau_f}{\tau_f - \tau_a} e^{-\frac{t}{\tau_f}} - \frac{\tau_a}{\tau_f - \tau_a} e^{-\frac{t}{\tau_a}}\right\}$$

But $$\tau_a = \frac{L_a}{R_a} = \frac{0.5 \times 10^{-3}}{0.1} = 5 \text{ m-sec}$$

and $$\tau_f = \frac{L_f}{R_f} = \frac{60}{100} = 0.6 \text{ sec}$$

Neglecting the effect of τ_a as it is very small, the approximate solution of $i_a(t)$ is as follows:

$$i_a(t) = \frac{\omega_r M_{af} V_f}{R_a R_f}\left\{1 - e^{-\frac{t}{\tau_f}}\right\} \text{ amp}$$

OBJECTIVE TYPE QUESTIONS

5.1 Armature conductors of dc machines generate:
(a) constant dc voltage
(b) alternating voltage
(c) rectangular waveform of voltage
(d) a voltage waveform that corresponds exactly to the flux-distribution waveform

5.2 A rotating commutator is used for
(a) transforming dc in conductor to ac in the external circuit
(b) transforming ac in conductor to dc in external circuit
(c) providing connection from rotating armature to stationary external circuit
(d) performing both function listed in (b) and (c)

A 12-pole dc machine has a lap winding runs at 1000 rpm. It has 288 conductors on the armature, a flux per pole of 0.12 Wb, and carries a total current of 500 amp. It has a brush drop of 2 volt and an armature resistance of 0.05 ohm. Answer the following questions for the machine:

5.3 The generated emf is:
(a) 388 V (b) 576 V
(c) 91.67 V (d) 3456 V

5.4 Torque developed by the machine is:
(a) 2750 N-m (b) 1375 N-m
(c) 16500 N-m (d) None of the above

5.5 When working as a motor the applied voltage should be:

(a) 549 V (b) 603 V
(c) 315 V (d) 261 V

5.6 The current in each parallel path in armature is:

(a) 6000 amp (b) 250 amp
(c) 41.67 amp (d) 83.3 amp

5.7 When functioning as a generator, the terminal voltage is:

(a) 549 V (b) 603 V
(c) 315 V (d) 261 V

5.8 When working as a motor, the terminal voltage is reduced to 400 volt, field flux is lowered to 0.1 wb and the armature current is still at 500 amp. The new speed of the machine is:

(a) same as before (b) 600 rpm
(c) 776 rpm (d) 1200 rpm

Answers

5.1 (d) **5.2** (d) **5.3** (b) **5.4** (a) **5.5** (b) **5.6** (c)
5.7 (a) **5.8** (c)

NUMERICAL PROBLEMS

5.1 A 1000 kW dc generator develops 500 volt at 300 rpm and has 10 poles, with a lap winding. The average flux-density over a pole-pitch is 0.7 tesla, with armature diameter of 1.45 m and core length of 0.35 m. Estimate

(a) flux per pole (b) number of conductors in armature (c) the torque developed

[**Ans.** (a) ϕ_p = 0.1116 Wb; (b) Z = 8.96; (c) $10^5/\pi$ N-m]

5.2 A separately excited dc generator develops a no-load voltage of 250 volt at the rated speed. The values of the resistance parameters are as follows:
The armature resistance, r_a = 0.05 ohm; inter-pole resistance, r_i = 0.05 ohm.
The machine produces voltage drop of 2.0 volts at full-load of 50 amp at the brush contacts, and the brush contact drops may be assumed to be proportional to the load current.

At full-load of 50 amp, the armature reactance is observed to weaken the main-axis flux by 10 per cent, while the demagnetizing effect of armature reaction may be taken as proportional to the armature current at other intermediate loads.
Determine the terminal voltage at current intervals of 10 amp and plot the external characteristics up to 120 per cent of full-load current.

5.3 A shunt-type dc generator has the following specifications:

I_f (in amp)	0.0	0.05	0.1	0.2	0.4	0.6	0.8	0.9	1.0	1.2	1.5
E_g (in volt)	20	24	35	70	140	210	253	265	275	290	300

(a) Estimate the values of the **critical resistance** and the value of the total field reseistnce needed to generate the rated voltage of 240 volt at 1200 rpm.
(b) If the value of the field resistance is maintained as in (a) above, and the machine is now driven at a reduced speed of 1000 rpm, will the machine voltage build up? Give justifiable reasons for your conclusions.
(c) Determine the reduced value of the field resistance to develop 240 volt at no-load when driven at the reduced speed of 1000 rpm using the open-circuit characteristics.

[**Ans.** (a) Critical resistance at 1200 rpm, $r_{f\,1200}$ = 350 ohm; r_f corresponding to 240 volts at 1200 rpm = 333 ohm;
(b) Critical resistance at 1000 rpm, $r_{f\,1000}$ = 291.07 ohm. The machine will not be able to build up with field resistance of 333 ohm
(c) r_f = 205.13 ohms; r_f = 291.67 ohm]

5.4 The dc machine in the previous exercise is, now, operated as a separately excited dc generator. The armature resistance, r_a = 0.2 ohm and its inter-polar resistance, r_i = 0.2 ohm. A brush contact drop of 2.0 volt occurs at its rated current. If the armature supplies a current of 20 amp at a terminal voltage of 240 volt, the field current required is 0.9 amp.

Estimate the demagnetizing effect of armature reaction in terms of the effective shunt field current and the effective shunt field ampere-turns, if the number of turns in the shunt field is 500 per pole .

[**Ans.** Effective reduction of shunt field current due to armature reaction, I_f = 0.125 amp Demagnetizing ampere-turns in shunt field equivalent = 62.5 amp-turns per pole. This is 13.89 per cent of the main field ampere-turns]

5.5 A dc series motor operates at a terminal voltage of 240 volt. If the effective armature circuit resistance, which comprises the resistance of the armature, r_a, the inter-polar resistnace, r_i, the series field resistance, r_{ae}, and also the effect of brush contact drop, is altogether 1.0 ohm, estimate the speed and torque developed, if the armature current is reduced to 10.0 amp.

It is given that the flux per pole is 60 per cent of that at 10.0 amps when compared with the flux per pole at 20 amp in the series field.

[**Ans.** Speed at 10 amps, N_{10} = 2090.91 rpm; Developed torque, T_{d10} = 10.5 N-m.
Speed at 20 amps, N_{20} = 1200 rpm; Developed torque, T_{d20} = 35.014 N-m.
The two torques are in the ratio of T_{d10}/T_{d20} = 30 per cent
Power ratings, P_{d10} = 2.3 kW and P_{d20} = 4.4 kW]

CHAPTER 6

Three-phase Systems and Their Transformations

6.1 THREE-PHASE SYSTEMS WITH DISTRIBUTED WINDINGS—GENERAL

In this chapter, we concentrate on consolidation of relevant information relating the 3-phase systems, which we are likely to encounter in our attempt in modelling of electro-mechanical systems.

Almost all the electric power in the commercial domain is generated and transmitted in the form of 3-phase voltage systems today. The balanced 3-phase system of voltages is developed by generating three single-phase voltages by a common rotating flux field in three electrical winding arrangements, which are electrically isolated and are identical in all respects, excepting for their spatial separation of effectively $\left(\frac{2\pi}{3}\right)$ electrical radians between two adjacent phases.

A typical 3-phase system of windings, connected in star, is shown in Figure 6.1. The winding could also be connected in delta to form the desired 3-phase connection.

We could write down the voltage/current relations for the circuit by introducing the concept of inductances offered by the voltage windings. It takes the general shape as follows:

$$v_a = (R_a i_a) + \frac{d}{dt}(L_a i_a) + \frac{d}{dt}(M_{ba} i_b) + \frac{d}{dt}(M_{ca} i_c)$$

$$v_b = \frac{d}{dt}(M_{ab} i_a) + (R_b i_b) + \frac{d}{dt}(L_b i_b) + \frac{d}{dt}(M_{cb} i_c) \qquad \text{...(6.1)}$$

$$v_c = \frac{d}{dt}(M_{ac} i_a) + \frac{d}{dt}(M_{cb} i_b) + R_c i_c + \frac{d}{dt}(L_c i_c)$$

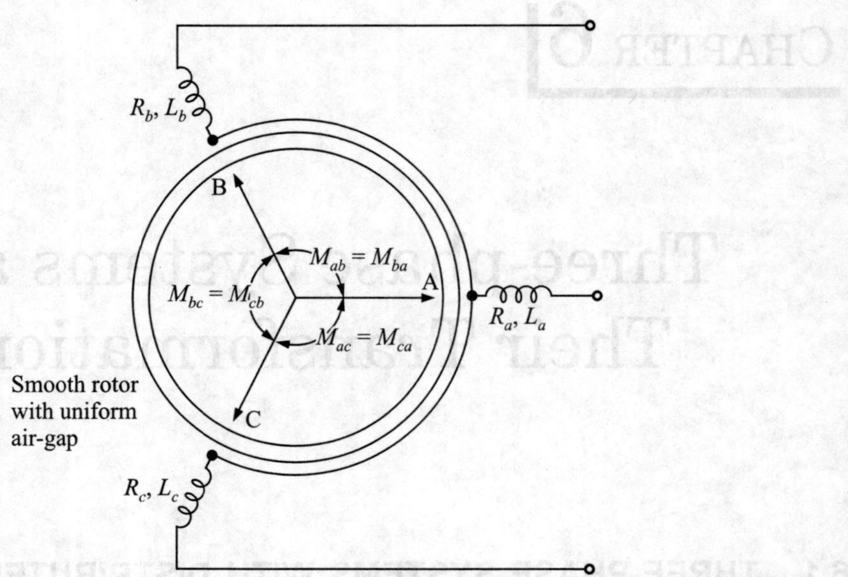

FIGURE 6.1 A balanced 3-phase supply system feeding a set of 3-phase machine windings, having smooth air-gap.

We would encounter such a winding arrangement or its equivalent in balanced 3-phase reactor, primary and the secondary of a 3-phase transformer, slip-ring induction motor, armature winding of a smooth air-gap, non-salient pole synchronous machine, or even in the stator winding of a poly-phase commutator motor—to identify a few of the standard machine configurations. Under balanced conditions, such a winding exhibits a resistance R, a coefficient of self-inductance L—the same value in each of the constituent phases, as well as the same value of the coefficient of mutual inductance between any two phases because of the spatial symmetry, M. The voltage–current relationship of the balanced 3-phase winding, therefore, simplifies to the following form:

$$\begin{bmatrix} v_a \\ v_b \\ v_c \end{bmatrix} = \begin{bmatrix} (R+Lp) & Mp & Mp \\ Mp & (R+Lp) & Mp \\ Mp & Mp & (R+Lp) \end{bmatrix} \begin{bmatrix} i_a \\ i_b \\ i_c \end{bmatrix} \quad \text{...(6.2)}$$

In the above expression, the separation of the quantity $\left(\dfrac{di}{dt}\right)$ into apparent product of Heaviside's differential parameter p and i components is artificial, and the time varying inductance components L and M work with i integrally, but are used to formalize the appearance of the matrices, in order to enable separation of the impedance terms and for executing inversion of matrices. This practice is adopted in this book as a standard practice. Thus, whenever the differential operator p appears in an impedance matrix, it should be evident that p operates on all entities associated with the concerned terms that appear to the right of p in the impedance matrix itself.

In practice, it is commonly accepted that with ac machines the voltages and currents can be taken as sinusoidal. We use complex numbers and phasor concepts to represent operation of such machines under steady-state condition. Under such conditions, pi is replaced by $j\omega i$, and the inductance terms are replaced by the corresponding reactance values, $\omega L = X$, and $\omega M = X_m$. Thus, the apparent restriction introduced with p form of the voltage equations in Eq. (6.2) associated with the impedance matrix automatically disappears with the voltage equations. It assumes a further simplicity and it can be written down with a labelled impedance matrix as follows:

$$\begin{bmatrix} v_a \\ v_b \\ v_c \end{bmatrix} = \begin{bmatrix} (R+jX) & jX_m & jX_m \\ jX_m & (R+jX) & jX_m \\ jX_m & jX_m & (R+jX) \end{bmatrix} \begin{bmatrix} i_a \\ i_b \\ i_c \end{bmatrix} \quad \text{...(6.3)}$$

The above 3 by 3 impedance matrix is open for inversion. It is a practice to simplify the process of inversion by introduction of suitable transformation methodologies that we shall take up subsequently.

6.2 TRANSFORMATIONS EMPLOYED IN GENERALIZED THEORY OF ELECTRO-MECHANICAL ENERGY CONVERSION

One of the basic aspects of modelling of electrical machines is the concept of transformations. The idea and need for transformation are not new or typical of generalized theory of electrical machines alone. Its usage is common to all branches of applied mathematics—from the simple case of change of axis in co-ordinate geometry, to the concept of logarithms and Laplace transforms, to symmetrical components transformation for solving problems of unbalance in electrical systems and power system, to direct- and quadrature-axis transformation in the analysis of salient-pole synchronous machines. Vector transformation plays a steller role in both classical and relativistic electro-magnetism—to indicate a few instances.

Transformations are used to simplify problems and in arriving at solutions with relative ease. In unified theory of machines, there are broadly two types of transformations used—the active and the passive—for arriving at a strategy of reconnecting the individual circuits constituting a system into simpler inter-connection, thereby, leading to a more simplistic approach or for simplification and reduction of problem to less laborious terms for solution. These strategies were initially proposed by W.G. Gibbs in 1752, in connection with Tensor Analysis of machines, for bringing electrical machines of different configurations under a common umbrella.

Before we embark into the detailed exercise associated with the two types of transformations—the active and the passive—we intend to spend a little more time in enumerating the main purpose of poly-phase transformations. The basic premise of transformation applied to poly-phase systems is to show that the stationary and rotating three-phase windings commonly in vogue in poly-phase electrical machines can be represented by a primitive machine with a 2-axis representation along stationary d- and q-axis. With this achieved, the $[L]$ and $[G]$ matrices can be formed by inspection to provide typical voltage and torque relations for 3-phase ac machines and dc machines in general.

In forming an equivalent *d-q* axis representation, identified as *winding transformation*, the necessary condition is that the space-distributed mmfs of the original 3-phase version and its transform are identical at all instants of time. In order to establish this equivalence of mmfs, it is a common practice to assume that the phase mmf's of the 3-phase winding are the fundamental crest values F_a, F_b and F_c at their respective phasor axes. Resolving F_a, F_b and F_c along two stationary (or rotating) orthogonal axes, *d* and *q*, and aggregating these resolved parts yield the crest values of F_d and F_q. These quantities in combination describe identically the resultant mmf distribution of the original 3-phase system. When the equivalence of mmfs established between the *d-q* and the 3-phase systems, the original mmf can be considered to originate from the equivalent *d-q* coils appropriately excited by equivalent i_d, i_q currents in these coils.

Now, as *d-q* axis coil arrangement is a 2-phase winding, this transformation leads to a change from a stationary or rotating 3-phase winding to a stationary 2-phase winding equivalent and vice versa. This transformation between 3-phase and 2-phase differs depending on whether the winding is stationary or moving, as in one case, the resolved angles are time invariant and in the other case, time dependent. The strategy adopted in the formation of both transformation is to establish a transformation between a 3-phase (*a-b-c*) and a 2-phase (α-β) winding, in which the phase separation between α-β axes and the *a-b-c* axes is maintained constant, irrespective of rotation of the windings at the same speed.

If the 3-phase (*a-b-c*) winding is stationary, the α-β and the *d-q* axes are assumed coincident and overlapping, but with rotation, a second transformation is necessitated between the rotating 2-phase α-β axes and the stationary 2-phase *d-q* windings.

The above elaboration is primarily related to equivalence of mmfs.

We recognise that in order to obtain the voltage and torque relations in terms of *d-q* axes entities, transformations are needed to be developed for equivalent voltage, current and impedances. These are, therefore, necessary exercises to be undertaken in subsequent sections.

But we can go back a little, and consider the active transformation first.

6.2.1 Active Transformation [C]

In the beginning of Section 6.2, it is mentioned that introduction of transformation to the modelling of electrical machines literally uses two distinct types of transformation—the active and the passive—with distinct purposes:

1. To achieve a strategy to reconnect a general set of circuits formed from the primitive unified system, with the ultimate aim of obtaining more simplistic approach.
2. to simplify problems to achieve solutions of problems with relative ease.

The active transformation involves a change in connections between two and more independent electrical circuits to form an inter-connected network.

In any network, the voltage sources can always be modified in such a way that the individual branches may be shorted or separated without affecting the current or power.

In the Generalized Machine Theory, the 'Active type of Transformation' is always used in the *reversed direction*, which signifies that the actual new network is always built up by suitably connecting the elements of the independent old circuits. [*C*] is the *connection* or *transformation matrix* whose appearance is a characteristic of Generalized Machine Theory. [*C*] always has the number of rows as there are circuits in the old system, and the same number of columns as there are number of meshes in the new system after transformation. Its elements are all either 0, +1 or –1. The connection matrix of the above transformation has more rows than columns. It is always rectangular and never square. Such matrices are comparatively rare and are identified as *singular* as they cannot have any inverse. *The physical significance of singularity is that the cyclic currents cannot be expressed in terms of the primitive currents.* That is, the converse relationship exists,

or $$[I] = [C]\,[I'] \qquad ...(6.4)$$

But, $$[I'] \neq [C]^{-1}\,[I]$$

as $[C]^{-1}$ is not mathematically tenable. The voltage relationship can also be shown as represented in the following matrix form:

$$[V'] = [C_t^*]\,[V] \qquad ...(6.5)$$

The above relationships in Eqs. (6.4) and (6.5) are said to be contragradient in the matrix parlance. The derivation of the current and voltage relationships has arisen naturally. While it was possible only to express old currents $[I]$ in terms of new currents $[I']$, the converse relationship held for the voltages and it was only possible to express the new voltages $[V']$ in terms of the old voltages $[V]$. But the basic laws of transformation matrix are closely associated in that the same transformation matrix $[C]$, transposed in the second case $[C_t]$, appears in the voltage relations. It can be shown that for invariant power, $P = P'$, for the old case and the transformed case, when the current relationship

$$[I] = [C]\,[I']$$

is developed, the voltage relationship

$$[V'] = [C_t^*][V]$$

follows automatically.

The correlation is most conveniently remembered in the starred form as follows:
V dashed matrix equals C transposed starred V matrix
The voltage relationships shall be is as follows:

$$[V'] = [C_t^*][V]$$
$$= [C_t^*][Z][I]$$
$$= [C_t^*][Z][C][I']$$

and since matrix multiplication is associative, therefore,

$$[V'] = [Z']\,[I']$$

We have:

$$[Z'] = [C_t^*][Z][C] \qquad ...(6.6)$$

Thus, the three relationships of prime importance in active transformation are as follows:

$$\left.\begin{aligned}[I] &= [C]\,[I']\\ [V'] &= [C_t^*][V]\\ [Z'] &= [C_t^*][Z][C]\end{aligned}\right\} \quad ...(6.7)$$

The purpose of active transformation in unified machines theory is that, when the circuit of any electrical machine is separated into primitive circuits, then the circuits of all electrical machines assume the same form and their primitive voltage and torque equations have the same form too, the only distinction manifests itself and appears:

1. In the manner in which the various windings are inter-connected to each other
2. In the algebraic form and the numerical values of their parameters
3. In the type of voltage source to which the machine is connected

An active transformation may be used to inter-connect complex networks. Transformations in cascade can be presented by a single transformation, which is the product of individual transformations, that is:

$$\left.\begin{aligned}&[Z'] = [C_{at}^*][Z][C_a]\\ \text{and}\quad &[Z''] = [C_{bt}^*][Z'][C_b]\\ \text{then,}\quad &[Z''] = [C_{bt}^*][C_{at}^*][Z][C_a][C_b]\end{aligned}\right\} \quad ...(6.8)$$

Two transformations in cascade equal a single transformation which is equal to the product of the individual transformations, since,

$$[C_{bt}^*][C_{at}^*] = [C_t]^* \quad \text{and} \quad [C_a][C_b] = [C]$$

Application of active transformation can be illustrated by taking up a suitable dc circuit chosen at random.

Illustrative example   The circuit is shown in Figure 6.2(a).

The resistance and voltage matrices can be written down by inspection. Applying Maxwell's cyclic current rule for writing down the voltage/current equations of the chosen circuit, by visual inspection, we have:

$$[R] = \begin{bmatrix} 7 & -1 \\ -1 & 8 \end{bmatrix}; \quad [V] = \begin{bmatrix} 8 \\ 6 \end{bmatrix}$$

so that,

$$[I] = \begin{bmatrix} I_1 \\ I_2 \end{bmatrix} = [R][V] = \begin{bmatrix} 1.5 \\ 2.5 \end{bmatrix}$$

The resulting branch–current distribution is detailed in Figure 6.2(b).

We note that the three voltage distributions and the two cyclic voltages correspond to the three original equations and two unknowns. We also note that the voltage in the middle branch of the network may be eliminated as shown in Figure 6.2(c). We can workout the general case, keeping the signs in mind, as illustrated in Figure 6.2(d). The resulting voltage matrix remains

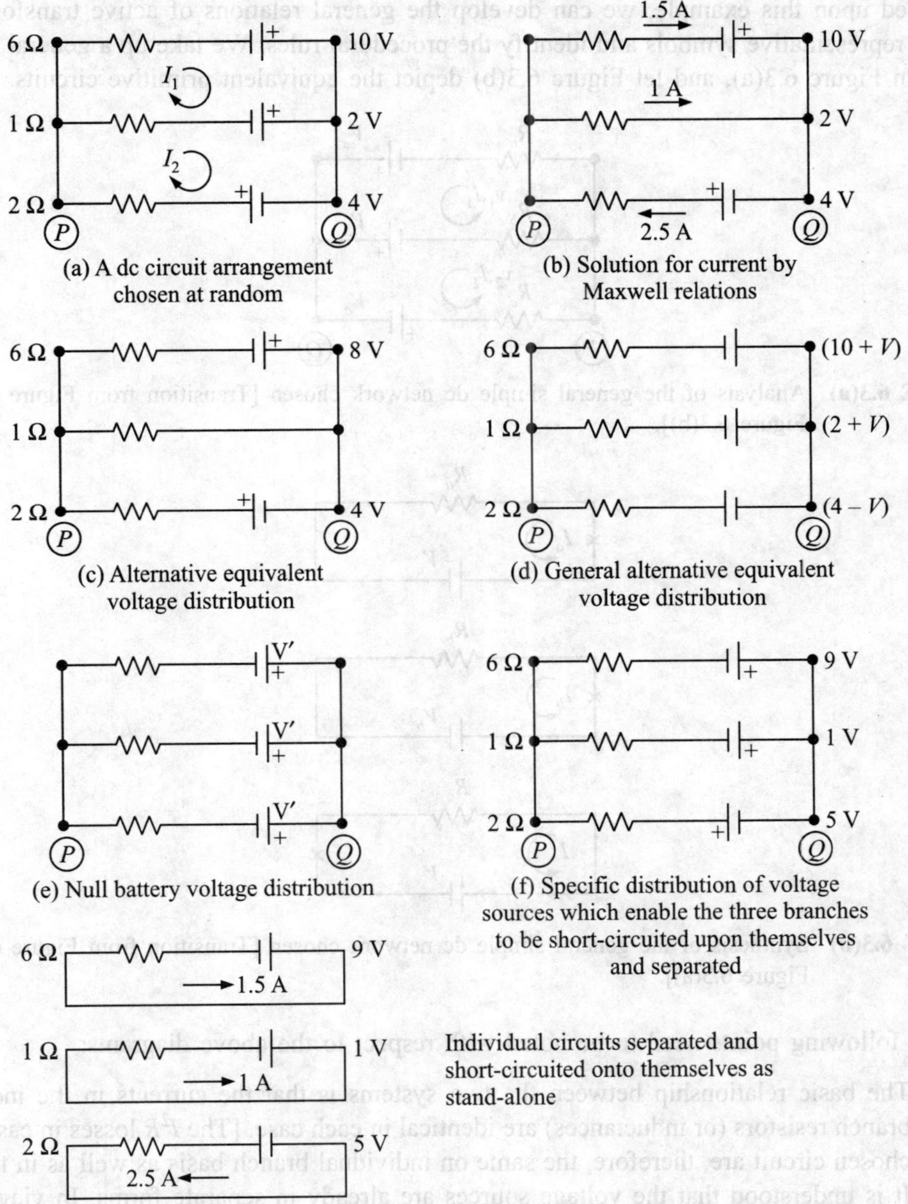

(a) A dc circuit arrangement chosen at random

(b) Solution for current by Maxwell relations

(c) Alternative equivalent voltage distribution

(d) General alternative equivalent voltage distribution

(e) Null battery voltage distribution

(f) Specific distribution of voltage sources which enable the three branches to be short-circuited upon themselves and separated

FIGURE 6.2 A simple dc circuit chosen at random to illustrate the application of active transformation.

the same as in Figure 6.2(a). Finally, we could arrive at the condition of null battery voltage distribution such that the potential difference across each branch is the same. This is shown by Figure 6.2(e). Figure 6.2(f) indicates a particular case with voltage $V = -1$ volt. The special property of this case is that the drop across each resistance branch equals the corresponding voltage source. The connecting bars P and Q in all the circuits of Figure 6.2 can now be replaced by short-circuiting individual branches and making each one of them self-sufficient independent circuits, without effecting any change in the branch currents.

Based upon this example, we can develop the general relations of active transformation. We use representative symbols and identify the procedural rules. We take up a general network shown in Figure 6.3(a), and let Figure 6.3(b) depict the equivalent primitive circuits.

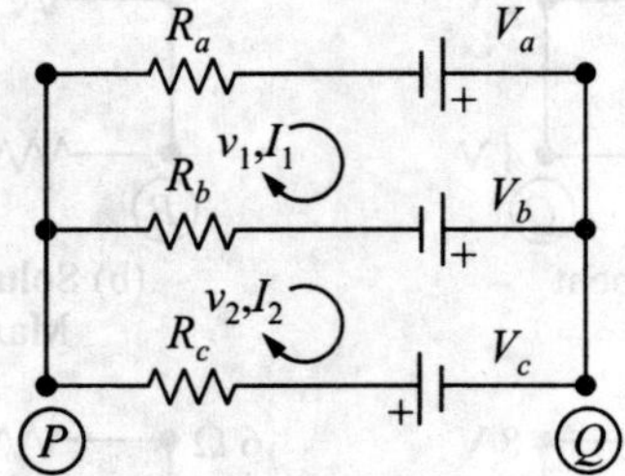

FIGURE 6.3(a) Analysis of the general simple dc network chosen [Transition from Figure 6.3(a) to Figure 6.3(b)].

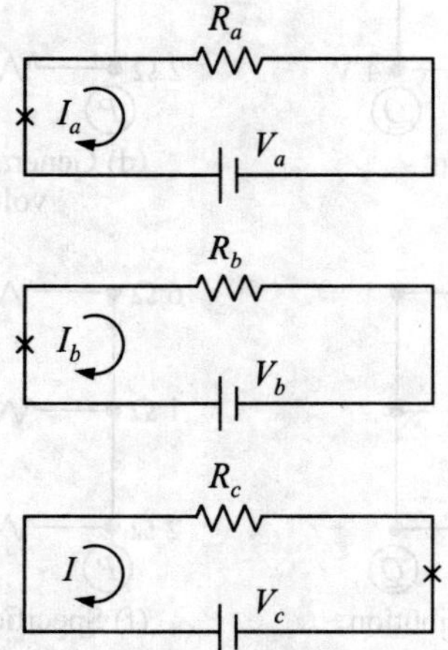

FIGURE 6.3(b) Synthesis of the general simple dc network chosen [Transition from Figure 6.3(b) to Figure 6.3(a)].

The following points need to be noted with respect to the above diagrams:

1. The basic relationship between the two systems is that the currents in the individual branch resistors (or inductances) are identical in each case. [The I^2R losses in case of the chosen circuit are, therefore, the same on individual branch basis as well as in totality].
2. It is understood that the voltage sources are already in separate forms. In view of the cluster of individual circuits of Figure 6.3(b), being quite independent, if required, their performance can be fully represented by a single matrix voltage equation of the form:

$$[V] = [Z]\,[I]$$

as given below:

$$\begin{bmatrix} V_a \\ V_b \\ V_c \end{bmatrix} = \begin{bmatrix} R_a & 0 & 0 \\ 0 & R_b & 0 \\ 0 & 0 & R_c \end{bmatrix} \begin{bmatrix} I_a \\ I_b \\ I_c \end{bmatrix} \qquad \text{...(6.9)}$$

Thus, whenever the resistance (or inductance) matrix consists of terms along the leading diagonal (or main diagonal only from left to right) only, then it denotes a number of electrically independent circuits. If we use primes—shown with dashes on top—to distinguish quantities in the new network of Figure 6.3(a), we may write a single matrix relationship of the form:

$$[V'] = [Z'] [I']$$

We can, now, write the elements of the individual matrices by inspection of the circuit in Figure 6.3(a), where, V_1 and V_2 are the cyclic voltages. Thus:

$$[V'] = \begin{bmatrix} V_1 \\ V_2 \end{bmatrix}; \; [I'] = \begin{bmatrix} I_1 \\ I_2 \end{bmatrix} \quad \text{and} \quad [Z'] = \begin{bmatrix} Z_1 & Z_2 \\ Z_3 & Z_4 \end{bmatrix}$$

where

$$Z_1 = R_a + R_b; \quad Z_2 = -R_b; \quad Z_3 = -R_b \quad \text{and} \quad Z_4 = R_b + R_c$$

The elements of the $[Z']$ matrix are written down on the basis of Maxwell's clockwise current rules.

In correlating the two circuits, special care is needed in choosing the signs of the elements. The batteries in the original network of Figure 6.3(a) are all uniformly connected to send currents in the clockwise direction. The new network may be obtained if:

1. The first two circuits of Figure 6.3(b) are disconnected on the left as indicated by the cross (×) on the vertical connecting lines and the batteries are treated as rotated clockwise to lie in line with the resistors.
2. The third independent circuit of Figure 6.3(b) is similarly disconnected on the right as indicated by the cross (×) on the vertical connector, and the resistor (or inductance) rotated to be co-linear with the battery.
3. The three opened up branches, suitably arranged along a vertical stack, are then reconnected by vertical connectors P and Q.

Since the currents in the individual resistor branches should remain the same, we can write down the relationship between the old and the new currents by inspection as follows:

$$I_a = +I_1; \; I_b = -I_1 + I_2; \; I_c = +I_2$$

Hence, the old and the new currents may be related in the matrix form as follows:

$$\begin{bmatrix} I_a \\ I_b \\ I_c \end{bmatrix} = \begin{bmatrix} +1 & 0 \\ -1 & +1 \\ 0 & +1 \end{bmatrix} \begin{bmatrix} I_1 \\ I_2 \end{bmatrix}$$

Following the earlier convension of representation, it is the same thing as saying:

$$[I] = [C] [I']$$

where, $[C] = \begin{bmatrix} +1 & 0 \\ -1 & +1 \\ 0 & +1 \end{bmatrix}$ is the *connection matrix* or the *transformation matrix*.

We can also get the voltage equation of the circuit in Figure 6.3(a), based on Maxwell's clockwise current law by inspection, which is:

$$V_1 = V_a - V_b; \quad \text{and} \quad V_2 = V_b + V_c$$

which leads to the matrix relationship as follows:

$$\begin{bmatrix} V_1 \\ V_2 \end{bmatrix} = \begin{bmatrix} +1 & -1 & 0 \\ 0 & +1 & +1 \end{bmatrix} \begin{bmatrix} V_a \\ V_b \\ V_c \end{bmatrix}$$

The voltage connection matrix is thus,

$$\begin{bmatrix} +1 & -1 & 0 \\ 0 & +1 & +1 \end{bmatrix} = \left[C_t^* \right]$$

Note that the star (*) marked on top of the connection (or transform) matrix denotes the conjugate and the suffix t stands for the transpose.

We may, now, repeat the purpose of active transformation once again for an unambiguous physical interpretation.

The use of active transformation is to enable the characteristic equations of the many different types of commutator machines that can be derived from the equations of one basic machine—the commutator primitive machine, proposed by Gabrail Kron. This, generally, involves the physical reconnection of one or more sets of windings in series.

When two windings are connected in series, the same current flows through both constituent windings, and the number of currents or dependent variables in the characteristic equations naturally reduce by one. The appearance of the above transformation is, therefore, typified by the fact that its connection matrix $[C]$ is rectangular and not square. The other distinguishing feature is that the elements of $[C]$ are always 0, +1 or –1. Since the connection matrix $[C]$ in active transformation is rectangular, it cannot have any inverse, and *the law of transformation for current, voltage and impedance are one directional only for the active transformation.*

6.2.2 Passive Transformations

'Passive transformations are applied to electro-mechanical energy converters with a purely mathematical objective of simplifying and then solving the performance equations of machines with phase windings'—basically like induction and synchronous machines, which contain poly-phase windings in their armatures.

When the voltage equations of poly-phase machines, like a poly-phase induction motor is written down for instance from first principles, it consists of six component equations—three for the stator member and three for the rotor phase windings. Since these equations have Heaviside's differential operator, $p\left(=\dfrac{d}{dt}\right)$, these six simultaneous equations are actually differential equations. Also, as impedance terms in the impedance matrix are dependent on the rotor angle, θ_r—and therefore time—these are non-linear differential equations.

Further, as the operator p, applied on the product of impedances—both self and mutual—and currents, solution of such equations poses special problems. The object of applying transformation is to linearize the equations by eliminating the rotor angle θ_r. This requires application of transformation in two stages in cascade.

The first stage consists of applying a phase transformation to both stator and rotor windings. This is known as Edith Clarke's *a-b-c* to α-β-γ transformation, which enables effective conversion of a standard 3-phase system into an effective 2-phase α-β-γ system. It is a dead transformation and all its elements are constants. It affords a simple change of variables. The actual machine currents are replaced by suitably chosen fictitious currents. The effect of applying Clarke's transformation is to reduce the original impedance matrix from 36 non-zero terms to only 14 terms with significant simplicity, reducing the remaining 22 terms to zero, while appreciably simplifying the remaining 14 terms in the process. Although Clarke's transformation can effect this significant simplification, it has the limitation that this transformation, by itself, cannot eliminate the variable rotor angle θ_r. This necessitates the use of a second transformation for the rotor windings, called the R.H. Park's transformation. It tends to transform the resulting *a-b-c* winding of the original 3-phase transformation into α-β-γ, that is equivalent to 2-phase system by Clarke's transformation earlier and then into a 2-phase *d-q-r* system. The final form of the Park's *d-q-r* system is, therefore, a cascaded system—from 3-phase *a-b-c* system to 3-phase *d-q-r* system via a 2-phase α-β-γ system to *d-q-r* system. The elements of Park's transformation are themselves functions of θ_r, and hence time. It is primarily a mathematical substitution but notionally it replaces an actual rotor distributed winding by its commutator winding equivalent.

6.2.3 Condition of Orthogonality of Passive Transformations

Since the two passive transformations discussed so far correspond mathematically to a change of variables, the matrices associated with passive transformations should always be square, and the number of new variables should always equal the number of old variables.

Although we identified that the significance of active and passive transformations are different, the laws of current, voltage, and impedance transformations have identical forms that is for each case, there is a specific relationship between one set of variables and the others. Further, when these relationships are expressed in matrix form, say, $[I] = [C]\,[I']$, where the new ones are marked with dashes ($'$), they have the same form whether $[C]$ denotes an active or a passive transformation, and the three non-reversible laws, detailed in Eq. (6.7), apply both to active and passive type of transformations. However, as pointed out earlier, as $[C]$ for passive transformations are mandatorily square types, these laws in the passive case are suitable for inversion. The full set of six passive transformations is, therefore, of the following form:

$$\left.\begin{aligned} &[I] = [C]\,[I']; &&[I'] = [C]^{-1}\,[I] \\ &[V'] = [C_t^*][V]; &&[V] = [C_t^*]^{-1}[V'] \\ &[Z'] = [C_t^*][Z][C]; &&[Z] = [C_t^*]^{-1}\,[Z'][C]^{-1} \end{aligned}\right\} \qquad ...(6.10)$$

Of these, the first set of laws, enumerated for $[I]$, $[V']$ and $[Z']$, are common to both active and passive transformations. The laws of current and voltage transformations are different. However, it would be especially advantageous if these laws were to be the same, as then only our memory is half loaded, that is, we have to remember only two such relationships.

We observe that $[I'] = [C]^{-1}[I]$ and $[V'] = [C_t^*][V]$ are identical, if:

$$[C]^{-1} = [C_t^*] \qquad \text{...(6.11)}$$

Again, $[I] = [C]\,[I']$ and $[V] = [C_t^*]^{-1}[V']$ are identical, if:

$$[C] = [C_t^*]^{-1} \qquad \text{...(6.12)}$$

However, using matrix multiplication rules,

$$[C][C_t^*] = [1], \text{ and } [C_t^*][C] = 1 \qquad \text{...(6.13)}$$

[1] being the unit matrix as defined in Chapter 2. The above condition for identical laws for voltage and current transformation as enumerated by Eq. (6.13) is called the *condition of orthogonality*. If the above condition is satisfied by a set of transformations, it is said to be *orthogonal transformation*. The above condition of orthogonality can now be stated as follows: *The inverse of an orthogonal matrix equals the conjugate of its transverse.*

All passive transformations discussed in this chapter, including their cascade combinations can be expressed in their orthogonal forms. The same is used in this book, unless otherwise indicated clearly.

6.3 TRANSFORMATION BETWEEN A 3-PHASE *a-b-c* WINDING AND A 2-PHASE (α, β) WINDING

Let Figure 6.4 denote a symmetrical 2-pole, 3-phase, distributed winding on a cylindrical magnetic structure, duly represented by three equivalent concentrated coils (see Section 4.5.2, under ac windings or multi-phase windings), with each coil of N_3 effective turns mutually displaced by $\left(\frac{2\pi}{3}\right)$ electrical radians as shown.

The mutually perpendicular α-β axes, fixed on the same magnetic structure, are also indicated in an arbitrary position by choosing the effect of the α-axis of the equivalent 2-phase system to be coincident with the *A*-axis of the 3-phase system. The position of the *q*-axis is in the same direction towards the *B*-phase although the two axes—β and *B*—have an angle of $\left(\frac{\pi}{12}\right)$ between them.

We wish to express the terminal currents in the coils of one set in terms of those of the other, assuming the current distribution—and therefore the air-gap mmf and flux distribution produced—as the same for both. We consider instantaneous currents first and introduce

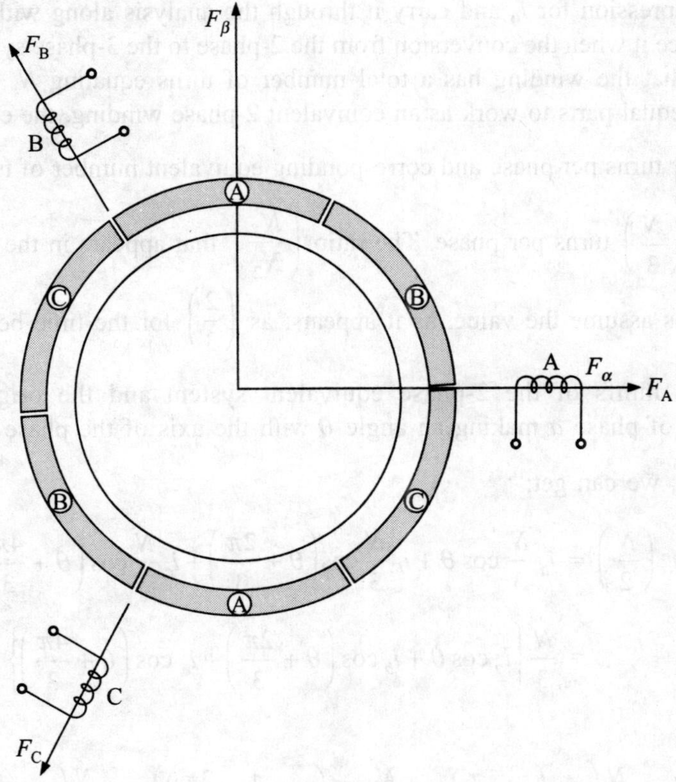

FIGURE 6.4 A symmetrical 2-pole, 3-phase distributed winding laid on a cylindrical magnetic structure.

rms currents later. It is straightforward to transform a given set of 3-phase terminal currents to the equivalent 2-phase currents; but we also wish to carryout inverse transformation—that is to produce a set of 3-phase currents equivalent to a given set of 2-phase currents. In doing this, we note that although there is one unique pair of 2-phase currents, i_α and i_β, which produce a given air-gap mmf pattern, there is no unique set of 3-phase currents, i_a, i_b and i_c, which produces the same pattern. To appreciate this, imagine that an identical value of current is passed through each of the phase windings—for example by connecting them in series. There will not be any effective air-gap mmf or flux produced because the three equal space-sinusoidal mmfs produced individually by three phase at $\left(\dfrac{2\pi}{3}\right)$ to each other cancel at every point around the air-gap. Thus, to each of the three unequal currents in the three phases which produce the desired air-gap mmf and flux pattern, we can always add an equal component of any magnitude, giving a new set of 3-phase currents without affecting the overall pattern. To find the value of this component from an arbitrary set of currents, we add them phasorically and divide the sum by 3. This component is called i_0 and it produces no effective air-gap flux. It, however, produces leakage flux. While resolving the 3-phase currents, i_a, i_b and i_c into the 2-phase currents, i_α and i_β, we automatically eliminate i_0. In performing the process we, therefore, write down an expression for i_0 and carry it through the analysis along with i_α and i_β, so that we can re-introduce it when the conversion from the 2-phase to the 3-phase system is carried out.

We assume that the winding has a total number of turns equaling N. When the same is divided into two equal parts to work as an equivalent 2-phase winding, the effective number of turns equals $\left(\frac{N}{2}\right)$ turns per phase and corresponding equivalent number of turns in the 3-phase system becomes $\left(\frac{N}{3}\right)$ turns per phase. The ratio $\left(\frac{N_3}{N_2}\right)$ that appears in the voltage equations, is arbitrary. Let us assume the value, as it appears, as $\left(\frac{2}{3}\right)$ for the time being.

Equating the mmfs of the 2-phase equivalent system and the original 3-phase *abc* systems, the axis of phase a making an angle θ with the axis of the phase α and equating i_0 to $\frac{1}{3}(i_a + i_b + i_c)$, we can get:

$$i_\alpha\left(\frac{N}{2}\right) = i_a\frac{N}{3}\cos\theta + i_b\frac{N}{3}\cos\left(\theta + \frac{2\pi}{3}\right) + i_c\frac{N}{3}\cos\left(\theta + \frac{4\pi}{3}\right)$$

$$= \frac{N}{3}\left\{i_a\cos\theta + i_b\cos\left(\theta + \frac{2\pi}{3}\right) + i_c\cos\left(\theta + \frac{4\pi}{3}\right)\right\}$$

Similarly,

$$i_\beta\left(\frac{N}{2}\right) = i_a\frac{N}{3}\left(\cos\left(\theta + \frac{\pi}{2}\right) + i_b\frac{N}{3}\cos\left(\theta + \frac{\pi}{2} + \frac{2\pi}{3}\right)\right) + i_c\frac{N}{3}\left(\theta + \frac{\pi}{2} + \frac{4\pi}{3}\right)$$

$$= \frac{N}{3}\left\{i_a\sin\theta + i_b\sin\left(\theta + \frac{2\pi}{3}\right) + i_c\sin\left(\theta + \frac{4\pi}{3}\right)\right\}$$

Also,

$$i_0\left(\frac{N}{2}\right) = \frac{1}{3}\frac{N}{3}(i_a + i_b + i_c)$$

Rearranging the three equations positionally and expressing them in matrix from, denoting $\frac{N/3}{N/2} = \frac{2}{3}$, we get:

$$\begin{bmatrix} i_0 \\ i_\alpha \\ i_\beta \end{bmatrix} = \frac{2}{3}\begin{bmatrix} \frac{1}{2} & \frac{1}{2} & \frac{1}{2} \\ \cos\theta & \cos\left(\theta + \frac{2\pi}{3}\right) & \cos\left(\theta + \frac{4\pi}{3}\right) \\ \sin\theta & \sin\left(\theta + \frac{2\pi}{3}\right) & \sin\left(\theta + \frac{4\pi}{3}\right) \end{bmatrix}\begin{bmatrix} i_a \\ i_b \\ i_c \end{bmatrix} \quad \text{...(6.14)}$$

By solving these equations for i_a, i_b and i_c, we develop the corresponding inverse expression as follows:

$$\begin{bmatrix} i_a \\ i_b \\ i_c \end{bmatrix} = \begin{bmatrix} 1 & \cos\theta & \sin\theta \\ 1 & \cos\left(\theta + \dfrac{2\pi}{3}\right) & \sin\left(\theta + \dfrac{2\pi}{3}\right) \\ 1 & \cos\left(\theta + \dfrac{4\pi}{3}\right) & \sin\left(\theta + \dfrac{4\pi}{3}\right) \end{bmatrix} \begin{bmatrix} i_0 \\ i_\alpha \\ i_\beta \end{bmatrix} \qquad \text{...(6.15)}$$

We can use the two current matrices to express the components of current distribution vector with respect to the two reference frames. The square matrices of coefficients transform the current distribution vector expressed in terms of a, b, c components into some quantities expressed in α, β, 0 components and vice-versa.

In order to complete the analysis, the relationships between two sets of voltages at the terminals of the 3-phase and the equivalent 2-phase windings are also to be obtained. Both windings—the original 3-phase winding and the equivalent 2-phase system—are linked by the same set of fluxes in the air-gap, produced by identical current patterns. If the total sinusoidally distributed flux per pole is ϕ, then it may be considered as split into two components—ϕ_α and ϕ_β, along the axes of the α-winding and the β-windings respectively. The voltages, we are seeking, are equal to the rates of change of flux linkage with the corresponding coils. We can calculate the flux linkages and their rates of change as follows:

Both the components of flux link each of the three coils, therefore:

$$v_a = N_3 \frac{d}{dt}(\phi_\alpha \cos\theta + \phi_\beta \sin\theta)$$

$$v_\alpha = N_2 \frac{d}{dt}(\phi_\alpha)$$

$$v_\beta = N_2 \frac{d}{dt}(\phi_\beta)$$

Thus,

$$v_a = \frac{N_3}{N_2}(v_\alpha \cos\theta + v_\beta \sin\theta)$$

$$= \frac{2}{3}(v_\alpha \cos\theta + v_\beta \sin\theta)$$

Similarly,

$$v_b = \frac{2}{3}\left\{v_\alpha \cos\left(\theta + \frac{2\pi}{3}\right) + v_\beta \sin\left(\theta + \frac{2\pi}{3}\right)\right\} \qquad \text{...(6.16)}$$

and

$$v_c = \frac{2}{3}\left\{v_\alpha \cos\left(\theta + \frac{4\pi}{3}\right) + v_\beta \sin\left(\theta + \frac{4\pi}{3}\right)\right\}$$

To the above voltages, we have to add the components of the induced voltages v_0, in order to obtain the correct terminal voltages. This voltage is produced by the leakage flux associated with the equal currents, i_0, which is present in each of the three phases. The algebraic sum of the three terminal voltages is equal to three times the value of this voltage, v_0. Thus,

$$\begin{bmatrix} v_a \\ v_b \\ v_c \end{bmatrix} = \frac{2}{3}\begin{bmatrix} \frac{3}{2} & \cos\theta & \sin\theta \\ \frac{3}{2} & \cos\left(\theta+\frac{2\pi}{3}\right) & \sin\left(\theta+\frac{2\pi}{3}\right) \\ \frac{3}{2} & \cos\left(\theta+\frac{4\pi}{3}\right) & \sin\left(\theta+\frac{4\pi}{3}\right) \end{bmatrix}\begin{bmatrix} v_0 \\ v_\alpha \\ v_\beta \end{bmatrix} \quad ...(6.17)$$

and by solving for v_0, v_α and v_β, we obtain:

$$\begin{bmatrix} v_0 \\ v_\alpha \\ v_\beta \end{bmatrix} = \begin{bmatrix} \frac{1}{3} & \frac{1}{3} & \frac{1}{3} \\ \cos\theta & \cos\left(\theta+\frac{2\pi}{3}\right) & \cos\left(\theta+\frac{4\pi}{3}\right) \\ \sin\theta & \cos\left(\theta+\frac{2\pi}{3}\right) & \sin\left(\theta+\frac{4\pi}{3}\right) \end{bmatrix}\begin{bmatrix} v_a \\ v_b \\ v_c \end{bmatrix} \quad ...(6.18)$$

The above square matrices of coefficients are the two voltage transformations. Since we have ignored losses, the total instantaneous power passing into the terminals with either of the two windings considered flows into the electro-magnetic field. As the field is the same—being produced by the same current pattern around the air-gap—the total instantaneous power should be invariant, whether it is expressed in the 3-phase or in the 2-phase system. For the 3-phase system, the power is as follows:

$$p = v_a i_a + v_b i_b + v_c i_c \quad ...(6.19)$$

If we express these voltages and currents in terms of 0, α and β values, using the derived relations, we can find the power for the 2-phase system. Since, the purely leakage flux associated with i_0 and its voltage, v_0, in the 3-phase winding cannot be produced by currents in the 2-phase winding, we shall, therefore, expect to find a term which corresponds to the power flow into this leakage field in the 3-phase form. Substituting 0, α and β quantities from Eqs. (6.15) and (6.17) into Eq. (6.19), we have:

$$p = 3v_0 i_0 + v_\alpha i_\alpha + v_\beta i_\beta \quad ...(6.20)$$

Hence, the power flow into the electro-magnetic field is invariant. The middle and the last term on the right hand side of Eq. (6.20) express in 2-phase terms, the important air-gap power along with a small leakage flux power common to the two systems, while the first term indicates in 3-phase terms, a purely 3-phase leakage flux power. i_0 and v_0 are the phase components in *a-b-c* system.

Balanced sinusoidal conditions

We considered only instantaneous conditions of operation so far. It is useful to discuss balanced sinusoidal conditions of operation as well. In Section 4.5 (under ac winding or multi-phase winding), we have seen that a balanced 3-phase winding fed with balanced 3-phase currents produces a fundamental pattern of mmf of constant amplitude, rotating at a uniform speed.

The same is true for a similar 2-phase winding fed from a 2-phase current system. We have also seen that the peak value of the 3-phase equals $\left(\frac{3}{2}\right)$ of that produced by one of the three phases, while that of the 2-phase mmf equals the peak mmf produced by one of the two phase windings. The equality of mmf for the 3-phase and the 2-phase systems is most simply expressed for this case by:

$$\frac{3}{2}\left(\sqrt{2}I_{ph}\right)N_3 \text{ for 3-phase system} = \left(\sqrt{2}I_{ph}\right)N_2 \text{ for 2-phase system}$$

using the rms currents.

or
$$\frac{3}{2}\left(\sqrt{2}I_{ph\,\text{3-phase}}\right)\frac{N_3}{N_2} = \sqrt{2}I_{ph\,\text{2-phase}}$$

Since $\frac{N_3}{N_2}$ is taken as $\frac{2}{3}$, we get:

$$I_{ph} \text{ for 3-phase} = I_{ph} \text{ for 2-phase}$$

And, as both 3-phase and 2-phase windings are linked by the same fundamental air-gap flux, which changes at the same rate, the voltages are in the same ratio as the number of turns. In other words, the two phase rms voltage is $\frac{3}{2}$ times the three phase value.

A critical scrutiny of the transformation of current and voltage reveals that the transformations between the 3-phase and 2-phase are not the same for current and voltage. In other words, the transformations are not in orthogonal form. However, it is convenient to imagine that when the 3-phase currents and voltage are expressed in terms of 2-phase and zero components, a change in the number of active conductors occurs in the ratio of $\sqrt{3}$ to $\sqrt{2}$. We also redefine i_0 and v_0 as $\frac{1}{\sqrt{3}}(i_a + i_b + i_c)$ and $\frac{1}{\sqrt{3}}(v_a + v_b + v_c)$ respectively. In other words, $\left(\frac{N_3}{N_2}\right)$ changes from $\left(\frac{3}{2}\right)$ to $\sqrt{\frac{3}{2}}$. In this process the voltage and current transformations assume identical expressions. Also, the expression for instantaneous power becomes the same in both 3-phase *abc* system and the 2-phase 0-α-β system. The results now become:

$$\begin{bmatrix} i_0 \\ i_\alpha \\ i_\beta \end{bmatrix} = \sqrt{\frac{2}{3}} \begin{matrix} \\ \\ \alpha \\ \beta \end{matrix} \begin{matrix} a & b & c \\ \end{matrix} \begin{bmatrix} 0 & \frac{1}{\sqrt{2}} & \frac{1}{\sqrt{2}} & \frac{1}{\sqrt{2}} \\ \alpha & 1 & -\frac{1}{2} & -\frac{1}{2} \\ \beta & 0 & \frac{\sqrt{3}}{2} & -\frac{\sqrt{3}}{2} \end{bmatrix} \begin{bmatrix} i_a \\ i_b \\ i_c \end{bmatrix} \quad \text{or} \quad [i_{0\alpha\beta}] = [C_1]\,[i_{abc}] \qquad \text{...(6.21)}$$

and

$$\begin{bmatrix} i_a \\ i_b \\ i_c \end{bmatrix} = \sqrt{\frac{2}{3}} \begin{matrix} & 0 & \alpha & \beta \\ a & \frac{1}{\sqrt{2}} & 1 & 0 \\ b & \frac{1}{\sqrt{2}} & -\frac{1}{2} & \frac{\sqrt{3}}{2} \\ c & \frac{1}{\sqrt{2}} & -\frac{1}{2} & -\frac{\sqrt{3}}{2} \end{matrix} \begin{bmatrix} i_0 \\ i_\alpha \\ i_\beta \end{bmatrix} \quad \text{or} \quad [i_{abc}] = [C_{1t}]\,[i_{0\alpha\beta}] \qquad \text{...(6.22)}$$

and identically for voltages,

$$[v_{0\alpha\beta}] = [C_1]\,[v_{abc}] \qquad \text{...(6.23)}$$

$$[v_{abc}] = [C_t]\,[v_{0\alpha\beta}] \qquad \text{...(6.24)}$$

Here,

$$[C_1] = \sqrt{\frac{2}{3}} \begin{bmatrix} \frac{1}{\sqrt{2}} & \frac{1}{\sqrt{2}} & \frac{1}{\sqrt{2}} \\ 1 & -\frac{1}{2} & -\frac{1}{2} \\ 0 & \frac{\sqrt{3}}{2} & -\frac{\sqrt{3}}{2} \end{bmatrix} \quad \text{and} \quad [C_{1t}] = \sqrt{\frac{2}{3}} \begin{bmatrix} \frac{1}{\sqrt{2}} & 1 & 0 \\ \frac{1}{\sqrt{2}} & -\frac{1}{2} & \frac{\sqrt{3}}{2} \\ \frac{1}{\sqrt{2}} & -\frac{1}{2} & -\frac{\sqrt{3}}{2} \end{bmatrix}$$

6.4 TRANSFORMATION BETWEEN ROTATING AND FIXED AXES

In Section 6.3, we considered transformation of voltages and currents between axes fixed to the winding conductors. We noted that the two reference frames—the 3-phase *abc* system and the 2-phase 0-α-β system—are stationary with respect to each other, although both of them can either be at rest or are rotating at the same speed in the same direction. If rotating, then the connections would have to be made through suitable number of slip rings.

In this section, we consider another important transformation. This is between rotating and fixed axes or vice versa.

Let us connect a set of active conductors to form a 3-phase winding or a 2-phase winding, and feed it with balanced 3-phase or 2-phase current respectively. Neglecting the effect of high-order space harmonics, a sinusoidally distributed pattern of air-gap flux is produced rotating at a constant speed and having constant magnitude.

Now, consider a simple dc armature winding with a commutator. Let the winding also have taps at three selected points $\frac{2\pi}{3}$ apart for a 2-pole configuration connected to suitable slip rings, so that the same winding can form a 2-pole, 3-phase balanced winding. Let us also tap the same winding at four points $\left(\frac{\pi}{2}\right)$ electrical radians apart to form a balanced 2-phase winding. This is indicated in Figure 6.5. With the 2-phase winding, the two phases use the same active conductors, but their electric and magnetic effects are unaffected by each other. The effects are as if the phases of the 3-phase and the 2-phase systems are separate but superposed on top of each other.

If the 3-phase taps or the 2-phase taps of the stationary armature are fed with balanced 3-phase or 2-phase currents, then the rotating current, mmf and flux patterns discussed earlier are produced. These two alternative sets of currents produce identical current, mmf and flux patterns, provided that the currents have appropriately adjusted values. ***These are only two different ways of expressing the same current distribution vectors***. Similar remarks also apply to the voltages and the voltage distribution vectors. The two sets of terminals are each stationary relative to each other.

Let us consider a particular instant of time and let us imagine that two pairs of brushes marked d and q, make contact with the commutator segments at the same points as the 2-phase taps. At this instant, conditions are duplicated if direct currents are passed through the stationary winding between opposite brushes of the pairs, the values of those currents being equal to the instantaneous values of the 2-phase currents at the instant chosen. By keeping the brush currents constant and steadily rotating the four brushes together around the commutator, making one revolution in the time of one cycle of the previously applied alternating current, we establish conditions identical to those obtained previously—that is embodying the same current, mmf and flux patterns moving at the same speed as we obtained when applying the 2-phase excitation to the stationary winding. The four brushes form a further set of terminals from which the same current pattern may be supplied, these terminals are moving at a constant speed relative to the conductors of the winding.

The advantage of supplying the desired current pattern through the moving brush contacts is that for balanced conditions simulating and producing balanced 3-phase (or 2-phase) supplies to the taps, we need only steady direct currents.

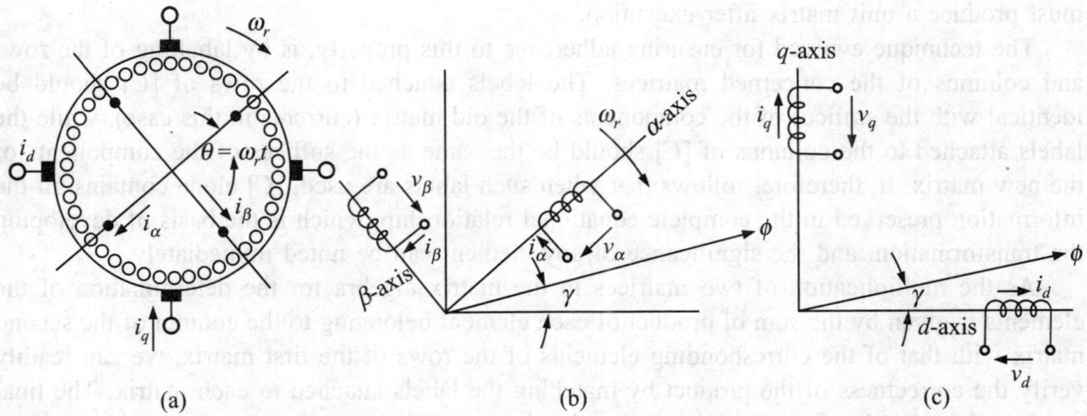

FIGURE 6.5 Two terminal-arrangements of commutator winding—2-phase and 2-axis.

It is simple to transform the 3-phase or 2-phase terminal currents, which express the current distribution vector along axis stationary with respect to the winding, into direct currents, which express the vector relative to axes rotating with respect to the winding and stationary with respect to the current distribution vector itself.

The three sets of currents (and voltages)—3-phase, equivalent 2-phase, and direct—thus form three descriptions of the same physical system, which are invariant among themselves.

Let us now suppose that the armature revolves at the same speed as the current, mmf and flux patterns—namely at synchronous speed—but in the opposite direction. These patterns are then at rest. The four brushes are stationary, and the 3-phase and 2-phase taps have to be joined to slip rings. Figure 6.5 shows the stationary flux pattern centred at an angle γ to the axis of the d-brushes. We now derive the transformation to change currents from the 2-phase to the dc reference system, with the former moving at synchronous speed with the dc axes held stationary. Figures 6.5(b) and 6.5(c) show the instantaneous conditions in the manner discussed earlier. We indicate the 2-phase currents and voltages fed to the 2-phase taps by suffices α and β, while the direct currents and voltages fed to the brushes by suffices d and q. If the armature winding has N effective turns or $\left(\frac{N}{2}\right)$ effective turns in each of the two parallel paths that carry half the terminal current in each parallel circuit, then in the single-circuit depicted in Figures 6.5(b) and 6.5(c), the coils have $\left(\frac{N}{2}\right)$ effective turns each.

Before writing down the equations, however, we should ensure that the order of appearance of the equations is in a manner that will ultimately ensure maintenance of orthogonality of the resulting matrix. Or, in other words, we should especially ensure that the transformation in its matrix form must ensure the property of its orthogonality, that is,

$$\left[C_t^*\right][C]$$

must produce a unit matrix after execution.

The technique evolved for ensuring adherence to this property, is by labelling of the rows and columns of the concerned matrices. The labels attached to the rows of $[C]$ should be identical with the suffices of the components of the old matrix (currents in this case), while the labels attached to the columns of $[C]$ should be the same as the suffices of the components of the new matrix. It, therefore, follows that when such labels are used, $[C]$ alone contains all the information preserved in the complete equational relationship, which is the basis of developing the transformation, and the significance of any element can be noted immediately.

As the multiplication of two matrices in the matrix algebra for the determination of the elements is given by the sum of product of each element belonging to the column of the second matrix with that of the corresponding elements of the rows of the first matrix, we can readily verify the correctness of the product by matching the labels attached to each matrix. The final result has the labels of rows of the first and columns of the second.

We are now ready to write down the instantaneous mmf equations along the axes of the brush pairs shown in Figure 6.5(a). We equate the mmfs along the two brush pairs and measure the time t from the instant when the α-taps and the d-axis are aligned. We have:

$$i_q\left(\frac{N}{2}\right) = i_\alpha\left(\frac{N}{2}\right)\sin\omega t + i_\beta\left(\frac{N}{2}\right)\sin(\omega t + 90°)$$

and
$$i_d\left(\frac{N}{2}\right) = i_\alpha\left(\frac{N}{2}\right)\cos\omega t + i_\beta\left(\frac{N}{2}\right)\cos(\omega t + 90°)$$

The above equations, when depicted in matrix form have the shape:

$$\begin{bmatrix} i_q \\ i_d \end{bmatrix} = \begin{matrix} \\ q \\ d \end{matrix}\begin{matrix} \alpha & \beta \\ \begin{bmatrix} \sin\omega t & \cos\omega t \\ \cos\omega t & -\sin\omega t \end{bmatrix} \end{matrix}\begin{bmatrix} i_\alpha \\ i_\beta \end{bmatrix} \qquad \text{...(6.25)}$$

and by solving for i_α and i_β:

$$\begin{bmatrix} i_\alpha \\ i_\beta \end{bmatrix} = \begin{matrix} \\ \alpha \\ \beta \end{matrix}\begin{matrix} q & d \\ \begin{bmatrix} \sin\omega t & \cos\omega t \\ \cos\omega t & -\sin\omega t \end{bmatrix} \end{matrix}\begin{bmatrix} i_q \\ i_d \end{bmatrix} \qquad \text{...(6.26)}$$

We note that:

$$[i_{qd}] = [C_{2t}]\,[i_{\alpha\beta}] \quad \text{and} \quad [i_{\alpha\beta}] = [C_2]\,[i_{qd}]$$

$$[C_{2t}] = [C_2]$$

and
$$[C_{2t}]\,[C_2] = \begin{bmatrix} 1 & 0 \\ 0 & 1 \end{bmatrix} = \text{Unit matrix}$$

$\therefore$ $[C_2]$ is orthogonal.

Since, i_α and i_β are instantaneous values of balanced 2-phase sinusoidal currents I. With i_α at its maximum value when $\omega t = \gamma$, and for a rotational frequency equal to the frequency of the currents ($\omega_r = \omega$), the order of rise of currents (or the phase rotation) is i_β followed by i_α to give a stationary current pattern. Using Eq. (6.25), we have:

$$\left.\begin{aligned} i_d &= \sqrt{2}\,I\{\cos(\omega t - \gamma)\cos\omega t + \cos(\omega t - \gamma + 90°)\cos(\omega t + 90°)\} \\ &= \sqrt{2}I\cos\gamma = I_d \\ \text{and similarly,} \\ i_q &= \sqrt{2}I\sin\gamma = I_q \end{aligned}\right\} \qquad \text{...(6.27)}$$

Now, we derive the transformations for the voltages which accompany these currents. We consider again the rotating armature supplied with 2-phase currents shown in Figure 6.5. A pair of 2-phase voltages v_α and v_β exists between the α and β pairs of terminals. We need to determine the corresponding voltages between the d and q pairs of terminals. Under balanced conditions, a steady stationary flux pattern is produced as already discussed. We can think of this pattern as formed by two components—one produced on the d-axis by the current i_d and the other, on the q-axis by the current i_q. The d and q pairs of terminals serve to define pseudo-stationary coils centred on the lines of the brushes. The coils are said to be pseudo-stationary because, when currents i_d and i_q pass, stationary current and flux patterns are produced, although the coils actually consist of conductors which are continuously changing. The rate of change of the flux-linkage due to the stationary flux-pattern with the moving coils instantly joined to the pair of d (or q) brushes is evidently constant. A steady direct voltage therefore appears between the brushes. To deduce the magnitude of this voltage, we observe that it equals the instantaneous voltage at the α-terminals when these are in line with the d (or q) brushes and therefore, in contact with them. Hence, the voltage transformations can be derived as follows. Figure 6.5 represents a typical instantaneous condition. The current pattern, denoted by the current vector, which is expressed as i_α and and i_β at the 2-phase taps or i_d and i_q at the brushes, produces a flux pattern centred at an angle γ to the d-axis. The pattern is stationary and constant in magnitude for balanced poly-phase conditions with constant frequency and synchronous speed. If the poly-phase currents are not balanced but the frequency and speed remain constant, the flux pattern wobbles a little. Voltages due to the latter are small and we can safely ignore it. The movement of the α and β coils through the flux pattern induces instantaneous voltages equal to v_α and v_β. In order to find their values, the flux linkages, ψ_α and ψ_β, with the moving coils α and β may be differentiated with respect to time. Now,

$$\psi_a = \frac{N}{2}\phi \cos(\omega t - \gamma)$$

Hence,

$$\left.\begin{aligned} v_\alpha &= -\frac{N}{2}\phi\omega \sin(\omega t - \gamma) \\ &\text{and similarly,} \\ v_\beta &= -\frac{N}{2}\phi\omega \sin(\omega t - \gamma + 90°) \end{aligned}\right\} \quad ...(6.28)$$

The situation regarding the d- and q-pseudo-stationary coils is as follows. As the commutator segments pass under the brushes, the brushes make contact with the d and q coils, which are identical with the α and β coils, but have only a small angular movement at the position of the brush axes. The movement is at the angular frequency of ω of the α and β coils. When the small angular movement producing the voltages v_d and v_q has been accomplished, the adjacent commutator segments connect a similar coil arrangement to the brushes, replacing the previous commutator sets and the connected coil, and the movement is repeated. The voltage produced between the opposite commutator segments due to the movement, therefore appears constantly between the brushes. Thus,

$$\psi_d = \frac{N}{2}\phi \cos(\omega t - \gamma)$$

and at $\omega t = 0$

$$\left.\begin{aligned} v_d &= \frac{N}{2}\phi\omega \sin\gamma = \psi_q\omega \\ &\text{and similarly,} \\ v_q &= -\frac{N}{2}\phi\omega \cos\gamma = -\psi_d\omega \end{aligned}\right\} \quad ...(6.29)$$

where, ψ_q and ψ_d are the flux-linkages with the pseudo-stationary coils on the q-axis and the d-axis respectively.

Taking the expressions for v_α and v_β from relations (6.28), we can write:

$$v_\alpha \cos \omega t + v_\beta \cos(\omega t + 90°) = -\frac{N}{2}\phi\omega \sin(\omega t - \gamma)\cos \omega t$$

$$-\frac{N}{2}\phi\omega \sin(\omega t - \gamma + 90°)\cos(\omega t + 90°)$$

$$= \frac{N}{2}\phi\omega \sin\gamma = v_d$$

and in the same way

$$v_\alpha \sin \omega t + v_\beta \sin(\omega t + 90°) = -\frac{N}{2}\phi\omega \cos\gamma = v_q$$

Also,

$$v_d \cos \omega t + v_q \sin \omega t = v_\alpha$$

and

$$v_d \cos(\omega t + 90°) + v_q \sin(\omega t + 90°) = v_\beta$$

The transformations for voltages are, therefore, identical with those previously found for currents, that is:

$$\begin{bmatrix} v_d \\ v_q \end{bmatrix} = \begin{matrix} \\ d \\ q \end{matrix}\overset{\begin{matrix}\alpha & \qquad\qquad \beta\end{matrix}}{\begin{bmatrix} \cos \omega t & \cos(\omega t + 90°) \\ \sin \omega t & \sin(\omega t + 90°) \end{bmatrix}} \begin{bmatrix} v_\alpha \\ v_\beta \end{bmatrix} \quad \text{...(6.30)}$$

and

$$\begin{bmatrix} v_\alpha \\ v_\beta \end{bmatrix} = \begin{matrix} \\ \alpha \\ \beta \end{matrix}\overset{\begin{matrix}d & \qquad\qquad q\end{matrix}}{\begin{bmatrix} \cos \omega t & \sin \omega t \\ \cos(\omega t + 90°) & \sin(\omega t + 90°) \end{bmatrix}} \begin{bmatrix} v_d \\ v_q \end{bmatrix} \quad \text{...(6.31)}$$

The voltages and currents expressing the voltage and current distribution vectors in the 2-phase (α, β) and the 2-phase (d, q) reference frames are summarised below for reference:

$$i_\alpha = \sqrt{2}I \cos(\omega t - \gamma) \qquad i_d = I_d = \sqrt{2}I \cos(-\gamma)$$
$$= \sqrt{2}I \cos(\gamma)$$

$$i_\beta = \sqrt{2}I \cos(\omega t - \gamma + 90°) \qquad i_q = I_q = \sqrt{2}I \cos(-\gamma + 90°)$$
$$= \sqrt{2}I \sin\gamma$$

$$v_\alpha = -\frac{N}{2}\phi\omega \sin(\omega t - \gamma) \qquad v_d = V_d = -\frac{N}{2}\phi\omega \sin(-\gamma)$$

$$= -\psi\omega \sin(\omega t - \gamma) \qquad = \frac{N}{2}\phi\omega \sin\gamma$$
$$= \psi_q\omega$$

$$v_\beta = -\frac{N}{2}\phi\omega \sin(\omega t - \gamma + 90°) \qquad v_q = V_q = -\frac{N}{2}\phi\omega \sin(-\gamma + 90°)$$

$$= -\frac{N}{2}\phi\omega \cos\gamma$$

$$= \psi_d \omega$$

Each of the 2-phase currents i_α and i_β, is seen to lag in phase by $\left(\frac{\pi}{2}\right)$ behind its driving voltage, the latter being equal to the induced emf without inclusion of the resistance element. The two voltages v_d and v_q, are reactance voltages. The reactance of each phase winding using rms values, is given by:

$$x = \frac{V}{I} = \frac{\frac{1}{\sqrt{2}} \times \frac{N}{2} \times \phi\omega}{I} = \frac{\left(\frac{N}{2}\right)\phi\omega}{\sqrt{2}I} = L\omega$$

as the maximum current $\sqrt{2}I$ in one of the phase windings itself produces the flux linkage, ψ, equalling $\left(\frac{N}{2}\right)\phi$ with that winding.

The alternating reference voltages v_α and v_β, at the rotating reference axes are transformed to steady rotational voltages, V_d and V_q at the stationary reference axes. The rotational voltage on one axis equals the product of the flux-linkage referred to the other axis and the angular speed. The sign convention adopted gives a positive voltage if that voltage, when applied to the terminal externally, would result in a positive current. A positive current is a current flowing in a direction to produce a positive flux, that is the one direction in the positive x or y directions. Thus, V_q is negative and V_d is positive.

The currents, I_d and I_q are steady direct currents for a balanced poly-phase system in the other (or α-, β-) reference frame. We transform the alternating currents, i_α and i_β, to direct currents by observing the stationary current distribution vector in the stationary reference system. In the absence of resistance, the driving voltages of I_d and I_q are zero. The voltages V_d and V_q are, clearly, rotational voltages. The space quadrature relationship of the current producing the flux-linkage on one axis and the rotational voltage on the other corresponds, in the stationary d, q reference system, to the time quadrature relationship indicated between the alternating currents and reactance voltages in the rotating α, β reference system.

We can imagine the stationary flux, ϕ, to be resolved into two component fluxes—ϕ_d along the d-axis and ϕ_q along the q-axis. Each flux component is associated with a proportional steady direct currents, I_d and I_q respectively, which produce it. Each flux component is also the cause of flux-linkage, equal to the product of its magnitude and the effective turns of the pseudo-stationary coil on its axis which carries the associated current. Each flux component also produces the steady rotational voltage on the other axis.

Here, we may note the consequences of arranging the conditions, so that the stationary flux pattern is on the d-axis, making $\gamma = 0$ in Figure 6.5. These are as follows:

$$i_\alpha = \sqrt{2}I \cos \omega t \qquad i_d = I_d = \sqrt{2}I$$
$$i_\beta = \sqrt{2}I \cos (\omega t + 90°) \qquad i_q = I_q = 0$$
$$v_\alpha = -\psi\omega \sin \omega t \qquad v_d = V_d = 0$$
$$v_\beta = -\psi\omega \sin (\omega t + 90°) \qquad v_q = V_q = -\psi\omega$$

If the flux is arranged entirely on the *d*-axis, then I_d, which flows in the pseudo-stationary *d*-coil, is equal to the maximum value of the 2-phase current and it produces the flux, ϕ. There is no voltage in this coil in the absence of resistance. The flux ϕ, in which the windings are rotating, produces the steady rotational voltage V_q in the other axis. So far, we have dealt with the balanced poly-phase conditions and rotation at synchronous speed, where the pseudo-stationary coils have only rotational voltages induced in them. In general, however, such pseudo-stationary coils have induced in them both transformer and rotational voltages. Transformer voltages are induced by changing currents in other coils which are magnetically linked with the pseudo-stationary coil—that is by coils which are on the same magnetic axis or on an axis at any other angle excluding 90°. Rotational voltages are induced by all components of flux at 90°, and such voltages are at every instant proportional to the product of the magnitude of the flux-linkage with the pseudo-stationary coil on the other axis and the speed. The self-inductance voltage in a pseudo-stationary coil, when the current in it changes, is a special case of the transformer voltage.

6.5 TRANSFORMATION BETWEEN ROTATING 3-PHASE *abc* SYSTEM AND STATIONARY 2-PHASE 0*dq* SYSTEM

Now, we shall undertake to develop a suitable transformation of a 3-phase *abc* system, which produces a synchronously rotating magnetic field with the winding configuration held at standstill when suitably excited with a balanced set of 3-phase currents, into a stationary 2-phase 0*dq* system.

This is achieved by first transforming the original 3-phase system into an equivalent 2-phase $0\alpha\beta$ system. This produces a rotating magnetic field of identical magnitude as the original 3-phase *abc* system, and the produced rotating magnetic field revolves in the same direction and at the same speed as the rotating magnetic field produced by the original 3-phase *abc* system by the procedure laid down in Section 6.3. The transformed 2-phase $0\alpha\beta$ system is then, further transformed in the 2-phase 0*dq* system in accordance with the methodology developed in Section 6.4.

Let Figure 6.5(a) indicate a 3-phase winding again in the form of a continuous winding with three taps separated by $\left(\frac{2\pi}{3}\right)$ electrical radians for every 2-pole winding configuration. In addition, let there be two pairs of stationary brushes for the 2-pole configuration shown. In this form, the 3-phase terminal currents would be different from the phase currents as the phase windings are effectively connected in delta. Figure 6.6(b) shows the phases separately, and the following analysis deals with phase currents and voltages.

Let us also include briefly our remarks concerning i_0 and v_0. As we had seen earlier, they were not related to the air-gap flux, being associated only with an exclusively 3-phase leakage flux. They must, therefore, be separated at the outset from the phase currents and voltages. We can, therefore, carry along with the desired 2-axis values which are associated with the remainder of the flux, especially the air-gap flux. They are eliminated automatically when the 2-axis quantities are derived as pointed out earlier.

Examination of Figures 6.6(b) and 6.6(c) show that they are identical with Figure 6.4 in principle when θ is replaced by $\omega_r t$, the arrow indicating rotation having been deleted, and the 2-phase $\alpha\beta$ quantities being changed to 2-phase *d-q* quantities.

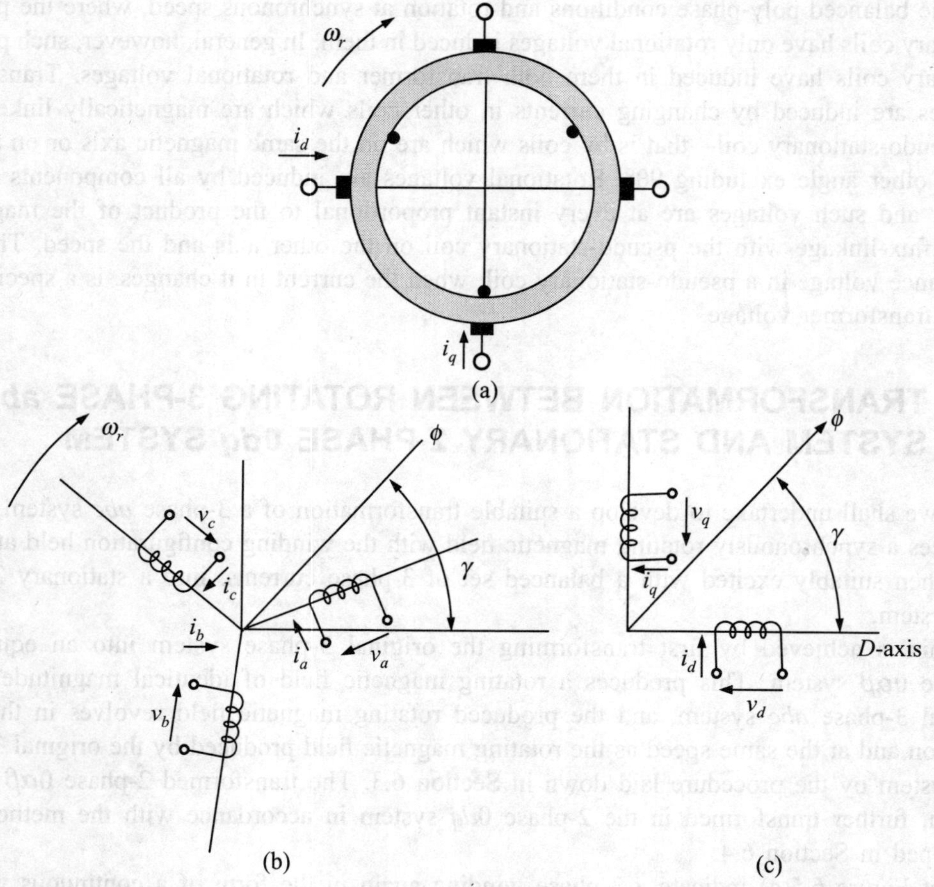

FIGURE 6.6 Two terminal-arrangements of commutator winding—3-phase and 2-phase.

By equating instantaneous mmfs along the axes of the brush pairs, we obtain relations similar to Eqs. (6.15) and (6.16).

$$\begin{bmatrix} i_0 \\ i_d \\ i_q \end{bmatrix} = \frac{2}{3}\begin{bmatrix} \frac{1}{2} & \frac{1}{2} & \frac{1}{2} \\ \cos\omega t & \cos\left(\omega t + \frac{2\pi}{3}\right) & \cos\left(\omega t + \frac{4\pi}{3}\right) \\ \sin\omega t & \sin\left(\omega t + \frac{2\pi}{3}\right) & \sin\left(\omega t + \frac{4\pi}{3}\right) \end{bmatrix}\begin{bmatrix} i_s \\ i_b \\ i_c \end{bmatrix} \quad \text{...(6.32)}$$

and

$$\begin{bmatrix} i_a \\ i_b \\ i_c \end{bmatrix} = \begin{bmatrix} 1 & \cos\omega t & \sin\omega t \\ 1 & \cos\left(\omega t + \frac{2\pi}{3}\right) & \sin\left(\omega t + \frac{2\pi}{3}\right) \\ 1 & \cos\left(\omega t + \frac{4\pi}{3}\right) & \sin\left(\omega t + \frac{4\pi}{3}\right) \end{bmatrix}\begin{bmatrix} i_0 \\ i_d \\ i_q \end{bmatrix} \quad \text{...(6.33)}$$

We assume that i_a, i_b and i_c are instantaneous values of balanced 3-phase sinusoidal currents and the time t is measured from the instant of alignment of the a terminals with the d brushes. We also assume that i_a attains its maximum value when $\omega t = \gamma$. For synchronous speed (when the actual rotational frequency equals the frequency of the currents and with a phase rotation of i_b followed by i_a followed by i_c, we have a stationary current phasor. Under this condition:

$$i_d = \frac{2}{3}\sqrt{2}I\left\{\cos(\omega t - \gamma)\cos\omega t + \cos\left(\omega - \gamma + \frac{2\pi}{3}\right)\cos\left(\omega t + \frac{2\pi}{3}\right)\right.$$

$$\left. + \cos\left(\omega - \gamma + \frac{4\pi}{3}\right)\cos\left(\omega t + \frac{4\pi}{3}\right)\right\}$$

$$= \sqrt{2}I\cos\gamma$$

and on similar lines,

$$i_q = \sqrt{2}I\sin\gamma$$

The derivation of the voltage transformation is similar to that for the 2-phase system, and remarks similar to those recorded in preceeding Eq. (6.28) are also valid here. For this reason, we do not provide the full details of the application to the 3-phase system. The final expressions have the following form:

$$\begin{bmatrix} v_0 \\ v_d \\ v_q \end{bmatrix} = \begin{bmatrix} \frac{1}{3} & \frac{1}{3} & \frac{1}{3} \\ \cos \omega t & \cos\left(\omega t + \frac{2\pi}{3}\right) & \sin\left(\omega t + \frac{2\pi}{3}\right) \\ \sin \omega t & \sin\left(\omega t + \frac{4\pi}{3}\right) & \sin\left(\omega t + \frac{4\pi}{3}\right) \end{bmatrix} \begin{bmatrix} v_a \\ v_b \\ v_c \end{bmatrix} \qquad \text{...(6.34)}$$

and

$$\begin{bmatrix} v_a \\ v_b \\ v_c \end{bmatrix} = \frac{2}{3} \begin{bmatrix} \frac{3}{2} & \cos \omega t & \sin \omega t \\ \frac{3}{2} & \cos\left(\omega t + \frac{2\pi}{3}\right) & \sin\left(\omega t + \frac{2\pi}{3}\right) \\ \frac{3}{2} & \cos\left(\omega t + \frac{4\pi}{3}\right) & \sin\left(\omega t + \frac{4\pi}{3}\right) \end{bmatrix} \begin{bmatrix} v_0 \\ v_d \\ v_c \end{bmatrix} \qquad \text{...(6.35)}$$

These instantaneous voltage equations are similar to Eqs. (6.17) and (6.18), if we make the changes indicated immediately before Eq. (6.32). The voltages and currents expressing the voltage and current distribution vectors in the 3-phase (*abc*) and the 2-phase (0-*d*-*q*) reference frame, when the poly-phase system is balanced, may be summarised as given below:

$$i_a = \sqrt{2}I \cos(\omega t - \gamma) \qquad i_d = I_d = \sqrt{2}I \cos \gamma$$

$$i_b = \sqrt{2}I \cos\left(\omega t - \gamma + \frac{2\pi}{3}\right) \qquad i_q = I_q = \sqrt{2}I \sin \gamma$$

$$i_c = \sqrt{2}I \cos\left(\omega t - \gamma + \frac{4\pi}{3}\right)$$

$$v_a = -\frac{N}{3}\phi\omega \sin(\omega t - \gamma) \qquad v_d = V_d = \frac{N}{2}\phi\omega \sin \gamma$$

$$= -\frac{2}{3}\psi\omega \sin(\omega t - \gamma) \qquad = \psi_q \omega$$

$$v_b = -\frac{2}{3}\psi\omega \sin\left(\omega t - \gamma + \frac{2\pi}{3}\right) \qquad v_q = V_q = -\frac{N}{2}\phi\omega \cdot \cos \gamma$$

$$v_c = -\frac{2}{3}\psi\omega \sin\left(\omega t - \gamma + \frac{4\pi}{3}\right) \qquad = -\psi_d \omega$$

It is observed that the differences between these expressions for the 3-phase/2-phase transformations and those for the 2-phase ($0\alpha\beta$)/2-phase ($0dq$) transformations are accounted for by the effective turns per phase of the 3-phase winding which is $\left(\frac{2}{3}\right)$ of that for the 2-phase winding. The d and q pseudo-stationary windings have the same numbers of effective turns as the phase windings of the 2-phase system.

We illustrated $\alpha\beta/dq$ and $abc/0dq$ transformations by a continuous commutator winding with taps connected to slip rings, because we observed the same current and voltage distribution patterns represented by vectors, from the different sets of terminals. However, the vectors are the same if phase windings are used. In such cases, for $0dq$ system, the commutator and brushes are hypothetical. The transformation provides a more convenient way of describing the rotating poly-phase system, by expressing it as a pseudo-stationary dc system. A machine consisting of stationary stator windings and pseudo-stationary rotor windings on the same axis as the stator windings, with the two sets of windings on each member mutually perpendicular so that they have no mutual inductance, gives equations which are much simpler, than those of a machine having three stator windings at $\left(\frac{2\pi}{3}\right)$ and also rotor winding of the same type. All such windings have both self- and mutual-inductances, the latter having a continuous variation caused by the relative motion between each stator and each rotor phase windings. An additional advantage is that we can use the same 2-axis diagram or a simplified similar diagram for the analysis of almost all electrical machines. If the currents and voltages of the simplified analysis are not required, the terminal currents and voltages of the windings of the machine to be analysed, we may easily transform them back to the actual reference frame.

6.5.1 Clarification about 3-phase *A-B-C* Windings

A 3-phase *a-b-c* system of windings is shown in Figure 6.1. When the system is balanced and is operating without any flow of zero-sequence current through its phase layouts (that is, with $I_0 = 0$), we can describe the 3-phase operation by the following relations:

$$\left.\begin{aligned} i_a &= I_{\max}\cos(\omega t - \phi) & v_a &= V_{\max}\cos\omega t \\ i_b &= I_{\max}\cos\left(\omega t - \frac{2\pi}{3} - \phi\right) & v_b &= V_{\max}\cos\left(\omega t - \frac{2\pi}{3}\right) \\ i_c &= I_{\max}\cos\left(\omega t - \frac{4\pi}{3} - \phi\right) & v_c &= V_{\max}\cos\left(\omega t - \frac{4\pi}{3}\right) \end{aligned}\right\} \quad \text{...(6.36)}$$

and the transformed currents and voltages of the 2-phase equivalent 0-α-β system using $[C_1]$ transformations are given by:

$$\left.\begin{aligned} i_0 &= 0 & v_0 &= 0 \\ i_\alpha &= \sqrt{\frac{3}{2}}I_{\max}\cos(\omega t - \phi) & v_\alpha &= \sqrt{\frac{3}{2}}V_{\max}\cos\omega t \\ i_\beta &= \sqrt{\frac{3}{2}}I_{\max}\sin(\omega t - \phi) & v_\beta &= \sqrt{\frac{3}{2}}V_{\max}\sin\omega t \end{aligned}\right\} \quad \text{...(6.37)}$$

The above expressions can also be written as:

$$\left.\begin{aligned} I_a &= I_a & V_a &= V_a \\ I_b &= a^2 I_a & V_b &= a^2 V_a \\ I_c &= a I_a & V_c &= a V_a \end{aligned}\right\} \qquad ...(6.38)$$

The labelled form of $[C_1]$ implies that the current and voltage transformations are governed by the following relations:

$$\left.\begin{aligned} i_a &= \sqrt{\frac{2}{3}}\left(\frac{1}{\sqrt{2}}i_0 + i_\alpha\right) & v_a &= \sqrt{\frac{2}{3}}\left(\frac{1}{\sqrt{2}}v_0 + v_\alpha\right) \\ i_b &= \sqrt{\frac{2}{3}}\left(\frac{1}{\sqrt{2}}i_0 - \frac{1}{2}i_\alpha + \frac{\sqrt{3}}{2}i_\beta\right) & v_b &= \sqrt{\frac{2}{3}}\left(\frac{1}{\sqrt{2}}v_0 - \frac{1}{2}v_\alpha + \frac{\sqrt{3}}{2}v_\beta\right) \\ i_c &= \sqrt{\frac{2}{3}}\left(\frac{1}{\sqrt{2}}i_0 - \frac{1}{2}i_\alpha - \frac{\sqrt{3}}{2}i_\beta\right) & v_c &= \sqrt{\frac{2}{3}}\left(\frac{1}{\sqrt{2}}v_0 - \frac{1}{2}v_\alpha - \frac{\sqrt{3}}{2}v_\beta\right) \end{aligned}\right\} \qquad ...(6.39)$$

and the inverse current and voltage transformations, $I' = [C_t^*][I]$ and $[V'] = [C_t^*][V]$ may be found using transpose of $[C_1]$. If we expand the matrix forms:

$$\left.\begin{aligned} i_0 &= \frac{1}{\sqrt{3}}(i_a + i_b + i_c) & v_0 &= \frac{1}{\sqrt{3}}(v_a + v_b + v_c) \\ i_\alpha &= \sqrt{\frac{2}{3}}\left(i_a - \frac{1}{2}i_b - \frac{1}{2}i_c\right) & v_\alpha &= \sqrt{\frac{2}{3}}\left(v_a - \frac{1}{2}v_b - \frac{1}{2}v_c\right) \\ i_\beta &= \frac{1}{\sqrt{2}}(0 + i_b - i_c) & v_\beta &= \frac{1}{\sqrt{2}}(0 + v_b - v_c) \end{aligned}\right\} \qquad ...(6.40)$$

Thus, by relating the above equations, we can get the vector form of equivalent 2-phase relations:

$$\left.\begin{aligned} I_0 &= 0 & V_0 &= 0 \\ I_\alpha &= \sqrt{\frac{3}{2}}I_a & V_\alpha &= \sqrt{\frac{3}{2}}V_a \\ I_\beta &= -j\sqrt{\frac{3}{2}}I_a & V_\beta &= -j\sqrt{\frac{3}{2}}V_a \end{aligned}\right\} \qquad ...(6.41)$$

6.6 TWO PHASE SYMMETRICAL COMPONENTS TRANSFORMATIONS

The concept of symmetrical components was first proposed by Charles L. Fortescue in 1918. The principle is well known and it is presumed that the reader is already acquainted with the principle underlying the basis of the transformation. Although the principle is general, it has been extensively used in the analysis of unbalanced conditions in 3-phase systems.

The phase transformation $[C_1]$ replaces the balanced 3-phase supply voltages, for which $v_\beta = jv_\alpha$, and in determining the values of i_α and i_β, the voltages appear in combination of $v_\alpha + jv_\beta$ and $v_\alpha - jv_\beta$. This suggests that the analysis would be simplified, if substitutions are made in the form:

$$v_+ = K(v_\alpha + jv_\beta) \quad \text{and} \quad v_- = K(v_\alpha - jv_\beta)$$

where, K is an arbitrary constant, introduced to maintain orthogonality. With this substitution introduced and with $K = \dfrac{1}{\sqrt{2}}$, $[C_3]$ is in orthogonal form, and

$$[C_3] = \begin{matrix} & \begin{matrix} + & - \end{matrix} \\ \begin{matrix} \alpha \\ \beta \end{matrix} & \begin{bmatrix} 1 & 1 \\ -j & j \end{bmatrix} \end{matrix}$$

This is a standard transformation matrix for a two-by-two transformation for a 2-phase analysis. But as we primarily deal with 3-phase systems the general case of $[C_3]$ transformation should be of three-by-three form. The problem is tackled in the conventional way by incorporating the additional row and column for the zero-sequence current, which is to remain unchanged by the transformation. We use constant multipliers as before, to make orthogonality possible. Let the $[C_3]$, in 2-phase form, may be denoted as in Eq. (6.42), after including zero-sequence components as follows:

$$[C_3] = m\begin{bmatrix} n & 0 & 0 \\ 0 & 1 & 1 \\ 0 & -j & j \end{bmatrix} \qquad ...(6.42)$$

The old variables are the i_0, i_α and i_β, and $[C_3]$ represents the forward transformation. We should note that the nature of the elements of $[C_3]$ are imaginary and conjugate form should be used in the assurtion of orthogonality. Then,

$$[C_t^*][C] = m\begin{bmatrix} n & 0 & 0 \\ 0 & 1 & j \\ 0 & 1 & -j \end{bmatrix} m\begin{bmatrix} n & 0 & 0 \\ 0 & 1 & 1 \\ 0 & -j & j \end{bmatrix}$$

$$= m^2\begin{bmatrix} n^2 & 0 & 0 \\ 0 & 2 & 0 \\ 0 & 0 & 2 \end{bmatrix}$$

For orthogonality in this case,

$$m^2 n^2 = 1 \quad \text{and} \quad n^2 = 2$$

Hence, $m = \dfrac{1}{\sqrt{2}}$ and $n = \sqrt{2}$

and the final form of $[C_3]$, for the 2-phase system, is as follows:

$$[C_3] = \frac{1}{\sqrt{2}} \begin{array}{c} \\ 0 \\ \alpha \\ \beta \end{array} \begin{array}{c} \begin{array}{ccc} 0 & + & - \end{array} \\ \begin{bmatrix} \sqrt{2} & 0 & 0 \\ 0 & 1 & 1 \\ 0 & -j & j \end{bmatrix} \end{array} \qquad \text{...(6.43)}$$

6.6.1 3-phase Symmetrical Components Transformation $[C_{13}]$

The conventional 3-phase symmetrical component transformation is, according to our representation $[C_{13}]$, which is the same as follows:

$$[C_{13}] = [C_{1t}]\,[C_3]$$

Direct multiplication requires the use of the 3×3 form of $[C_3]$ contained in Eq. (6.43) and it leads to:

$$[C_{13}] = [C_{1t}]\,[C_3] = \sqrt{\frac{2}{3}} \begin{array}{c} \\ a \\ b \\ c \end{array} \begin{array}{c} \begin{array}{ccc} 0 & \alpha & \beta \end{array} \\ \begin{bmatrix} \frac{1}{\sqrt{2}} & 1 & 0 \\ \frac{1}{\sqrt{2}} & -\frac{1}{2} & \frac{\sqrt{3}}{2} \\ \frac{1}{\sqrt{2}} & -\frac{1}{2} & -\frac{\sqrt{3}}{2} \end{bmatrix} \end{array} \frac{1}{\sqrt{2}} \begin{array}{c} \\ 0 \\ \alpha \\ \beta \end{array} \begin{array}{c} \begin{array}{ccc} 0 & + & - \end{array} \\ \begin{bmatrix} \sqrt{2} & 0 & 0 \\ 0 & 1 & 1 \\ 0 & -j & j \end{bmatrix} \end{array}$$

$$= \frac{1}{\sqrt{3}} \begin{bmatrix} 1 & 1 & 1 \\ 1 & a^2 & a \\ 1 & a & a^2 \end{bmatrix} \qquad \text{...(6.44)}$$

where, $a = -\frac{1}{2} + j\frac{\sqrt{3}}{2}$ and $a^2 = -\frac{1}{2} - j\frac{\sqrt{3}}{2}$.

The symmetrical components transformation use the cube roots of unity as denoted by the standard nomenclature, as follows:

$$a = -\frac{1}{2} + j\frac{\sqrt{3}}{2}; \quad a^2 = -\frac{1}{2} - j\frac{\sqrt{3}}{2} \text{ and } a^3 = 1$$

These roots can also be denoted in the polar form as:

$$1 = e^{j0} = e^{j360°} = e^{j720°} = \ldots \text{ so that}$$

$$\sqrt[3]{1} = e^{j0},\ e^{j360°},\ e^{j720°} \ \ldots \text{ and}$$

$$a = e^{j120°},\ e^{j480°},\ e^{j840°} \ \ldots$$

$$a^2 = e^{j240°},\ e^{j600°},\ e^{j960°} \ \ldots$$

We can write down the relationships between the sequence currents and the phase currents as follows:

$$\left.\begin{aligned} &I_{0a} = I_{0b} = I_{0c} = I_0 \\ &I_{+a} = I_+,\ I_{+b} = a^2 I_+,\ I_{+c} = aI_+ \\ &I_{-a} = I_-,\ I_{-b} = aI_-,\ I_{-c} = a^2 I_- \end{aligned}\right\} \qquad ...(6.45)$$

and also,

$$\left.\begin{aligned} &I_a = I_0 + I_+ + I_- \\ &I_b = I_0 + a^2 I_+ + aI_- \\ &I_c = I_0 + aI_+ + a^2 I_- \end{aligned}\right\} \qquad ...(6.46)$$

If we carefully study the expression for $[C_{13}]$ in Eq. (6.38), we observe some striking revelations. $[C_{13}]$ is orthogonal, but the orthogonality is achieved by introducing a constant common factor of $\left(\frac{1}{\sqrt{3}}\right)$, in the transformation of phase currents to sequence currents. Since the conjugates of a, that is $a^* = a^2$ and $(a^2)^* = a$, the inverse relations read:

$$\left.\begin{aligned} &i_0 = \frac{1}{\sqrt{3}}(i_a + i_b + i_c) && v_0 = \frac{1}{\sqrt{3}}(v_a + v_b + v_c) \\ &i_+ = \frac{1}{\sqrt{3}}(i_a + a i_b + a^2 i_c) && v_+ = \frac{1}{\sqrt{3}}(v_a + a v_b + a^2 v_c) \\ &i_- = \frac{1}{\sqrt{3}}(i_a + a^2 i_b + a i_c) && v_- = \frac{1}{\sqrt{3}}(v_a + a^2 v_b + a v_c) \end{aligned}\right\} \qquad ...(6.47)$$

And, in the particular case, when the 3-phase currents (and voltages) are balanced,

$$i_b = a^2 i_a \quad \text{and} \quad i_c = a\, i_a$$

Hence, under balanced conditions:

$$\left.\begin{aligned} &I_0 = \frac{1}{\sqrt{3}}(1 + a^2 + a)I_a = 0 \\ &i_+ = \frac{1}{\sqrt{3}}(1 + a^3 + a^3)I_a = \sqrt{3}\, I_a \\ &i_- = \frac{1}{\sqrt{3}}(1 + a^4 + a^2)I_a = 0 \end{aligned}\right\} \qquad ...(6.48)$$

6.7 PHYSICAL SIGNIFICANCE OF SEQUENCE IMPEDANCES

6.7.1 Zero-sequence Impedance, Z_0

In order to appreciate the physical significance of Z_0, we must first arbitrarily create a situation where a pure zero-sequence current only can flow in a 3-phase winding configuration, such that co-phasal current of equal magnitude flows in a balanced 3-phase winding system. This can be easily done by connecting the three phase-windings either in series or in parallel as shown in Figure 6.7, with application of single-phase voltages across the open terminals of the combination.

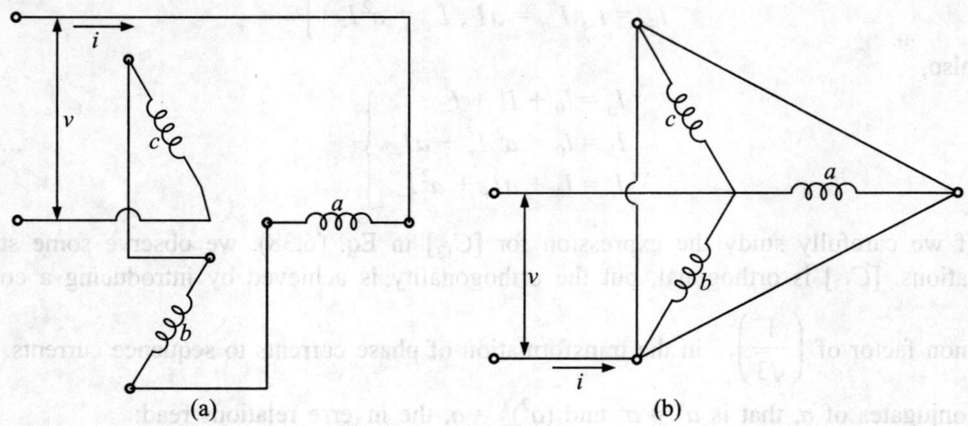

FIGURE 6.7 (a) Series and (b) parallel connections of 3-phase balanced windings to carry pure zero-sequence currents in phases.

By visual inspection of the circuit arrangement in Figure 6.7(a), we can write,

$$i_a = i_b = i_c = i$$

where, i is the measured value of the current in the series circuit.

But from Eq. (6.41), we observe that:

$$i_0 = \frac{1}{\sqrt{3}}(i_a + i_b + i_c) = \frac{3}{\sqrt{3}}i$$

While,

$$i_+ = \frac{1}{\sqrt{3}}(i_a + a\,i_b + a^2\,i_c) = 0, \text{ and}$$

$$i_- = \frac{1}{\sqrt{3}}(i_a + a^2 i_b + a\,i_c) = 0$$

Hence, $i_0 = \sqrt{3}\,i$ or $i = \dfrac{i_0}{\sqrt{3}}$

Also, $v_a + v_b + v_c = 3v$, as $v_a = v_b = v_c = v$

By a similar exercise, we find that,

$v = \sqrt{3}v_0$ while, $v_+ = 0$ and $v_- = 0$

Since, we define the impedance offered by the circuit per phase, $Z_0 = \dfrac{v_0}{i_0}$ and the measured impedance offered by the circuit,

$$\frac{v}{i} = \frac{\sqrt{3}v_0}{\dfrac{i_0}{\sqrt{3}}} = 3\frac{v_0}{i_0} = 3Z_0$$

or the value of Z_0 is one-third of the impedance measured by this method. A similar conclusion can be arrived at for the circuit arrangement in Figure 6.7(b). From visual inspection of the arrangement, we can conclude that:

$$v_a = v_b = v_c = v$$

and using expression (6.47), we have

$$v_0 = \frac{1}{\sqrt{3}}(v_a + v_b + v_c) = \sqrt{3}v$$

while,

$v_+ = \dfrac{1}{\sqrt{3}}(v_a + av_b + a^2v_c) = 0$, and $v_- = \dfrac{1}{\sqrt{3}}(v_a + a^2v_b + av_c)$ is also equal to zero.

Hence, the single-phase voltage applied across one of the phase terminals $a - n$ of the balanced 3-phase winding in the circuit arrangement of Figure 6.7(b) amounts to:

$$v = \frac{v_0}{\sqrt{3}} \quad \text{or} \quad v_0 = \sqrt{3}v$$

Also, $i_0 = \dfrac{v_0}{Z_0} = \dfrac{\sqrt{3}v}{Z_0}$, while, as before, $v_+ = v_- = 0$.

Again, from visual inspection of the circuit arrangement, $i_a = i_b = i_c = \dfrac{v}{Z_0} = \dfrac{i}{3}$.

But from Eq. 6.47, we have

$$i_0 = \frac{1}{\sqrt{3}}(i_a + i_b + i_c) = \sqrt{3}\,i$$

Hence,

$$Z_0 = \frac{v_0}{i_0} = \frac{\sqrt{3}v}{\sqrt{3}i}$$

In other words, the zero-sequence offered by the circuit per phase is three times the single phase impedance measured by the above circuit arrangement.

The result bears out that since in Figure 6.7(a), the value of Z_0 appears in series, while in Figure 6.7(b), Z_0 offers itself in parallel, the above conclusions are in order.

6.7.2 Comparative Quantification of Z_0

The magnitude of Z_0, when compared with the self-inductance of the winding, is ordinarily small at normal saturation levels. In other words:

$$X_L > X_0$$

The value of X_L appears in the relation,

$$Z = R + jX_L$$

and stands for the inductive reactance $j\omega L$ of the individual phase windings, offering impedance Z to a single-phase supply of frequency f. L_L denotes the self-inductance of the winding due to its own current flow. X_0 similarly represents the zero-sequence reactance, $X_0 = j\omega l$ per phase. X_0 is also the leakage reactance per phase of the winding arrangement as seen by the excitation voltage effective across each phase. Hence, l can also be identified as the leakage inductance of the winding.

To physically appreciate why l is identified as the *leakage inductance* of a winding, let us investigate the flux-distribution scenario in the magnetic field in a poly-phase transformer and a poly-phase rotating machine schematic. We consider that the poly-phase winding arrangements are inter-connected between their phases according to Figure 6.7(a). Let Figure 6.8(a) depict a poly-phase winding arrangement and the magnetic circuit schematic of a poly-phase transformer along with the directions of flux-distribution when the windings are connected according to Figure 6.7(a), and energised through a 1-phase supply system. The three sets of ampere-conductor distributions are instantaneously established in the respective core limbs as denoted by the arrows. These have equal values and are in time-phase, all acting in the same direction in space. Under such a condition, no mutual flux—that is flux components linking two phase

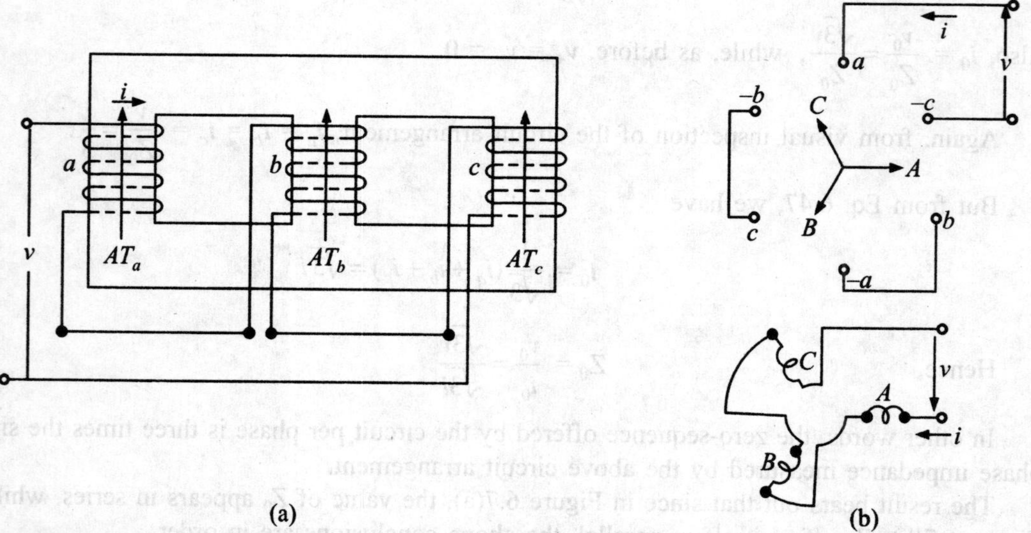

FIGURE 6.8 Balanced 3-phase winding schematic and ampere–conductor/mmf distribution pattern produced in the magnetic circuit when inter-connected phase-to-phase as per Figure 6.7(a). (a) A three-phase transformer/inductor with inter-connections as per Figure 6.7(a) showing directions of instantaneous ampere–conductor distributions along each magnetic limb, when energised with with 1-phase balanced supply. (b) A smooth air-gap 3-phase rotating machine with series inter-connection of the phases as per Figure 6.7(a), showing relative instantaneous directions of phasor mmfs.

windings—exists effectively. The only flux linking each winding is the flux which, while passing through the magnetic core of a winding has to find a return path all the way through the air. As the reluctance of the air-path is high, the consequent inductance and reactances offered by the winding are low.

Figure of 6.8(b) similarly depicts the instantaneous mmf distributions inside a 2-pole rotating electrical machine with smooth air-gap surface. We have chosen to denote the phase windings by single coils only per phase, with alignments such that the phase mmf's are $\left(\frac{2\pi}{3}\right)$ electrical angle apart in space. When the coils are inter-connected in accordance with Figure 6.7(a), as before, the ampere-turns produced are equal in magnitude and are in time-phase, so that they add up to zero. Under this condition, no mutual-flux between two phase windings can exist and only leakage flux of relatively small magnitude can exist in each phase, as shown in Figure 6.8(b).

6.7.3 Positive Sequence Impedance, Z_1

We have found that a balanced system of 3-phase currents and voltages are the same things as the equivalent positive sequence systems. The resulting depiction of a balanced positive sequence system on a per phase basis can be described by the following voltage equation:

$$v_a = Z_1 i_a$$

In this relationship, the positive sequence impedence is the effective impedance offered by each phase, when a balanced 3-phase circuit is fed from a balanced 3-phase source, as shown in Figure 6.9.

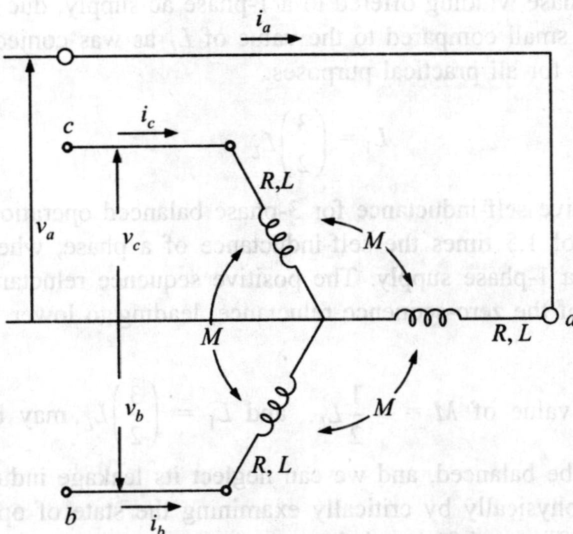

FIGURE 6.9 A smooth air-gap balanced 3-phase machine winding of a rotating machine fed from a balanced 3-phase supply system.

6.7.4 Quantitative Relationship among Self-, Mutual- and Leakage-inductances

A relationship among the self-, mutual- and leakage-inductances can be broadly laid down from analytical considerations based on physical analysis arrived at in Sections 6.7.1 and 6.7.2. Thus,

$$\left.\begin{aligned} X_0 &= X_L + 2X_m \\ \text{that is,} \quad l &= L_L + 2M \\ \text{and} \quad M &= -\frac{1}{2}(L_L - l) \end{aligned}\right\} \qquad \text{...(6.49)}$$

$$\left.\begin{aligned} \text{Also,} \quad X_1 &= X_L - X_m \\ \text{that is,} \quad L_1 &= L_L - M \\ &= L_L + \frac{1}{2}(L_L - l) \end{aligned}\right\} \qquad \text{...(6.50)}$$

Using Eq. (6.50), we have

$$L_1 = \frac{3}{2}L_L - \frac{1}{2}l \qquad \text{...(6.51)}$$

In other words, using Eq. (6.51), which needs to be physically authenticated, the positive sequence inductance of a phase winding is predicted to be slightly less than $\left(\frac{3}{2}\right)$ times the self-inductance of a phase winding offered to a 1-phase ac supply, due to its own flux. And if l is neglected as very small compared to the value of L_L as was conjectured in Section 6.7.2, we may conclude that for all practical purposes:

$$L_1 = \left(\frac{3}{2}\right)L_L$$

That is, the effective self-inductance for 3-phase balanced operation on a phase-by-phase basis is of the order of 1.5 times the self-inductance of a phase, when measured with only the phase excited by a 1-phase supply. The positive sequence reluctance is thus, in general, lesser than the value of the zero-sequence reluctance, leading to lower values of X_0, the zero-sequence reactance.

The approximate value of $M = -\frac{1}{2}L_L$, and $L_1 = \left(\frac{3}{2}\right)L_L$, may be surmised when the system is assumed to be balanced, and we can neglect its leakage inductance. We can verify the above conjecture physically by critically examining the state of operation of the internal magnetic circuit using Figure 6.8(a) and (b).

If only one of the three-phase windings in Figure 6.8(a) or (b) is energised, the flux flowing in one direction through the magnetic core of the concerned phase divides equally so that only one-half of the outgoing flux returns in the *reverse direction* through each of the remaining two core limbs in Figure 6.8(a). As in a balanced operation, the number of turns is equal in each phase winding, the coefficients of self- and mutual-inductance are proportional to the amount of flux linked by the other phases. But as the self-inductance is always positive, we may conclude that the nature of M is equal $-\frac{1}{2}L_L$.

The ampere-turns of the energised phase must develop the full-flux in the core of the concerned phase and only half of this flux value flows in the other two cores in parallel and effective along the other two winding axes or phases. Hence, the effective permeance of this magnetic circuit has to be 1.5 times the permeance of the single core.

For a balanced 3-phase operation, the sum of the instantaneous currents must always be zero, and the sum of the instantaneous fluxes is also zero. If follows that the current in the winding mounted on any one limb is just sufficient to set up the flux in that concerned limb. The effective reluctance of this connection is, thus, the reluctance of a single core. Since the coefficient of self-inductance is inversely proportional to the values of the reluctance, the result, $L_1 = \left(\frac{3}{2}\right)L_L$ is validated.

With reference to Figure 6.8(b), our assumption of a balanced magnetic circuit, which is only approximately so for the case in Figure 6.8(a), is more valid, and the conclusion that

$$M = -\frac{1}{2}L_L$$

is more justified. This result is clearly manifest from test results also. The nature of the windings, their layouts and the magnetic circuit are such that the reluctances of each phase are balanced.

6.8 CASCADED TRANSFORMATIONS IN THE 3-PHASE FORM

In Section 6.6.1 we had described the rudiments of one of the three-by-three forms of cascaded transformations—the transformation $[C_{13}]$. Its applications are discussed in greater detail when we take up the modelling of a 3-phase wound-rotor induction motor and 3-phase smooth air-gap synchronous machines.

In the present section, we intend to describe the basis of the two more compound transformations in their 3-by-3 forms—the transformations $[C_{12}]$ and $[C_{123}]$. We follow the procedure laid down in the case of transformation matrix $[C_{13}]$, which involves multiplication of the transformation matrices, i.e.,

$$[C_{12}] = [C_{1t}]\,[C_{2t}]$$

and

$$[C_{123}] = [C_{12}]\,[C_3]$$

We borrow the matrix values of $[C_{1t}]$ from Eq. (6.22) in Section 6.3 and modified version of $[C_2]$ from Eq. (6.25), for obtaining the matrix form of $[C_{12}]$.

6.8.1 Cascaded Transformation $[C_{12}]$ in Three-by-Three Form

As pointed out:

$$[C_{12}] = [C_{1t}]\,[C_2] = \sqrt{\frac{2}{3}} \begin{array}{c} \\ a \\ b \\ c \end{array}\!\!\begin{array}{c} \begin{array}{ccc} 0 & \alpha & \beta \end{array} \\ \begin{bmatrix} \frac{1}{\sqrt{2}} & 1 & 0 \\ \frac{1}{\sqrt{2}} & -\frac{1}{2} & \frac{\sqrt{3}}{2} \\ \frac{1}{\sqrt{2}} & -\frac{1}{2} & -\frac{\sqrt{3}}{2} \end{bmatrix} \end{array} \begin{array}{c} \\ 0 \\ \alpha \\ \beta \end{array}\!\!\begin{array}{c} \begin{array}{ccc} 0 & q & d \end{array} \\ \begin{bmatrix} 1 & 0 & 0 \\ 0 & \sin\theta & \cos\theta \\ 0 & \cos\theta & -\sin\theta \end{bmatrix} \end{array}$$

$$= \sqrt{\frac{2}{3}} \begin{array}{c} \\ a \\ b \\ c \end{array}\!\!\begin{array}{c} \begin{array}{ccc} 0 & q & d \end{array} \\ \begin{bmatrix} \frac{1}{\sqrt{2}} & \sin\theta & \cos\theta \\ \frac{1}{\sqrt{2}} & \sin\theta_2 & \cos\theta_2 \\ \frac{1}{\sqrt{2}} & \sin\theta_3 & \cos\theta_3 \end{bmatrix} \end{array}$$

where, $$\theta_2 = \theta + \left(\frac{2\pi}{3}\right) \text{ and } \theta_3 = \theta + \left(\frac{4\pi}{3}\right) \qquad ...(6.52)$$

In the above transformation θ_2 and θ_3 have been introduced and the following trigonometric identities are used for simplification:

$$-\frac{1}{2}\sin\theta + \frac{\sqrt{3}}{2}\cos\theta = \sin\theta_2 = \sin\left(\theta + \frac{2\pi}{3}\right)$$

and $$-\frac{1}{2}\sin\theta - \frac{\sqrt{3}}{2}\cos\theta = \sin\theta_3 = \sin\left(\theta + \frac{4\pi}{3}\right)$$

Also, $$-\frac{1}{2}\cos\theta - \frac{\sqrt{3}}{2}\sin\theta = \cos\theta_2 = \cos\left(\theta + \frac{2\pi}{3}\right)$$

and $$-\frac{1}{2}\cos\theta + \frac{\sqrt{3}}{2}\sin\theta = \cos\theta_3 = \cos\left(\theta + \frac{4\pi}{3}\right) \qquad ...(6.53)$$

6.8.2 Cascaded Transformation $[C_{123}]$ in Three-by-Three Form

The complete cascaded transformation $[C_{123}]$ for the armature of a synchronous machine with 3-phase windings can be realised by multiplying the result contained in Eq. (6.52) by the three-by-three form of symmetrical components transformation derived in Eq. (6.43) in Section 6.7.

Thus,

$$[C_{123}] = [C_{12}]\,[C_3] = \sqrt{\frac{2}{3}} \begin{array}{c} \\ a \\ b \\ c \end{array} \begin{array}{c} \begin{array}{ccc} 0 & q & d \end{array} \\ \begin{bmatrix} \frac{1}{\sqrt{2}} & \sin\theta & \cos\theta \\ \frac{1}{\sqrt{2}} & \sin\theta_2 & \cos\theta_2 \\ \frac{1}{\sqrt{2}} & \sin\theta_3 & \cos\theta_3 \end{bmatrix} \end{array} \frac{1}{\sqrt{2}} \begin{array}{c} \\ 0 \\ d \\ q \end{array} \begin{array}{c} \begin{array}{ccc} 0 & + & - \end{array} \\ \begin{bmatrix} \sqrt{2} & 0 & 0 \\ 0 & 1 & 1 \\ 0 & -j & j \end{bmatrix} \end{array}$$

$$= \frac{1}{\sqrt{3}} \begin{array}{c} \\ a \\ b \\ c \end{array} \begin{array}{c} \begin{array}{ccc} 0 & + & - \end{array} \\ \begin{bmatrix} 1 & e^{j\theta} & e^{-j\theta} \\ 1 & e^{j\theta_2} & e^{-j\theta_2} \\ 1 & e^{j\theta_3} & e^{-j\theta_3} \end{bmatrix} \end{array} \qquad \text{...(6.54)}$$

The relation used for simplification from the trigonometric form to complex form is:

$$\cos\theta \pm j\sin\theta = e^{\pm j\theta}$$

So, the relation between the phase currents and voltages and their sequence components are as follows:

$$\left.\begin{array}{ll} I_a = \dfrac{1}{\sqrt{3}}[I_0 + e^{-j\theta}I_+ + e^{j\theta}I_-] & I_0 = \dfrac{1}{\sqrt{3}}[I_a + I_b + I_c] \\ I_b = \dfrac{1}{\sqrt{3}}[I_0 + e^{-j\theta_2}I_+ + e^{j\theta_2}I_-] & I_+ = \dfrac{1}{\sqrt{3}}[e^{j\theta}I_a + e^{j\theta_2}I_b + e^{j\theta_3}I_c] \\ I_c = \dfrac{1}{\sqrt{3}}[I_0 + e^{-j\theta_3}I_+ + e^{j\theta_3}I_-] & I_- = \dfrac{1}{\sqrt{3}}[e^{-j\theta}I_a + e^{-j\theta_2}I_b + e^{j\theta_3}I_c] \end{array}\right\} \qquad \text{...(6.55)}$$

and

$$\left.\begin{array}{ll} V_a = \dfrac{1}{\sqrt{3}}[V_0 + e^{-j\theta}V_+ + e^{j\theta}V_-] & V_0 = \dfrac{1}{\sqrt{3}}[V_a + V_b + V_c] \\ V_b = \dfrac{1}{\sqrt{3}}[V_0 + e^{-j\theta_2}V_+ + e^{j\theta_2}V_-] & V_+ = \dfrac{1}{\sqrt{3}}[e^{j\theta}V_a + e^{j\theta_2}V_b + e^{j\theta_3}V_c] \\ V_c = \dfrac{1}{\sqrt{3}}[V_0 + e^{-j\theta_3}V_+ + e^{j\theta_3}V_-] & V_- = \dfrac{1}{\sqrt{3}}[e^{-j\theta}V_a + e^{-j\theta_2}V_b + e^{-j\theta_3}V_c] \end{array}\right\} \qquad \text{...(6.56)}$$

6.9 CONSOLIDATION OF DERIVED TRANSFORMATIONS

We record here the relations so far derived under transformations among all frames of reference.

1. For transformation from 3-phase *abc* system to 2-phase $0\alpha\beta$ system in 3-by-3 form:

$$[C_1] = \sqrt{\frac{2}{3}} \begin{array}{c} \\ a \\ b \\ c \end{array} \begin{array}{c} \begin{matrix} 0 & \alpha & \beta \end{matrix} \\ \begin{bmatrix} \frac{1}{\sqrt{2}} & 1 & 0 \\ \frac{1}{\sqrt{2}} & -\frac{1}{2} & \frac{\sqrt{3}}{2} \\ \frac{1}{\sqrt{2}} & -\frac{1}{2} & -\frac{\sqrt{3}}{2} \end{bmatrix} \end{array} \qquad [C_1]^{-1} = [C_t] = \sqrt{\frac{2}{3}} \begin{array}{c} \\ 0 \\ \alpha \\ \beta \end{array} \begin{array}{c} \begin{matrix} a & b & c \end{matrix} \\ \begin{bmatrix} \frac{1}{\sqrt{2}} & \frac{1}{\sqrt{2}} & \frac{1}{\sqrt{2}} \\ 1 & -\frac{1}{2} & -\frac{1}{2} \\ 0 & \frac{\sqrt{3}}{2} & -\frac{\sqrt{3}}{2} \end{bmatrix} \end{array}$$

For mmf transformation:

$$[F_{0\alpha\beta}] = \frac{\sqrt{3}}{2}[C_{1t}][F_{abc}] \qquad [F_{abc}] = \sqrt{\frac{2}{3}}[C_1][F_{0\alpha\beta}]$$

For current transformation:

$$[i_{0\alpha\beta}] = [C_{1t}]\ [i_{abc}] \qquad [i_{abc}] = [C_1]\ [i_{0\alpha\beta}]$$

For flux-linkage transformation:

$$[\psi_{0\alpha\beta}] = [C_{1t}]\ [\psi_{abc}] \qquad [\psi_{abc}] = [C_1]\ [\psi_{0\alpha\beta}]$$

For voltage transformation:

$$[v_{0\alpha\beta}] = [C_{1t}]\ [v_{abc}] \qquad [v_{abc}] = [C_1]\ [v_{0\alpha\beta}]$$

For inductance transformation:

$$[L_{0\alpha\beta}] = [C_{1t}]\ [L_{abc}]\ [C_1] \qquad [L_{abc}] = [C_1]\ [L_{0\alpha\beta}]\ [C_{1t}]$$

For impedance transformation:

$$[Z_{0\alpha\beta}] = [C_{1t}]\ [Z_{abc}]\ [C_{11}] \qquad [Z_{abc}] = [C_1]\ [Z_{0\alpha\beta}]\ [C_{1t}]$$

2. For transformation from 2-phase ($0\alpha\beta$) system to ($0dq$) system in 3-by-3 form:

$$[C_2] = \begin{array}{c} \\ 0 \\ \alpha \\ \beta \end{array} \begin{array}{c} \begin{matrix} 0 & q & d \end{matrix} \\ \begin{bmatrix} 1 & 0 & 0 \\ 0 & \sin\theta & \cos\theta \\ 0 & \cos\theta & -\sin\theta \end{bmatrix} \end{array} \qquad [C_{2t}] = \begin{array}{c} \\ 0 \\ q \\ d \end{array} \begin{array}{c} \begin{matrix} 0 & \alpha & \beta \end{matrix} \\ \begin{bmatrix} 1 & 0 & 0 \\ 0 & \sin\theta & \cos\theta \\ 0 & \cos\theta & -\sin\theta \end{bmatrix} \end{array}$$

In final 3-by-3 form of $[C_{12}]$

$$[C_{12t}] = \sqrt{\frac{2}{3}}\begin{array}{c} \\ a \\ b \\ c \end{array}\begin{array}{c} \begin{array}{ccc} 0 & q & d \end{array} \\ \begin{bmatrix} \dfrac{1}{\sqrt{2}} & \dfrac{1}{\sqrt{2}} & \dfrac{1}{\sqrt{2}} \\ \sin\theta & \sin\left(\theta + \dfrac{2\pi}{3}\right) & \sin\left(\theta + \dfrac{4\pi}{3}\right) \\ \cos\theta & \cos\left(\theta + \dfrac{2\pi}{3}\right) & \cos\left(\theta + \dfrac{4\pi}{3}\right) \end{bmatrix} \end{array}$$

Hence,

$$[C_{12}] = \sqrt{\frac{2}{3}}\begin{array}{c} \\ 0 \\ q \\ d \end{array}\begin{array}{c} \begin{array}{ccc} a & b & c \end{array} \\ \begin{bmatrix} \dfrac{1}{\sqrt{2}} & \sin\theta & \cos\theta \\ \dfrac{1}{\sqrt{2}} & \sin\left(\theta + \dfrac{2\pi}{3}\right) & \cos\left(\theta + \dfrac{2\pi}{3}\right) \\ \dfrac{1}{\sqrt{2}} & \sin\left(\theta + \dfrac{4\pi}{3}\right) & \cos\left(\theta + \dfrac{4\pi}{3}\right) \end{bmatrix} \end{array}$$

3. For transformation from 2-phase $(0\alpha\beta)$ to 2-phase symmetrical component in $(0+-)$ system in 3-by-3 form and (abc) to $(0+-)$ system:

$$[C_3] = \frac{1}{\sqrt{2}}\begin{array}{c} \\ 0 \\ \alpha \\ \beta \end{array}\begin{array}{c} \begin{array}{ccc} 0 & + & - \end{array} \\ \begin{bmatrix} \sqrt{2} & 0 & 0 \\ 0 & 1 & 1 \\ 0 & -j & j \end{bmatrix} \end{array} \qquad [C_{3t}] = \frac{1}{\sqrt{2}}\begin{array}{c} \\ 0 \\ + \\ - \end{array}\begin{array}{c} \begin{array}{ccc} 0 & \alpha & \beta \end{array} \\ \begin{bmatrix} \sqrt{2} & 0 & 0 \\ 0 & 1 & -j \\ 0 & 1 & j \end{bmatrix} \end{array}$$

$$[C_{13}] = \sqrt{\frac{1}{3}}\begin{array}{c} \\ a \\ b \\ c \end{array}\begin{array}{c} \begin{array}{ccc} 0 & + & - \end{array} \\ \begin{bmatrix} 1 & 1 & 1 \\ 1 & a^2 & a \\ 1 & a & a^2 \end{bmatrix} \end{array} \qquad \text{where, } a = -\frac{1}{2} + j\frac{\sqrt{3}}{2} \text{ and } a^2 = -\frac{1}{2} - j\frac{\sqrt{3}}{2}$$

$$[C_{13t}] = \frac{1}{\sqrt{3}}\begin{array}{c} \\ 0 \\ + \\ - \end{array}\begin{array}{c} \begin{array}{ccc} a & b & c \end{array} \\ \begin{bmatrix} 1 & 1 & 1 \\ 1 & a^2 & a \\ 1 & a & a^2 \end{bmatrix} \end{array}$$

$$[C_{123}] = \frac{1}{\sqrt{3}} \begin{matrix} \\ a \\ b \\ c \end{matrix} \begin{matrix} 0 & + & - \\ \begin{bmatrix} 1 & e^{j\theta} & e^{-j\theta} \\ 1 & e^{j\theta_2} & e^{-j\theta_2} \\ 1 & e^{j\theta_3} & e^{-j\theta_3} \end{bmatrix} \end{matrix}$$

where, $\theta_2 = \theta + \frac{2\pi}{3}$ and $\theta_3 = \theta + \frac{4\pi}{3}$

OBJECTIVE TYPE QUESTIONS

6.1 Linear transformation from an original to a new system of variables and vice versa is governed by linear equations. Consider the following statements:

(i) Application of logarithm for multiplication of two numbers is a linear-transformation.
(ii) Laplace transform, in transforming time-domain equations to s-domain equations or vice versa is a linear transformation.
(iii) Symmetrical components of currents applied to three-phase currents are linear transformation.
(iv) the process of referring secondary quantities to primary or its reverse is a linear transformation.
(v) A transformation may be arbitrary and chosen purely on mathematically based grounds to reduce/simply performance equations, and have physical interpretation, although the latter is not absolutely necessary.

Choose the correct alternative:

(a) All above statements are correct
(b) Only statements (i), (ii), (iii) and (iv) are true
(c) Statements (ii) and (iii) are not valid
(d) Only statements (i) and (v) are correct.

6.2 In machine analysis, currents are transformed under condition of invariance, as the transformed system of currents must produce mmfs identical with that of the original system of currents, so that the new system must be in time and space invariance with the original:

(a) The above statement is not true
(b) The statement is only partially true in that the time invariance only should be maintained
(c) The above statement correctly describe the process of transformation
(d) The requirement of space invariance only needs to be satisfied

6.3 Transformation between three-phase system to an equivalent two-phase system is frequently desired, because:

(i) The process of analysis becomes elegent
(ii) The number of variables in equations is significantly reduced by the transformation

(iii) mutual coupling between constituent windings in the two-phase transformed system is absent.

(iv) a reverse transformation to the original three-phase system is not required only if power invariant condition is imposed.

(v) concept of two phase depiction is easier to purceive.

(vi) pseudo-stationary coil in the representation of rotor armature, forms the basis of denoting circuits of rotating machines.

Identify the correct alternative:

(a) Statements (i) and (ii) are only true

(b) Statements (iii), (iv), (v) and (vi) are valid

(c) All statements are valid

(d) Statements (i) and (v) are not valid.

Answers

6.1 (a) **6.2** (c) **6.3** (d)

NUMERICAL PROBLEMS

6.1 A 3-phase induction machine has the following per phase parameters referred to the stator:

Stator resistance — 0.6 ohm

Rotor resistance — 0.76 ohm

Stator and rotor leakage reactance — 4.5 ohm

Magnetizing reactance — 120 ohm

Obtain the values of the parameters of the equivalent 2-phase induction machine, if tis per-phase turns are:

(a) the same as that of its 3-phase counterpart.

(b) 1.5 times that of the 3-phase induction machine

(c) $\sqrt{1.5}$ times that of the 3-phase counterpart.

[**Ans.** (a) 0.4, 0.5, 3.0, 80 ohm; (b) 0.9, 1.25, 6.75, 180 ohm (c) Same as those of the 3-phase machine]

6.2 A 2-pole synchronous machine has its 3-phase armature mounted on its rotor frame. When fed from a 3-phase supply, the armature currents are given by:

$$i_a = I_m \cos \omega t$$

$$i_b = I_m \cos\left(\omega t - \frac{2\pi}{3}\right)$$

$$i_c = I_m \cos\left(\omega t - \frac{4\pi}{3}\right)$$

At time $t = 0$, the rotor phase a-axis is aligned to the axis of the field winding. Show that the direct and quadrature axes current components are given by

$$i_d = \sqrt{\frac{3}{2}} I_m \quad \text{and} \quad i_q = 0.$$

6.3 A 3-phase, 6 pole, 25 Hz squirrel cage induction motor delivers a torque of 400 N-m to a load at a rated voltage of 250 volt. The stator winding of the machine is replaced by a 2-phase stator, with number of poles, the effective turns per phase and the number of conductors maintained the same as the 3-phase stator. Compute:

(a) the rated phase voltage of the 2-phase machine.

(b) the torque, if the 2-phase induction motor is supplied at the rated value of the phase voltage obtained from part (a) above from a 2-phase 50 Hz supply. Please ignore the losses occurring in the machine.

(c) the kW rating of the machine, when working on the 3-phase, 25 Hz, as well as the 2-phase, 50 Hz mains.

[**Ans.** (a) 166.67 volt; (b) T_{e2} = 88.89 N-m
(c) P_3 = 31.42 kW, P_2 = 13.96 kW]

CHAPTER 7

Analysis of Actual Machine Configurations

7.1 CONVENTIONS ADOPTED FOR GENERAL ANALYSIS

Before commencing the study of motor and generator action of actual configurations, it becomes absolutely necessary to elaborate the conventions adopted in this book about the flow of current and the resulting electro-magnetic torque. It is also important for us to indicate clearly the basis on which these assumptions are adopted. A coil is invariably represented by a cluster of conductors with a specified axis of its influence along the axis of its symmetry. When current flows through such a coil, an mmf is developed along this axis and flux is produced.

A positive mmf/flux is produced when it is along the radial direction and is pointed radially outwards away from the centre of the representative diagram. This means that the direction of *turning* of the coil is in the counter-clockwise sense when viewed from the top and the pole formation at the point of entry of the current (at the end nearer to the centre of the diagram) is South Pole. The positive direction of the current flow in the coil is therefore at the end of the coil nearer to the centre of the diagram, which is taken as positive. To this end, the instantaneous positive voltage is taken as applied and this terminal is taken to be at a higher potential. Since the applied voltage is taken as positive and causes a positive current flow into a coil, an electric power flow into the electrical circuit is taken as positive, that is, motoring action is assumed as positive in our convention.

The phase layout of an ac winding having *a-b-c* phase sequence, presents the axis of symmetry of the phase *a* followed by axis of symmetry of *b*-phase, a $\left(\dfrac{2\pi}{3}\right)$ electrical behind in the clockwise direction, followed by *c*-phase, a $\left(\dfrac{4\pi}{3}\right)$ electrical behind the phase *a* axis. The rotating magnetic field produced in such a winding fed by a 3-phase ac supply with *a-b-c* phase sequence revolves from $a \rightarrow b \rightarrow c$ along the clockwise direction. We assume the clockwise direction of rotation to be the positive direction of rotation. However, we should always keep in the back of our minds that the positive angle is measured along the counter-clockwise direction.

In a primitive machine representation, the reference currents are always shown to be positive, that is, there are shown as entering the coils at the positive potential ends. As has already been pointed out, the end of the coils where the reference currents enter are considered as developing south poles. If then, for a particular case, the current is found from the analysis as leaving the positive terminal, the coil current denoted by a different symbol is taken as negative. The two positive alternative cases are indicated in the reference diagram in Figure 7.1.

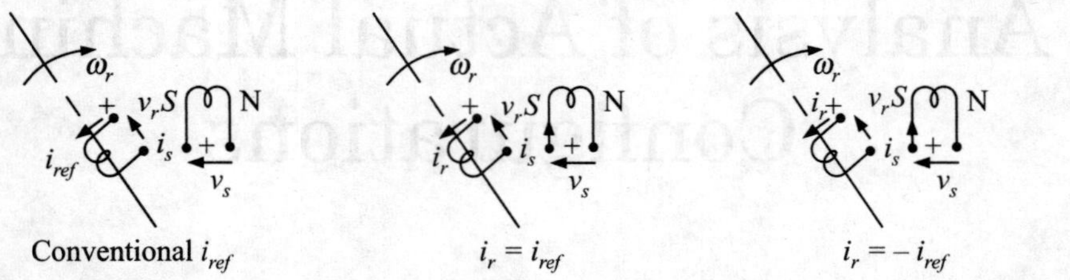

Conventional i_{ref} $\qquad$ $i_r = i_{ref}$ $\qquad$ $i_r = -\,i_{ref}$

FIGURE 7.1 Coil current and polarity convention adopted in present text.

A similar convention is also introduced for the electro-magnetic torque. The reference torque is always taken as positive when the same is tending to rotate the rotating member along the positive direction of rotation as assumed, that is, in the clockwise direction. The convention is indicated in Figure 7.2, where the difference between the two constituent diagrams is indicated by the current flow directions in the pseudo-stationary coils mounted on the rotors of the commutator windings.

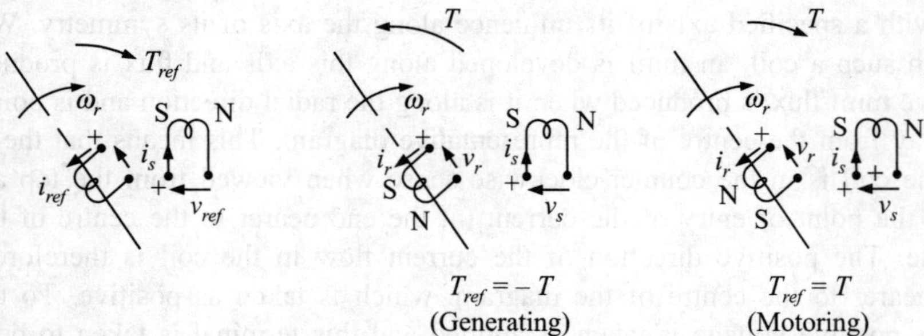

FIGURE 7.2 Motor and generator action indicated for the torque convention on the basis of direction of current in rotor pseudo-stationary coils.

A significant group of ac and dc machines uses the commutating windings on the rotor member. A simple form of a 2-pole commutator machine with salient pole stator structure is considered in Figures 7.3(a) and (b) and is adopted as a model for obtaining the performance characteristics of commutator machines.

For rendering sufficient generality in the subsequent analysis, it is assumed that the voltages, and therefore, the currents into the stationary coil *s*, that is, v_s and i_s, as well as the voltage and current in the pseudo-stationary coil, *r*, that is, v_r and i_r, are time-varying in nature. The uniformly distributed commutator winding in Figure 7.3(a) is being connected to the supply with the help of stationary brush-pair laid on an axis at θ_r to the flux-axis (that is, *D*-axis) as elaborated in Section 4.5.1. The commutator action effectively divides the commutator

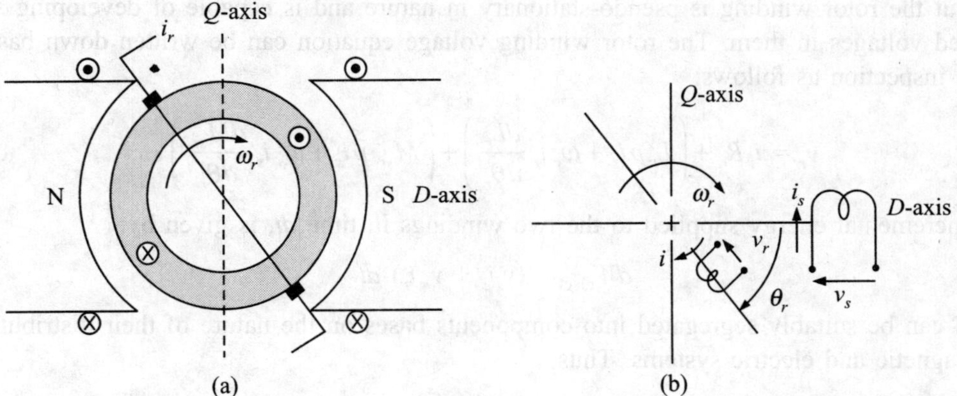

FIGURE 7.3 (a) A 2-pole schematic of a commutator machine, and (b) Its equivalent primitive depiction.

winding along the brush axis to form two uniform current sheets of opposite polarities. The commutator winding, therefore, behaves as a stationary coil of N_r concentrated turns only from the view point of the axis of the mmf produced by it, but the coils are attributed with special property of being capable of inducing rotational voltages in them—a characteristic that renders the name *pseudo-stationary*. The representation is shown in Figure 7.3(b). The two stationary coils depicted in the figure are considered to produce sinusoidally distributed air-gap mmfs. The space layouts of these mmfs are separated by an angle θ_r, while their time separation is decided by the respective currents.

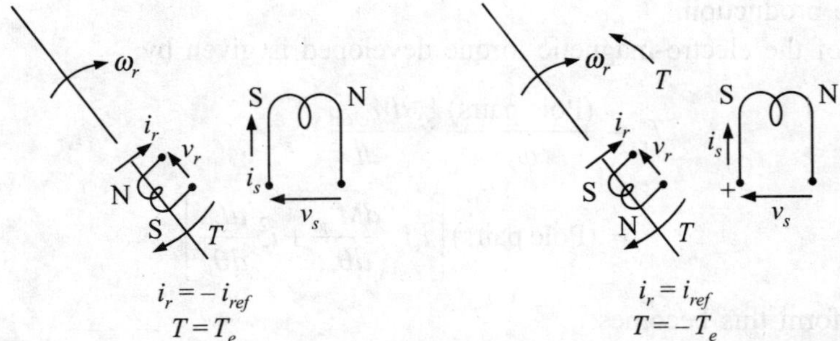

FIGURE 7.4 Motoring and generating action consolidated by specifying rotor current and torque relationships with reference rotor current and torque for clockwise rotation.

A commutator machine clearly consists of two windings—one on the stator and the other denoted by pseudo-stationary coil on the rotor. Each of these windings has [L] matrices of similar form, but the [G] matrix for these two windings is different. The [G] matrix, therefore, needs to be especially identified before developing the voltage and torque equations.

There is no rotational voltages associated with the stator coil, it being stationary in space. Hence, for a cylindrical rotor structure:

$$v_s = i_s R_s + L_s p i_s + M_{sr}\, p i_r \qquad \text{...(7.1)}$$

where, R_s, L_s, M_{sr} and p have the usual meanings as discussed earlier.

But the rotor winding is pseudo-stationary in nature and is capable of developing speed-induced voltages in them. The rotor winding voltage equation can be written down based on visual inspection as follows:

$$v_r = i_r R_r + \left(L_r p i_r + \omega_r i_r \frac{dL_r}{d\theta_r} \right) + \left(M_{sr} p i_s + \omega_s i_s \frac{dM_{sr}}{d\theta_r} \right) \quad ...(7.2)$$

The incremental energy supplied to the two windings in time dt, is given by:

$$dW_{\text{elect}} = (v_s i_s + v_r i_r)\, dt$$

which can be suitably segregated into components bases on the nature of their distribution in the magnetic and electric systems. Thus,

$$v_s i_s\, dt + v_r i_r\, dt = \underbrace{(i_s^2 R_s + i_r^2 R_r)\, dt}_{\text{First term}} + \underbrace{\left[L_s p \frac{1}{2} (i_s)^2 + L_r p \frac{1}{2} (i_r)^2 + M_{sr} p (i_s i_r) \right] dt}_{\text{Second term}}$$

$$+ \underbrace{\omega_r \left[i_s i_r \frac{dM_{sr}}{d\theta_r} + (i_r)^2 \frac{dL_r}{d\theta_r} \right] dt}_{\text{Third term}} \quad ...(7.3)$$

The first term in the above equation denotes the incremental energy loss in winding resistance, followed by incremental increase in field storage energy in the second term. This is followed by the incremental energy storage for electro-mechanical energy conversion and resulting torque production.

The value of the electro-magnetic torque developed is given by,

$$T_e = \frac{\text{(Pole pairs)}}{\omega_r} \times \frac{dW_{\text{elect}}}{dt}$$

$$= \text{(Pole pairs)} \left[i_s i_r \frac{dM_{sr}}{d\theta_r} + i_r^2 \frac{dL_r}{d\theta_r} \right]$$

and, in matrix form this becomes:

$$T_e = \text{(Pole pairs)}\, [i_s \quad i_r] \begin{bmatrix} 0 & 0 \\ \dfrac{dM_{sr}}{d\theta_r} & \dfrac{dL_r}{d\theta_r} \end{bmatrix} \begin{bmatrix} i_s \\ i_r \end{bmatrix} \quad ...(7.4)$$

$$= \text{(Pole pairs)}\, [i_t]\, [G]\, [i] \quad ...(7.5)$$

Equation (7.5), when compared with Eq. (3.20) shows that the former is one-half of the latter.

For the commutator machine with stationary s winding,

$$[R] = \begin{bmatrix} R_s & 0 \\ 0 & R_r \end{bmatrix}; \; [L] = \begin{bmatrix} L_s & M_{sr} \\ M_{sr} & L_r \end{bmatrix} \quad \text{and} \quad [G] = \begin{bmatrix} 0 & 0 \\ \dfrac{dM_{sr}}{d\theta_r} & \dfrac{dL_r}{d\theta_r} \end{bmatrix}$$

so that the voltage equation is of the form:

$$\begin{bmatrix} v_s \\ v_r \end{bmatrix} = \left[\begin{bmatrix} R_s & 0 \\ 0 & R_r \end{bmatrix} + \begin{bmatrix} L_s & M_{sr} \\ M_{sr} & L_r \end{bmatrix} p + \omega_r \frac{\partial}{\partial\theta_r} \begin{bmatrix} 0 & 0 \\ M_{sr} & L_r \end{bmatrix}\right] \begin{bmatrix} i_s \\ i_r \end{bmatrix} \quad ...(7.6)$$

And, with a brush displacement of θ_r from the flux or direct axis:

L_s = Constant (as the air-gap is smooth),

$$L_r = L_{r1} + L_{r2} \cos 2\theta_r \qquad \therefore \qquad \frac{dL_r}{d\theta_r} = -2L_{r2} \sin 2\theta_r$$

$$M_{sr} = M_m \cos \theta_r \qquad \frac{dM_{sr}}{d\theta_r} = -M_m \sin \theta_r$$

so that the general expression for torque becomes:

$$T_e = \text{(Pole-pairs)} \, [i_s \;\; i_r] \begin{bmatrix} 0 & 0 \\ -M_m \sin\theta_r & -2L_{r2} \sin 2\theta_r \end{bmatrix} \begin{bmatrix} i_s \\ i_r \end{bmatrix} \quad ...(7.7)$$

The final form of the voltage equation, then becomes:

$$\begin{bmatrix} v_s \\ v_r \end{bmatrix} = \left\{\begin{bmatrix} R_s & 0 \\ 0 & R_r \end{bmatrix} + \begin{bmatrix} L_s & M_m \cos\theta_r \\ M_m \cos\theta_r & L_r \end{bmatrix} + \omega_r \begin{bmatrix} 0 & 0 \\ -M_m \sin\theta_r & -2L_{r2} \sin 2\theta_r \end{bmatrix}\right\} \begin{bmatrix} i_s \\ i_r \end{bmatrix} \quad ...(7.8)$$

And with brushes along the quadrature-axis, that is, with $\theta_r = \frac{\pi}{2}$,

$$-M_m \sin \theta_r = M_m \quad \text{and} \quad 2L_{r2} \sin 2\theta_r = 0$$

Hence,

$$\left.\begin{aligned} T_e &= \text{(Pole pairs)}\, i_s\, i_r\, M_m \\ v_s &= i_s R_s + L_s p\, i_s \\ \text{and, } v_r &= i_r R_r + L_r p\, i_r + \omega_r M_m\, i_s \end{aligned}\right\} \quad ...(7.9)$$

Under steady-state conditions with a non-varying voltage at the terminals (that is with direct voltages):

$$p = 0, \quad v_s = V_f \quad \text{and} \quad v_r = V_a, \quad M_m = M_a$$

Then for motoring action:

$i_s = I_f$, $i_r = I_a$, and

$$\left.\begin{aligned} \text{Torque} &= \text{(Pole pairs)}\, I_f\, I_a\, M_m \\ V_f &= I_f R_f \\ V_a &= I_a R_a + \omega_r I_f I_a M_a \end{aligned}\right\} \quad ...(7.10a)$$

While for generator action:

$i_s = I_f, \quad i_r = -I_a$

$\therefore$

$$\left.\begin{aligned} T_e &= -\text{ (Pole pairs) } I_f\, I_a\, M_a \\ V_f &= I_f\, R_f \\ V_a &= -I_a\, R_a - \omega_r\, M_a\, I_f \end{aligned}\right\} \qquad \text{...(7.10b)}$$

In the above expressions, M_m is the mutual inductance between the field winding and the pseudo-stationary rotor equivalent coil, replacing the commutator winding with its axis coinciding the quadrature axis. The above conditions, therefore, represent the general operation of dc motor and generator systems.

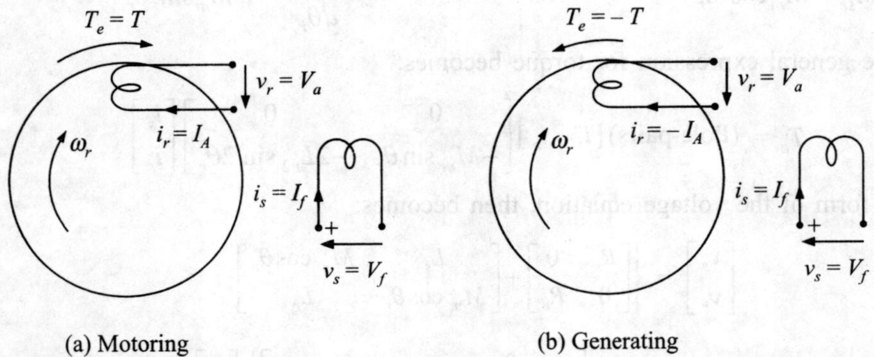

FIGURE 7.5 Equivalent steady-state circuit depiction for commutator machine operating on dc supply.

7.2 NATURE OF REPRESENTATION OF CIRCUIT PARAMETERS IN MULTI-WINDING MACHINES

A transformation of significance is the one in which the machine windings are transformed into *d*-axis and *q*-axis arrays of stator circuits and rotor pseudo-stationary coils. As such, this arrangement agrees with a two-windings commutator machine indicated in Figure 7.3, and the equations developed for its analysis are closely relevant. These can be used to determine the performance characteristics of the transformed machine and hence the original machine itself.

The name *primitive machine* is attributed to the *d*-*q*-array of coils, and each coil is used in the stand-alone mode. Such a machine has been nomenclatured as the *KRON PRIMITIVE*.

A Kron Primitive with two orthogonally placed pseudo-stationary rotor coils and two orthogonal stator coils is shown in Figure 7.6. Each coils is marked with two suffixes at the bottom on the right, so that the coil under reference may be identified precisely. The convention adopted for identification is the axis of influence (*d* or *q*) as the first of the suffixes and the component number of the machine to which it belongs (*s* or *r*, for stator or rotor member, as the case may be). Thus, on diagram, the coil A_{ds} is the coil located with its action of influence along the *D*-axis and placed on the stator, and the coil C_{qr} is a pseudo-stationary rotor coil with its axis of influence along the *Q*-axis.

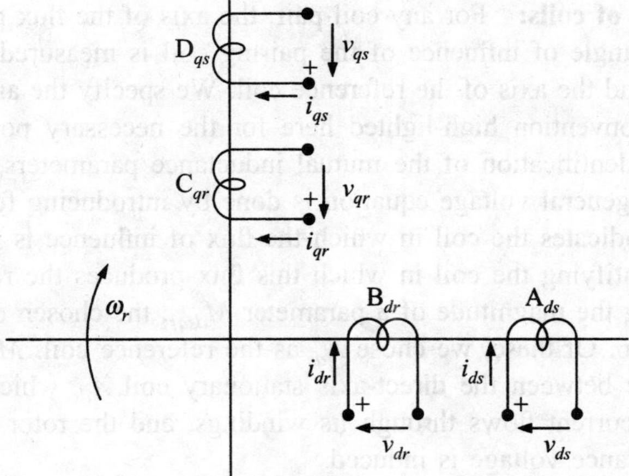

FIGURE 7.6 *D-Q* axis primitive machine configuration represented with one coil each shown along the direct- and quadrature-axis on both the rotor and stator members.

7.2.1 Matrix Form of Voltage Coefficients

Let us concentrate on the formation of the $[L]$ and $[G]$ matrices for the primitive machine depicted in Figure 7.6, on the basis of the above conventions. These can be written down based on the voltage equations derived in expression (7.8) for the two-winding commutator machine.

From Eq. (7.8), we have

$$[L] = \begin{bmatrix} L_s & M_m \cos\theta_r \\ M_m \cos\theta_r & L_r \end{bmatrix}$$

The mutual inductance coefficient term is:

$$M_{sr} = M_{rs} = M_m \cos\theta_r$$

When $\theta_r = 0$, $M_{sr} = M_m$

But with $\theta_r = \pm\dfrac{\pi}{2}$ $M_{sr} = 0$

Hence, the mutual inductance elements are present only when the two concerned coils share a common axis.

At this stage, it has become necessary to lay down rules for attaching suffices to the mutual inductance parameters and pairing coils to specify the basis of our convention. The issues involved in specifying the nature of mutual inductance between two coils are as follows:

1. Precise identification of the coils involved in determining a mutual—Induction parameter;
2. The relative position of the two concerned coils;
3. The magnitude of the parameter.

All the above considerations require precise pairing of the coils. We specify a rule for allotting suffices to the mutual-inductance parameters of coil-pairs first. The basis of our adopted conventions are laid down below:

Precise identification of coils: For any coil-pair, the axis of the flux producing coil is taken as reference and the angle of influence of the pairing coil is measured between the angle of the pairing coil axis and the axis of the reference coil. We specify the angle with a sign taking due care about the convention high-lighted here for the necessary positive angle along the clockwise direction. Identification of the mutual inductance parameters of the coil pairs used in $[L]$ and $[G]$ of the general voltage equation is done by introducing four suffices in all. The first pair of suffices indicates the coil in which the flux of influence is produced, followed by a subsequent pair identifying the coil in which this flux produces the reaction voltage. As an example, in specifying the magnitude of a parameter M_{dsqr}, the chosen coil-pair is A_{ds} and C_{qr} as shown in Figure 7.6. Of these, we chose A_{ds} as the reference coil. M_{dsqr}, therefore, denotes the mutual inductance between the direct-axis stationary coil A_{ds} which is taken as the flux producing coil when current flows through its windings, and the rotor pseudo-stationary coil C_{qr} in which the reactance voltage is induced.

Relative position of the two concerned coils: The angle between the axes of the concerned coils A_{ds} (in the above example is $\frac{\pi}{2}$). It has to be reassured after attributing a sign (positive or negative). We consider the flux axis of the reference coil A_{ds} as fixed and unmoving while the axis of the influenced coil C_{qr} is notionally moved and brought in line with the reference axis by the shorter of the two alternative paths between coils C_{qr} and A_{ds}. This is depicted in Figure 7.6. If the direction of the notional displacement is along the positive direction of the envisaged motion, we add a negative sign to the absolute value of the angle. This will mean that the axis of the coil under reference makes an angle of $(-\pi/2)$ with the axis of the flux produced by A_{ds}.

Magnitude of the parameter: The value of the parameter is the magnitude of physical coupling between the two coils, obtained after the influenced coil is notionally brought to share the same magnetic axis as the reference coil. Hence, the magnitude of M_{dsqr} is the mutual inductance between two coils placed along the direct axis—coil A_{ds} and B_{dr}, since B_{dr} and C_{qr} are supposed to be identical in all respects but their position, the coefficient of M_{dsqr} equals the magnitude of M_{dsdr}.

The *d*-*q* axis coil array can consist of a number of individual coil-pairs. Of these, if we choose a *d*-axis coil as a flux producing reference coil, pairs are formed with coils displaced with angles of $\theta_r = 0$ on the direct axis and with angles of $\theta r = -\pi/2$ on the quadrature axis, and relative to a *q*-coil pairs will be formed with coils displaced by angles of $\theta_r = 0$ and $\theta_r = +\pi/2$. Thus for all coil pairs the displacement angles of concern will be $\theta_r = 0$ and $\theta_r = \pm\pi/2$. We had seen at the beginning of this section for the $[L]$-matrix. If the matrix is expressed in a labelled form it will assume the following shape:

If the matrix is expressed in a labelled form, it assumes the following shape:

Reference coils

$$[L] = \text{coils influenced by reference coils} \downarrow \begin{array}{c} \\ ds \\ dr \\ qs \\ qr \end{array} \begin{array}{cccc} ds(A) & dr(B) & qs(D) & qr(C) \\ \end{array} \begin{bmatrix} L_{ds} & M_{drds} & 0 & 0 \\ M_{dsdr} & L_{dr} & 0 & 0 \\ 0 & 0 & L_{qs} & M_{qrqs} \\ 0 & 0 & M_{qsqr} & L_{qr} \end{bmatrix} \quad ...(7.11a)$$

Also, from Eq. (7.8); we have

$$[G] = \begin{bmatrix} 0 & 0 \\ -M_m \sin\theta_r & -2L_{r2} \sin 2\theta_r \end{bmatrix} \qquad \text{...(7.11b)}$$

In this case, $G_{sr} = -M_m \sin \theta_r$

and when $\theta_r = 0, \qquad G_{sr} = 0$

while with $\theta_r = \pm\dfrac{\pi}{2} \qquad G_{sr} = \mp M_m$

We observe that for determining the relevant G matrix elements, we have to identify only coil-pairs formed between a pseudo-stationary coil on the rotor—that is coils B_{dr} and C_{qr} (as per Figure 7.6)—and four flux-producing developing coils—two on the stationary member, coils A_{ds} and D_{qs} and two pseudo-stationary coils B_{dr} and C_{qr}. Further, the relative positions of the pseudo-stationary coils influenced by the flux should be at right angles to the flux-axis.

Thus, the stationary flux producing coil, A_{ds} couples up with C_{qr}, and D_{qs} with B_{dr} in contributing to G matrix terms, while the flux producing pseudo-stationary coil B_{dr} couples up with C_{qr}, and reciprocally C_{qr} ties up with B_{dr} to contribute to G matrix coefficients. Therefore, the mutual inductance terms M_{dsqr}, M_{qsdr}, M_{drqr} and M_{qrdr} constitute reluctance terms in the formation of the G-matrix coefficients.

For a smooth air-gap profile without any saliency (like the cases of induction or cylindrical rotor synchronous machines), some of the mutual elements have the same values; these are as follows:

$$L_{ds} = L_{qs} = L_{dr} = L_{qr}$$

$$M_{dsdr} = M_{drds} = M_{qsqr} = M_{qrqs}$$

But for salient structure (like dc machines and salient-pole synchronous machines):

$$L_{ds} \neq L_{qs}, \quad L_{dr} \neq L_{qr} \quad \text{and} \quad M_{dsdr} \neq M_{qsqr}$$

Further, we observe that the direction of notional rotation of the affected pseudo-stationary coils to align with the axis of the respective flux-producing coils, has an intimate connection with the sign of the movement. To clarify this point, let us first take coils A_{ds} as flux-producing coils which interact individually with the pseudo-stationary coil C_{qr}.

It will be noted that as per chosen convention high-lighted in the paragraph under the heading "Relative positon of the two concerned coils" in the present sub-section 7.2.1, the angle of notional displacement is $(-\pi/2)$, so that the G coefficient of C_{rq} with source voltage in stator coil A_{ds} will be:

$$G_{dsqr} = -M_{dsqr} \sin\left(-\frac{\pi}{2}\right) = M_{sdqr}$$

The corresponding angle of notional movement for the coil B_{dr}, with source of flux in the stator coil D_{qr}, will be:

$$G_{qsdr} = -M_{qsdr} \sin\left(+\frac{\pi}{2}\right) = -M_{qsdr}$$

Thus, the G coefficients with stator coils as source of flux, will work out to be

$$G_{dsqr} = M_{dsqr} \quad ; \quad G_{qsdr} = -M_{qsdr}$$

We can, now, consider the coefficients related to the two pseudo-stationary coils, B_{dr} and C_{qr}, taken as flux producing coils, for the effective angle of notional movement of the affected coils C_{qr} and B_{dr} respectively. Their values are found to be $(+\pi/2)$ and $(-\pi/2)$, in that order, so that:

$$G_{qrdr} = -M_{qrdr} \sin\left(+\frac{\pi}{2}\right) = -M_{qrdr}$$

and

$$G_{drqr} = -M_{drqr} \sin\left(-\frac{\pi}{2}\right) = M_{drqr}$$

However, the coil C_{qr} shares the flux producing coil B_{dr} along the d-axis, while the coil B_{dr} shares the flux-producing coil C_{qr} along the q-axis. For all practical purposes the coils B_{dr} and C_{qr} have identical shapes in terms of the effective number of turns and layout. They will offer identical self-inductance when a unit current flows through each coil. Besides the mutual inductance offered by the two identical coils will be denoted by replacing the mutual inductance term by the product of the self inductance terms, L_{dr} and L_{qr}, as shown. Hence,

$$G_{drqr} = -M_{drdr} \sin\left(-\frac{\pi}{2}\right) = -M_{drdr}$$

so that,

$$M_{drdr} = L_{dr}L_{dr} = L_{dr}$$

And

$$G_{qrdr} = -M_{qrdr} \sin\left(+\frac{\pi}{2}\right) = -M_{qrdr}$$

so that,

$$M_{qrdr} = -M_{qrqr} = -L_{qr}L_{qr} = -L_{qr}$$

We are, now, ready to consolidate the information arrived at in connection with the G-matrix coefficients.

(a) For coefficients with stator coils as the source of flux:

$G_{dsqr} \rightarrow$ angle of notional displacement is negative $(-\pi/2)$.

And $\quad G_{dsqr} = M_{dsqr}$

Also, $G_{qsdr} \rightarrow$ angle of notional displacement is positive $(+\pi/2)$

so that, $\quad G_{qsdr} = -M_{qsqr}$

(b) Coefficients involving two pseudo-stationary coils:

$G_{qrdr} \rightarrow$ angle of notional displacement is positive $(+\pi/2)$

so that, $\quad G_{qrdr} = -M_{qrdr} = -M_{qrqr} = -L_{qr}$

Also, $G_{drqr} \rightarrow$ angle of notional displacement is negative $(-\pi/2)$.

so that, $G_{drqr} = M_{drqr} = M_{drdr} = L_{dr}$

M_{drdr} is the self-inductance of the pseudo-stationary coil when placed along the direct axis, and M_{qrqr} is the self-inductance of the coils when placed with their axis coincident with the quadrature axis as already explained.

We, now, can obtain the G-matrix in general form:

$$[G] = \text{Labelled coils influenced by reference coils} \downarrow \quad \overset{\text{Reference coils}}{\begin{array}{c} \\ ds \\ dr \\ qs \\ qr \end{array}\begin{array}{cccc} ds & dr & qs & qr \\ \end{array}} \begin{bmatrix} 0 & 0 & 0 & 0 \\ 0 & 0 & M_{qsdr} & M_{qrdr} \\ 0 & 0 & 0 & 0 \\ M_{dsqr} & M_{drqr} & 0 & 0 \end{bmatrix} \quad ...(7.12)$$

by appropriate substitutions derived above as follows:

$$[G] = \text{Coils influenced by reference coils} \downarrow \quad \overset{\text{Reference coils}}{\begin{array}{c} \\ ds \\ dr \\ qs \\ qr \end{array}\begin{array}{cccc} ds & dr & qs & qr \\ \end{array}} \begin{bmatrix} 0 & 0 & 0 & 0 \\ 0 & 0 & -M_{qsdr} & -L_{qr} \\ 0 & 0 & 0 & 0 \\ M_{dsqr} & L_{dr} & 0 & 0 \end{bmatrix} \quad ...(7.13)$$

This leads us to the following general voltage relationships

$$[v] = \{[R]+[L]p+\omega_r[G]\}[i]$$

by substituting the matrix elements derived in Eqs. (7.11) and (7.13) in Eq. (7.6), we get

$$\begin{bmatrix} v_{ds} \\ v_{dr} \\ v_{qs} \\ v_{qr} \end{bmatrix} = \begin{array}{c} \\ ds \\ dr \\ qs \\ qr \end{array} \overset{\begin{array}{cccc} ds & dr & qs & qr \end{array}}{\begin{bmatrix} (R_{ds}+L_{ds}p) & M_{drds} & 0 & 0 \\ M_{dsdr} & (R_{dr}+L_{dr}p) & -\omega_r M_{qsqr} & -\omega_r L_{qr} \\ 0 & 0 & (R_{qs}+L_{qs}p) & M_{qrqs}p \\ \omega_r M_{dsdr} & \omega_r L_{dr} & M_{qsqr}p & (R_{qr}+L_{qr}p) \end{bmatrix}} \begin{bmatrix} i_{ds} \\ i_{dr} \\ i_{qs} \\ i_{qr} \end{bmatrix} \quad ...(7.14)$$

7.2.2 Expression for Torque

The expression for torque can be derived now by substituting Eq. (7.13) in equation (7.05). This leads to:

$$[T_e] = (\text{Pole pairs})\,[i_{ds} \quad i_{dr} \quad i_{qs} \quad i_{qr}]$$

$$\begin{array}{c} \\ ds \\ dr \\ qs \\ qr \end{array} \begin{array}{c} \begin{array}{cccc} ds & dr & qs & qr \end{array} \\ \begin{bmatrix} 0 & 0 & 0 & 0 \\ 0 & 0 & -M_{qsqr} & -L_{qr} \\ 0 & 0 & 0 & 0 \\ +M_{dsdr} & +L_{ds} & 0 & 0 \end{bmatrix} \end{array} \begin{bmatrix} i_{ds} \\ i_{dr} \\ i_{qs} \\ i_{qr} \end{bmatrix} \quad ...(7.15)$$

When simplified and expressed in expanded form, we get

$$T_e = (\text{Pole pairs})\ [M_{dsdr}\ i_{ds}\ i_{qr} - M_{qsqr}\ i_{qs}\ i_{qr} + (L_{ds} - L_{qr})\ i_{ds}\ i_{qr}] \quad ...(7.16)$$

In the above expanded form, the torque component $(L_{ds} - L_{qr})\ i_{ds}\ i_{qr}$ is present only in salient-pole machines and is identified as the torque component due to the effect of saliency.

The remaining two components in Eq. (7.16) are contributed as a result of uniform air-gap and are nomenclatured as electro-magnetic torque due to cylindrical effect of the two components, the positive is the motoring torque and the negative is the generating torque.

The net sum of all the components determines the machine action—motoring or generating.

7.3 ANALYSIS OF SIMPLE DIRECT CURRENT MACHINES

We are now ready to take up the analysis of simple dc machines by using the procedure developed for the commutator machine primitive in Section 7.2. A simple dc machine can be denoted by a primitive machine with one stator coil for the main field and a pseudo-stationary coil along the quadrature axis on the rotor, as already pointed out at the end of Section 7.1 and Section 7.2 and is duly supported by mathematical relations (7.10a) and (7.10b). We represent the primitive of a basic dc machine again in Figure 7.7 along with the adopted convention marked on the diagram. We can always add more coils along appropriate axes in the stator and the rotor to consider additional flux sources subsequently as the need arises.

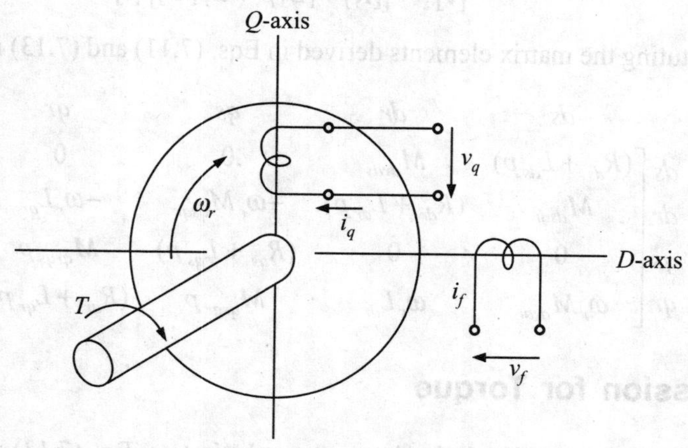

FIGURE 7.7 Primitive machine depiction of a simple dc machine with two coils.

We arrive at the equations relating the terminal voltages and currents, the torque, speed of the machine in terms of self-, mutual- and rotational-inductance of the machine, which also show the shaft on the rotating member of the diagram.

We note that the basic convention adopted is such that all powers—both mechanical and electrical—have positive sign when flowing into the device and a negative sign when flowing out of it. When arriving at the equations, the field power flow is always taken as positive. Also, the quadrature-axis electrical power and the mechanical power are always with opposite signs. Figure 7.7 gives voltage equations having the form:

$$v_f = (R_f + L_f\,p)\,i_f$$

$$v_q = (V \text{ due to rotation}) + (R_q + L_q p)\,i_q$$

$$= i_f\,M_f\,\omega_r + (R_q + L_q p)\,i_q$$

In matrix form, the above set of equations is expressed as follows:

$$\begin{bmatrix} v_f \\ v_q \end{bmatrix} = \begin{bmatrix} (R_f + L_f p) & 0 \\ M_f \omega_r & (R_q + L_q p) \end{bmatrix} \begin{bmatrix} i_f \\ i_q \end{bmatrix} \qquad \text{...(7.17)}$$

which agrees with Eqs. (7.10a) and (7.10b), fully.

To obtain the steady-state equations, we put the operator p as equal to zero in the above equation. If the field winding is excited along the adopted positive direction (which is the case) as shown in Figure 7.7, it establishes a flux pattern as per the adopted convention from left to right as shown, where the lower brush placed on the quadrature axis of the rotor denotes positive terminal, with the upper brush as negative. If the machine is motoring, the current in the quadrature winding is positive, thus giving a positive electrical power input.

Let us spend some more time on the conceptual issue. Voltages are induced in conductors when they are located in a flux-distribution pattern that pulsates or moves relatively with them. If the fundamental component of the flux is denoted by a phasor, we call such a phasor as the current distribution/voltage distribution phasor (as the case may be); for such a winding distribution Figure 7.8(a) represents, belts marked at centres by dots and crosses, the pattern of current flow in the conductor arrangement around the air-gap.

The mmf produced by the current flow is along the horizontal centre line. The current distribution is indicated by phasor I in the diagram, positioned along the mmf pattern. With commutator provided to the winding, the phasor I will also be positioned along the axis of brushes that supply the current.

In Figure 7.8(b), the belts representing the dots and crosses shown at the belt centres, indicate the voltages generated in the conductors when they are rotating in a stationary flux configuration of constant magnitude centred around the horizontal axis. The rotational voltage could also be denoted by a voltage phasor v_{rot} along the axis of the brushes between which the maximum value of the voltage appears if the winding were fitted with such brushes and commutator.

Figure 7.8(c), similarly indicates the voltage induced in the same conductors when they are stationary with respect to the field pattern laid along the horizon axis as shown, and the mmf decreasing in magnitude. This transformer voltage distribution has been indicated by a voltage phasor v_{tr} on the diagram.

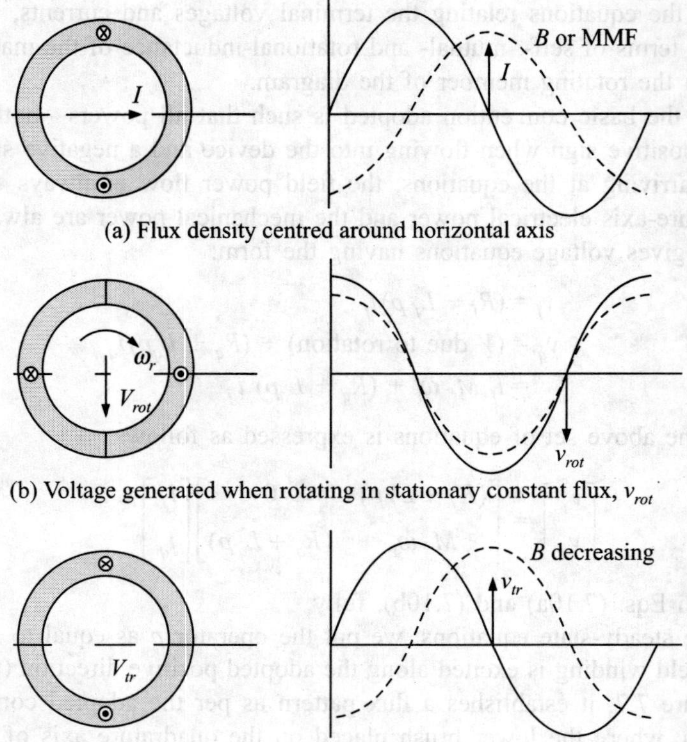

FIGURE 7.8 Production of current and voltage distribution patterns in conductors.

If the field winding is excited in the positive direction as shown in Figure 7.8, establishing a flux distribution directed from left to right, then with the positive direction of rotation taken as clockwise, the generated speed induced voltage is positive. As shown in Figure 7.8(b), the lower brush placed on the quadrature axis is positive and the upper brush is negative. If the machine is motoring, the current in the quadrature axis is also positive, giving positive input power.

With a positive I_f and a positive value of I_q, the torque aids rotation. The machine is, therefore, in motoring mode with the mechanical power flow in the negative direction.

An easier way of remembering this result is to examine how the pseudo-stationary coil, in which the speed-induced voltage is being developed, is brought under the direct influence of the flux responsible for production of the said voltage. If the pseudo-stationary coil is observed to be moving in the clockwise direction to traverse the shortest of the two routes, thus aiding the conventional positive direction of the armature, *then the speed-induced voltage assumes a positive sign*.

Referring to Figure 7.9, the shortest of the two routes for the notional traverse of q pseudo-stationary coil to D-axis (the flux responsible for the production of the speed-induced voltage is in the clockwise direction along q-a-d and not along q-b-d), and the purported movement aids the conventional positive direction of rotation. Hence, the speed-induced voltage in Q-axis pseudo-stationary coil assumes a positive sign. We had come to the same conclusion as in Eq. (7.17). If there were a pseudo-stationary coil in the representative diagram with its axis of influence along the direct axis, then the speed-induced voltage developed in it due to flux in quadrature coils would have negative signs as per the above consideration. This is bourn out by Eq. (7.14).

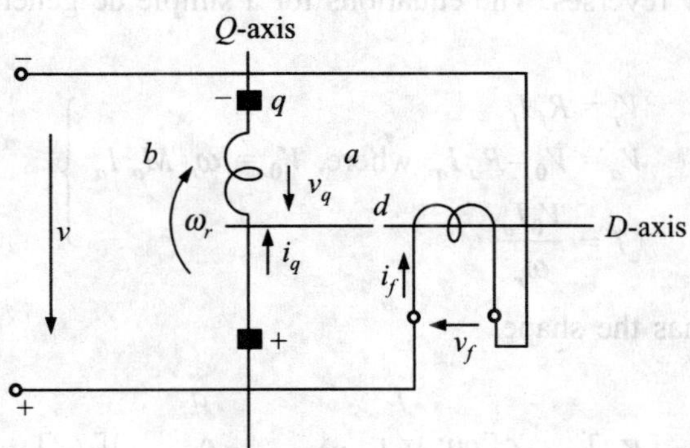

FIGURE 7.9 A dc shunt motor representation.

Simple dc machines include the separately-excited, shunt- and series-excited motors and generators and also compound wound machines. Compound machines have both shunt and series field windings, but in effect these machines have effectively two windings only—the field and the armature. In field windings, the shunt and the series are both mounted on the stationary member and jointly contribute to the production of the total magnetic field, while the armature is mounted on the rotor. The circuit arrangement shown in Figure 7.7 can, therefore, denote a dc machine quite satisfactorily. In general, suffices of the primitive machine are modified to suit a particular machine being considered and here *D*-axis stator winding is denoted by *F* and the quadrature-axis rotor winding *q* is normally replaced by *a* for the armature. The suffices of all elements of the voltage equation are also modified accordingly. By removing the two unwanted winding elements in the four winding primitive machine of Figure 7.6, that is, Eq. (7.14) in Section 7.3, we can arrive at a simplified equation:

$$\begin{bmatrix} V_f \\ V_a \end{bmatrix} = \begin{bmatrix} R_f & 0 \\ +\omega_r M_m & R_a \end{bmatrix} \cdot \begin{bmatrix} I_f \\ I_a \end{bmatrix} \quad ...(7.18)$$

The above relation denotes the steady-state form of Eq. (7.14), by putting the terms with operator *p* as zero. The sign of the generated voltage term is arbitrarily taken as positive for convenience. In the expanded form, the first equation

$$V_f = R_f I_f$$

is given by Ohm's law; in other words the armature and the field circuit do not have an interactive property. The second equation,

$$V_a = \omega_r M\, I_f + R_a I_a$$

indicates that V_a is utilized in supplying the armature resistance drop $I_a R_a$ and in countering the generated voltage (also called the back induced emf, $E = \omega_r L I_f$), for constant values of I_1 and ω_r.

The only difference between the generating and the motoring action is that the armature current in the generating condition flows from the positive terminals outwards, while that in motoring operation flows from the supply system into the positive terminal. In other words, the current direction of flow reverses. The equations for a simple dc generator, therefore, have the following form:

$$\left.\begin{aligned} V_r &= R_f I_f \\ V_a &= V_0 - R_a I_a, \text{ where, } V_0 = \omega_r M_a I_a \\ T &= \frac{V_0 I_a}{\omega_r} \end{aligned}\right\} \qquad \text{...(7.19)}$$

and, in matrix form it has the shape,

$$\begin{bmatrix} V_f \\ V_a \end{bmatrix} = \begin{matrix} \\ f \\ a \end{matrix} \overset{\begin{matrix} f & & & a \end{matrix}}{\begin{bmatrix} (R_f + L_f p) & 0 \\ \omega_r M_m & (R_a + L_a p) \end{bmatrix}} \begin{bmatrix} I_f \\ I_a \end{bmatrix} \qquad \text{...(7.20)}$$

or the shape of the impedance matrix in the voltage equation remains unchanged due to the change in action between the motor and generator operation as expected.

7.3.1 Separately Excited dc Generator

In this configuration, the field winding is supplied from a separate supply source. Equations (7.19) and (7.20) apply. The performance of the machine can be described by the characteristic curves—the open-circuit and the load characteristics.

A separately excited dc generator can operate at any no-load voltage between zero and the rated value, through variation of excitation current in the field circuit, and once the value of the field current has been set and the armature is rotated at a constant speed, the terminal voltage remains almost constant over the range of load currents.

7.3.2 Shunt dc Generator

It is an accepted procedure to connect the main field winding across the terminals of the armature as shown in Figure 7.9. This makes the field winding as a shunt to the armature; so the machine derives its excitation from its own armature. The accompanying changes in the voltage equations at steady-state, now becomes:

$$\begin{aligned} V_a &= (R_f + R_{rh}) I_f \\ V_a &= \omega_r M_m I_f - R_a I_a \end{aligned} \qquad \text{...(7.21)}$$

and, $\qquad I_{\text{eff}} = I_a - I_f$

At no load, condition, $\qquad I_a = I_f$

Also, $V_a = (R_f + R_{rh})I_f$ and, $V_a = \omega_r M_m I_f - R_a I_f$

which can be combined to write,

$$(R_f + R_{rh} + R_a)I_f = \omega_r M_m I_f \quad ...(7.22)$$

In the above equations, R_{rh} is an external resistance connected in series with the field system.

Mathematically, Eq. (7.22) can lead to solution depending on whether:

1. $(R_f + R_{rh} + R_a) > \omega_r M_m$, leading to $I_f = 0$
2. $(R_f + R_{rh} + R_a) = \omega_r M_m$; this leads to values of I_f between zero and positive.
3. $(R_f + R_{rh} + R_a) < \omega_r M_m$; leading to infinite values of I_f, amounting to armature short-circuit.

Physical interpretation of the above observations are as follows:

1. So long as the total resistance of the circuit $(R_f + R_{rh} + R_a)$ is greater than the *critical value*, $\omega_r M_m$, the field current I_f and the armature voltage V_a, remain zero, that is, the machine is not self-excited.
2. If the resistance of the field rheostat, R_{rh}, is now reduced, as soon as the critical total field resistance $(R_f + R_{rh} + R_a)$ equals $\omega_r M_m$, the values of I_f and V_a start rising indefinitely. This ideal situation *fortunately* does not occur in practice, because the value of M_m, the mutual inductance between the field winding and the quadrature axis pseudo-stationary coil, does not remain constant due to saturation effect of the magnetic circuit. M_m decreases with increase of I_f. Saturation of the *D*-axis circuit containing the field winding, thus, plays a very significant role in the performance of the machine, and especially in case of the shunt generator operation. Without natural saturation as an integral behavioural property of the magnetic circuit, a stable operation of the shunt-machine would not have been possible. It becomes a matter of great importance to study the influence of I_f on the value of M_m (and, therefore, the maximum value of M_m). This is done by reconnecting the machine in a separately-excited mode, running it as a generator with external excitation at a pre-determined speed (usually the roted generator speed) and plotting its no-load characteristic. Taking Eq. (7.22),

$$(R_f + R_{rh} + R_a)I_f = \omega_r M_m$$

we can obtained the desired no-load voltage, E_0, within the working range. It is necessary to note the value of I_f from the open-circuit characteristics and set the value of R_{rh}, so that the equity equation is satisfied:

$$(R_f + R_{rh} + R_a)I'_f = E'_0$$

Graphically, the no-load voltage is given by the point of intersection of the open-circuit characteristics curve

$$E_0 = \omega_r M_m I_f$$

with the resistance line, that is, the straight line passing through the origin and denoted by the equation:

$$E'_0 = (R_f + R_{rh} + R_a)I_f$$

having a slope of $(R_f + R_{rh} + R_a)$. As R_{rh} is changed, the slope of the resistance line changes, and this action effectively shifts the point of intersection of the resistance line with the open-circuit characteristic. Increasing the value of R_{rh} effectively reduces the no-load terminal voltage. The movement of this point of intersection shows that the more pronounced the saturation, the more stable the operation of the machine is. In addition to this, the presence of residual magnetism in the pole and yoke structure of the field system is an essential condition for self-excitation. Also, the direction of rotation of a shunt generator should be such that the initial value of small no-load voltage generated due to residual magnetism, should aid the building up of the voltage in the generator. The change in terminal voltage from full-load to no-load is usually rather large for shunt machines when compared to separately excited machines.

7.3.3 Compound-wound dc Generators

A simple series generator is hardly ever used, but the addition of a series field winding to a shunt-field machine, to produce a compound field machine is found to be quite in vogue.

We consider a configuration with one brush-pair along the quadrature axis and two sets of stator coils along the direct axis—one for the separately excited or the shunt field and the remaining for the series field. The new arrangement can be derived from the two-winding primitive machine configurations is given in Figure 7.6 and is shown in Figure 7.10.

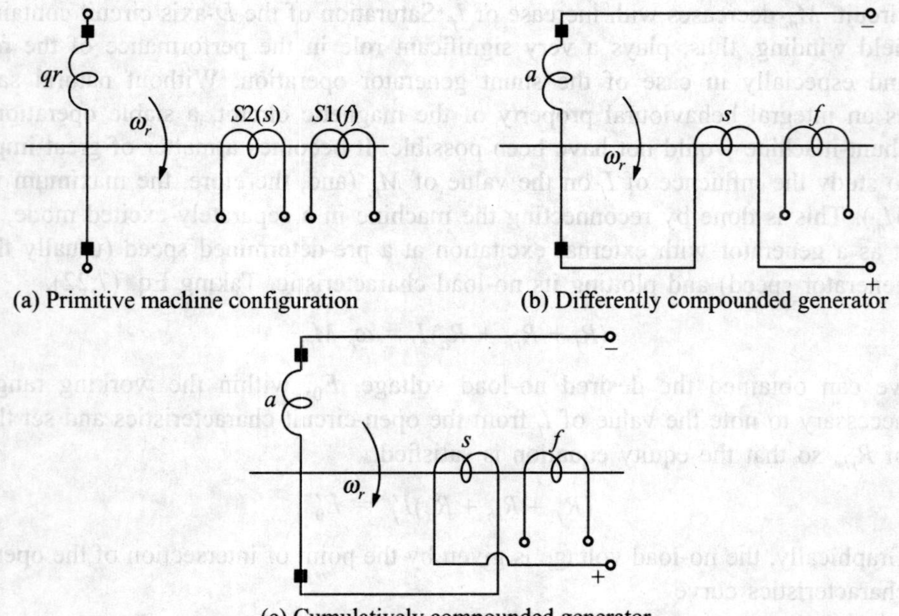

FIGURE 7.10 The equivalent primitive commutator machine configuration with added coils along *D*-axis, and different field circuit connections for compound generators.

The machine voltage and torque equations may be written down by inspection of the primitive machine equivalent based on the conventions already highlighted in Section 7.1 and elaborated in Section 7.3. Equation (7.14) can accordingly be modified as follows:

$$\left.\begin{array}{l} \begin{bmatrix} v_{s1} \\ v_{s2} \\ v_{qr} \end{bmatrix} = \begin{array}{c} \\ s1 \\ s2 \\ qr \end{array} \begin{array}{c} \begin{array}{ccc} s1 & s2 & qr \end{array} \\ \begin{bmatrix} (R_{s1}+L_{s1}p) & M_{s2s1}p & 0 \\ M_{s1s2}p & (R_{s2}+L_{s2}p) & 0 \\ \omega_r M_{ds1qr} & \omega_r M_{ds2qr} & (R_{qr}+L_{qr}p) \end{bmatrix} \end{array} \begin{bmatrix} i_{s1} \\ i_{s2} \\ i_{qr} \end{bmatrix} \\ \text{and} \qquad T_e = (\text{pole pairs})\ i_a\ \{G_{ds1dr}\ i_{s1} + G_{ds2dr}\ i_{s2}\} \end{array}\right\} \quad ...(7.23)$$

where, $s1$ stands for the separately excited or shunt field and $s2$ represents the series field while the element with suffix qr denotes the quantities pertaining to the rotor pseudo-stationary coils.

As already elaborated, we label only the mutual inductance and G matrix terms by four suffices and choose to denote the two field windings by f, indicating the main separately excited or shunt field and s denoting the series field respectively, with quadrature axis pseudo-stationary coil by a.

In Eq. (7.23), the rotational inductances may be replaced by equivalent mutual inductances $M_{f\text{-}am}$ and $M_{s\text{-}am}$, and it takes the form:

$$\left.\begin{array}{l} \begin{bmatrix} v_f \\ v_s \\ v_a \end{bmatrix} = \begin{array}{c} \\ f \\ s \\ a \end{array} \begin{array}{c} \begin{array}{ccc} f & s & a \end{array} \\ \begin{bmatrix} (R_f+L_f p) & M_{sf}p & 0 \\ M_{sf}p & (R_s+L_s p) & 0 \\ \omega_r M_{f\text{-}am} & \omega_r M_{s\text{-}am} & (R_a+L_a p) \end{bmatrix} \end{array} \begin{bmatrix} i_f \\ i_s \\ i_a \end{bmatrix} \\ T_e = (\text{Pole pairs})\ i_a\ (M_{f\text{-}am}\ i_f + M_{s\text{-}am}\ i_s) \end{array}\right\} \quad ...(7.24)$$

The above equations are applied to steady state and transient analysis. At steady-state, as already mentioned, we put the differential operator p as zero.

The series and the separately excited or the shunt field may be connected to yield mmfs in the same or opposite directions to each other. When both mmfs are additive, the configuration is a cumulative system, the system being differential in the subtractive arrangement. The two alternative connections are depicted in the generating mode in Figure 7.10(b) and (c). Reversal of armature current due to change over from generator to motor mode also reverses the mmf direction of the series field. Hence, a differentially compounded generator acts as a cumulatively compounded motor and a cumulatively compound generator functions as a differentially compounded motor. The primitive and actual machine currents and voltages can, therefore, be related as follows:

Direct axis, $i_f = I_f$	Direct axis $V_{s1} = V_f$
Direct axis, $i_s = \pm\, i_a$	Direct axis $V_{s2} = \pm V_s$
Quadrature axis, $i_{qr} = i_a$	Quadrature axis $V_{qr} = V_a$

In all cases, current i_a is shown as entering the terminal of C_{qr} at the nearest end of the centre of the diagram, which is assumed to be positive. The end is marked by a dot. For motoring action, i_a is a negative quantity. The armature circuit voltage:

$$V = V_{qr} \pm V_{ds2} \quad \text{or} \quad V = V_a \pm V_s$$

If we substitute for the primitive voltages and currents combining V_{qr} and V_{ds2}, and also write $R = (R_a + R_s)$ and $L = (L_a + L_s)$, matrix equation (7.24) assumes the form:

$$\begin{bmatrix} V_f \\ V \end{bmatrix} = \begin{bmatrix} (R_f + Lp) & M_{sf}\,p \\ (\omega_r M_{f\text{-}am} + M_{fa}\,p) & (R + Lp + \omega_r M_{s\text{-}am}) \end{bmatrix} \begin{bmatrix} i_f \\ i_a \end{bmatrix} \qquad \text{...(7.25)}$$

and

$$T_e = (\text{Pole pairs})\ \{i_a\, M_{f\text{-}am}\, i_f \pm i_a^2 M_{s\text{-}am}\}$$

In Eq. (7.25), the alternative sign of i is positive for a cumulatively compounded motor and a differentially compounded generator and negative for a differentially compounded motor or a cumulatively compounded generator.

7.3.4 Series dc Machine

Removing the separately excited winding from consideration and the corresponding terms from matrix Eq. (7.25), lead to steady-state equation for series motor, expressed in expanded form as follows:

$$\left.\begin{aligned} V &= (R + \omega_r\, M_{s\text{-}am}) I_a \\ T_e &= (\text{Pole pairs})\ \ (I_a^2 M_{s\text{-}am}) \end{aligned}\right\} \qquad \text{...(7.26)}$$

The characteristics of series motor displaces high starting torque and should always be started with initial load.

7.4 MACHINES WITH POLY-PHASE ALTERNATING CURRENT DISTRIBUTED WINDING ON THE ARMATURE

We now, take up the discussion of the two major forms of commutator-free machines in extensive use and lay down the general rules for the analysis of machines belonging broadly to this family. In this category, we are taking up the 3-phase synchronous machines and the poly-phase induction motor for discussion. These machines are also broadly categorised as slip-ring machines, and have many common conditions among themselves.

A 3-phase synchronous machine employs two main windings—a field winding energised from a direct current source, and a 3-phase distributed winding energised with applied (for motoring condition) or induced (for generating), alternating currents. To develop a steady torque as a motor (or a counter torque as a generator) the two mmf distribution at the intervening air-gap, must be maintained at a constant torque angle. This constant phase displacement of mmfs is achieved by using two alternative forms of machine construction in synchronous machines with the armature of the machine housed on the stator—a common feature in construction strategy for large synchronous machines—the field winding is mounted on the rotating member and is revolved along the direction of the phase sequence of rotation from $a \to b \to c$ at synchronous speed, $\omega_r = \omega_s$. Alternatively, with the 3-phase armature winding mounted on the rotating member, the rotor is revolved such that it causes the appearance of the phasor axes of the armature phases according to the sequence of phase rotation—from $a \to b \to c$ with respect to the stationary field axis. With the assumption of positive direction of rotation as clockwise, the positioning of the phase windings should be $a \to b \to c$ along the counter-clockwise direction for the rotating magnetic field configuration to be stationary with the fixed field axis. To achieve this, the rotor is required to be revolved in the clockwise direction. Such a rotation makes the counter-clockwise phase distribution of the rotor to align along the stationary field axis according to the clockwise rotating phases $a \to b \to c$. This arrangement is in keeping with the convention adopted in case of dc commutator machines, where the field system is stationary.

Let us now compare the situation prevailing in case of the induction machines. For a positive direction of rotation taken as clockwise, the excitation field produced in induction machine is usually the stator member in which a rotating magnetic field at synchronous speed and constant magnitude is produced by 3-phase alternating current excitation. An induction machine does not have a clearly defined axis of symmetry. It is of annular construction, having a cylindrical stator and rotor arrangement, separated by a relatively small air-gap. It is usually possible to house the primary and the secondary windings on the stator or the rotor member respectively, but in practice the stator almost invariably houses the primary. With such an arrangement, the secondary housed on the rotor consists of a squirrel-cage winding type with stout bars, short-circuited at both ends. Such a machine is very robust, almost maintenance-free and is cheap. However, if the starting torque and the starting current are needed to be controlled, the secondary circuit on the rotor needs to be of the slip-ring type. The poly-phase primary winding usually mounted on the stator is fed from a poly-phase supply system, and it produces a synchronously rotating magnetic field pattern with respect to the primary winding. The poly-phase secondary winding mounted on the rotor is also linked by this rotating field system and a poly-phase voltage is induced in it. This secondary winding is either a 3-phase winding arrangement normally connected in star and the winding terminals brought out through three slip-rings or a squirrel-cage winding arrangement already discussed above. The nature of current induced in the short-circuited squirrel-cage winding is identical with that of the primary. Due to this, the cage-rotor winding can also be looked upon as a symmetrical 3-phase winding arrangement like a normal poly-phase winding. Short-circuiting of the secondary winding causes the secondary current to flow, and it produces a second component of field which also rotates as the first field in the same direction. The angle separating the two field patterns develops a torque and the rotor accelerates in a direction tending to reduce the rate of change of flux-linkage of the primary winding with the secondary system as per Lenz's law.

As the rotor gains speed, the frequency of the secondary current starts diminishing and so also does the field pattern produced by it, when compared with the previous amount as seen by the secondary. However, the reduction of the relative speed of the rotor with the rotating magnetic field in the gap of the machine, effectively results in a gain of the actual speed of the rotor. A secondary field distribution pattern is now developed which, relative to the primary winding, is of constant speed. If no retarding torque is applied on the rotor, the speed of the rotor tends to rise to be equal to the synchronous speed of the system. At this speed, however, the secondary winding is stationary with respect to the field produced by the primarily winding. There is no secondary voltage induced at this condition and the secondary current and field pattern in the secondary cease to exist. In actual practice, the secondary cannot attain this condition of functioning.

An induction machine in its motoring mode operates on full-load at a small value of slip. This slip speed equals the difference between the synchronous speed and the actual rotor speed. At the running speed, the strength of the primary and secondary components of the magnetic field and the torque angle between them provide the necessary torque to supply the shaft torque demand and the friction and windage torque, maintaining a near-constant speed. The schematic of a 3-phase induction machine is shown in Figure 7.11.

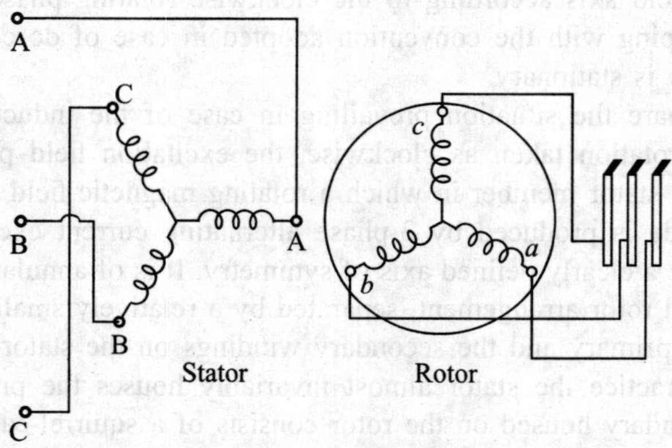

FIGURE 7.11 Schematic of a 3-phase induction machine.

The rotor speed is denoted by, $\omega_r = (1 - s)\omega_s$, where ω_s is the angular speed of supply frequency or the synchronous speed.

$$\text{Per-unit slip, } s = \frac{\omega_s - \omega_r}{\omega_s}$$

We have noted that in case of induction machines, for an assumed positive direction of rotation to be clockwise, the rotating magnetic field produced by the stationary stator member, when excited with a 3-phase alternating current excitation system with phase sequence of rotation to be $a \rightarrow b \rightarrow c$, should have a sequence of phase layout as $a \rightarrow b \rightarrow c$ along the clockwise direction. In other words, phase-b should lag a-phase positionally and further phase-c should lag phase-b positionally in the layout. Hence, when compared with the phase layout scenario in the armature of the 3-phase synchronous machine, the exciting winding in induction machine should have the axis of b- and c-phases, both for the stator and rotor winding arrangements interchanged.

In case of induction machine, therefore, the initial phase layout should lead to oppositely rotating magnetic field in the rotor when excited by 3-phase excitation with the same phase sequence.

7.4.1 Sequence of Analysis

For both types of slip-ring machines—the poly-phase synchronous as well as poly-phase induction machines—we need to subject the analytical equations to c_1 transformation in order to incorporate simplification in the voltage/current relationship of 3-phase *abc* system to equivalent 2-phase $0\alpha\beta$ system in 3-by-3 form. However, while applying the said transformation, we have to keep in mind two very important points which have a strong bearing on the final outcome of the machine analysis. These are as follows:

1. The phase layouts that we encounter in the 3-phase distributed windings in the original machine have altered arrangements in the armatures of the two machines:

   If it is spread along the *counter-clockwise* direction in sequence $a \to b \to c$, it is the case of synchronous machines, while it has a sequence of phase layout according to $a \to b \to c$ along the *clockwise* direction for the induction machines. Both machines are assumed to be rotated in positive sense along the clockwise direction.
2. In case of both machines, the exciting field is assumed to be generated on the stator. In synchronous machines, the excitation is developed in the stationary field windings, while the primary 3-phase armature winding produces the rotating magnetic field, revolving at synchronous speed and constant amplitude in the air-gap in case of induction machines.

Let us critically examine the implications of the consideration 1 above, first.

We have noted earlier, that in applying transformation c_1, the physical positioning of the β-phase of the equivalent $0\alpha\beta$ system depends on the actual layout of the relative phase-spread positioning of the original 3-phase *abc* system. We have concluded that the β-phase lies in the direction of *b*-phase and the axis of β-phase is separated from the axis of the *b*-phase by 30° electrical (as *b*-phase is separated from *a*-phase effectively by 120° electrical, while β-phase is separated from α-phase by 90°, and we assumed *a*-phase as well as the α-phase as coincident).

We also observe that in case of synchronous machines, *b*-phase leads *a*-phase by $\left(\frac{2\pi}{3}\right)$, while β-phase lies ahead of α-phase effectively by $\left(\frac{\pi}{2}\right)$. But in case of induction machines, the situation has changed—*b*-phase lagged *a*-phase effectively by $\left(\frac{2\pi}{3}\right)$ and by implication β-phase also lags α-phase by $\left(\frac{\pi}{2}\right)$. Hence, we observe that β-phase in the two cases is located in phase-opposition to each other. This has been depicted clearly in Figure 7.12.

The effect of this phenomenon is to consider reversal of all terms derived for β-phase in applying transformation c_1 in one of the machines, say synchronous machine analysis. This permits the use of a single common voltage equation for both machines, and we can apply the reversal of signs subsequently.

The rotor winding arrangements of both machines need linearization by applying a second 3-phase, *abc* transformation to 2-phase 0*dq* transformation in 3-by-3 form, c_2. The synchronous machine and the induction machine equations then become identical in form with those of commutator machines, and analysis of all different categories of machines broadly assumes a common form.

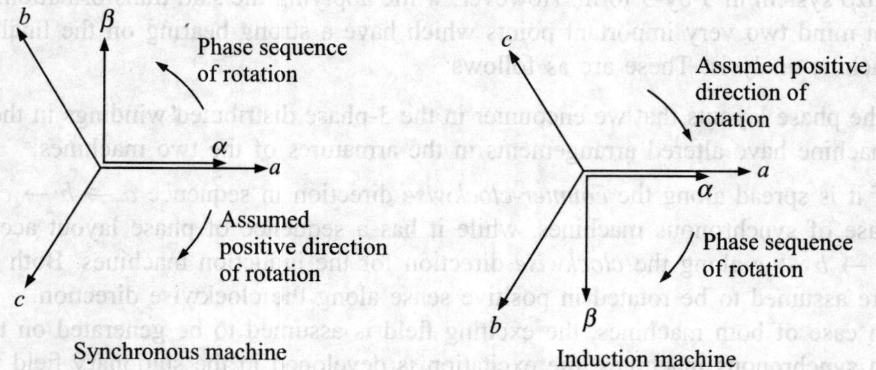

FIGURE 7.12 Relative positions of β-phase in case of synchronous and induction machines with respect to common α-phase, for applying c_1 transformation.

We shall consider the application of transformation c_1 and c_2 in separate chapters for both these machines.

OBJECTIVE TYPE QUESTIONS

7.1 Armature conductors of a dc machine, shall develop in them:

(a) dc voltage of constant magnitude
(b) alternating voltage of frequency corresponding to the number of poles and speed of rotation of armature
(c) a rectangular waveform pulsating voltage
(d) a voltage waveform corresponding exactly to the flux-distribution waveform.

7.2 A rotating commutator provides:

(a) transformation of dc in conductors
(b) continuous contact from rotating armature to stationary external circuit
(c) frequency changer action by transforming ac in the conductors to dc in external circuit
(d) both functions listed under (b) and (c) above

7.3 A 12-pole dc machine is provided with a simplex lap winding, and is driven at 1200 rpm. It has 288 conductors in its armature and the flux per pole is 0.12 Wb. It carries a total current of 500 amps. If the brush contact drop is taken as 2.0 volts, and the armature resistance is 0.05 ohm:

(a) the generated voltage available at the brushes shall be:

(i) 288 volt (ii) 91.67 volt
(iii) 576 volt (iv) 3456 volt

(b) When functioning as a motor, the applied voltage must be:

(i) 549 volt (ii) 603 volt
(iii) 315 volt (iv) 261 volt

(c) Current in each parallel path is:

(i) 6000 amp (ii) 250 amp
(iii) 83.3 amp (iv) 41.67 amp

(d) If the field current is 10 amp at full-load in its shunt-field winding, and there is no series field, when the above machine is functioning as a motor and the armature is drawing 500 amp from supply, the efficiency is:

(i) 95% (ii) 85%
(iii) 93.65% (iv) 89%

7.4 A 600 volt, 1000 rpm, 500 amp motor has a voltage drop of 2.0 volt at the brush contact. It has an armature circuit resistance of 0.02 ohm including the series field winding. In order to limit the starting current to 1000 amp, an additional resistance has to be inserted in series.

(a) Its value shall be:

(i) 5 ohm (ii) 0.3878 ohm
(iii) 0.4 ohm (iv) 10 ohm

(b) The starting torque of the motor shall be:

(i) 8422.5 N-m (ii) 4211.25 N-m
(iii) Zero (iv) Cannot be obtained with given data

7.5 In a short-shunted compounded dc generator the shunt-field is connected across:

(a) the load terminals (b) the armature terminals
(c) the series field terminals (d) none of the above given alternatives

7.6 In a long-shunt compound dc motor, the shunt-field winding is connected across:

(a) the load terminals (b) the armature terminals
(c) the series field terminals (d) none of the above given options

7.7 A dc generator is driven by a 3-phase induction motor with a negligible effective resistance in its stator circuit. It is found that the drive unit takes power inputs from the ac supply of:

(a) 1000 Watts with the generator uncoupled
(b) 2000 Watts with the generator coupled but unexcited
(c) 3000 Watts with the generator excited

The excitation loss of the dc generator is:

(i) 2000 Watts (ii) 500 Watts
(iii) 1000 Watts (iv) 3000 Watts

Answers

7.1 (d) **7.2** (d) **7.3** (a) (iii); (b) (ii); (c) (iv); (d) (iii)
7.4 (a) (ii); (b) (i) **7.5** (b) **7.6** (a) **7.7** (iii)

NUMERICAL PROBLEMS

7.1 A separately excited dc generator has a lap-connected 4-pole winding in its armature, with 400 conductors. If the armature is carrying a total current of 100 amps. Under steady-state operation, draw the mmf and flux-density diagrams. Neglect fringing of flux.

(a) Estimate the magnetizing (*d*-axis) and cross-magnetizing (*q*-axis) components of mmf in amp-turns, if the rotor brushes are displaced by 10° electrical from the neutral magnetic axis.

(b) Analysis the phenomenon of brush-shifting in a generator.

[**Ans.** (a) F_d = 1231 Amp-turns; (b) F_q = 217 Amp-turns]

7.2 An ideal dc motor with a negligible power loss operates at no-load with a fixed value of field excitation supplied from a separate source. Obtain an expression for the current drawn when the armature is subjected to a varying voltage V_t.

Derive the constant in the expression if the moment of inertia of the armature is 0.75 kg-m^2 and the armature develops 150 volts at a speed of 750 rpm.

[**Ans.** J/Km^2 = 0.2056]

CHAPTER 8

Application of Transformations C_1 and C_2 to Slip-ring Windings

8.1 SLIP-RING INDUCTION MACHINE

We start by considering the 3-phase armature windings arrangement in case of slip-ring machines encountered in the more straightforward case of wound-rotor induction machine, first. In a typical case, both stator and rotor are cylindrical in shape, with a small but uniform air-gap intervening the two members, which carry uniformly distributed poly-phase windings along their respective surfaces.

The magnitude of resistance *R* and coefficient of self-inductance, *L* of both the stator and the rotor windings, as well as the coefficients of mutual-inductance between the phases of the same member—the stator or the rotor—remain constant even with change in rotor angular position between the rotor and the stator members when the machine rotates. However, the primary and the secondary phase windings continue to present mutual inductances, which are functions of their relative angular positions. In fact, it is on these variations that the operation of induction machines are governed.

We could derive the performance of a machine by denoting the mutual inductance as a function of the angle between the concerned windings, and taking the axis of one of the phases, say phase *A*, as the direct axis. We can then write down the instantaneous currents and flux-linkage expressions, obtain the voltages and transform the equations in *D-Q* terms using transformation matrices already developed in Chapter 6. We have to undertake the step integrally for the stator and the rotor arrangement simultaneously. In other words, we end up with six non-linear, simultaneous, differential voltage/current relationships—three for the stator and three for the rotor phases. The resulting relationship is indicated qualitatively in the descriptive representative expression (8.1).

The task at hand to conceive and formulate the impedance matrix for writing down the voltage/current relationship by inspection mathematically in exact form is unconceivable because the stator-to-stator and rotor-to-rotor mutual inductances between phases, which contribute only to transformer induced voltages and field energy storage (and not torque production) are not

	AS	BS	CS	ar	br	cr
AS	$R_A + L_A p$	Transformer effect of M_{BA}	Transformer effect of M_{CA}	Transformer and speed effect of M_{aA}	Transformer and speed effect of M_{bA}	Transformer and speed effect of M_{cA}
BS	Transformer effect of M_{AB}	$R_B + L_B p$	Transformer effect of M_{CB}	Transformer and speed effect of M_{aB}	Transformer and speed effect of M_{bB}	Transformer and speed effect of M_{cB}
CS	Transformer effect of M_{AC}	Transformer effect of M_{BC}	$R_C + L_C p$	Transformer and speed effect of M_{aC}	Transformer and speed effect of M_{bC}	Transformer and speed effect of M_{cC}
ar	Transformer and speed effect of M_{Aa}	Transformer and speed effect of M_{Ba}	Transformer and speed effect of M_{Ca}	$r_a + L_a p$	Transformer effect of M_{ba}	Transformer effect of M_{ca}
br	Transformer and speed effect of M_{Ab}	Transformer and speed effect of M_{Bb}	Transformer and speed effect of M_{Cb}	Transformer effect of M_{ab}	$r_b + L_p p$	Transformer effect of M_{cb}
cr	Transformer and speed effect of M_{Ac}	Transformer and speed effect of M_{Bc}	Transformer and speed effect of M_{Cc}	Transformer effect of M_{ac}	Transformer effect of M_{bc}	$r_c + L_c p$

...(8.1)

conceivable mathematically. Also, it is totally obscure and impossible to physically conceive the interaction pattern of stator to rotor mutual inductances, which contribute to field energy storage as well as torque production. We can, at best, identify these terms descriptively as indicated in Eq. (8.1). Fortunately for us, it has been possible to develop effective simplifying tools which require our denoting the impedance in matrix form.

First, we recapitulate the details of our identification convention again, even at the cost of repetition. We have the windings of the 3-phase machines shown in Figure 8.1. The phase-sequence of the supply is assumed to be $A \rightarrow B \rightarrow C$ and the stator phases are represented by *AS*, *BS* and *CS* (or for brevity simply by *A*, *B*, *C*), while the rotor phases are depicted by *ar*, *br* and *cr* (or simply as *a*, *b*, *c*). Clockwise direction of rotation is assumed as positive. This let to six sets of voltage equations—three for the stator and three for, the rotor member—with thirty-six qualitatively described elements in the matrix representation of the impedance matrix. We may, therefore, use the compound form of matrix for simplification of the voltage equations, by dividing the equations into components by drawing double-lines so that the stator and rotor quantities may be segregated. The compound matrix of the voltage/current relationship, therefore, assumes the following form:

$$\begin{bmatrix} V_1 \\ \hline V_2 \end{bmatrix} = \left[\begin{array}{c|c} Z_{11} & Z_{12} \\ \hline Z_{21} & Z_{22} \end{array}\right] \begin{bmatrix} I_1 \\ \hline I_2 \end{bmatrix} \qquad \text{...(8.2)}$$

Each of the elements, V_1, ..., Z_{11}, ..., and I_1, ..., are matrices themselves, for which it is permitted to use separate voltage equations of the form:

$$[V_1] = [Z_{11}]\ [I_1] \qquad \text{...(8.3)}$$

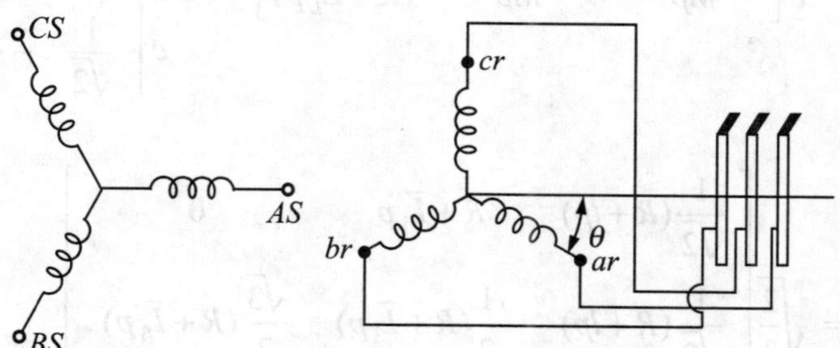

FIGURE 8.1 The schematic of a 3-phase induction motor.

Both stator and rotor elements of the above set of voltage equations shall first be transformed by $[C_1]$, so that:

$$[C] = \begin{bmatrix} [C_1] & 0 \\ 0 & [C_1] \end{bmatrix} \qquad \text{...(8.4)}$$

We now, apply transformation matrix $[C_1]$ on the impedance matrix. We have seen in Eq. (6.10) that to convert 3-phase $[Z]$ matrix to 2-phase form, we require to apply:

$$[Z'] = [C^*_{1t}][Z][C_1] \qquad \text{...(8.5)}$$

where, $[Z']$ is the new form of the matrix, while $[Z]$ is in the old form.

In expanding expression (8.4), we observe that $[C_1]$ is a 'passive transformation' with only real constant terms as its elements have no complex numbers. Hence, $[C_{1t}^*] = [C_{1t}]$, and we get from expression (8.5):

$$[Z'] = \begin{bmatrix} [C_{1t}][Z_{11}][C_1] & [C_{1t}][Z_{12}][C_1] \\ [C_{1t}][Z_{21}][C_1] & [C_{1t}][Z_{22}][C_1] \end{bmatrix} \qquad \text{...(8.6)}$$

so that each element of the impedance matrix transforms in the same manner.

While discussing the 3-phase windings in Chapter 6, we noted that the impedance matrix in expression (6.2) was derived as:

$$[Z] = \begin{bmatrix} (R+\bar{L}_L p) & \bar{M}p & \bar{M}p \\ \bar{M}p & (R+\bar{L}_L p) & \bar{M}p \\ \bar{M}p & \bar{M}p & (R+\bar{L}_L p) \end{bmatrix} \qquad \text{...(8.7)}$$

Hence, in obtaining the new transformed value of $[Z']$, we commence from the backside by first multiplying $[Z]$ by $[C_1]$, leading to:

$$[Z][C_1] = \begin{matrix} & a & b & c \\ a & (R+\bar{L}_L p) & \bar{M}p & \bar{M}p \\ b & \bar{M}p & (R+\bar{L}_L p) & \bar{M}p \\ c & \bar{M}p & \bar{M}p & (R+\bar{L}_L p) \end{matrix} \; \sqrt{\frac{2}{3}} \begin{matrix} & 0 & \alpha & \beta \\ a & \frac{1}{\sqrt{2}} & 1 & 0 \\ b & \frac{1}{\sqrt{2}} & -\frac{1}{2} & \frac{\sqrt{3}}{2} \\ c & \frac{1}{\sqrt{2}} & \frac{1}{2} & -\frac{\sqrt{3}}{2} \end{matrix}$$

$$= \sqrt{\frac{2}{3}} \begin{bmatrix} \frac{1}{\sqrt{2}}(R+lp) & R+\bar{L}_1 p & 0 \\ \frac{1}{\sqrt{2}}(R+lp) & -\frac{1}{2}(R+\bar{L}_1 p) & \frac{\sqrt{3}}{2}(R+\bar{L}_1 p) \\ \frac{1}{\sqrt{2}}(R+lp) & -\frac{1}{2}(R+\bar{L}_1 p) & -\frac{\sqrt{3}}{2}(R+\bar{L}_1 p) \end{bmatrix} \qquad \text{...(8.8)}$$

where, the value of l, the leakage reactance, $\bar{L}_1$, the self-inductance are $\bar{M}$, the mutual inductances between phases are as shown in Eqs. (6.43) and (6.44) in Section (6.6.3) to be:

$$l = L_L + 2M \quad \text{and} \quad L_1 = \bar{L}_L - \bar{M}$$

where L_L is the self-inductance of a phase winding and L_1 is the effective self-inductance for 3-phase balanced operational, and the value of M is negative. The primed values $\bar{L}_L$ and $\bar{M}$ are the winding parameters with a single-phase excitation applied to the corresponding phase. All other terminals are kept open circuited during determination of these parameters.

Since,
$$[C_{1t}^*] = [C_{1t}] = \sqrt{\frac{2}{3}}\begin{bmatrix} \frac{1}{\sqrt{2}} & \frac{1}{\sqrt{2}} & \frac{1}{\sqrt{2}} \\ 1 & -\frac{1}{2} & -\frac{1}{2} \\ 0 & \frac{\sqrt{3}}{2} & -\frac{\sqrt{3}}{2} \end{bmatrix}$$

The final product

$[Z'] = [C_{1t}]\ [Z]\ [C_1]$ has the following form:

$$[Z'] = \sqrt{\frac{2}{3}}\begin{bmatrix} (R+lp) & 0 & 0 \\ 0 & (R+L_1 p) & 0 \\ 0 & 0 & (R+L_1 p) \end{bmatrix} \qquad ...(8.9)$$

We can physically interpret from the shape of the impedance matrix of expression (8.9), and arrive at a very significant conclusion. We elaborated in Section 6.2.3, under the conditions of orthogonality of passive transformations, that the laws governing the active and passive transformations were the same, although the purpose of their application to the analysis of electro-mechanical devices were quite different. We have also seen in Section 6.2.1 in connection with application of 'active transformations' by way of the solved example, immediately after Eq. (6.9) that whenever an inductance matrix is composed of terms along the main diagonal only—that is the diagonal from left to right—then it signifies a number of totally independent electrical circuits with no inter-dependent effects. The zero-sequence component of the impedance matrix of 3-phase systems, as presented in Eq. (8.9), in its new transformed form, satisfies this condition. It, therefore, implies that the zero-sequence component of the circuit forms a completely electrically independent circuit and any current flowing in it does not, in any way, affect or contribute to operating parameters of positive sequence circuit element. We can, therefore, keep the effect of zero-sequence terms in reserve for later use during inverse transformation from 2-phase to 3-phase stage in the 0*dq* to *abc* stage.

Now, we are ready to take up the voltage/current relations of individual component matrices of Eq. (8.2).

The relationship connecting the stator voltages and stator currents, considering the rotor open-circuited—that is with $[I_2]$ equal to zero—have been derived in Eqs. (8.7) and (8.8). We use these expressions for developing stator-to-stator impedance matrix $[Z_{11}]$, so that the relation can be written down as given below, using the adopted suffix convention of 1 for the stator and 2 for the rotor.

For the 3-phase *A*, *B*, *C* system, the voltage/current relationship is shown as follows:

$$\begin{bmatrix} V_A \\ V_B \\ V_C \end{bmatrix} = \begin{bmatrix} (R_A+\bar{L}_A p) & \bar{M}_{BA} p & \bar{M}_{CA} p \\ \bar{M}_{AB} p & (R_B+\bar{L}_B p) & \bar{M}_{CB} p \\ \bar{M}_{AC} p & \bar{M}_{BC} p & (R_C+\bar{L}_C p) \end{bmatrix}\begin{bmatrix} I_A \\ I_B \\ I_C \end{bmatrix}$$

In this expression, $\bar{L}_A, \bar{L}_B$ and $\bar{L}_C$ are self-inductance for single-phase excitation and

$\bar{M}_{AB} = \bar{M}_{BA}$, $\bar{M}_{BC} = \bar{M}_{CB}$ and $\bar{M}_{CA} = \bar{M}_{AC}$, with L_1 and M_1 denoting the stator-to-stator self- and mutual-inductances.

The stator-to-stator impedance matrix element $[Z_{11}]$:

$$[Z_{11}] = \begin{matrix} & A & B & C \\ A & (R_1 + L_1 p) & M_1 p & M_1 p \\ B & M_1 p & (R_1 + L_1 p) & M_1 p \\ C & M_1 p & M_1 p & (R_1 + L_1 p) \end{matrix} \quad ...(8.10)$$

derived in Eq. (8.7).

And the transformed value of stator-to-stator impedance matrix, also derived in Eq. (8.9) is:

$$[Z'_{11}] = \begin{matrix} & 0 & \alpha & \beta \\ 0 & (R_1 + L_{01} p) & 0 & 0 \\ \alpha & 0 & (R_1 + L_1 p) & 0 \\ \beta & 0 & 0 & (R_1 + L_1 p) \end{matrix} \quad ...(8.11)$$

where, $L_{01} = \bar{L}_L + 2\bar{M}$, and $L_1 = \bar{L}_L - \bar{M}$ and $\bar{M}$ is negative.

The rotor-to-rotor impedance matrix element, $[Z_{22}]$: The rotor-to-rotor impedance matrix element, $[Z_{22}]$ can be written down similarly, because of symmetry of rotor construction and winding layouts directly, following expressions (8.7) and (8.9). These have following shapes:

$$[Z_{22}] = \begin{matrix} a \\ b \\ c \end{matrix} \begin{bmatrix} (R_2 + \bar{L}_2 p) & \bar{M}_2 p & \bar{M}_2 p \\ \bar{M}_2 p & (R_2 + \bar{L}_2 p) & \bar{M}_2 p \\ \bar{M}_2 p & \bar{M}_2 p & (R_2 + \bar{L}_2 p) \end{bmatrix} \quad ...(8.12)$$

where, the elements with suffix 2 are rotor quantities, and $\bar{L}_2$ and $\bar{M}_2$ denote the per phase values as were listed earlier in connection with $[Z_{11}]$ for stator component matrix. We, therefore, obtain:

$$[Z'_{22}] = [C_{1t}]\,[Z_{22}]\,[C_1]$$

$$= \begin{matrix} & 0 & \alpha & \beta \\ 0 & (R_2 + L_{02} p) & 0 & 0 \\ \alpha & 0 & (R_2 + L_2 p) & 0 \\ \beta & 0 & 0 & (R_2 + L_2 p) \end{matrix} \quad ...(8.13)$$

where, $L_{02} = \bar{L}_2 + 2\bar{M}_2$ and $L_2 = \bar{L}_2 - \bar{M}_2$ and $\bar{M}_2$ is negative.

The rotor-to-stator impedance matrix element, $[Z_{21}]$: The coefficient of mutual-inductance for the particular case of a-phase of the rotor and A-phase of the stator windings can be given as follows:

$$M_{aA} = \bar{M} \cos\theta + \bar{M}_3 \cos 3\theta$$

where, θ is the angular displacement between the axis of a-phase and A-phase, and M_3 is the third harmonic component. M_{aA} has the same value as M_{bB} and M_{cC}. It can also be conceived that M_{bA}, M_{cB} and M_{aC} also yield equal results, with θ changed to $\theta_2 = \left(\theta + \frac{2\pi}{3}\right)$, and for M_{cA}, M_{aB} and M_{bC} in which case θ is replaced by $\theta_3 = \left(\theta + \frac{4\pi}{3}\right)$. Thus,

$$M_{bA} = \bar{M} \cos \theta_2 + \bar{M}_3 \cos 3\theta_2$$

and,

$$M_{cA} = \bar{M} \cos \theta_3 + \bar{M}_3 \cos 3\theta_3$$

The complete impedance matrix can be written down as follows:

$$[Z_{21}] = \begin{bmatrix} (\bar{M}p \cdot \cos\theta & (\bar{M}p \cdot \cos\theta_3 & (\bar{M}p \cdot \cos\theta_2 \\ +\bar{M}_3 p \cdot \cos 3\theta) & +\bar{M}_3 p \cdot \cos 3\theta_3) & +\bar{M}_3 p \cdot \cos 3\theta_2) \\ (\bar{M}p \cdot \cos\theta_2 & (\bar{M}p \cdot \cos\theta & (\bar{M}p \cdot \cos\theta_3 \\ +\bar{M}_3 p \cdot \cos 3\theta_2) & +\bar{M}_3 p \cdot \cos 3\theta) & +\bar{M}_3 p \cdot \cos 3\theta_3) \\ (\bar{M}p \cdot \cos\theta_3 & (\bar{M}p \cdot \cos\theta_2 & (\bar{M}p \cdot \cos\theta \\ +\bar{M}_3 p \cdot \cos 3\theta_3) & +\bar{M}_3 p \cdot \cos 3\theta_2) & +\bar{M}_3 p \cdot \cos 3\theta) \end{bmatrix} \quad ...(8.14)$$

We apply the associative law for executing the multiplication, to which matrices are amenable, and write:

$$[Z_{21}]\,[C_1] = \begin{matrix} \\ a \\ b \\ c \end{matrix} \begin{matrix} A & B & C \\ \bar{M}p \cos \theta & \bar{M}p \cos \theta_3 & \bar{M}p \cos \theta_2 \\ \bar{M}p \cos \theta_2 & \bar{M}p \cos \theta & \bar{M}p \cos \theta_3 \\ \bar{M}p \cos \theta_3 & \bar{M}p \cos \theta_2 & \bar{M}p \cos \theta \end{matrix} \;\sqrt{\frac{2}{3}} \begin{matrix} \\ A \\ B \\ C \end{matrix} \begin{bmatrix} 0 & \alpha & \beta \\ \frac{1}{\sqrt{2}} & 1 & 0 \\ \frac{1}{\sqrt{2}} & -\frac{1}{2} & \frac{\sqrt{3}}{2} \\ \frac{1}{\sqrt{2}} & -\frac{1}{2} & -\frac{\sqrt{3}}{2} \end{bmatrix}$$

$$= \sqrt{\frac{3}{2}}\bar{M} \begin{matrix} \\ a \\ b \\ c \end{matrix} \begin{bmatrix} 0 & \alpha & \beta \\ 0 & p\cos\theta & p\sin\theta \\ 0 & p\cos\theta_2 & p\sin\theta_2 \\ 0 & p\cos\theta_3 & p\sin\theta_3 \end{bmatrix} \quad ...(8.15)$$

By using the trigonometrical identities:

1. $\cos \theta + \cos \theta_2 + \cos \theta_3 = 0$
2. $\cos \theta_3 - \cos \theta_2 = \sqrt{3} \sin \theta$
3. $\cos \theta - \cos \theta_3 = \sqrt{3} \sin \theta_2$
4. $\cos \theta_2 - \cos \theta = \sqrt{3} \sin \theta_3$

we can achieve the simplified form shown in Eq. (8.15).

We now take up the multiplication of the second part—that of the third harmonic part. This part is simple in form. We observe that $3\theta_2$, $3\theta_3$ and 3θ have the same contribution. It adds a constant factor of $M_3 p \cos 3\theta$, which adds a matrix of the form:

$$M_3 p \cos 3\theta \begin{matrix} \\ a \\ b \\ c \end{matrix} \begin{matrix} A & B & C \\ \begin{bmatrix} 1 & 1 & 1 \\ 1 & 1 & 1 \\ 1 & 1 & 1 \end{bmatrix} \end{matrix}$$

so that,

$$[Z_{21}]\,[C_1] = \bar{M}_3 p \cos 3\theta \begin{bmatrix} 1 & 1 & 1 \\ 1 & 1 & 1 \\ 1 & 1 & 1 \end{bmatrix} \sqrt{\frac{2}{3}} \begin{bmatrix} \frac{1}{\sqrt{2}} & 1 & 0 \\ \frac{1}{\sqrt{2}} & -\frac{1}{2} & \frac{\sqrt{3}}{2} \\ \frac{1}{\sqrt{2}} & -\frac{1}{2} & -\frac{\sqrt{3}}{2} \end{bmatrix}$$

$$= \sqrt{3}\,\bar{M}_3 p \cos 3\theta \begin{matrix} \\ a \\ b \\ c \end{matrix} \begin{matrix} 0 & \alpha & \beta \\ \begin{bmatrix} 1 & 0 & 0 \\ 1 & 0 & 0 \\ 1 & 0 & 0 \end{bmatrix} \end{matrix}$$

Then, the complete transformed matrix has the following form:

For the first part:

$$[Z'_{21}] = [C_{1t}]\,[Z_{21}]\,[C_1]$$

$$= \sqrt{\frac{2}{3}} \begin{matrix} \\ 0 \\ \alpha \\ \beta \end{matrix} \begin{matrix} a & b & c \\ \begin{bmatrix} \frac{1}{\sqrt{2}} & \frac{1}{\sqrt{2}} & \frac{1}{\sqrt{2}} \\ 1 & -\frac{1}{2} & -\frac{1}{2} \\ 0 & \frac{\sqrt{3}}{2} & -\frac{\sqrt{3}}{2} \end{bmatrix} \end{matrix} \begin{matrix} \\ a \\ b \\ c \end{matrix} \begin{matrix} 0 & \alpha & \beta \\ \begin{bmatrix} 0 & p\cos\theta & p\sin\theta \\ 0 & p\cos\theta_2 & p\sin\theta_2 \\ 0 & p\cos\theta_3 & p\cos\theta_3 \end{bmatrix} \end{matrix} \sqrt{\frac{3}{2}}\bar{M}$$

$$= \bar{M} \begin{matrix} \\ 0 \\ \alpha \\ \beta \end{matrix} \begin{matrix} 0 & \alpha & \beta \\ \begin{bmatrix} 0 & 0 & 0 \\ 0 & p\cos\theta & p\sin\theta \\ 0 & -p\sin\theta & p\cos\theta \end{bmatrix} \end{matrix}$$

in which, $M = \frac{3}{2}\bar{M}$, since the identity, $\sin\theta_2 - \sin\theta_3 = \sqrt{3}\cos\theta$.

Also, for the second stage, that is, the 3rd harmonic component:

$$[Z'_{21}] = [C_{1t}]\,[Z_{21}]\,[C_1]$$

Hence,

$$[Z'_{21}] = \sqrt{\frac{2}{3}}\begin{matrix} \\ 0 \\ \alpha \\ \beta \end{matrix}\begin{matrix} a & b & c \\ \left[\begin{matrix} \frac{1}{\sqrt{2}} & \frac{1}{\sqrt{2}} & \frac{1}{\sqrt{2}} \\ 1 & -\frac{1}{2} & -\frac{1}{2} \\ 0 & \frac{\sqrt{3}}{2} & -\frac{\sqrt{3}}{2} \end{matrix}\right] \end{matrix} \sqrt{3}\bar{M}_3 p \cos 3\theta \begin{matrix} \\ a \\ b \\ c \end{matrix}\begin{matrix} 0 & \alpha & \beta \\ \left[\begin{matrix} 1 & 0 & 0 \\ 1 & 0 & 0 \\ 1 & 0 & 0 \end{matrix}\right] \end{matrix}$$

$$= \bar{M}_3 p \cos 3\theta \begin{matrix} \\ 0 \\ \alpha \\ \beta \end{matrix}\begin{matrix} 0 & \alpha & \beta \\ \left[\begin{matrix} 1 & 0 & 0 \\ 0 & 0 & 0 \\ 0 & 0 & 0 \end{matrix}\right] \end{matrix}$$

where,

$$M_3 = 3\bar{M}_3$$

Then, addition of third harmonic component to the first component, gives the complete transformed matrix component $[Z'_{21}]$, which has the final shape:

$$[Z'_{21}] = \begin{matrix} \\ 0 \\ \alpha \\ \beta \end{matrix}\begin{matrix} 0 & \alpha & \beta \\ \left[\begin{matrix} \bar{M}_3 p \cos 3\theta & 0 & 0 \\ 0 & Mp\cos\theta & Mp\sin\theta \\ 0 & -Mp\sin\theta & Mp\cos\theta \end{matrix}\right] \end{matrix}$$

Stator-to-rotor impedance matrix element, $[Z_{12}]$: This can be developed similarly and we observe that it has the form:

$$[Z_{12}] = [Z_{21t}] \qquad \text{...(8.16)}$$

so that,

$$[Z'_{12}] = [Z'_{21t}] \qquad \text{...(8.17)}$$

that is, their transformed values also match.

It is possible to combine the four transformed component matrices, to arrive at the total transformed matrix, as follows:

$$[Z'] = \begin{array}{c} \\ 0 \\ \alpha(A) \\ \beta(B) \\ 0 \\ \alpha(a) \\ \beta(b) \end{array} \begin{array}{c} \begin{array}{cccccc} 0 & \alpha(A) & \beta(B) & 0 & \alpha(a) & \beta(b) \end{array} \\ \begin{bmatrix} (R_1 + L_{01}p) & 0 & 0 & M_3 p \cos 3\theta & 0 & 0 \\ 0 & (R_1 + L_1 p) & 0 & 0 & Mp\cos\theta & -Mp\sin\theta \\ 0 & 0 & (R_1 + L_1 p) & 0 & Mp\sin\theta & Mp\cos\theta \\ M_3 p\cos 3\theta & 0 & 0 & (R_2 + L_{02}p) & 0 & 0 \\ 0 & Mp\cos\theta & Mp\sin\theta & 0 & (R_2 + L_2 p) & 0 \\ 0 & -Mp\sin\theta & Mp\cos\theta & 0 & 0 & (R_2 + L_2 p) \end{bmatrix} \end{array} \quad ...(8.18)$$

We can rearrange the above matrix by moving the fourth row and column to second positions of rows and columns respectively, so that it can have a new form:

$$\begin{array}{c} \\ 0 \\ 0 \\ \alpha(A) \\ \beta(B) \\ \alpha(a) \\ \beta(b) \end{array} \begin{array}{c} \begin{array}{cccccc} 0 & 0 & \alpha(A) & \beta(B) & \alpha(a) & \beta(b) \end{array} \\ \left[\begin{array}{cc|cccc} (R_1 + L_{01}p) & M_3 p\cos 3\theta & 0 & 0 & 0 & 0 \\ M_3 p\cos 3\theta & (R_2 + L_{02}p) & 0 & 0 & 0 & 0 \\ \hline 0 & 0 & (R_1 + L_1 p) & 0 & Mp\cos\theta & -Mp\sin\theta \\ 0 & 0 & 0 & (R_1 + L_1 p) & Mp\sin\theta & Mp\cos\theta \\ 0 & 0 & Mp\cos\theta & Mp\sin\theta & (R_2 + L_2 p) & 0 \\ 0 & 0 & -Mp\sin\theta & Mp\cos\theta & 0 & (R_2 + L_2 p) \end{array}\right] \end{array} \quad ...(8.19)$$

The above matrix could be looked upon in the compound matrix form by de-partitioning it after drawn vertical and horizontal lines below the second column and the next, as also the second row and the next in the impedance matrix $[Z']$ as shown in Eq. (8.19).

We observe that the above matrix in its new form assumes the shape:

$$[Z'] = \begin{bmatrix} Z_0 & 0 \\ 0 & Z'' \end{bmatrix}$$

so as to yield a new set of voltage/current matrix relationships:

$$[V_0] = [Z_0]\ [I_0] \quad \text{and} \quad [V''] = [Z'']\ [I'']$$

The two relationships above are totally independent of each other with no interaction between them. The part of the matrix component:

$$\begin{bmatrix} V_{01} \\ V_{02} \end{bmatrix} = \begin{bmatrix} (R_1 + L_{01}p) & M_3 p\cos 3\theta \\ M_3 p\cos 3\theta & (R_2 + L_{02}p) \end{bmatrix} \begin{bmatrix} I_{01} \\ I_{02} \end{bmatrix} \quad ...(8.20)$$

has no effect on the other elements as discussed in Sections 6.2.1 and 6.2.3.

The physical significance of the above sets of equations is that, under certain unusual conditions of operation, it indicates production of torque, but for this to happen, the zero-sequence current must flow. Under usual conditions, neither I_{01} nor I_{02} flows in the machine.

This component matrix must, therefore, be safely taken out of the transformed impedance matrix, leaving the positive sequence impedance matrix representing the remaining part, denoted by $[Z'']$ only. This reduced from of independence matrix is fully capable of providing the analytical equations of poly-phase slip-ring machines of the induction machine type with full impunity. It has the final form:

$$[Z''] = \begin{array}{c} \\ \alpha(A) \\ \beta(B) \\ \alpha(a) \\ \beta(b) \end{array} \begin{array}{c} \begin{array}{cccc} \alpha(A) & \beta(B) & \alpha(a) & \beta(b) \end{array} \\ \begin{bmatrix} (R_1 + L_1 p) & 0 & Mp\cos\theta & -Mp\sin\theta \\ 0 & (R_1 + L_1 p) & Mp\sin\theta & Mp\cos\theta \\ Mp\cos\theta & Mp\sin\theta & (R_2 + L_2 p) & 0 \\ -Mp\sin\theta & Mp\cos\theta & 0 & (R_2 + L_2 p) \end{bmatrix} \end{array} \quad ...(8.21)$$

so that,

$$[I''] = [Z_t''][V'']$$

Mathematical significance of the transformation is that the form of impedance matrix has been drastically simplified and the 3-phase angles θ_2 and θ_3 have been eliminated. Due to presence of θ in the impedance expression, however, its form is still non-linear.

We now, hold the final form of expression (8.21) in reserve and come back to it at the time of analysing induction machines. It is now time for us to develop similar strategies for transforming impedance matrix for salient-pole synchronous machines following similar lines, to be able to compare the nature of impedance matrices in the two cases and arrive at conclusions.

8.2 SALIENT-POLE SYNCHRONOUS MACHINE

In this section, we pursue a similar exercise in transforming matrix of a poly-phase synchronous machine.

In Section 7.4, we have touched upto the general features of poly-phase alternating current distributed windings mounted on the armature of the main-frame of slip-ring machines. We have spent some time on the general features of synchronous machines. We have also noted the special features of induction machines with the help of Figure 7.11 and have outlined our strategies for analysing this family of machines in Section 7.4.1, after noting that there existed significant similarity in the analysis of synchronous and induction machines.

Let Figure 8.2 depict the schematic of a salient-pole synchronous machine with the existing field and the 'damper' arrangement shown on the stationary member. From design considerations in salient-pole synchronous machines, the field windings are placed on the salient-poles and the air-gap surface is cylindrical on the armature side only.

In the schematic diagram, the field system is indicated by the field winding, F on the stator member. In addition, the short-circuited windings on the stator, marked D and Q, represent the damper windings of the machine. These comprise squirrel-cage bars laid along the pole-faces and are short-circuited at both ends through end rings. Under balanced steady-state conditions, the flux in the poles and yoke is stationary with respect to the damper windings. Thus, there is no current in damper windings as they are in-operative. There is also no

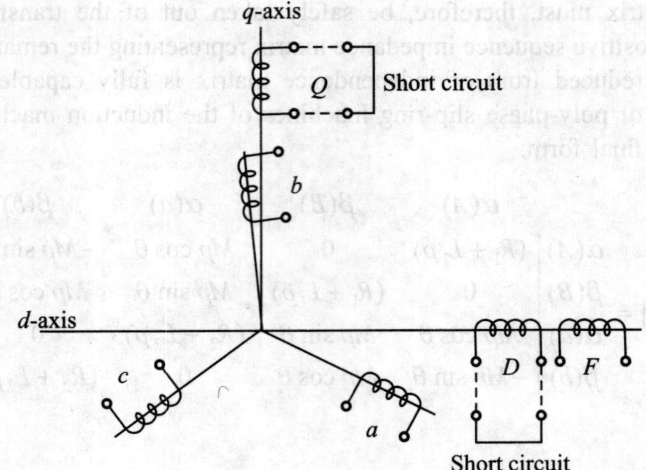

FIGURE 8.2 Schematic of a 3-phase, salient-pole synchronous machine.

eddy-current in the field structure. Only when there occurs a sudden change in the balanced currents due to change in load or due to a fault or a steady unbalanced current flow in the machine armature, the flux in the pole changes and the damper system becomes operative.

The general voltage equation of the machine may be written in compound matrix form with suffices (1) and (2) indicating the field and armature respectively as follows:

$$\begin{bmatrix} V_1 \\ V_2 \end{bmatrix} = \begin{bmatrix} Z_{11} & Z_{12} \\ Z_{21} & Z_{22} \end{bmatrix} \begin{bmatrix} I_1 \\ I_2 \end{bmatrix}$$

The next stage is to replace the actual 3-phase armature by an equivalent 2-phase system using phase transformation $[C_1]$. The field system being already two-axis in nature, needs no alteration. This amounts to applying the first transformation to the impedance matrix in the voltage relation by a transformation $[C]$ of the following form:

$$[C] = \begin{array}{c} \\ 1 \\ 2 \end{array} \begin{array}{c} \begin{array}{cc} 1 & 2 \end{array} \\ \begin{bmatrix} 1 & 0 \\ 0 & C_1 \end{bmatrix} \end{array}$$

so that the transformed impedance matrix assumes the form:

$$[Z'] = [C_{1t}]\ [Z]\ [C_1] = \begin{bmatrix} [Z_{11}] & [Z_{12}][C_1] \\ [Z_{21}][C_1] & [C_{1t}][Z][C_1] \end{bmatrix}$$

8.2.1 Idealised Reactances of Synchronous Machines

Now, we spend time on the deviation of expressions on self- and mutual-inductances for the salient-pole synchronous machines. The derivations are based on the idealised simplifying assumptions listed in Section 4.6.

(a) Self- and mutual-inductance of the stationary member housing field and damper windings: Figure 8.2 indicates that in addition to the field winding, there are additional short-circuited windings *D* and *Q*, identified as damper windings on the stator.

As pointed out, the rotor is taken to be of cylindrical structure, and the mmfs of the stator windings—*F*, *D* and *Q*—are exposed to magnetic paths of constant permeance for all values of θ_r. The self-inductance, L_F, L_D and L_Q and the mutual-inductance M_{FD} therefore, have constant values independent of θ_r. Further, there is no coupling between the field axis or *D*-axis winding and *Q*-axis winding. The general form of the field impedance matrix component, $[Z_{11}]$, is therefore denoted in the matrix form as follows:

$$[Z_{11}] = \begin{matrix} & F & D & Q \\ F & (R_F + L_F p) & M_{FD} p & 0 \\ D & M_{DF} p & (R_D + L_D p) & 0 \\ Q & 0 & 0 & (R_Q + L_Q p) \end{matrix}$$

We may introduce a simplified assumption that the effect of *D*-axis damper winding can be coupled with the main field winding for the time-being, only to be separated by re-introducing separate row and column in the impedance matrix in the identical form as due to the field, when the effect of damper winding needs to be accounted for. Under this simplified form, the impedance matrix assumes to form:

$$[Z_{11}] = \begin{matrix} & F(D) & Q \\ F(D) & (R_F + L_F p) & 0 \\ Q & 0 & (R_Q + L_Q p) \end{matrix}$$

It needs to be emphasised that we have not neglected *D*-axis damper winding and have only taken *Q*-axis damper into consideration. *D*-axis damper has a coupling with the field winding, *F*. This has a similar form, although different in value from *F* winding with other windings on the rotor.

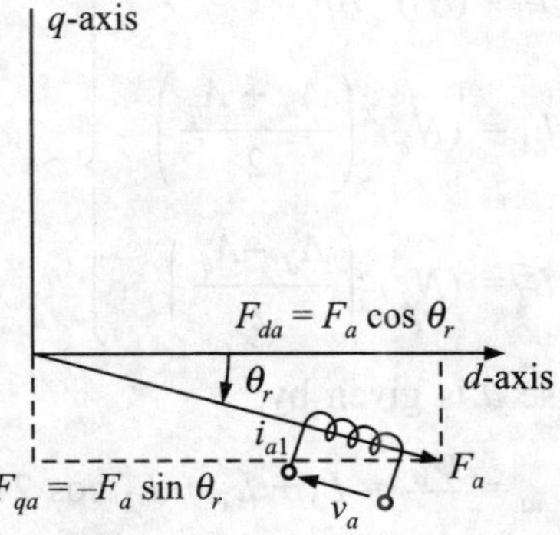

FIGURE 8.3 Phase *a*, *d*-*q* axis mmf components.

(b) Rotor self-inductance: Referring to Figure 8.3, the instantaneous phase mmf F_a is resolved into two components along *d*-axis and *q*-axis. The values are:

$$F_{da} = F_a \cos\theta_r = N_2 i_a \cos\theta_r$$
$$F_{qa} = -F_a \sin\theta_r = -N_2 i_a \sin\theta_r \qquad ...(8.22)$$

F_{da} and F_{qa} may be assumed to act respectively on the main gap permeances, Λ_d and Λ_q.

The corresponding values of flux-linkages along these two axes—*D*-axis and *Q*-axis—are:

$$\left.\begin{aligned}\psi_{Da} &= N_2\Lambda_d F_{Da} = (N_2)^2\Lambda_d i_a \cos\theta_r \\ \psi_{Qa} &= -(N_2)^2\Lambda_q i_a \sin\theta_r\end{aligned}\right\} \qquad ...(8.23)$$

The gap flux-linkage for the *a*-phase is given by:

$$\psi_{ga} = \psi_{da}\cos\theta_r - \psi_{qa}\sin\theta_r \qquad ...(8.24)$$

which works out to be:

$$\psi_{ga} = (N_2)^2 i_a (\Lambda_d \cos^2\theta_r + \Lambda_q \sin^2\theta_r)$$

$$= (N_2)^2 i_a \left\{\frac{(\Lambda_d + \Lambda_q)}{2} + \frac{(\Lambda_d - \Lambda_q)}{2}\cos 2\theta_r\right\} \qquad ...(8.25)$$

The leakage flux of the rotor is virtually not affected by the rotor position. If we assume that the leakage flux remains constant and independent of θ_r:

$$\psi_{la} = N_2\Lambda_l F_a = (N_2)^2\Lambda_l i_a \qquad ...(8.26)$$

The total flux-linkage for the phase-*a* is, therefore, given by:

$$\psi_a = \psi_{la} + \psi_{ga} \qquad ...(8.27)$$

Hence,

$$\psi_a = (N_2)^2 i_a \left\{\Lambda_l + \frac{(\Lambda_d + \Lambda_q)}{2} + \frac{(\Lambda_d - \Lambda_q)}{2}\cos 2\theta_r\right\} \qquad ...(8.28)$$

In the above expressions, N_2 stands for the total effective turns per rotor phase. If we define:

$$\left.\begin{aligned}L_l &= (N_2)^2\,\Lambda_l \\ L_1 &= (N_2)^2\left(\frac{\Lambda_d + \Lambda_q}{2}\right) \\ L_2 &= (N_2)^2\left(\frac{\Lambda_d - \Lambda_q}{2}\right)\end{aligned}\right\} \qquad ...(8.29)$$

then the self-inductance of phase *a* is given by:

$$L_{aa} = \frac{\psi_a}{i_a} = L_l + L_1 + L_2\cos 2\theta_r \qquad ...(8.30)$$

Similarly, the self-inductance of phases *b* and *c* are as follows:

$$L_{bb} = \frac{\psi_b}{i_b} = L_l + L_1 + L_2\cos 2\left(\theta_r + \frac{2\pi}{3}\right) \qquad ...(8.31)$$

and $$L_{cc} = \frac{\psi_c}{i_c} = L_l + L_1 + L_2 \cos 2\left(\theta_r + \frac{4\pi}{3}\right) \quad ...(8.32)$$

(c) Rotor mutual-inductance: The mutual inductance between any two pairs of rotor coils may be obtained from the mutual flux-linkage obtained with one coil excited and the other left open circuited.

Let us consider phase coils a and b, with coil a excited and coil b held open-circuited. If we resolve coil a flux-linkage into two components—ψ_{da} and ψ_{qa}—from expression (8.24) along the axis of b-phase, to result in total flux-linkage with b-phase, then:

$$\psi_{ba} = \psi_{qa} \cos\left(\theta_r + 90° + \frac{2\pi}{3}\right) + \psi_{da} \cos\left(\theta_r + \frac{2\pi}{3}\right)$$

$$= -\psi_{qa} \sin\left(\theta_r + \frac{2\pi}{3}\right) + \psi_{da} \cos\left(\theta_r + \frac{2\pi}{3}\right) \quad ...(8.33)$$

Now, substituting Eq. (8.23) in Eq. (8.33), we get

$$\psi_{ba} = (N_1)^2 i_a \left\{\Lambda_q \sin\left(\theta_r + \frac{2\pi}{3}\right) \sin\theta_r + \Lambda_d \cos\left(\theta_r + \frac{2\pi}{3}\right) \cos\theta_r\right\}$$

or $$\psi_{ba} = (N_1)^2 i_a \left\{-\left(\frac{\Lambda_d + \Lambda_q}{2}\right) + \left(\frac{\Lambda_d - \Lambda_q}{2}\right) \cos 2\left(\theta_r + \frac{\pi}{3}\right)\right\} \quad ...(8.34)$$

We can afford to neglect the small amount of leakage flux, which does not cross the air-gap. Hence, by substituting Eq. (8.29) in Eq. (8.34), the mutual inductance coefficient between phases a and b is:

$$L_{ba} = L_{ab} = \frac{\psi_{ba}}{i_a} = -\frac{L_1}{2} + L_2 \cos 2\left(\theta_r + \frac{\pi}{3}\right) \quad ...(8.35)$$

Expressions for L_{cb} and L_{ac} can be written down directly by inspection by noting that in Eq. (8.35), the angle $\left(\theta_r + \frac{\pi}{3}\right)$ is the angle measured from the d-axis to the line bisecting the angle between axis of phases a and b. Hence, for phases c and b, the above angle is $(\theta_r + \pi)$ and for phases c and a, it should be $\left(\theta_r + \frac{5\pi}{3}\right)$. Hence:

$$L_{bc} = L_{cb} = \frac{\psi_{cb}}{i_b} = -\frac{L_1}{2} + L_2 \cos 2(\theta_r + \pi) \quad ...(8.36)$$

and

$$L_{ca} = L_{ac} = \frac{\psi_{ca}}{i_c} = -\frac{L_1}{2} + L_2 \cos 2\left(\theta_r + \frac{5\pi}{3}\right) \quad ...(8.37)$$

(d) Stator to rotor mutual inductances: We can write down expressions for the mutual-inductances between rotor phases and the field winding on d-axis stator by inspection as follows:

$$\left.\begin{aligned} L_{aF} = L_{Fa} &= M_{rF}\cos\theta_r \\ L_{bF} = L_{Fb} &= M_{rF}\cos\left(\theta_r + \frac{2\pi}{3}\right) \\ L_{cF} = L_{Fc} &= M_{rF}\cos\left(\theta_r + \frac{4\pi}{3}\right) \end{aligned}\right\} \qquad ...(8.38)$$

where, M_{rF} is the maximum value of mutual-inductance coefficient between a rotor phase and the field winding F.

We may also write (when we take its effect separately from the combined field and d-axis damper), that:

$$\left.\begin{aligned} L_{aD} = L_{Da} &= M_{rD}\cos\theta_r \\ L_{bD} = L_{Db} &= M_{rD}\cos\left(\theta_r + \frac{2\pi}{3}\right) \\ L_{cD} = L_{Dc} &= M_{rD}\cos\left(\theta_r + \frac{4\pi}{3}\right) \end{aligned}\right\} \qquad ...(8.39)$$

where, M_{rD} is the maximum value of mutual-inductance between a rotor phase and the damper winding, D.

Similarly,

$$\left.\begin{aligned} L_{aQ} = L_{Qa} &= L_{rQ}\cos\left(\theta_r + \frac{\pi}{2}\right) = -M_{rQ}\sin\theta_r \\ L_{bQ} = L_{Qb} &= L_{rQ}\cos\left(\theta_r + \frac{\pi}{2} + \frac{2\pi}{3}\right) = -M_{rQ}\sin\left(\theta_r + \frac{2\pi}{3}\right) \\ \text{and} \quad L_{cQ} = L_{Qc} &= L_{rQ}\cos\left(\theta_r + \frac{\pi}{2} + \frac{4\pi}{3}\right) = -M_{rQ}\sin\left(\theta_r + \frac{4\pi}{3}\right) \end{aligned}\right\} \qquad ...(8.40)$$

where, M_{rQ} is the maximum value of the mutual-inductance coefficient between a rotor phase and Q-axis damper winding.

We have to spend a little more time in discussing the representation of the general form of inductance matrix between the members of the machine. Let us start with armature to field.

1. Armature to field: Between the armature and the field, the general form of impedance matrix is written as follows:

$$[Z_{21}] = \begin{array}{c} \\ a \\ b \\ c \end{array}\begin{array}{c} \begin{array}{cc} F & Q \end{array} \\ \begin{bmatrix} p\bar{M}_{aF} & p\bar{M}_{aQ} \\ p\bar{M}_{bF} & p\bar{M}_{bQ} \\ p\bar{M}_{cF} & p\bar{M}_{cQ} \end{bmatrix} \end{array} \qquad ...(8.41)$$

and we have found that:

$$M_{aF} = \bar{M}_F \cos \theta_r$$

$$M_{bF} = \bar{M}_F \cos\left(\theta_r + \frac{4\pi}{3}\right) = \bar{M}_F \cos \theta_3$$

$$M_{cF} = \bar{M}_F \cos\left(\theta_r + \frac{2\pi}{3}\right) = \bar{M}_F \cos \theta_2$$

We have retained the original values of θ_2 and θ_3 to be the same as those used for the induction machine in this case also, to establish possible parity between the two end results of transformation by $[C_1]$.

Similarly, we assume that the mutual-inductances of the armature phases with Q-axis dampers differ only in magnitude and 90° phases displacement.

Hence,

$$\left.\begin{aligned} M_{aQ} &= \bar{M}_Q \sin \theta \\ M_{bQ} &= \bar{M}_Q \sin \theta_3 \\ M_{cQ} &= M_Q \sin \theta_2 \end{aligned}\right\} \qquad \text{...(8.42)}$$

so that the complete matrix component of expression (8.41) can be written as:

$$[Z_{21}] = \begin{matrix} \\ a \\ b \\ c \end{matrix} \begin{matrix} F & Q \\ \left[\begin{matrix} \bar{M}_F p \cos \theta \\ \bar{M}_F p \cos \theta_3 \\ \bar{M}_F p \cos \theta_2 \end{matrix}\right. & \left.\begin{matrix} \bar{M}_Q p \sin \theta \\ \bar{M}_Q p \sin \theta_3 \\ \bar{M}_Q p \sin \theta_2 \end{matrix}\right] \end{matrix} \qquad \text{...(8.43)}$$

and, the transformed value of $[Z_{21}]$ becomes:

$$[Z'_{21}] = [C_{1t}]\,[Z_{21}] = \sqrt{\frac{3}{2}} \begin{bmatrix} \bar{M}_F p \cos \theta & \bar{M}_Q p \sin \theta \\ \bar{M}_F p \sin \theta & -\bar{M}_Q p \cos \theta \end{bmatrix} \qquad \text{...(8.44)}$$

where

$$M_F = \sqrt{\frac{3}{2}}\,\bar{M}_F \quad \text{and} \quad M_Q = \sqrt{\frac{3}{2}}\,\bar{M}_Q$$

2. Field to armature: Since mutual-inductance effects are reciprocal, it is natural to assume that Z_{12} is simply reciprocal of Z_{21}. The appropriate value of Z'_{12} therefore, is:

$$Z'_{12} = \begin{matrix} \\ F \\ Q \end{matrix} \begin{matrix} \alpha & \beta \\ \left[\begin{matrix} M_F p \cos \theta \\ M_Q p \sin \theta \end{matrix}\right. & \left.\begin{matrix} M_F p \sin \theta \\ M_Q p \cos \theta \end{matrix}\right] \end{matrix} \qquad \text{...(8.45)}$$

3. Armature to armature: For a salient-pole synchronous machine, the main source of difficulty is posed by the form of the coefficients of self- and mutual-inductances of the armature winding, because the air-gap opposite to the armature is not cylindrical as in case of induction machines. In general, the inductance matrix component has the form:

$$Z_{22} = \begin{array}{c} \\ a \\ b \\ c \end{array}\begin{array}{c} \begin{array}{ccc} a & b & c \end{array} \\ \begin{bmatrix} (R_a + pL_a) & p\,M_{ba} & p\,M_{ca} \\ p\,M_{ab} & (R_b + pL_b) & p\,M_{cb} \\ p\,M_{ac} & p\,M_{bc} & (R_c + pL_c) \end{bmatrix} \end{array} \qquad \text{...(8.46)}$$

Effecting the transformation of $[Z_{22}]$ by $[C_1]$ and $[C_{1t}]$ to yield $[Z_{22}]$ is extremely involved. The first stage given as $[Z_{22}]\,[C_1]$ results in:

$$[Z_{22}]\,[C_1]$$

$$= \begin{bmatrix} \frac{1}{\sqrt{2}}\{R_a + p(L_a + M_{ab} + M_{ac})\} & \left\{R_a + p\left(L_a - \frac{1}{2}M_{ab} - \frac{1}{2}M_{ac}\right)\right\} & \frac{\sqrt{3}}{2}\{p(M_{ab} - M_{ac})\} \\ \frac{1}{\sqrt{2}}\{R_b + p(L_b + M_{ab} + M_{bc}\} & \left\{-\frac{1}{2}R_b + p\left(-\frac{1}{2}L_b - M_{ab} - \frac{1}{2}M_{bc}\right)\right\} & \frac{\sqrt{3}}{2}\{R_b + p(L_b - M_{ac})\} \\ \frac{1}{\sqrt{2}}\{R_c + p(L_c + M_{ac} + M_{bc}\} & \left\{-\frac{1}{2}R_c + p\left(-\frac{1}{2}L_c - M_{ac} - \frac{1}{2}M_{bc}\right)\right\} & -\frac{\sqrt{3}}{2}\{R_c + p(L_c - M_{ac})\} \end{bmatrix} \qquad \text{...(8.47)}$$

And if the completely transformed matrix is written in the form:

$$Z'_{22} = [C_{1t}]\,[Z_{22}]\,[C_1] = \begin{bmatrix} Z_0 & Z_{0\alpha} & Z_{0\beta} \\ Z_{0\alpha} & Z_\alpha & Z_{\alpha\beta} \\ Z_{0\beta} & Z_{\alpha\beta} & Z_\beta \end{bmatrix}$$

where,

$$Z_0 = R_2 + \frac{1}{3}p(L_a + L_b + L_c + 2M_{ab} + 2M_{ac} + 2M_{bc})$$

$$Z_{0\alpha} = \frac{\sqrt{2}}{3}p\left(L_a - \frac{1}{2}L_b - \frac{1}{2}L_c + \frac{1}{2}M_{ab} + \frac{1}{2}M_{ac} - M_{bc}\right)$$

$$Z_{0\beta} = \frac{1}{\sqrt{6}}p(L_b - L_c + M_{ab} - M_{ac})$$

$$Z_\alpha = R_2 + \frac{2}{3}p\left(L_a + \frac{1}{4}L_b + \frac{1}{4}L_c - M_{ab} - M_{ac} + \frac{1}{2}M_{bc}\right)$$

$$Z_\beta = R_2 + \frac{1}{2}p(L_b + L_c - 2M_{bc})$$

and
$$Z_{\alpha\beta} = \frac{1}{\sqrt{3}}p\left(-\frac{1}{2}L_b + \frac{1}{2}L_c + M_{ab} - M_{ac}\right) \quad \text{...(8.48)}$$

The finally transformed matrix, after substituting the values of phase self- and mutual-inductances derived in parts (a) to (d) in Section 8.2, leads to the following expressions for the completed transformation:

$$Z'_{22} = \begin{array}{c} \\ 0 \\ \alpha \\ \\ \beta \\ \end{array} \overset{\begin{array}{ccc} 0 & \alpha & \beta \end{array}}{\begin{bmatrix} (R_2 + L_{0a}p) & p(L_{0b}\cos 2\theta_r + L_{0c}\cos\theta_r) & p(-L_{0b}\sin 2\theta_r + L_{0c}\sin 4\theta_r) \\ p(L_{0b}\cos 2\theta_r + L_{0c}\cos 40_r) & R_2 + p(L_0 + L_b\cos 2\theta_r + L_c\cos 4\theta_r) & p(L_b\sin 2\theta_r - L_{0c}\sin 4\theta_r) \\ p(L_{ab}\sin 2\theta_r + L\theta_{0c}\sin 4\theta_r) & p(L_b\sin 2\theta_r - L_c\sin 4\theta_r) & R_2 + p(L_a - L_b\cos 2\theta_r + L_c\cos 4\theta_r) \end{bmatrix}} \quad \text{...(8.49)}$$

where, L_a, L_b, L_c and M_{bc}, M_{ac} and M_{ab} are of the following forms:

$$L_a = \bar{L}_a + \bar{L}_b\cos 2\theta_r \qquad L_a = \bar{L}_a - \bar{M}_a \qquad L_{0a} = \bar{L}_a + 2\bar{M}_a$$

$$L_b = \bar{L}_a + \bar{L}_b\cos 2\theta_{3r} \qquad L_b = \frac{1}{2}\bar{L}_b + \bar{M}_b \qquad L_{0b} = \frac{1}{\sqrt{2}}(\bar{L}_c - \bar{M}_c)$$

$$L_c = \bar{L}_a + \bar{L}_b\cos 2\theta_{2r} \qquad L_c = \frac{1}{2}\bar{L}_c + \bar{M}_c \qquad L_{0c} = \frac{1}{\sqrt{2}}(\bar{L}_c - \bar{M}_c)$$

$$M_{bc} = \bar{M}_a + \bar{M}_b\cos 2\theta_r$$
$$M_{ca} = \bar{M}_a + \bar{M}_b\cos 2\theta_{3r}$$
$$M_{ab} = \bar{M}_a + \bar{M}_b\cos 2\theta_{2r}$$

In pages 168–169 under Section Armature/Armature, C.V. Jones, in his book on *Unified Theory of Electrical Machines*, has shown that the effect of applying transformation by $[C_1]$ on 3-phase *a-b-c* system in a salient-pole configuration to 2-phase 0-α-β system does not lead to any tangible simplification, but at first sight, causes prohibitively intricate result. However, under normal operating conditions, when looked at under idealized conditions experimentally, the machine displays negligible effect on the mutual inductance M_{0b} and M_{0c}. No zero-sequence voltage is generated in the armature as a result of its reaction with the field system[1].

When the armature carries sinusoidal currents, the in-phase third harmonic voltages appears as generated not by the field but by reactions between the armature currents themselves.

Test results show that the magnitude and effects of third harmonic voltage in phases of the armature are trivial. It is, therefore, legitimate to ignore and omit the zero-sequence row and

[1] Jones, C.V., *The United Theory of Electrical Machines,* pp. 168–169, Butterworths, London, 1967.

column from the transformed impedance matrix. Test results further show that the values of the fourth harmonic terms L_c and M_c in the phase impedances are significant but their combinations,

$$L_c = \left(\frac{1}{2}\right)L_c + M_c$$

In the transformed matrix is negligible. The terms containing L_c may therefore be neglected[2].

The final value of Z'_{22}, thus, reduces to a two-by-two matrix eliminating the zero-sequence row and column to a new form:

$$Z'_{22} = \begin{matrix} \\ \alpha \\ \beta \end{matrix} \begin{matrix} \alpha & \beta \\ \left[\begin{matrix} R_2 + p(L_a + L_b \cos 2\theta_r) \\ pL_b \sin 2\theta_r \end{matrix}\right. & \left.\begin{matrix} pL_b \sin 2\theta_r \\ R_2 + p(L_a - L_b \cos 2\theta_r) \end{matrix}\right] \end{matrix} \qquad \text{...(8.50)}$$

The four components derived above and depicted in Eqs. (7.41), (8.44), (8.45) and (8.50) may be combined to yield a completely transformed impedance matrix.

$$Z'' = \begin{matrix} \\ F \\ Q \\ \alpha \\ \beta \end{matrix} \begin{matrix} F & Q & \alpha & \beta \\ \left[\begin{matrix} (R_F + pL_F) \\ 0 \\ M_F p \cos\theta_r \\ M_F p \sin\theta_r \end{matrix}\right. & \begin{matrix} 0 \\ (R_Q + pL_Q) \\ M_Q p \sin\theta_r \\ -M_Q p \cos\theta_r \end{matrix} & \begin{matrix} M_F p \cos\theta_r \\ M_Q p \sin\theta_r \\ R_2 + p(L_a + L_b \cos 2\theta_r) \\ p(L_b \sin 2\theta_r) \end{matrix} & \left.\begin{matrix} M_F p \sin\theta_r \\ -M_Q p \cos\theta_r \\ p\, L_b \sin 2\theta_r \\ R_2 + p(L_a - L_b \cos 2\theta_r) \end{matrix}\right] \end{matrix}$$

...(8.51)

The difference in position between the β rotor phases in the induction machine and the synchronous machine was discussed at length in Section 7.4.1. In order to align the equations of the two types of machines, it is desirable to eliminate this difference. The easiest way to achieve this is to express the transformed impedance matrix of Eq. (8.50) not in terms of β but in terms of ($-\beta$) as already indicated in Section 7.4.1. Physically, this implies reversal of connection for the β-phase. Mathematically, this indicates that the signs of all elements in the fourth row and column of $[Z']$ need be reversed excepting for the element (4, 4). The final impedance matrix is therefore of the form:

$$[Z'] = \begin{matrix} \\ F(D) \\ Q \\ \alpha \\ \beta \end{matrix} \begin{matrix} F(D) & Q & \alpha & \beta \\ \left[\begin{matrix} (R_F + L_F p) \\ 0 \\ M_F p \cos\theta_r \\ -M_F p \sin\theta_r \end{matrix}\right. & \begin{matrix} 0 \\ (R_Q + L_Q p) \\ M_Q p \sin\theta_r \\ M_Q p \cos\theta_r \end{matrix} & \begin{matrix} M_F p \cos\theta_r \\ M_Q p \sin\theta_r \\ R_2 + p(L_a + L_b \cos 2\theta_r) \\ -pL_b \sin 2\theta_r \end{matrix} & \left.\begin{matrix} -M_F p \sin\theta_r \\ M_Q p \cos\theta_r \\ -pL_b \sin 2\theta_r \\ R_2 + p(L_a - L_b \cos 2\theta_r) \end{matrix}\right] \end{matrix} \qquad \text{...(8.52)}$$

Let us compare the two alternative forms of relationships of $[Z']$ for the induction machines in Eq. (8.21) and the synchronous machines in Eq. (8.52). The result is very revealing in that it shows that if we substitute in Eq. (8.52) the following modifications:

[2] Ibid.

$$\left.\begin{aligned} R_F = R_Q &= R_1 \\ L_F = L_Q &= L_1 \\ M_F = M_Q &= M \\ L_a &= L_2, \text{ and} \\ L_b &= 0 \end{aligned}\right\}$$

both the above matrix relations assume identical forms. $[C_1]$ phase transformation, therefore, results in voltage equations of synchronous and induction machines belonging to slip-ring machine family of similar form. The difference between the two machines is thus not a fundamental one but are only in relative values of their inductance parameters and the nature of applied voltage.

Mathematically, the matrix voltage equations form four simultaneous partial differential equations. Since θ_r is present, these equations are non-linear. Heaviside's differential operator, $p\left(=\dfrac{d}{dt}\right)$, appearing in the expressions of Z causes the product of functions of θ_r and current, both of which are functions of time, contributes to non-linearity in determining the inverse and poses special difficulties for determination of i. We eliminate the function of θ_r—the rotor angle—by applying a second transformation, $[C_2]$ to the $[Z]$ matrix, thereby, linearizing the non-linear form of the impedance matrix given in Eqs. (8.21) and (8.52).

Since $[C_2]$ itself includes θ_r—a function of time—it signifies a 'live' transformation. Especial care is required for its application. It has to be applied on the 'rotor' of both the machines and we can include its application by considering the voltage equations of the equivalent 2-phase machines, already derived in Eqs. (8.21) and (8.52). For this, we use the equations in the compound matrix form, which yield:

$$\begin{bmatrix} V_1 \\ V_2 \end{bmatrix} = \begin{bmatrix} Z_{11} & Z_{12} \\ Z_{21} & Z_{22} \end{bmatrix} \begin{bmatrix} I_1 \\ I_2 \end{bmatrix}$$

and, the transformation is of the following form:

$$[C] = \begin{matrix} 1 \\ 2 \end{matrix}\begin{bmatrix} 1 & 0 \\ 0 & [C_2] \end{bmatrix}$$

with the result that the transformed impedance matrix:

$$[Z'] = [C_{2t}]\,[Z]\,[C_2] = \begin{bmatrix} [Z_{11}] & [Z_{12}][C_2] \\ [C_{2t}][Z_{21}] & [C_{2t}][Z_{22}][C_2] \end{bmatrix} \qquad \text{...(8.53)}$$

as before.

We now, execute the transformation of Eq. (6.51) first, taking each element of the transformed matrix in turn.

It is clear that $[Z'_{11}] = [Z_{11}]$ and the stator-to-stator element remains unaffected by this transformation.

The value of $[C_2]$, as was derived in Eq. (6.26) in Section 6.4 has the form:

$$[C_2] = \begin{matrix} & q & d \\ \alpha & \sin\theta & \cos\theta \\ \beta & \cos\theta & -\sin\theta \end{matrix}$$

Stator-to-stator element $[Z_{11}]$:

$$[Z'_{11}] = \begin{matrix} & F & Q \\ F & (R_F + L_F p) & 0 \\ Q & 0 & (R_Q + L_Q p) \end{matrix} \qquad \text{...(8.54)}$$

Stator-to-rotor element $[Z_{12}]$: In connection with the impedance component $[Z_{12}]$ also there is no special precaution needed to be taken for applying the live transformation $[C_2]$ since the Heaviside's differential operator p in $[Z_{12}]$ acts upon the products of all terms to its right only. Therefore:

$$[Z'_{12}] = [Z_{12}]\,[C_2] \begin{matrix} & \alpha & \beta \\ F & M_F p\cos\theta & -M_F p\sin\theta \\ Q & M_Q p\sin\theta & M_Q p\cos\theta \end{matrix} \begin{matrix} & q & d \\ \alpha & \sin\theta & \cos\theta \\ \beta & \cos\theta & -\sin\theta \end{matrix}$$

$$= \begin{matrix} q & d \\ 0 & M_F p \\ M_Q p & 0 \end{matrix} \qquad \text{...(8.55)}$$

Rotor-to-stator element $[Z_{21}]$: We have to treat this transformation with special care.

$$[Z'_{21}] = [C_{2t}]\,[Z_{21}] = \begin{matrix} & \alpha & \beta \\ q & \sin\theta & \cos\theta \\ d & \cos\theta & -\sin\theta \end{matrix} \begin{matrix} & F & Q \\ \alpha & M_F p\cos\theta & M_Q p\sin\theta \\ \beta & -M_F p\sin\theta & M_Q p\cos\theta \end{matrix}$$

In this expression the differential operator p occurs in the middle of the transformation and does not operate upon the functions of θ, which happens to its left. To obtain element Z_{11} of this transformed matrix.

Z_{11} has an expanded form:

$$Z_{11} = M_F\,(\sin\theta p\cos\theta - \cos\theta p\sin\theta)$$

This is not zero as the differential operator p in Z actually operates on the original value of $[Z]$ jointly with terms i. $[Z]$ matrix is only a part of the original expression Zi $(= v)$; the segregation is artificially undertaken for neatness of the final expression. The correct expression for this term should have been:

$$M_F\,\{\sin\theta p(\cos\theta i) - \cos\theta p(\sin\theta i)\}$$

which requires the use of partial derivatives of the products following p individually. The result of the final expression in expanded form should have been:

$$Z'_{11} = M_F \{\sin\theta \cos\theta(pi) - i \sin\theta \sin\theta(p\theta) - \cos\theta \sin\theta(pi) - i \cos\theta \cos\theta(p\theta)\}$$

$$\therefore \quad Z'_{11} = -M_F \cdot i(p\theta)$$

Since, we have conceptually borrowed the matrix element i from the current matrix, we could return it to its original place.

Hence, we could conclude that the appearance of the expression:

$$\sin\theta p \cos\theta - \cos\theta p \sin\theta \quad \rightarrow \quad -(p\theta) \qquad \text{...(8.56)}$$

and hence, the term Z'_{11} in the matrix component $[Z'_{21}]$ assumes a value of $(-M_F\, p\theta)$ after re-segregation of i.

The second term Z'_{12}: In a similar manner, the correct value of Z'_{12} should be:

$$Z'_{12} = M_Q \{\sin\theta p (\sin\theta i) + \cos\theta p (\cos\theta i)\}$$

whose expanded form yields:

$$Z'_{12} = M_Q[(i)\{\sin\theta \cos\theta(p\theta) - \cos\theta \sin\theta(p\theta)\} + \cos\theta \cos\theta(pi) + \sin\theta \sin\theta(pi)]$$

$$= M_Q p(i)$$

Hence,

$$\sin\theta p \sin\theta + \cos\theta p \cos\theta \quad \rightarrow \quad p \qquad \text{...(8.57)}$$

In a similar manner:

The third term Z'_{21}:

$$Z'_{21} = M_F\{\cos\theta p \cos\theta + \sin\theta p \sin\theta\} = M_F p$$

The fourth term Z'_{22}:

$$Z'_{22} = M_Q\{\cos\theta p \sin\theta - \sin\theta p \cos\theta\} = M_Q(p\theta)$$

The elements of the transformed matrix component $[Z'_{21}]$: We can now write down the final form of the transformed component matrix $[Z'_{21}]$, by collation of the different elements estimated. This has to form:

$$[Z'_{21}] = \begin{matrix} & F & Q \\ q & -M_F(p\theta) & M_Q p \\ d & M_F p & M_Q(p\theta) \end{matrix}$$

And, since $p\theta$ in this case is the angular frequency ω_s, we have:

$$[Z'_{21}] = \begin{matrix} & F & \\ q & -M_F\omega_s & M_Q p \\ d & M_F p & M_Q\omega_s \end{matrix} \qquad \text{...(8.58)}$$

It is interesting to note that although $[Z_{21}]$ is a transpose of $[Z_{21}]$ but $[Z'_{21t}] \neq [Z'_{12}]$. The lack of symmetry in the finally transposed impedance matrix to which it points may be considered later for physical interpretation.

The elements of the transposed matrix component, $[Z'_{22}]$: Before working out the expanded form of the compound matrix component $[Z'_{22}]$ in expression (8.53), it is desirable to separate the constant resistance component parts R_2 from the time dependent parts pL_2 which contains terms $\{R_2 + p(L_a + L_b \cos 2\theta)\}$ etc. in the matrix format separately as summation of matrices.

The $[Z_{22}]$ in the rotor-to-rotor impedance matrix prior to transformation to $0dq$ frame of reference could be denoted by:

$$[Z_{22}] = \begin{matrix} & \alpha & \beta \\ \alpha & \{R_2 + p(L_a + L_b \cos 2\theta)\} & -(L_b p \sin \theta) \\ \beta & -(pL_b \sin \theta) & \{R_2 + p(L_a - L_b \cos 2\theta)\} \end{matrix} \begin{matrix} & q & d \\ \alpha & \sin\theta & \cos\theta \\ \beta & \cos\theta & -\sin\theta \end{matrix}$$

$$= \begin{bmatrix} R_2 & 0 \\ 0 & R_2 \end{bmatrix} + \begin{bmatrix} p(L_a + L_b \cos 2\theta) & -(pL_b \sin 2\theta) \\ -(L_b p \sin 2\theta) & p(L_a - L_b \cos 2\theta) \end{bmatrix}$$

$$= [r] + p[L]$$

As before, transforming the impedance matrix component $[Z_{22}]$, we have relationship shown earlier where,

$$[r'] = [C_{2t}]\,[r]\,[C_2] + [C_{2t}] \begin{bmatrix} p(L_a + L_b \cos 2\theta) & -(pL_b \sin 2\theta) \\ -(pL_b \sin 2\theta) & p(L_a - L_b \cos 2\theta) \end{bmatrix} [C_2]$$

We take up the resistance component first.

$$[r'] = [C_{2t}]\,[r]\,[C_2]$$

where,

$$C_{2t} = \begin{matrix} & \alpha & \beta \\ q & \sin\theta & \cos\theta \\ d & \cos\theta & -\sin\theta \end{matrix} \quad \text{and} \quad [C_2] = \begin{matrix} & q & d \\ \alpha & \sin\theta & \cos\theta \\ \beta & \cos\theta & -\sin\theta \end{matrix}$$

We start by transforming from the back side:

$$[r]\,[C_2] = R_2 \begin{matrix} & \alpha & \beta \\ & 1 & 0 \\ & 0 & 1 \end{matrix} \begin{matrix} & q & d \\ \alpha & \sin\theta & \cos\theta \\ \beta & \cos\theta & -\sin\theta \end{matrix}$$

$$[r_{\text{int}}] = \begin{matrix} & q & d \\ \alpha & R_2 \sin\theta & R_2 \cos\theta \\ \beta & R_2 \cos\theta & -R_2 \sin\theta \end{matrix}$$

Hence,

$$[C_{2t}]\,[r_{\text{int}}] = \begin{matrix} & \alpha & \beta \\ q & \sin\theta & \cos\theta \\ d & \cos\theta & -\sin\theta \end{matrix} \begin{matrix} & q & d \\ \alpha & R_2\sin\theta & R_2\cos\theta \\ \beta & R_2\cos\theta & -R_2\sin\theta \end{matrix}$$

$$= R_2\begin{bmatrix} 1 & 0 \\ 0 & 1 \end{bmatrix} = \begin{bmatrix} R_2 & 0 \\ 0 & R_2 \end{bmatrix}$$

Thus, we conclude that transformation of a constant term without any differential operator associated with it does not affect the transformed value after completing the transformation. Hence, we can safely segregate the inductance matrix component from the constant resistance term before applying the transformation and include the same with respective inductance terms after transformation, for joint consideration in the impedance matrix. This practice is followed throughout the book. We can take up transforming the impedance component $p[L]$ separately, as the next step.

As has already been done for transforming $[r]$ previously, we work out:

$$[L_{\text{int}}] = [L]\,[C_2]$$

$$[L_{\text{int}}] = \begin{bmatrix} p(L_a + L_b\cos 2\theta) & -p(L_b\sin 2\theta) \\ -p(L_b\sin 2\theta) & p(L_a - L_b\cos 2\theta) \end{bmatrix} \begin{bmatrix} \sin\theta & \cos\theta \\ \cos\theta & -\sin\theta \end{bmatrix}$$

The product works out to be:

$$\begin{matrix} & q & d \\ \alpha & (L_a + L_b)p\sin\theta & (L_a + L_b)p\cos\theta \\ \beta & (L_a - L_b)p\cos\theta & -(L_a - L_b)p\sin\theta \end{matrix}$$

In the above matrix, we make the following substitutions:

$$L_d = (L_a + L_b) \quad \text{and} \quad L_q = (L_a - L_b)$$

and work out the final stage of transformation after including the resistance terms appropriately. The final result is:

$$[Z'_{22}] = [C_{2t}]\,[Z_{22}]\,[C_2]$$

$$= \begin{matrix} & \alpha & \beta \\ q & \sin\theta & \cos\theta \\ d & \cos\theta & -\sin\theta \end{matrix} \begin{bmatrix} (R_2 + L_q p)\sin\theta & (R_2 + L_d p)\sin\theta \\ (R_2 + L_q p)\cos\theta & (R_2 + L_d p)\cos\theta \end{bmatrix}$$

which leads to the following on the basis of relations (8.55) and (8.56):

$$[Z'_{22}] = \begin{matrix} & q & d \\ q & (R_2 + L_q p) & -L_d\omega \\ d & L_q\omega & (R_2 + L_d p) \end{matrix} \qquad \text{...(8.59)}$$

We are, now, ready to write the complete transformed matrix by putting together expressions obtained in Eqs. (8.54), (8.55), (8.58) and (8.59). It would assume the following form:

$$[Z'] = \begin{array}{c} \\ F(D) \\ Q \\ q \\ d \end{array} \begin{array}{c} \begin{array}{cccc} F(D) & Q & q & d \end{array} \\ \begin{bmatrix} (R_F + L_F p) & 0 & 0 & M_F p \\ 0 & (R_Q + L_Q p) & M_Q p & 0 \\ -M_F \omega & M_Q p & (R_2 + L_q p) & -L_d \omega \\ M_F p & M_Q \omega & L_q \omega & (R_2 + L_d p) \end{bmatrix} \end{array} \quad \text{...(8.60)}$$

In the above form, the field F is located on the direct axis; the suffix D has been kept aside for denoting the D-axis damper winding. We may note the order of the rows and columns is $DQqd$, in keeping with orthogonality of the matrix.

8.3 INDUCTION MACHINES

We now, take up the exercise of developing similar equation for the induction machines.

We have noted that in Eq. (8.21) when the 3-phase *abc* winding of a wound rotor induction machine was represented by an equivalent 2-phase $0\alpha\beta$ machine, the result of the impedance matrix is represented by:

$$\begin{array}{c} \\ \alpha(A) \\ \beta(B) \\ \alpha(a) \\ \beta(b) \end{array} \begin{array}{c} \begin{array}{cccc} \alpha(A) & \beta(B) & \alpha(a) & \beta(b) \end{array} \\ \begin{bmatrix} (R_1 + L_1 p) & 0 & Mp\cos\theta & -Mp\sin\theta \\ 0 & (R_1 + L_1 p) & Mp\sin\theta & Mp\cos\theta \\ Mp\cos\theta & Mp\sin\theta & (R_2 + L_2 p) & 0 \\ -Mp\sin\theta & Mp\cos\theta & 0 & (R_2 + L_2 p) \end{bmatrix} \end{array}$$

As before, we apply transformation $[C_2]$ to the rotor of the induction machine, which amounts to operation of a transformation $[C]$ of the form:

$$[C] = \begin{bmatrix} 1 & 0 \\ 0 & C_2 \end{bmatrix}$$

We have undertaken this operation in connection with salient pole synchronous machines and have obtained the complete transformed impedance matrix in Eq. (8.60). Exactly the same form of impedance matrix is found in case of wound-rotor induction machines as well, if we make the following changes by substituting

$$R_F = R_Q = R_1; \quad L_F = L_Q = L_1; \quad L_d = L_q = L_2;$$
$$M_F = M_Q = M \quad \text{and} \quad \omega = \omega_r$$

We also replace $F(D)$ by $\alpha(A)$ and Q by $\beta(B)$, being the representation of axes in equivalent 2-phase induction machine, as they denote the same parameters as a result of uniformity of air-gap around the periphery. Thus, the final form of the equivalent 2-phase machine is given in $0dq$ reference frame as follows:

$$\begin{array}{c} \\ \alpha(A) \\ \beta(B) \\ q \\ d \end{array} \begin{array}{c} \begin{array}{cccc} \alpha(A) & \beta(B) & q & d \end{array} \\ \begin{bmatrix} (R_1+L_1p) & 0 & 0 & Mp \\ 0 & (R_1+L_1p) & Mp & 0 \\ -M\omega_r & Mp & (R_2+L_2p) & -L_2\omega_r \\ Mp & M\omega_r & L_2\omega_r & (R_2+L_2p) \end{bmatrix} \end{array} \quad \text{...(8.61)}$$

In solving the general induction motor problems, since we know that the nature of the applied voltages solution of induction machine problems under any condition of operation need inversion of the above impedance matrix in Eq. (8.61). A solution for balanced normal operation can, however, be obtained more simply by making a few obvious assumptions consistent with the voltage equations. We may assume that:

1. The currents and voltages in both the stator and rotor are balanced and sinusoidal.
2. While considering the physical interpretation of applying commutator transformation $[C_2]$, it is justified that the rotor currents and voltages are transformed into quantities having the same magnitudes but with their frequency changed from slip to supply frequency. In the transformed voltage equation of the machine, all currents and voltages, therefore, are at supply frequency.
3. For consideration of balanced operation, therefore, Heaviside's differential operator, $p\left(=\dfrac{d}{dt}\right)$, can be legitimately replaced by $j\omega_1 = 2\pi f_1\omega$, where f_1 is the supply frequency.

We can, therefore, make the following replacements and denote the terms associated with inductances by corresponding reactances:

$$L_1p = jX_1; \quad L_2p = jX_2 \quad \text{and} \quad Mp = jX_m$$

and also the rotor angular velocity ω_r by the difference between the synchronous and slip speed:

$$\omega_r = \omega_1 - \omega_2 = \omega_1(1-s)$$

so that:

$$M\omega_r = (1-s)X_m \quad \text{and} \quad L_2\omega_r = (1-s)X_2$$

The final form of the voltage equation in Eq. (8.61), therefore, assume the form under balanced steady-state conditions as follows:

$$\begin{bmatrix} V_1 \\ -jV_1 \\ -jV_2 \\ V_2 \end{bmatrix} = \begin{array}{c} \\ \alpha(A) \\ \beta(B) \\ q \\ d \end{array} \begin{array}{c} \begin{array}{cccc} \alpha(A) & \beta(B) & q & d \end{array} \\ \begin{bmatrix} (R_1+jX_1) & 0 & 0 & jX_m \\ 0 & (R_1+jX_1) & jX_m & 0 \\ -(1-s)X_m & jX_m & (R_2+jX_2) & -(1-s)X_2 \\ jX_m & (1-s)X_m & (1-s)X_2 & (R_2+jX_2) \end{bmatrix} \end{array} \begin{bmatrix} i_1 \\ -ji_1 \\ -ji_2 \\ i_2 \end{bmatrix} \quad \text{...(8.62)}$$

If we write the first equation in expanded form:

$$V_1 = (R_1 + jX_1)I_1 + jX_mI_2$$

The second equation in expanded form,

$$-jV_1 = -j(R_1 + jX_1)I_1 - jX_m I_2$$

is exactly common to the first equation with a constant multiplier $(-j)$.

The expanded form of the fourth equation,

$$V_2 = jX_m I_1 - j(1-s)X_m I_1 - j(1-s)X_2 I_2 + (R_2 + jX_2)I_2$$
$$= jsX_m I_1 + (R_2 + jsX_2)I_2$$

is, similarly an exact replication of the third equation.

The second and the third equations can be safely omitted as repeatition of Eqs. (1) and (4). This merely shows that the q-axis equations are identical with their d-axis counterparts with a constant multiplier $(-j)$, indicating their relative positions $\left(\frac{\pi}{2}\right)$ electrical radians behind the d-axis components. Hence, it suffices to redraw the voltage equation by a two-by-two matrix:

$$\begin{bmatrix} V_1 \\ V_2 \end{bmatrix} = \begin{bmatrix} (R_1 + jX_1) & jX_m \\ jsX_m & (R_2 + jsX_2) \end{bmatrix} \begin{bmatrix} I_1 \\ I_2 \end{bmatrix}$$

which can be rewritten as:

$$\begin{bmatrix} V_1 \\ \left(\frac{V_2}{s}\right) \end{bmatrix} = \begin{bmatrix} (R_1 + jX_1) & jX_m \\ jX_m & \left(\frac{R_2}{s} + jX_m\right) \end{bmatrix} \begin{bmatrix} I_1 \\ I_2 \end{bmatrix} \qquad ...(8.63)$$

In normal balanced operation as a motor, V_2 and $\left(\frac{V_2}{s}\right)$ is zero, as the rotor terminals are short-circuited.

The actual rotor resistance can be separated from the term $\left(\frac{R_2}{s}\right)$ in the impedance matrix, so that:

$$\frac{R_2}{s} = R_2 + \frac{(1-s)}{s} R_2$$

thus showing that the short-circuit at the rotor terminals indicates that the rotor is de-facto close-circuited with an effective resistance $\frac{(1-s)}{s} R_2$. The equivalent circuit of the induction motor, therefore, can be likened with that of a transformer, whose secondary terminals are close-circuited with a resistance $\frac{(1-s)}{s} R_2$.

8.3.1 Expression for Torque in Induction Motor

As torque is a scalar quantity, invariant to transformations of currents and voltages, the transformed torque has the same value before and after transformation by $[C_2]$. The original expression for torque derived in Section 3.5.1 and contained in Eq. (3.20), can, therefore, be written in the form:

$$T = \frac{\text{pole pairs}}{2}[I'_t]\left[\frac{dL'}{d\theta}\right][I'] \qquad ...(8.64)$$

where

$[I] = [C]\,[i']$

$[I_t] = [i'_t][C_t]$

and $[L'] = [C_t]\,[L]\,[C]$

This basic form of torque in Eq. (8.64) is not applied any longer after the transformation $[C_2]$. Since $[L]$ is not a function of θ_r any more. Since Eq. (8.61) shows the form of impedance matrix after $[C_2]$ has been applied, the new form of torque expression is given by:

$$\text{Torque, } T_e = [i_t]\,[G]\,[i]$$

This gives the instantaneous torque in Newton-metre. When the current is denoted phasorically, then the expression for torque as shown in Section 2.2.7, Eq. (2.27), assumes the form:

Torque = Real part of the complex expression

$$[I_t^*][G][I]$$

where, I_t^* is the conjugate of the transpose of the current matrix.

The torque matrix $[G]$ is simply the set of coefficients of ω_r in Z' in Eq. (8.61). Its value is as follows:

$$[G'] = ([G]+[G_t]) = \begin{array}{c} \\ \alpha(A) \\ \beta(B) \\ q \\ d \end{array}\begin{array}{c} \begin{array}{cccc} \alpha(A) & \beta(B) & q & d \end{array} \\ \left[\begin{array}{cccc} 0 & & & \\ & & & \\ -M & & & -L_2 \\ & M & L_2 & \end{array}\right] \end{array} \qquad ...(8.65)$$

The expression for the torque can be written as follows:

$$T = \frac{1}{2}[i_t][G'][i] = \frac{1}{2}[i_t]\{[G]+[G_t]\}[i]$$

$$= \frac{1}{2}[i_t][G][i] + \frac{1}{2}[i_t][G_t][i] \qquad ...(8.66)$$

The torque and both the components are one-to-one matrices or a single scalar quantity. Their values are, therefore, identical and remain unchanged by transformation and:

$$\frac{1}{2}[i_t][G][i] = \frac{1}{2}([i_t][G_t][i])_t$$

$$= \frac{1}{2}[i_t][G][i]$$

Hence, $$T_e = [i_t]\,[G]\,[i] \qquad ...(8.67)$$

and, the factor $\left(\frac{1}{2}\right)$ in the general torque equation disappears. This point was documented in Section 2.2.7, in Eq. (2.27).

$$\therefore \qquad \text{Torque } T_e = \text{Real part of } \left\{-M(I_A I_q^* - I_B I_d^*)\right\} \qquad ...(8.68)$$

In the above expression the stator current, $I_{\alpha(A)}$, that is, I_A and $I_{\beta(B)}$, that is I_B are balanced 2-phase currents, related to 3-phase currents by expressions (6.41) in Section 6.5.1:

$$I_{\alpha(A)} = \sqrt{\frac{3}{2}}(I_1) \quad \text{and} \quad I_{\beta(B)} = -j\sqrt{\frac{3}{2}}(I_1) \qquad ...(8.69)$$

where, I_1 is the steady-state current in phase A. The values of I_d and I_q differ in frequency only and not in magnitude from the currents in the rotor of the 2-phase 0-α-β machine. Further, the 2-phase rotor currents differ from the current in a-phase of the 3-phase a-b-c machine rotor by a simple friction,

$$\left.\begin{aligned} I_d = I_\alpha &= \sqrt{\frac{3}{2}}\, I_2 \\ I_q = I_\beta &= -j\sqrt{\frac{3}{2}}\, I_2 \end{aligned}\right\} \qquad ...(8.70)$$

Substitution in the torque expression, thus, gives

$$\text{Torque, } T_e = \text{Real part of } (-j3MI_1 I_2^*)$$

In the above expression I_1 may be eliminated by using $V_2 = 0$, so that:

$$I_1 = -\left[\frac{(R_2 + jsX_2)}{jsX_m}\right] \qquad ...(8.71)$$

$$\therefore \qquad \text{Torque, } T_e = \text{Real part of } \left[\frac{3M}{X_m}\left\{\frac{R_2}{s} + jX_2\right\} I_2 I_2^*\right]$$

But $I_2 I_2^*$ is a real number denoting $(I_2)^2$, and the real part of the value of torque is given simply as the resistance term. Also, $X_m = \omega_1 M$.

Hence, $$\text{torque, } T_e = 3\frac{R_2}{s}\frac{1}{\omega_1}(I_2)^2 \text{ Nm}$$

When the actual rotor winding is imagined to be replaced by equivalent winding as in equivalent circuits, we find $R_2' = n^2 R_2$ and $I_2' = -\left(\frac{I_2}{n}\right)$, where n is the turns ratio, $\left(\frac{N_1}{N_2}\right)$, that is, the ratio of the effective number of turns in the stator to effective number of turns of rotor per phase, so that the torque equivalent is given by:

$$\text{Torque} = 3\frac{R_2'}{s}\left(\frac{1}{\omega_1}\right)(I_2')^2 \text{ Nm}$$

$$= 3\left(\frac{R_2'}{s}\right)(I_2')^2 \text{ Synchronous watt} \qquad \text{...(8.72)}$$

8.3.2 Physical Interpretation of Applying Commutator Transformation $[C_2]$ to Rotor of Induction Machine

When the rotor currents of the induction machine is subjected to commutator transformation $[C_2]$, the balanced 3-phase rotor currents are transformed into balanced 2-phase rotor currents, but now the second rotor winding is β, in which the current lags the current in α-winding by $\left(\frac{\pi}{2}\right)$ as was pointed out in Section 7.4.1. Also, the rotor currents are at slip frequency $\omega_2 = s\omega_1$ and not at supply frequency, ω_1. The 2-phase currents may be expressed as follows:

$$\left.\begin{aligned} i_\alpha &= I \cos \omega t \\ i_\beta &= I \sin \omega t \end{aligned}\right\} \qquad \text{...(8.73)}$$

where, I is the rms value of the current.

Application of $[C_2]$ yields:

$$\left.\begin{aligned} i_d &= I (\cos \omega t \cos \theta - \sin \omega t \sin \theta) \\ &= I \cos (\omega t + \theta) \\ \text{and} \quad i_q &= I (\cos \omega t \sin \theta + \sin \omega t \cos \theta) \\ &= I \sin (\omega t + \theta) \end{aligned}\right\} \qquad \text{...(8.74)}$$

Here, $\omega = \omega_2 = s\omega_1$ and $\theta = \omega_r t$.

The angular velocity of rotor, ω_r is the slip speed, or $\omega_r = \omega_1 - \omega_2 = (1 - s)\omega_1$.

The d- and q-currents are therefore

$$i_d = I \cos \omega_1 t; \quad i_q = I \sin \omega_1 t \qquad \text{...(8.75)}$$

The effect of applying $[C_2]$ transformation is clearly producing a change of frequency. The actual rotor currents were at slip frequency, ω_2. These are now replaced by current having the same magnitude as before but at supply frequency ω_1.

8.4 ANALYSING 3-PHASE INDUCTION MOTOR BY 2-AXIS THEORY—AN ALTERNATIVE TREATMENT

While introducing ourselves, we have observed that the induction machine does not have an identifiable axis of symmetry other than the axes of the 3-phase stator and rotor axes of windings. The flux axis being a rotating one, changes continuously while operating, and there are no self-evident direct and quadrature axes. We have also noted that the primary and secondary windings offer mutual inductances as functions of angular positions, and their variations cause the machine to function.

We could derive the operational performance by expressing the mutual inductances as functions of the angle between the concerned windings, and by selecting the axis of one of the phase—say phase *A*—as the direct axis. Then, we could write down the values of flux-linkages and currents and obtain thereform expressions of voltages. We also obtain the corresponding equations in the *d-q* format by employing the transformation. This strategy was adopted in the previous sections but it tended to obscure the physical picture of the machine operation and we ended up in the mathematical manipulations.

In this section, we see how the familiar two-axis equivalence already developed is gainfully employed for analysis of poly-phase induction machines.

We choose stationary axis with respect to the rotating pattern of flux produced by primary and secondary members of the machine. We consider that the machine with 3-phase primary and 3-phase secondary winding arrangement and the primary located on the stator, is fed from a balanced 3-phase source. The resulting rotating magnetic field produced is considered in the clockwise direction, that is, the assumed positive direction of its rotation, at synchronous speed.

We also imagine that the stator winding as reconnected to a hypothetical commutator carrying two pairs of mutually perpendicular brushes. If the brushes are rotated around the commutator at synchronous speed, the 3-phase alternating currents earlier considered supplied to the 3-phase winding through stationary terminals are now, replaced by direct current supplied to the rotating brushes. For the balanced 3-phase conditions, the brush currents and voltages or the 2-axis terminals conditions are identical.

In Figure 8.4, are shown two coils, D_s and Q_s, which with the 2-axis pattern, have been brought to rest by imagining the stator winding to rotate at synchronous speed in the *negative* (that is, counter-clockwise) direction. Coils D_s and Q_s have, therefore, been attributed the property of pseudo-stationary coils. A current in either coil produces a stationary flux pattern, in which the winding, rotating at synchronous speed, produces rotational voltages. This voltage appears between the terminals of the other coil. We can, now, consider the rotor or secondary winding on similar lines. The winding is also observed at terminals which are stationary, relative to the pattern of flux. It is the same thing as considering that the secondary winding is connected to a hypothetical commutator and is carrying two pairs of brushes like the stator. Further, the two separate pairs of brushes are in alignment with the other pairs for the stator member. When the induction motor runs unloaded at almost synchronous speed, the secondary winding, for all practical purposes, is stationary relative to the flux pattern and the hypothetical brushes, and the secondary voltages and currents are virtually zero. But when load torque is applied to the rotor shaft, the secondary winding starts revolving in the negative direction relative to the stationary flux pattern and brushes at a slip speed $s\omega_1$, and rotational voltages are induced in the pseudo-stationary secondary coils. As the secondary is short-circuited, these rotational voltages develop currents, which, in turn, produce a secondary component of flux.

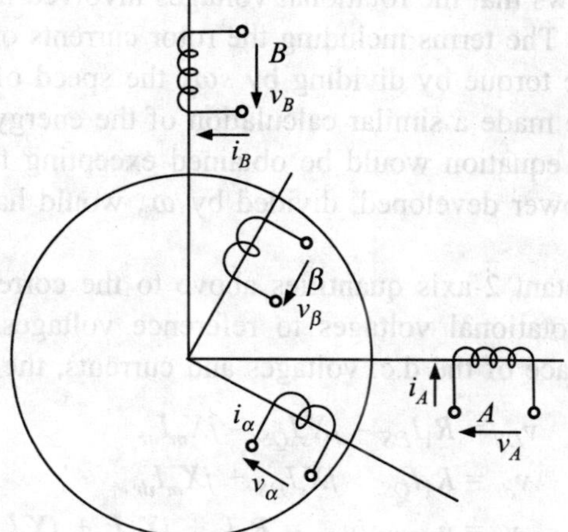

FIGURE 8.4 2-axis depiction of poly-phase induction motor.

The energy converted between the mechanical and electrical forms is given by the sum of the products of the rotational voltages and the currents in each coil. The rotational voltages are produced by the flux pattern of the other axis. Also, the two patterns produced by the secondary currents, develop rotational voltages in the pseudo-stationary coils in the primary (or stator) pseudo-stationary notional coils D_s and Q_s. This causes additional currents to flow to draw additional energy from the supply, which is converted in the other member. The additional current in the primary winding also restores the main flux to almost its original value, the small drop being accounted for due to the primary leakage flux and primary resistance, just like in transformer operation. The 2-axis equations for a balanced poly-phase induction motor in the steady-state, with the reference axis stationary relative to the flux pattern is shown in Figure 8.4.

Because of the inherent symmetry in induction machines, the parameters of the two windings on the same member are equal. Since we assume that the flux is sinusoidally distributed, the total self-inductance of a given winding equals the rotational inductance between that winding and the other similar winding on the same member. Also, the mutual inductance between windings belonging to the stator-rotor combination, is equal to the rotational inductance between a winding on one member and the winding of the other member on the other axis. Thus, we have:

$$\left.\begin{aligned} v_{DS} &= R_1 I_{DS} - \omega_s L I_{QS} - \omega_s M I_{qr} \\ v_{QS} &= R_1 I_{QS} + \omega_s L I_{DS} + \omega_s M I_{dr} \\ v_{dr} &= R_2 I_{dr} - s\omega_s L I_{qr} - s\omega_s M I_{QS} = 0 \\ v_{qr} &= R_2 I_{qr} + s\omega_s L I_{dr} + s\omega_s M I_{DS} = 0 \end{aligned}\right\}$$

and, the power converted, when expressed in terms of the rotor quantities,

$$\begin{aligned} P_d &= 3(-s\omega_s L I_{qr} - s\omega_s M I_{QS}) I_{dr} + (s\omega_s L I_{dr} + s\omega_s M I_{DS}) I_{qr} \\ &= 3s\omega_s\ [M I_{DS} I_{qr} - M I_{QS} I_{dr}] \end{aligned}$$

The above relation shows that the rotational voltages involved in the energy conversion are only due to stator currents. The terms including the rotor currents on the other axis cancel out. We obtain the value of the torque by dividing by $s\omega_s$, the speed of the rotor winding relative to the reference axis. If we made a similar calculation of the energy developed in terms of the stator quantities, the same equation would be obtained excepting for the appearance of ω_s in place of $s\omega_s$. The stator power developed, divided by ω_s, would have given torque developed in the above case.

If we change the constant 2-axis quantities above to the corresponding alternating phase quantities by converting rotational voltages to reference voltages, and if we enter the rms voltages and currents in place of the d.c. voltages and currents, the following equations result:

$$v_{DS} = R_1 I_{DS} - jX_s I_{QS} - jX_m I_{qr}$$

$$v_{dr} = R_1 I_{QS} + jX_s I_{DS} + jX_m I_{dr}$$

and
$$v_1 = v_{DS} - jv_{QS} = R_1 I_s + jX_m I_r + jX_s I_s \qquad \text{...(8.76)}$$

Also,

$$v_{dr} = R_2 I_{dr} - jsX_r I_{qr} - jsX_m I_{QS} = 0$$

$$v_{qr} = v_{dr} - jv_{qr} = R_2 I_2 + jsX_2 I_2 + jsX_m I_s = 0 \qquad \text{...(8.77)}$$

If Eq. (8.77) is divided by s, we get:

$$\left(\frac{R_2}{s}\right) I_r + jX_2 I_r + jX_m I_s = 0 \qquad \text{...(8.78)}$$

And
$$X_s = X_{ls} + X_m$$

when this equation becomes:

$$v_1 = R_1 I_s + j(X_{ls} + X_m) I_s + jX_m I_r$$

$$= R_1 I_s + jX_{ls} I_s + jX_m (I_s + I_r) \qquad \text{...(8.79)}$$

Also, we can write:

$$X_r = X_{lr} + X_m$$

and Eq. (8.79) becomes:

$$\left(\frac{R_r}{s}\right) I_r - j(X_{lr} + X_m) I_r - jX_m I_s = 0$$

or
$$\left(\frac{R_r}{s}\right) I_r - jX_{lr} I_r - jX_m (I_s + I_r) = 0 \qquad \text{...(8.80)}$$

Equations (8.79) and (8.80) indicate, when all quantities are referred to the primary winding and the positive direction of the secondary current is reversed (the secondary voltage is a generated voltage and it causes current to flow in reversed direction when compared with motoring current), that the circuit of the poly-phase induction motor under balanced operation is shown by Figure 8.5.

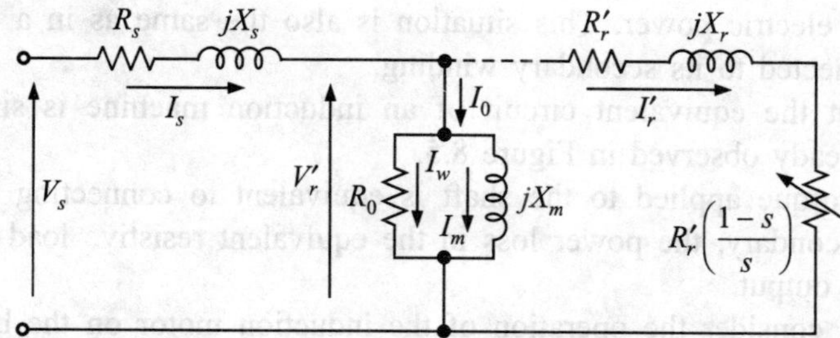

FIGURE 8.5 Equivalent circuit of induction motor.

In the above diagram, the iron loss component of resistance has not been included. The iron loss may be conveniently subtracted from the output power developed and friction and windage loss may be included with it.

The analysis of induction motor is very similar to that of a transformer. When it is operating without an applied shaft torque, the speed of the motor is almost equal to the synchronous speed. The secondary winding practically runs at synchronous speed, which is almost the same speed as the primary component of the flux pattern, and a very small current is induced in the secondary. The secondary winding at this point of operation has a near negligible effect on the primary winding and the situation resembles that of a transformer fed at normal rated voltage with an open-circuited secondary. The primary winding carries a magnetizing current to produce the mmf pattern, which develops a flux pattern of magnitude required to induce in the primary winding, a voltage almost equal to the applied voltage on the assumption that the primary resistance is negligibly small.

With a braking torque imposed on the motor shaft, the speed of the machine tends to reduce. A current now flows in the secondary winding and produces an mmf stationary with respect to the primary mmf component. This follows the development of the secondary flux pattern, and it produces a component voltage at supply frequency in the primary just like in the case of transformers. To effect an ampere-turn balance, the primary reacts by increasing the primary current, sufficient to overcome the mmf due to the secondary winding, thus replenishing the flux linking the primary winding to its former value. Thus, an induction motor also functions as a 'constant flux' device like a transformer.

The two components of magnetic flux are at a certain torque angle such that its value is just equal to the applied torque on the motor shaft. Thus, the increase in primary current, together with the applied effective voltage, provides electric power equalling the mechanical power output along with mechanical power losses in the machine. The resultant air-gap flux suffers very little change because the applied voltage is assumed constant. But actually, there is a small drop in air-gap flux linking the primary winding as a result of the presence of an increased leakage flux produced by increased primary current.

On increasing the braking torque further, the secondary and primary currents, the torque angle and input power all increase and the speed decreases, till the maximum torque is developed. Any further augmentation of braking torque causes the machine to enter an unstable mode and causes the machine to transit to stand-still condition, that is, at $s = 1$.

While accounting for the primary winding, we observe that the effect of the secondary winding is always to induce a supply frequency voltage at a phase angle. This causes the necessary flow of electric power. This situation is also the same as in a transformer with a resistive load connected to its secondary winding.

It follows that the equivalent circuit of an induction machine is similar to that of a transformer, as already observed in Figure 8.5.

The braking torque applied to the shaft is equivalent to connecting a resistive load to the transformer secondary, the power loss in the equivalent resistive load being equal to the mechanical power output.

We can, now, consider the operation of the induction motor on the basis of component current, voltage and flux-distribution pattern on which its performance depends. Figure 8.6(a) depicts the development of the rotating distribution pattern of the magnetizing current, I_m, and the corresponding mmf pattern, F_m when the primary winding is fed from a 3-phase supply. The horizontal lines in the diagram denote the periphery of the primary and secondary members on which the windings are housed. The small gap between the horizontal lines indicate the air-gap length. The mmf pattern also represents the flux-density pattern of distribution B_m, and this gives the flux-density distribution at points around the air-gap. Ignoring the small drop in flux due to leakage reactance and resistance of the primary while carrying load current, the Figure 8.6(a) also displays the resultant air-gap flux distribution at no load, as well as on-load conditions.

Figure 8.6(b), depicts the voltage distribution scenario V_r developed in the secondary winding by the flux-pattern, when the secondary is operating with a slip. Since the secondary is short-circuited and is inductive, the secondary produces a voltage distribution V_r and a corresponding short-circuit current distribution pattern I_r, which lags behind V_r by an impedance angle θ_r as shown in Figure 8.6(b). If the slip is small (as it is so even on full-load), as the secondary frequency is small, the secondary leakage reactance is small when compared with its effective resistance. We consider the secondary voltage as induced by the resultant air-gap flux and not the primary component of flux when load current flows, the secondary current being given by the ratio of the secondary voltage produced by the resultant flux and the combination of the resistance and effective leakage reactance of the secondary. The angle of lag ϕ_2 of the secondary current from the secondary voltage is, therefore, small at small value of slip. The secondary current distribution produces the secondary component of mmf, as indicated in Figure 8.6(b). The angle between the resultant flux pattern and the secondary flux pattern is nearly equal to the load angle δ, as indicated in the diagram. It is not exactly equal to the load angle since the leakage reactance voltage drop has been ignored. If we aggregate the reverse of the secondary component voltage drop equivalent flux to the resultant flux distribution, we get the primary component of flux distribution. This is indicated in Figure 8.6(c). The angle between the axis of the primary and secondary flux-distribution is the torque angle δ_T.

The secondary component of current distribution I_2 produces the secondary mmf, also indicated in Figure 8.6(b). The angle between the resultant flux-pattern and the secondary flux scenario is nearly equal to the load angle δ, as indicated in Figure 8.6(b). It is not exactly equal to the load angle since the leakage reactance voltage drop has been ignored as pointed out earlier.

The polarities of the primary and secondary members are indicated in Figure 8.6(d). We observe that the torque angle for an induction machine is significantly more than $\left(\frac{\pi}{2}\right)$. It is

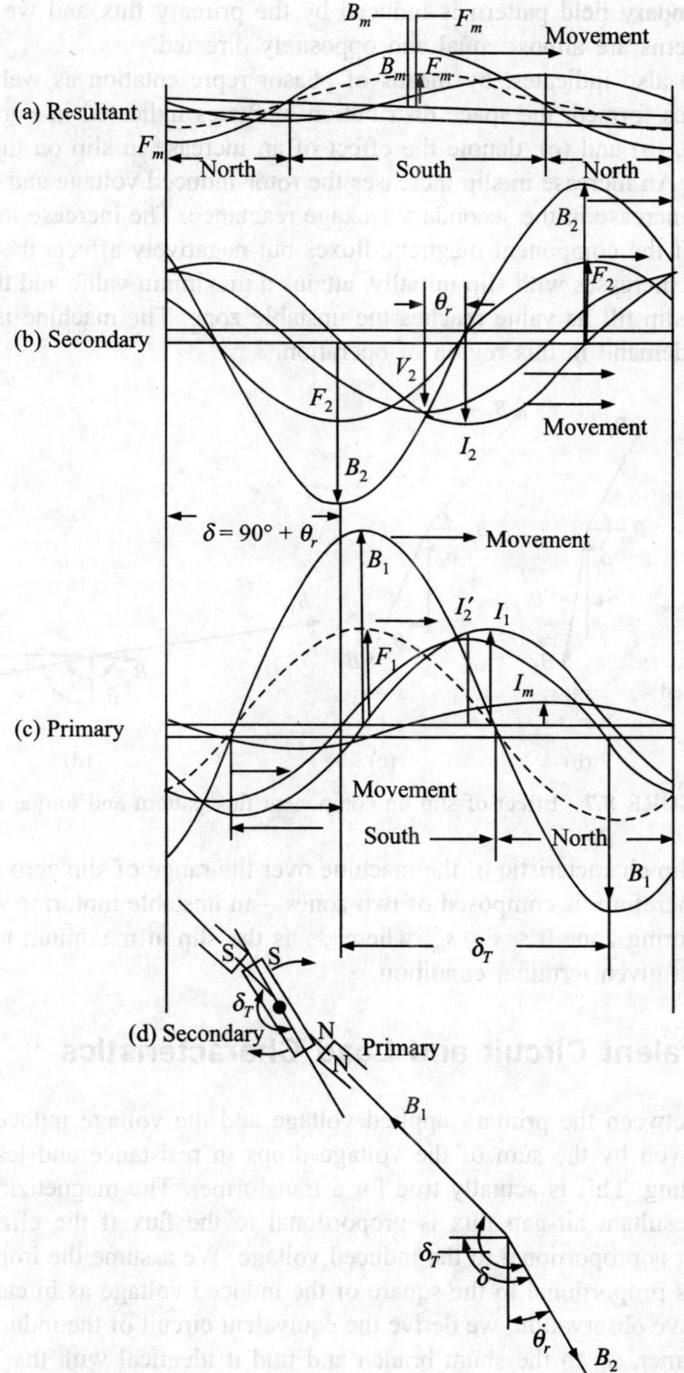

FIGURE 8.6 Development of component flux-pattern in induction motor with mechanical load.

because the secondary field pattern is induced by the primary flux and we can conceive that the two flux patterns are almost equal and oppositely directed.

Figure 8.6(d) also indicates, by means of phasor representation as well as representative magnetic polarities formed, the space distribution of fluxes indicated in Figure 8.6(a), (b) and (c). Figure 8.7(a), (b) and (c), denote the effect of an increase in slip on the flux components and torque angle. An increase in slip increases the rotor-induced voltage and the frequency, and consequently an increase in the secondary leakage reactance. The increase in the slip augments the magnitudes of the component magnetic-fluxes but negatively affects the torque angle. The torque, therefore, increases with slip initially, attains a maximum value and then starts reducing with increase of slip till its value reaches the unstable zone. The machine is unable to sustain the shaft torque demand in this region of operation.

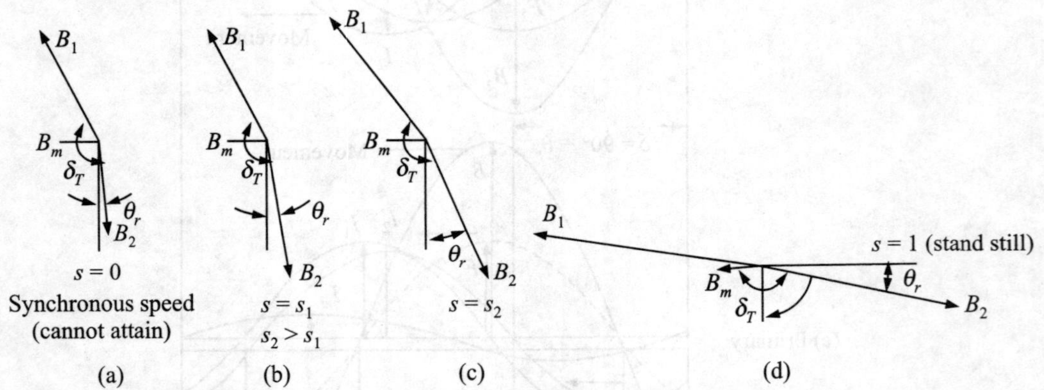

FIGURE 8.7 Effect of slip on component flux-pattern and torque angle.

The torque-slip characteristic of the machine over the range of slip zero to unity, called the motoring zone, therefore, is composed of two zones—an unstable motoring zone for $1 > s > s_m$ and a stable motoring zone $0 < s < s_m$, where, s_m is the slip at maximum torque produced by the machine for a given terminal condition.

8.4.1 Equivalent Circuit and Load Characteristics

The difference between the primary applied voltage and the voltage induced by the resultant air-gap flux is given by the sum of the voltage drops in resistance and leakage reactance of the primary winding. This is actually true for a transformer. The magnetizing current required to produce the resultant air-gap flux is proportional to the flux if the effect of saturation is ignored, that is, it is proportional to the induced voltage. We assume the iron losses in the core of the machine as proportional to the square of the induced voltage as in case of transformers.

From the above observation, we derive the equivalent circuit of the induction machine from that of a transformer, up to the shunt branch and find it identical with that of a transformer.

Now, we consider the portion of the equivalent circuit that represents the effect of the primary winding of what happens to the secondary. We discussed in the previous section details of mmf balance across the air-gap of the primary and secondary load currents. The situation is similar to that of a transformer. However, in case of the induction machine, the primary and secondary windings are distributed and the flux distribution is approximately sinusoidal. Hence, the load components of the primary and secondary currents are in inverse proportion to the number of effective turns per phase. Hence, we may refer the secondary phase resistance and reactance parameters to the primary winding for inclusion in the equivalent circuit by using the effective turns ratio, as we did in case of transformers. In the analysis, we assume that the primary and the secondary windings have the same number of phases, and the ratio of secondary to primary effective turns per phase is K, that is:

$$K = \frac{V_r}{V_s} = \frac{\text{Secondary voltage per phase at stand-still on open circuit}}{\text{Primary voltage per phase}}$$

and also:

$$s = \frac{\omega_s - \omega_r}{\omega_s}$$

The secondary winding impedance per phase at stand-still is taken as:

$$Z_r = R_r + jX_r$$

Now, at a slip s, the secondary winding impedance per phase is as follows:

$$Z_r' = R_r + jsX_r$$

and the secondary induced voltage per phase at the slip s is sV_r at a frequency sf. Thus, the short-circuit secondary current works out to be:

$$I_r \text{ at short-circuit} = \frac{sV_r}{R_r + jsX_r}$$

This current when referred to the primary, is given by:

$$I_r' = KI_r = K\frac{sV_r}{R_r + jsX_r}$$

The primary induced voltage per phase is:

$$V_r' = \frac{1}{K}V_r, \text{ that is, } V_r = KV_r'$$

If we replace voltage V_r by equivalent V_r' in the expression for I_r', we get:

$$I_r' = \frac{K^2 sV_r'}{R_r + jsX_r} = \frac{sV_r'}{\dfrac{R_r}{K^2} + js\dfrac{X_r}{K^2}}$$

In the above expression,

$$R_r' = \frac{R_r}{K^2} \quad \text{and} \quad X_r' = \frac{X_r}{K^2}$$

and they are respectively the secondary resistance and secondary leakage reactance referred to the primary winding, just as in case of transformers. The above expression may, now, be rewritten as follows:

$$I'_r = \frac{V'_r}{\frac{R'_r}{s} + jX'_r}$$

In the above expression, if we critically examine the right hand side, we observe that the frequency of the referred value of the secondary current has notionally changed to primary frequency, f.

To obtain the equivalent circuit depiction, we should add the no-load current I_0 to I'_r. I_0 consists of two components—the magnetizing component of current and the iron-loss component of current. Similarly, to V_r we add the voltage drops caused by the sum of I'_r and I_0 in the primary winding resistance and leakage reactance. The equivalent circuit, therefore, assumes the form shown in Figure 8.5 in Section 8.4. We observe that the impedance of the secondary circuit referred to the primary is $\left(\frac{R'_r}{s} + jX'_r\right)$. But the resistance of the secondary winding referred to the primary is only R'_r. The remainder of the total resistance, given by $\frac{R'_r}{s} - R'_r = \left(\frac{1-s}{s}\right)R'_r$. The loss in this resistor component equals the mechanical power developed and its value is a function of the slip.

Let us now consider the machine running as a motor. The total power input per phase is given by:

$$P_s = V_s I_s \cos\theta_s$$

where, $\cos\theta_s$ is the primary power factor.

Similarly, the pre-phase power crossing the air-gap to the secondary member is given by:

$$P_r = V'_r I'_r \cos\theta_r = (I'_r)^2\left(\frac{R'_r}{s}\right)$$

But, as the secondary copper loss per phase is given by $(I'_r)^2 R'_r$, the gross mechanical power developed has the form:

$$(I'_r)^2\left\{\frac{R'_r}{s} - R'_r\right\} = \frac{(1-s)}{s}(I'_r)^2 R'_r$$

and, the gross mechanical torque developed per phase

$$T_e = \frac{P_{\text{mech}}}{\omega_r} = \frac{P_{\text{mech}}}{2\pi n_s(1-s)} = \frac{P'_r}{2\pi n_s}$$

where, n_s is the primary synchronous speed in revolution per second. The gross mechanical torque is the same as the electromagnetic torque. The available torque is less than the above value because it includes the friction and windage torque being also supplied by the machine when functioning as a motor. The gross torque at a given slip is proportional to the secondary input power, which is the square of the secondary voltage. If we assume that V_r is approximately equal to V_s, the torque at a given slip is approximately proportional to the square of the applied voltage V_s.

We also observe that the ratio of the secondary input power per phase is such that the ratio of the secondary winding copper loss per phase to the total secondary power input per phase equals the value of the per-unit slip s, that is,

$$\frac{\text{Secondary copper loss per phase}}{\text{Secondary input power per phase}} = \frac{(I_r')^2 R_r'}{\dfrac{(I_r')^2 R_r'}{s}} = s$$

The secondary power input, P_r is termed as the *torque in synchronous watts* because if the actual torque at a slip s occurred at synchronous speed, it would do work equal to the total rotor power input including the rotor copper loss. We developed the equivalent circuit of an induction motor originally on the basis of its operational similarly with a transformer. This approach presented in Section 8.4 and depicted in Figure 8.5, includes a series-parallel equivalent network of lumped resistance-inductance elements, very similar to that of a transformer, with the secondary circuit close-circuited with a resistance equivalent of $\left(\dfrac{1-s}{s}\right)R_r'$. In an alternative approach to the development of the equivalent circuit, we could apply the circuit concept based on the total energy flow from the primary input terminals to the rotor, across the air-gap. At suitable steps, we reduce the input energy by the losses occurring at various points along the path of energy flow, like:

1. The stator—in the form of stator copper loss.
2. The fixed losses in stator at the interface at the air-gap, which include the core loss at constant voltage and frequency in the stator (core-loss in rotor is insignificant under operating conditions and has values only at stand-still conditions as the frequency of the rotating magnetic field equals that of the stator), the friction and windage losses in the rotor.
3. Available energy in rotor, which contributes to the production of rotor copper loss, and energy available for supply to the rotor shaft.

The energy flow-diagram on a per-phase basis is shown in Figure 8.8, and the equivalent circuit thus developed is presented in Figure 8.9.

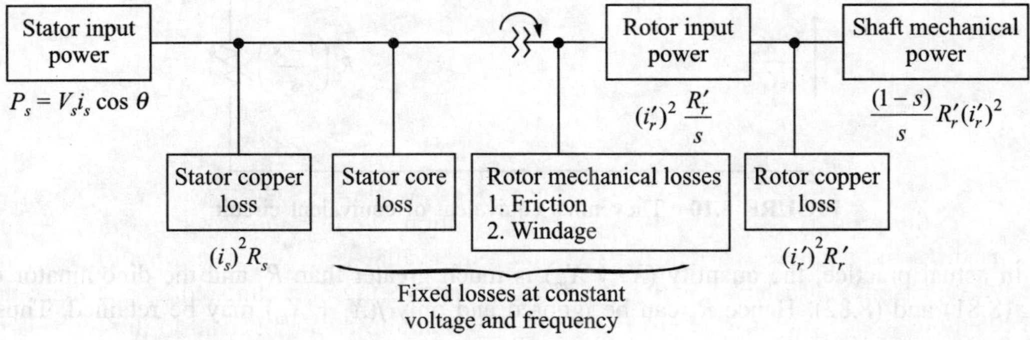

FIGURE 8.8 Power flow diagram of induction motor.

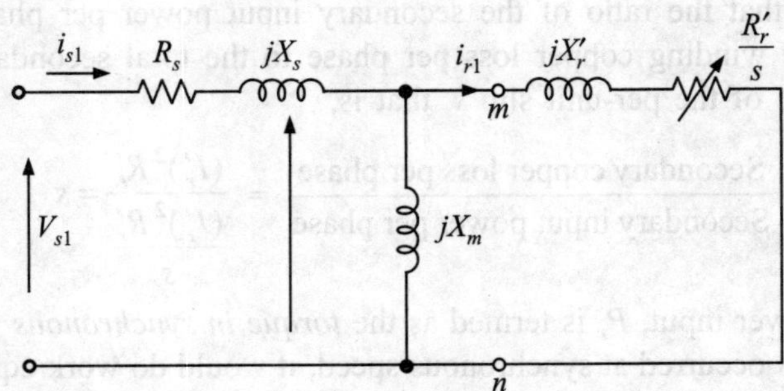

FIGURE 8.9 Equivalent circuit of induction machine based on energy-flow consideration.

Torque-slip and power-slip characteristics of 3-phase induction motor can be readily obtained from this form of its equivalent circuit. Simplification of this equivalent circuit using Thevenin's equivalence leads to easier methods of determining these characteristics.

We take two points *m* and *n* on the equivalent circuit shown in Figure 8.9. If we look into the equivalent circuit from points *m-n* towards the, stator side, then the part of the circuit constituting R_s, X_s and X_m and the input voltage source V_s, can be replaced by an equivalent voltage source V_e and an equivalent series impedance, $Z_e = R_e + jX_e$ as indicated by Figure 8.9. Referring to the Figures 8.9 and 8.10, we can write,

$$V_e = \frac{V_s \cdot jX_m}{R_s + j(X_s + X_m)} \quad \text{...(8.81)}$$

and

$$Z_e = \frac{(R_s + jX_s)(jX_m)}{R_s + j(X_s + X_m)} \quad \text{...(8.82)}$$

In the figure, V_e is the voltage appearing across the terminals *m-n*, with the rotor open-circuited.

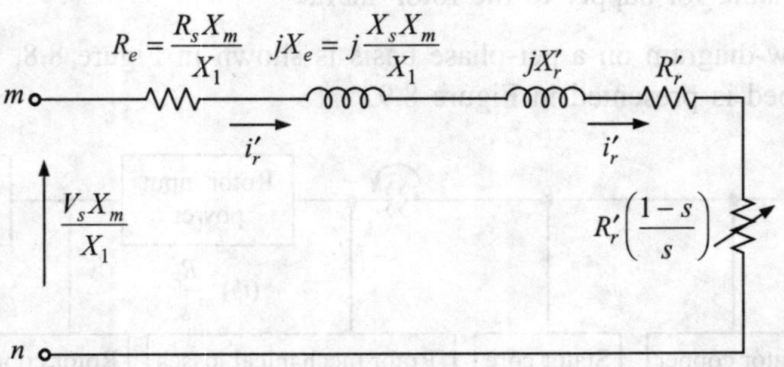

FIGURE 8.10 Thevenin's equivalent of equivalent circuit.

In actual practice, the quantity $(X_s + X_m)$ is much greater than R_s and the dinominator of Eqs. (8.81) and (8.82). Hence R_s can be ignored and only $j(X_s + X_m)$ may be retained. Thus:

$$V_e = \frac{V_s X_m}{X_s + X_m} = \frac{V_s X_m}{X_1} \quad \text{...(8.83)}$$

and
$$Z_e = R_e + jX_e$$
$$= \frac{R_s X_m}{X_s + X_m} + \frac{jX_s X_m}{X_s + X_m}$$
$$= \frac{R_s X_m}{X_1} + j\frac{X_s X_m}{X_1} \quad \text{...(8.84)}$$

In the above expressions,

$$X_1 = X_s + X_m$$

denotes the self-reactance of the stator windings per phase.

Thus, the Thevenin's equivalent of the equivalent circuit taken at points *m-n* (at the input terminals of the rotor) is given by a series impedance composed of effective resistance of $\left(\frac{R_s X_m}{X_1}\right)$ and an effective reactance of $j\left(\frac{X_s X_m}{X_1}\right)$, with an effective voltage of $\left(\frac{V_s X_m}{X_1}\right)$ across the *m-n* terminals as shown in Figure 8.10.

The rotor effective current as obtained from the equivalent circuit of Figure 8.10, is given by:

$$i_r = \frac{V_e}{\left\{R_e + \frac{R_r'}{s}\right\} + j(X_r' + X_e)} \quad \text{...(8.85)}$$

and the total torque in N-m is given by:

$$T = \frac{m}{\omega_s} \frac{V_e^2}{\left\{R_e + \frac{R_r'}{s}\right\}^2 + \{X_r' + X_e\}^2} \frac{R_r'}{s} \quad \text{...(8.86)}$$

where, m is the number of stator phases and ω_s is the synchronous speed of the stator rotating magnetic field.

Torque-speed (or torque-slip) characteristic of an induction motor working at a specified applied voltage and frequency can be obtained by substituting different values of slip s. A typical shape of the curve is shown in Figure 8.11. It functions as a motor when its slip is greater than zero (corresponding to synchronous speed of stator field and any positive value of s (denoting motor speed less than synchronous, including rotating in the reverse direction.

But operation of the machine in the slip range of $0 \le s \le 2$ is most significant. This regions is further divided into two discrete zones:

1. The stable motoring zone $0 \le s \le s_{mt}$ and
2. The unstable motoring zone $s_{mt} \le s \le 2$

where, s_{mt} is the slip at maximum torque or the pull-out torque.

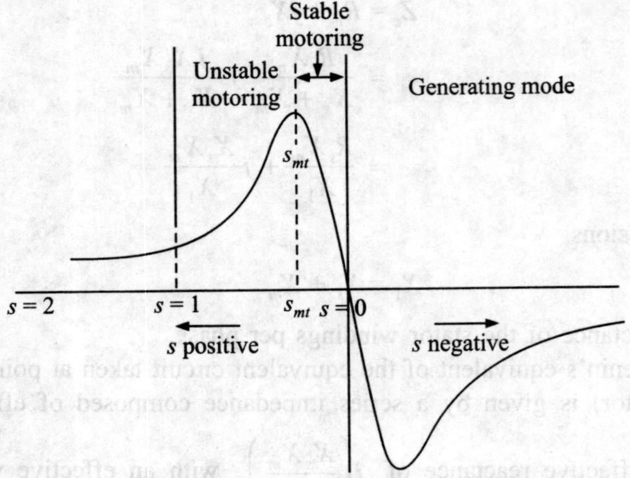

FIGURE 8.11 Typical torque-slip characteristic of an induction machine.

The machine is found to operate as an induction generator when operating at negative slip (corresponding to rotor speed driven at super-synchronous condition).

8.4.2 Maximum Internal Torque

This is identified as the *maximum torque* at slip $s = s_{mt}$, shown in Figure 8.11, or the pull-out or *stalling* torque and is especially identified as T_{mt}.

The internal torque is proportional to $(I_r')^2 \dfrac{R_r'}{s}$ in synchronous watts. It assumes a maximum value when the power delivered to the effective rotor resistance in the equivalent circuit of Figure 8.9, that is, $\left(\dfrac{R_r'}{s}\right)$ maximizes. We apply the maximum power transfer theorem to obtain the value of s_{mt} and consequently the value of T_{mt}.

The maximum power transfer occurs when the load impedance equals the source impedance. Thus, at operation $s = s_{mt}$.

$$Z_{\text{load}} = Z_{\text{source}}$$

Since the load in this case is $\dfrac{R_r'}{s} = \dfrac{R_r'}{s_{mt}}$, we get condition as:

$$\frac{R_r'}{s_{mt}} = \sqrt{R_e^2 + (X_r' + X_e)^2} \qquad \text{...(8.87)}$$

Hence, s_{mt} is obtained as:

$$s_{mt} = \frac{R_s'}{\sqrt{R_e^2 + (X_r' + X_e)^2}} \qquad \text{...(8.88)}$$

and, the value of the maximum torque is:

$$T_{mt} = \left[\frac{mV_e^2}{\omega_s}\right]\left[\frac{1}{\left\{R_e + \frac{R_r'}{s_{mt}}\right\}^2 + \{X_r' + X_e\}^2}\right]\left(\frac{R_r'}{s_{mt}}\right) \quad ...(8.89)$$

which works out to be:

$$T_{mt} = \frac{mV_e^2}{2\omega_s\left[R_e + \sqrt{R_e^2 + (X_r' + X_e)^2}\right]} \text{ Nm} \quad ...(8.90)$$

Examination of Eqs. (8.88) and (8.90) indicates that:

1. The slip at which the maximum torque occurs is directly proportional to the net value of the rotor resistance.
2. T_{mt} is directly proportional to the square of the stator applied voltage.
3. T_{mt} is reduced by an increase in stator resistance, R_s (that is R_e) and
4. T_{mt} is independent of the rotor circuit resistance R_r, while it reduces with increase of X_r' and X_s.

This leads us to conclude that:

1. Speed of motor can be controlled by changing the rotor-circuit resistance.
2. Starting torque may be varied by controlling the rotor-circuit resistance.
3. T_{mt} can be made to occur at $s = 1$ by altering the rotor-circuit resistance.

With wound-rotor induction motors, the shorting torque is increased by inserting suitable external resistance in the rotor circuit at start. As the motor starts and accelerates, the external resistance is cut till it is finally reduced to zero.

Starting torque

Initially, at stand-still, $s = 1$. Hence, the value of the starting torque,

$$T_{st} = \frac{mV_e^2}{\omega_s} \cdot \frac{R_r'}{(R_e + R_r')^2 + (X_s' + X_e)^2} \quad ...(8.91)$$

8.4.3 Torque Produced in the Motor in Terms of Pull-out Torque, T_{mt}

In order to obtain a simple approximate expression, let us neglect the stator resistance R_s, which is the same thing as neglecting the stator equivalent resistance R_e.

If we divide Eq. (8.86) by Eq. (8.89), we get

$$\frac{T}{T_{mt}} = \frac{2\left[R_e + \sqrt{R_e^2 + (X_r' + X_e)^2}\right]}{\left(R_e + \frac{R_r'}{s}\right)^2 + (X_r' + X_e)^2} \frac{R_r'}{s}$$

and since R_s and R_e are neglected:

$$\frac{T}{T_{mt}} = \frac{2\,(X_r' + X_e)}{\left(\dfrac{R_r'}{s}\right)^2 + (X_r' + X_e)^2} \frac{R_r'}{s} \qquad ...(8.92)$$

Also,

$$R_r' = s_{mt}(X_r' + X_e) \qquad ...(8.93)$$

as R_e is negligible.

Substitution of Eq. (8.93) in Eq. (8.92) leads to:

$$\frac{T}{T_{mt}} = \frac{2\,(X_r' + X_e)}{\left(\dfrac{s_{mt}X}{s}\right)^2 + X^2} \frac{s_{mt}X}{s}$$

where,

$$X = X_r' + X_e$$

or

$$\frac{T}{T_{mt}} = \frac{2}{\left(\dfrac{s_{mt}}{s}\right) + \left(\dfrac{s}{s_{mt}}\right)}$$

That is,

$$T = \frac{2T_{mt}}{\left(\dfrac{s_{mt}}{s}\right) + \left(\dfrac{s}{s_{mt}}\right)} \qquad ...(8.94)$$

8.4.4 Power-slip Characteristics

The total internal mechanical power developed in an induction motor at a slip s, is:

$$P_{\text{mech}} = m(i_r')^2 R_r' \left(\frac{1-s}{s}\right) \text{ watt}$$

or

$$P_{\text{mech}} = \left[\frac{mV_e^2}{\left(R_e + \dfrac{R_r'}{s}\right)^2 + X^2}\right]\left[\frac{(1-s)R_r'}{s}\right] \qquad ...(8.95)$$

We use the maximum power transfer theorem again but use the value given in Eq. (8.95) to determine the value of S_{mp}. Then,

$$\frac{R_r'(1-s_{mp})}{s_{mp}} = \sqrt{(R_r' + R_e)^2 + X^2}$$

or

$$s_{mp} = \left[\frac{R_r'}{R_r' + \sqrt{(R_e + R_r')^2 + X^2}} \right] \quad \text{...(8.96)}$$

Corresponding to s_{mp}, the value of maximum mechanical power $P_{m\ \text{mech}}$ can be shown:

$$P_{m\ \text{mech}} = \left\{ \frac{mV_e^2}{\left[R_e + R_r' + R_r'\left(\dfrac{1 - s_{mp}}{s_{mp}}\right)\right]^2 + X^2} \right\} \left\{ \frac{R_r'(1 - s_{mp})}{s_{mp}} \right\}$$

which simplifies to:

$$P_{m\ \text{mech}} = \frac{mV_e^2}{2\left[R_e + R_r' + \sqrt{(R_e + R_r')^2 + X^2} \right]} \quad \text{...(8.97)}$$

It may be noted that $P_{m\ \text{mech}}$ depends on the value of the rotor resistance R_r', whereas T_{mt} does not.

EXAMPLE 8.1 A 3-phase, star-connected, 400 volt, 50 Hz induction motor has the following circuit parameters in ohm per phase referred to stator:

R_s, the stator resistance per phase = 0.15 Ω
X_s, the stator reactance per phase when referred to stator side = 0.45 Ω
R_r', the rotor resistance per phase, when referred to stator side = 0.12 Ω
X_r', the rotor reactance per phase, when referred to stator side = 0.45 Ω
X_m, the magnetizing reactance of the machine as seen from the stator side = 28.5 Ω
Fixed losses (core, friction and windage losses) of the machine = 400 watt
Stator to rotor effective turns ratio = 1.2.

Obtain:

(a) the slip at which maximum torque occurs, the value of the maximum torque and the corresponding power developed;
(b) the rotor current and torque at start;
(c) the value of the external resistance to be inserted in the rotor circuit to produce the maximum torque at starting.

Solution: Using the equivalent circuit of the machine shown in Figure 8.9, we reduce the circuit to Thevenin's equivalent network shown in Figure 8.10. We get,

$$Z_e = \frac{(0.15 + j0.45)\,(j28.5)}{(0.15 + j28.95)}$$

$$\simeq \frac{(0.15 + j0.45)\,(j28.5)}{j28.95}$$

Hence, $$R_e = \frac{(0.15)(j28.5)}{j28.95} = 0.1476\,\Omega$$

and $$X_e = \frac{(0.45)(j28.5)}{j28.95} = 0.443\,\Omega$$

We write,

$$\frac{0.12}{s_{mt}} = 0.1476 + j(0.45 + 0.443)$$

(a) Hence, $$s_{mt} = \frac{0.12}{\sqrt{(0.1476)^2 + (0.893)^2}} = 0.1326$$

$$V_e = \frac{V_s}{\sqrt{3}}\frac{28.5}{28.95} = \frac{400}{\sqrt{3}}\frac{28.5}{28.95} = 227.4 \text{ volt/ph}$$

Then,

$$T_{mt} = \frac{3 \times (227.4)^2}{(50\pi) \times 2\left\{0.1476 + \sqrt{(0.1476)^2 + (0.893)^2}\right\}}$$

$$= 469.1 \text{ Nm}$$

Corresponding mechanical power developed at slip $s = 0.1326$, is:

$$P_{\text{shaft}} = 469.1\ (1 - 0.1326)\ 50\pi - 400$$

Shaft, $P = 63.5153$ kW

Maximum shaft torque developed:

$$= \frac{P_{shaft}}{\omega_s(1 - s_{mt})}\ \frac{63515.3}{(50\pi)\,(1 - .1326)}$$

$$= \frac{63515.3}{50\pi \times 0.8674} = 466.2 \text{ Nm}$$

(b) At start, $s = 1$, and rotor current at starting

$$I'_{r\,st} = \frac{V_e}{\sqrt{(R_e + R'_r)^2 + (X'_r + X_e)^2}}$$

$$= \frac{227.4}{\sqrt{(0.2676)^2 + (0.895)^2}}$$

$$= 243.94 \text{ amp}$$

$\therefore$ $$\text{Starting torque} = 3 \times (243.94)^2 \times (0.12)\,\frac{1}{50\pi}$$

$$= 136.4 \text{ Nm.}$$

(c) By inspecting Figure 8.9, we conclude that, at starting for maximum torque to occur:

$$\frac{R_r' + R_{\text{ext}}}{1} = |0.1476 + j0.893|$$

$$= \sqrt{(0.1476)^2 + (0.893)^2}$$

$$= 0.9051$$

∴ The external resistance referred to stator that must be inserted in rotor circuit

$$R_{\text{ext}} = 0.9051 - 0.12$$

$$= 0.7851 \text{ ohm}$$

∴ The actual resistance that must be connected to the rotor terminals for producing maximum torque at starting is given by:

$$\frac{0.7851}{(1.2)^2} = 0.5452 \text{ ohm}$$

per phase, star connected.

8.5 ANALYSIS OF POLY-PHASE SYNCHRONOUS MACHINES WITH UNIFORM AIR-GAP

In a poly-phase synchronous machine the armature is fitted with a poly-phase winding and carries poly-phase currents through it, while the field winding is fed from a direct current source. The armature produces a flux-pattern which rotates at constant synchronous speed, decided by the number of poles and the frequency of the current, while the d.c. winding produces a flux-distribution which remains stationary with respect to the field winding. It is required that the angle between the component fluxes should be constant in order to develop a constant torque. Only when such a condition is satisfied, that a steady energy-conversion between the mechanical and electrical form can come about. For this to happen, the rotor must rotate at synchronous speed such that the relative speed between the direct field (produced by the field arrangement) and the synchronously rotating field produced by the armature should be zero.

As pointed out earlier, synchronous machines are built in many forms. In large machines, the armature is housed in slots on the stationary or outer member and armature winding is accessed by rigid connections through stationary tapping parts.

The field system in such cases is mounted on the rotor with a smooth air-gap in slots or it can also be on projected structures (or salient poles) with lumped field windings. Alternatively, the field system may be mounted on the stationary (that is, outer member) with salient poles with polar projections while the armature is housed in slots of a smooth rotor. In this case the armature terminals are accessed by brush-slip ring interface. In either case, the schematic of a synchronous machine may be represented by Figure 8.2. We take up the smooth air-gap machines first. When the field winding is excited and a relative movement occurs between the field flux and the armature conductors due to rotation of one of the members, a balanced set of 3-phase voltages, V_0 is developed across the three phases of the armature, corresponding to the effective movement at synchronous speed. If this set of voltages is connected to 3-phase balanced impedances, a balanced set of 3-phase currents ensue. The terminal voltage is found to alter from the initial value of V_0 per phase, to a new value V_a per phase, because, as expected the winding presents resistance and reactance drops due to current flow through them. This change in terminal voltage occurs although the field current—and, therefore, the field flux—remains unaltered, and the speed of relative movement is kept constant. The current flowing in the armature winding produces the 'main flux' along with a small percentage of 'leakage flux'. The mmf component contributing to the main flux is called the armature reaction mmf and the corresponding flux component as the armature reaction flux. Leakage flux also exists and links with the turns of the armature winding predominantly along air paths. We can, therefore, divide the total reactance offered by 3-phase armature windings into main magnetizing reactance, and leakage reactance of the armature. We re-iterate that the air-gap flux-pattern produced by the poly-phase winding is stationary with respect to the field system, and the mmf referred above is super-imposed on the field mmf. The flux-distribution produced in the air-gap by the field system, induces a voltage V_0 in the armature phases which remains unaffected by any current flowing in the phase windings. However, the current in the phase windings induces super-posed voltages in them having magnitude of (jX_mI_a) per phase. Additionally, the leakage component of the flux also contributes an additional voltage induced of value (jX_lI_a). The terminal voltage is the phasor sum of the voltage V_0, produced by the field flux, the voltage (jX_mI_a) due to the main armature reaction flux and the voltage component (jX_lI_a) due to the leakage flux of the armature, in addition to any voltage drop due to the phase windings. The phasor diagram in Figure 8.12(a) denotes the sinusoidal currents and voltages induced in one phase of the armature winding, along with the drops indicated above.

The two component air-gap flux patterns are at the torque angle for a mechanical power intake sufficient to provide for the electrical power output and all the losses except for the loss incurred in the resistance of the field winding. The corresponding equivalent circuit of the synchronous machine is depicted in Figure 8.12(b). This is an idealized version of the equivalent circuit and assumes smooth air-gap between the armature and the field system. The voltage can be expressed by relations:

$$V_0 = V_a + \{R_a + j(X_m + X_l)\}I_a$$
$$= V_a + \{R_a + jX_s\}I_a$$

where, $X_s = (X_m + X_l)$, is the idealised value of the synchronous reactance.

The torque angle δ_T, between the component flux distributions developed by the d.c. excited field winding at the poly-phase member is not the angle between the phasors V_0 and V_a. This angle is commonly known as the 'load angle'. The angle δ_T, in this case, is greater than $\left(\frac{\pi}{2}\right)$ as shown in Figure 8.12(c), because V_0 is greater than V_a. This is the same as saying that the component flux distributions produced by the field winding and the poly-phase armature winding, inducing in the armature the component voltages V_0 and (jX_sI_a)—in that order—have components directed in opposition. Figure 8.12(c), indicates, with the help of equivalent magnets, the relative positions of these individual flux patterns.

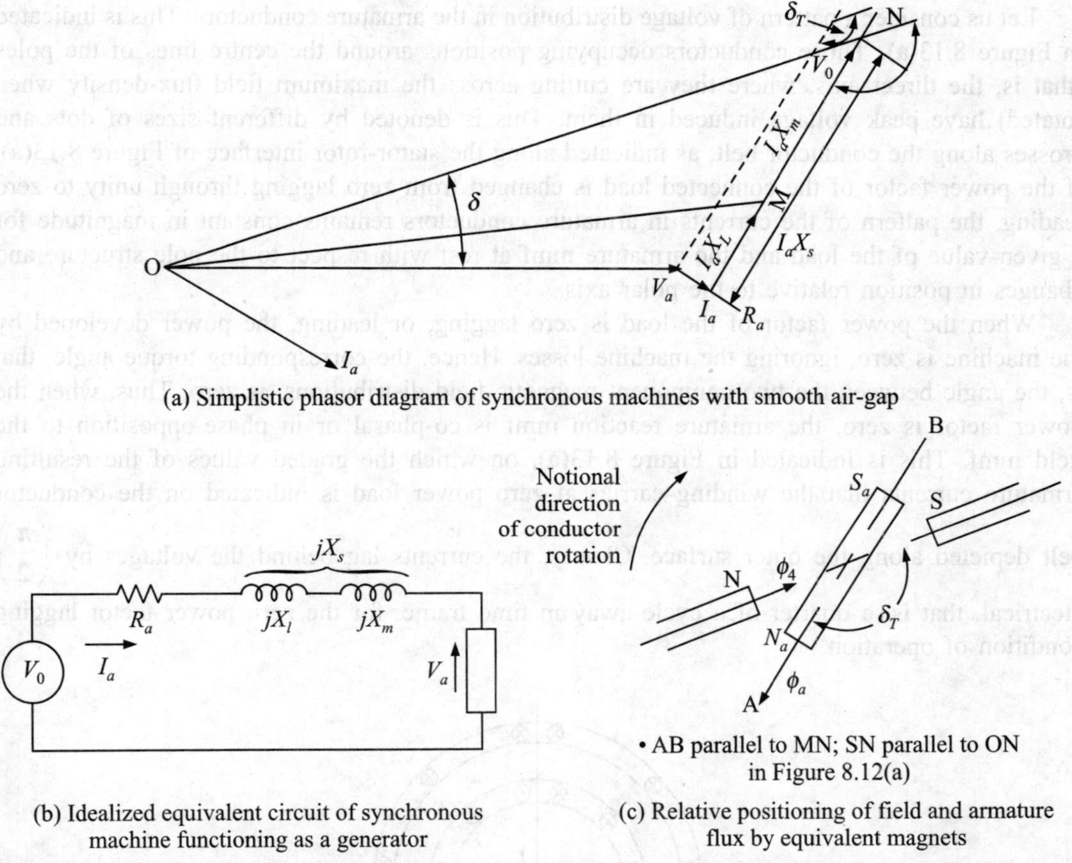

(b) Idealized equivalent circuit of synchronous machine functioning as a generator

(c) Relative positioning of field and armature flux by equivalent magnets

FIGURE 8.12 Idealized equivalent circuit and phasor diagram of poly-phase synchronous machine with uniform air-gap at steady-state.

The net voltage induced per phase is actually V_a', as indicated in Figure 8.12(a). The deviation in value between V_a' and the actual terminal voltage V_a is due to the joint effects of winding resistance and leakage reactance.

The magnitude of V_0 is always proportional to the value of the field excitation I_f for constant speed operation, with saturation in the field ignored. V_0 is identified as the 'internal voltage' or the 'excitation voltage' of the machine.

8.5.1 Variation of Position of Armature Reaction mmf with Power Factor

It may be worthwhile to examine the nature of mmf produced by the alternating current in the armature winding and flux it produces. The space distribution of the armature reaction mmf is almost sinusoidal, and for a given set of 3-phase balanced currents, it is constant in magnitude and at rest with respect to the field mmf. But its line of action varies in position relative to that of the field flux axis, and is dependent on the phase angle of the armature current with respect to the terminal voltage.

Let us consider a pattern of voltage distribution in the armature conductors. This is indicated in Figure 8.13(a). Those conductors occupying positions around the centre lines of the poles (that is, the direct axis, where they are cutting across the maximum field flux-density when rotated) have peak voltage induced in them. This is denoted by different sizes of dots and crosses along the conductor belt, as indicated along the stator-rotor interface of Figure 8.13(a). If the power factor of the connected load is changed from zero lagging through unity to zero leading, the pattern of the currents in armature conductors remains constant in magnitude for a given value of the load and the armature mmf at rest with respect to the pole structure and changes in position relative to the polar axis.

When the power factor of the load is zero lagging, or leading, the power developed by the machine is zero, ignoring the machine losses. Hence, the corresponding torque angle, that is, the angle between the two component magnetic field distributions, is zero. Thus, when the power factor is zero, the armature reaction mmf is co-phasal or in phase-opposition to the field mmf. This is indicated in Figure 8.13(a), on which the graded values of the resulting armature currents that the winding carries at zero power load is indicated on the conductor belt depicted along the outer surface. Clearly, the currents lag behind the voltages by $\left(\dfrac{\pi}{2}\right)$ electrical, that is, a quarter of a cycle away in time frame, for the zero power factor lagging condition of operation.

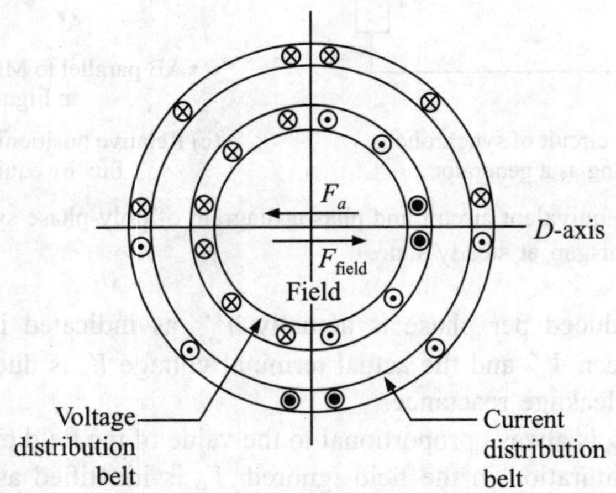

FIGURE 8.13(a) Voltage and current distribution in armature for zero power factor lagging operation.

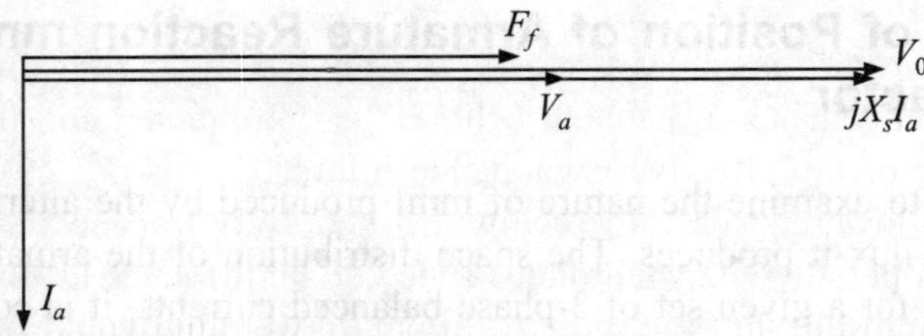

FIGURE 8.13(b) Phasor diagram of machine delivering load at zero power-factor lagging at constant load current.

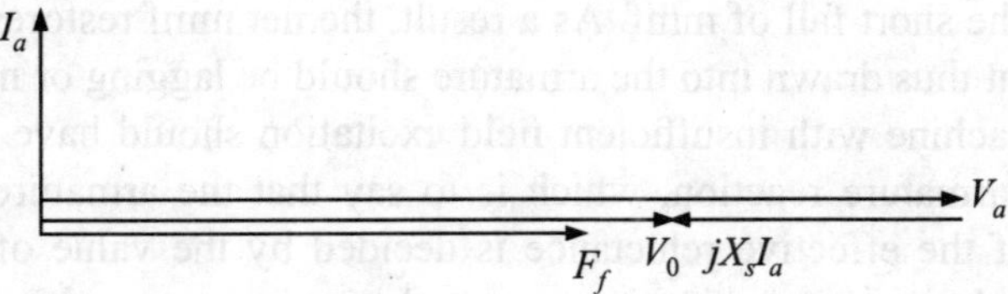

FIGURE 8.13(c) Phasor diagram of machine delivering load at zero power factor leading at constant load current.

The corresponding positions of the armature reaction mmf vis-a-vis the field mmf is indicated in Figure 8.13(b) for the zero power factor lagging and in Figure 8.13(c) for zero leading power factor respectively.

Thus the armature reaction mmf alters its relative position with respect to the field mmf from de-magnetizing, through cross-magnetizing to magnetizing, when the power factor of the load is changed from zero-lagging, through unity to zero-leading.

8.5.2 Poly-phase Synchronous Machine with Uniform Air-Gap Connected to Infinite Bus-Bar—The Development of Operational Chart

When the armature terminals of a poly-phase synchronous machine are connected to a large power system, the terminal voltage and the frequency of the machine are both constant. An independent operation of the machine is superseded and dictated upon by the large power system. The machine is, then, said to be connected to an infinite bus-bar.

In analysing the machine behaviour, let us first consider that the connected machine is floating on a power system, that is, both the current and the power flow in the machine are negligibly small. At this condition, the excitation voltage, V_0, and the terminal voltage V_a are equal. While the machine continues to float (that is, takes no part in power sharing), the excitation current in the field circuit is increased, the shaft torque continuing to be zero.

Since, however, the terminal voltage cannot be altered, a net current, flows in the armature winding in such a way that the armature reaction mmf will nullify the increase in the value of the field mmf. For this to happen, the component mmfs—the field mmf and the armature reaction mmf—should share the common direct axis of the machine. This statement is duly supported by the fact that the net torque on the shaft of the machine remains zero, and, thus, the torque angle is not developed.

Increase in field current forces the armature to draw leading current from the terminals (or a de-magnetizing current). Thus a machine behaves as a *capacitor* and the magnitude of the field decides the value of the effective capacitance offered.

If the excitation is lowered to a value below that necessary to generate the terminal voltage on open circuit, when operating at synchronous speed, to be the same as the terminal voltage of the system, the maximum flux-linkage with the armature winding would then be insufficient to develop a voltage equating the terminal supply voltage. The supply would provide a current in the armature to make up the short fall of mmf. As a result, the net mmf restores itself to the original mmf value. The current thus drawn into the armature should be lagging or magnetizing in nature.

A synchronous machine with insufficient field excitation should have its armature winding produce magnetizing armature reaction, which is to say that the armature is behaving like an inductor. The value of the effective reluctance is decided by the value of the field current.

Thus, when a synchronous machine is connected to a constant voltage system, the current taken is always sufficient to produce a terminal voltage equalling the system voltage.

We are, now, ready to consider a situation in which it is, once again, floating on the large power system, with the field current re-adjusted for a zero power input. We consider that a braking torque is applied to the machine shaft. The machine will immediately react by falling back in angular position relative to the synchronously rotating reference. The ensuing changes in the flux-linkage produced by the field winding with the armature winding are retarded in time phase. The supply system will produce an armature current to add an armature reaction mmf with appropriate phase relationship with the field mmf. The resultant will restore the net flux-linkage with the armature winding to the original value. The flux distribution produced by the armature current should be at an appropriate torque angle with the field mmf. This armature current is at a phase angle with the supply voltage, so as to provide an input electric power equalling the mechanical power output.

If the shaft torque is, now, reversed becoming positive, the whole situation reverses. As a result the rotor advances ahead of its no-load position. Mechanical power is, now, fed into the machine shaft and electrical power is fed to the supply system. For a given value of the shaft torque, the field current totally controls the value of the reactive power—magnetizing or de-magnetizing—from the supply system. This has no effect on the active power, which is totally determined by the mechanical torque. This happens only at the constant synchronous speed. If the applied torque is changed, without changing the field current, there ensues changes in both active and reactive power, because a change in torque angle alters the angle between the field and the armature reaction mmf, and thus their resultant value. This urges a change in the reactive component of armature current to restore the maximum armature flux-linkage to the previous value.

The complete operation of a synchronous machine on a system of constant voltage and frequency system or an infinite bus system, can be depicted by its Operation Chart, shown in Figure 8.14(a). In this figure, the resistance of the armature has been neglected and a phasor diagram corresponding to a smooth air-gap synchronous machine is drawn in chain-dot representation. Line AB is drawn proportional to the constant terminal voltage, V_a, which is chosen along the horizontal axis of the diagram. I_a denotes the armature current per phase and is shown at a lagging power factor angle of ψ, behind the constant terminal voltage V_0. BD shows the synchronous reactance drop $I_a(X_m + X_l)$ to the chosen scale of voltage, and the internal voltage V_0 is then given by AD.

A vertical line drawn perpendicular to AB at point B makes an angle ψ, so that BD makes an angle ψ with this vertical line, as indicated. As the terminal voltage V_a, is constant, the active power flow from the generator is given by the product ($V_0 I_a$ cos ψ) per phase. Hence, the vertical line GBJ at point B, when calibrated to a suitable power scale, also indicates the active power. By a similar reasoning, the horizontal line ABH can be calibrated to read reactive power in *kvt* units proportionately to the current, I_a by the same chosen power scale.

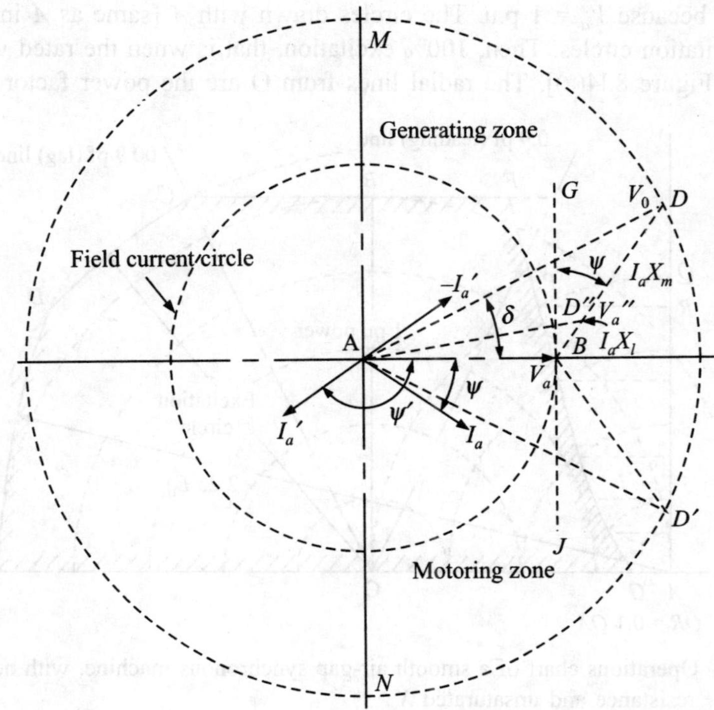

FIGURE 8.14(a) Operations chart of a uniform air-gap synchronous machine on infinite bus-bar.

The floating condition is indicated by point B, and it shows that a field current of I_{fp} would generate on open circuit at synchronous speed, a terminal voltage equalling the system voltage. Operation of the machine as synchronous reactor or synchronous capacitor, shall be represented by points on the horizontal axis to the left or to the right of point B, as marked on the diagram of operations chart. The values of the field current required for a particular lagging or leading reactive power values may be obtained by interpolation between the field current circles. If a positive torque is applied to the rotor and the field current is maintained constant, the operating point moves along a particular field excitation circle and the generated power can be read off the power scale calibrated along the line GBJ. Also any increase or reduction of field current at a constant torque will correspond to a horizontal traverse either to the right or left of the diagram.

The Operations Chart, indicating the zone of stable operation of a poly-phase synchronous machine, when working as a generator, is shown in Figure 8.14(b). For the machine chosen, it is assumed that X_s (unsaturated value) is 1.25 per unit and the infinite bus-bar voltage, V_a is 1 per unit. For zero field excitation, V_0 = zero, so that $I_a = \left(\frac{V_a}{X_s}\right) = 80\%$ or 0.8 pu and the point O [same as B in Figure 8.14(a)] extended to the right depicts VAR lagging. The vertical line drawn at O indicates active power in watts on the same chosen power scale. Circles drawn with O as centre, result in constant current circles with same scale as the volt-ampere reactive, because $V_a = 1$ p.u. The circles drawn with A [same as A in Figure 8.14(a)] as centre are excitation circles. Then, 100% excitation, that is when the rated voltage V_a is AO [same as AB in Figure 8.14(a)]. The radial lines from O are the power factor lines.

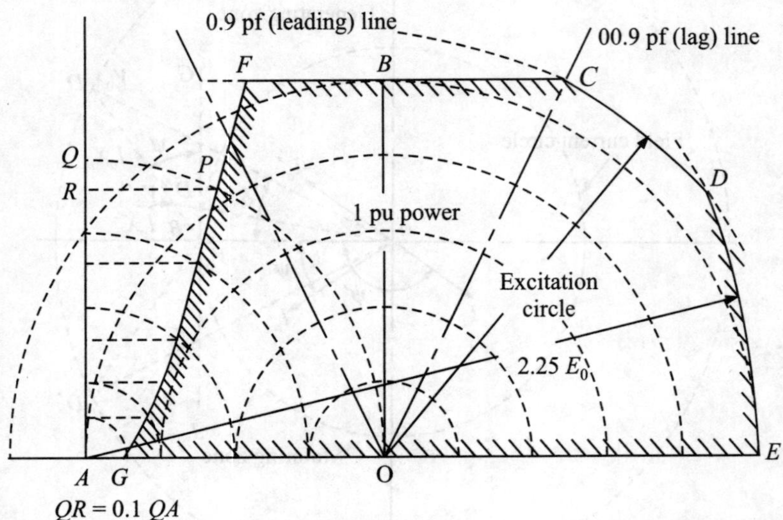

FIGURE 8.14(b) Operations chart of a smooth air-gap synchronous machine, with negligible armature resistance and unsaturated X_s.

The Operations Chart is drawn, taking 100% as the power limit of the prime mover. With this limit imposed from the prime mover, the vertical limit of the operating zone FBC is marked. Assuming 0.9 as the rated power factor, point C lies on the 0.9 pf (lagging) line to limit the stator or the armature current to its rated value. An arc CD is drawn with OC as radius. This is the limit line to maintain the armature current with limiting values. At D, the excitation current limit is attained. This happens at an assumed imposed limit of excitation voltage, V_0, of 2.25 pu. Arc DF is drawn to complete the portion of the operational limit FBCDG.

Line CB cannot be extended up to the theoretical stability limit, since at this location, any small perturbation leads to instability. It is usual to assume a safety limit of 10 per cent of full-load less than the theoretical maximum for a given excitation. Hence, for the 100% excitation, QR = 10% and PR drawn horizontally intersects 100% excitation circle at point P. Following this procedure, the complete stable working range HFBCDG for the machine is obtained when functioning as a generator. Any point within the marked area HFBCDG immediately indicates to the chosen scale, the volt–ampere, watts, volt–ampere reactive, value of current, power factor, d.c. excitation.

Let us refer once again to Figure 8.14(a). There is a maximum torque and a corresponding active power. If this value is crossed, the machine loses synchronism and runs above the synchronous speed. The maximum power is proportional to the field current corresponding to AM for the field current circle I_{FP}.

If a braking torque is applied to the shaft, the movement of the operating point is down a circle of constant field current into the lower half of the figure below the extended horizontal line AB. The maximum torque for a given field current, that is, the pull-out torque at which the machine loses synchronism and runs at sub-synchronous speed is also shown in the figure. The vertical length AN indicates the maximum motoring power with a field excitation of I_{fp}. Dividing this power with synchronous speed in radians per second leads to determination of the pull-out torque.

In Figure 8.14(a), the generated electrical power is plotted vertically upwards and is represented by positive sign. The received electric power, that equals, the mechanical power output, is therefore denoted by negative sign.

We could redraw the chain-dot line phasor diagram by placing the current phasor in the negative direction to denote the true position of the phasor, but if we do so, the torque angle becomes $(90° + \psi + \delta)$, which is less than 180° and is leading the terminal voltage phasor as indicated. In this case, all other phasors excepting phasor V_a need to be rotated clockwise. A motor which receives a leading current is, in effect, supplying a lagging or magnetizing current and the sign of the reactive power depends on the excitation current of the field only. This decides whether the machine functions as a motor or a generator.

8.5.3 Physical Interpretation of Applying Transformation [C_2] on Synchronous Machines

Let us consider the synchronous machine running at synchronous speed as a generator and supplying balanced 3-phase currents to a load out of its terminals.

The current per phase in phase-*a* may be expressed as follows:

$$i_a = \sqrt{2} I \cos \omega t$$

The effect of applying $[C_1]$ for transformation of the phases is to convert the 3-phase rotor currents into 2-phase currents, but the rotor winding β leads that of α by $\left(\frac{\pi}{2}\right)$. As discussed in Section 7.4.1, according to our convention, the balanced 2-phase currents can now be written down as follows:

$$I_\alpha = I' \cos \omega t \quad \text{and} \quad I_\beta = -I' \sin \omega t$$

where, $I' = \sqrt{3} I$. The direct- and quadrature-rotor currents is, therefore,

$$\begin{aligned} i_d &= i_\alpha \cos \theta - i_\beta \sin \theta \\ &= I' (\cos \omega t \cos \theta + \sin \omega t \sin \theta) \\ &= I' \cos (\theta - \omega t) \end{aligned}$$

and

$$\begin{aligned} i_q &= i_\alpha \sin \theta + i_\beta \cos \theta \\ &= I' (\cos \omega t \sin \theta - \sin \omega t \cos \theta) \\ &= I' \sin (\theta - \omega t) \end{aligned}$$

where, θ is the angle between the α-phase of the 2-phase $0\alpha\beta$ system and D-axis at $t = 0$.

At time $t = 0$, the current in *a*-phase attains positive maximum. If at this time instant, the angle of the rotor *a*-phase with *d*-axis stator is given by $\theta = \omega t + \delta$, then,

$$i_d = I' \cos \delta \quad \text{and} \quad i_q = I' \sin \delta$$

Thus, the rotor currents are no longer dependent on time; that is to say that they are 'direct currents'. As the phase transformation $[C_1]$ and the commutator transformation $[C_2]$ are both orthogonal, for balanced operation, the voltage also transforms into direct voltages.

The result of transforming by $[C_2]$ under balanced condition is, therefore, equivalent to that of a commutator machine operating on direct current. Balanced 3-phase currents produce rotating magnetic field in the rotor with constant magnitude, rotating at synchronous speed with respect to the rotor winding, the direction of rotation being decided by the phase sequence of rotation. If the winding itself is rotated at synchronous speed in the opposite direction, this rotating magnetic field is stationary in space. It can, therefore, be produced by direct current in a rotating commutator winding.

We have, so far, analysed the behaviour of synchronous machines with respect to its terminals, normally used to supply the machine with electrical power (in case of motoring) or from which the electric power is drawn (when operating as a generator), that is, with reference to the reference frames stationary with the machine. The analysis is significant if we employ the 3-phase to 2-phase transformation originally applied by Park.

Figure 8.15 shows the 2-axis phasor diagram of a salient-pole synchronous machine in which the armature resistance has been ignored. Current I_a, producing the armature mmf pattern centred at some point between the direct and quadrature axes, is shown split into two components I_d and I_q. This produces mmf patterns centred on the direct- and the quadrature-axes respectively. This splitting is shown in Figure 8.15(b). Also shown in the diagram are the mmf distribution depicted by phasor F and the component phasors F_d and F_q, so that:

$$\overline{F} = \overline{F}_d + \overline{F}_q$$

Each of these mmfs produces a component of fundamental flux distribution and consequently component voltages in the armature windings equal to the product of the component currents, I_d and I_q, and the corresponding magnetizing reactances. These reactances have been identified as the direct-axis magnetizing reactance, X_{ad} and the quadrature axis magnetizing reactance,

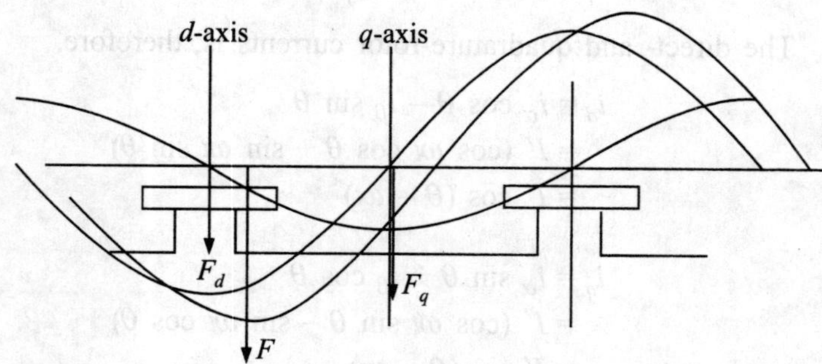

FIGURE 8.15(a) MMF, F, resolved into components F_d and F_q.

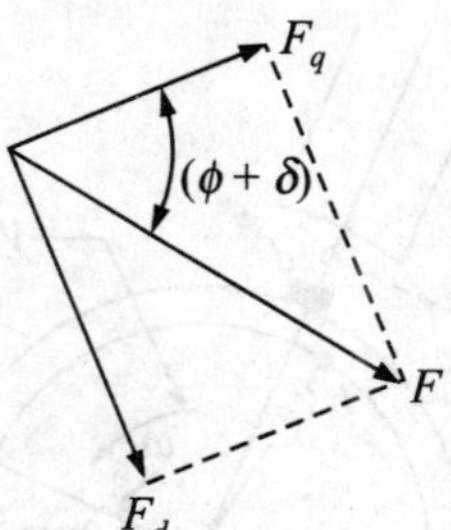

FIGURE 8.15(b) Phasor relation among F, F_d and F_q.

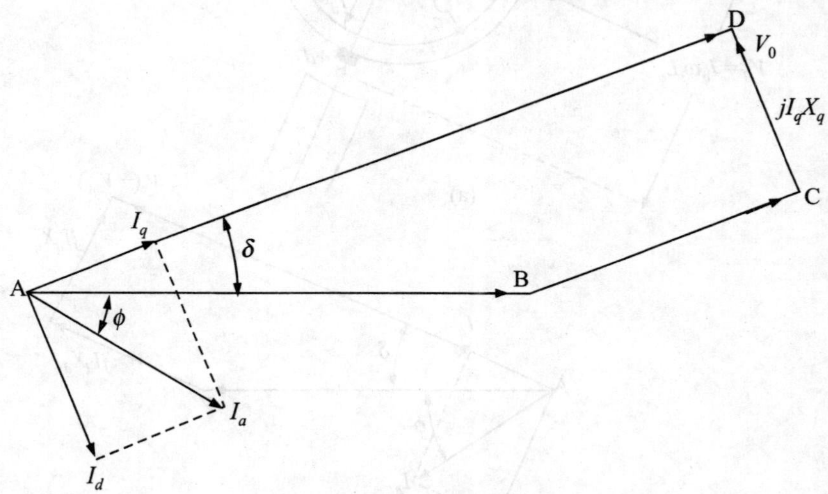

FIGURE 8.15(c) Two-axis phasor diagram of synchronous machine when operating at a lagging power factor as a generator, with ignored armature resistance.

X_{aq}. The direct- and the quadrature-axis synchronous reactances are obtained by adding leakage reactance component X_l, to each, that is:

$$X_d = X_{ad} + X_l$$
$$X_q = X_{ad} + X_l$$

Figure 8.16 is a reproduction of Figure 8.15, ignoring the effect of armature resistance. This is a 2-phase phasor diagram of the poly-phase synchronous machine. On top of the diagram is super-imposed (on the rotor member) a 3-phase winding rotating at synchronous speed and provided with a notional commutator with two sets of orthogonal brush-pairs located along planes in the field axis and at right angles to it, as shown. Alternatively, the d.c. field is considered to be moved to the stator member. The brush pairs, therefore, occupy the direct and the quadrature axes.

The phasor diagram is drawn for the standard direction of the 3-phase winding. This is the same thing as saying that the armature phase sequence is *a-b-c* in the counter-clockwise direction and the armature conductors are moving in the clockwise direction or the field system is effectively rotating in the counter-clockwise direction.

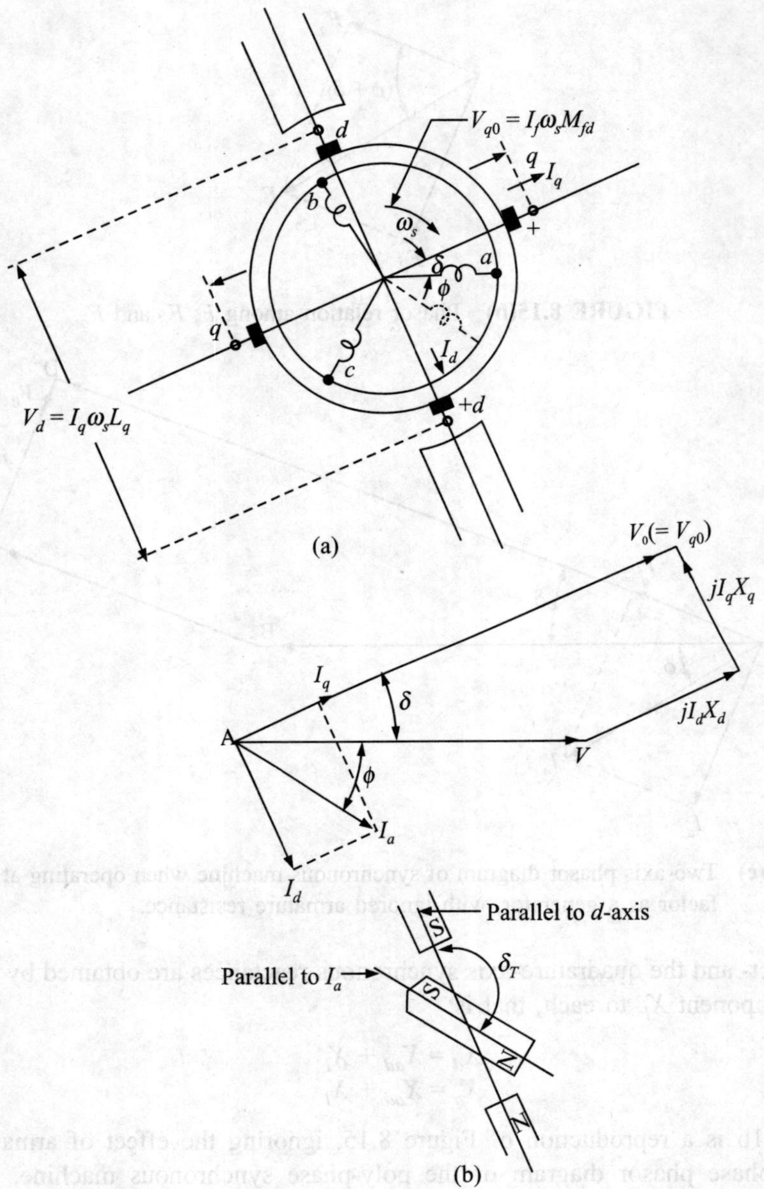

FIGURE 8.16 Synchronous machine in *a-b-c* and *d-q* reference frames.

We may, now, measure the stationary current and voltage distribution phasors across the *d*- and *q*-pair of brushes. We observe that these phasors and the phasor diagrams are vector diagrams representing the stationary voltage and current distribution. In this representation $\bar{I}_d$ and $\bar{I}_q$ are, to an appropriate scale I_d and I_q—steady direct, time-invariant currents flowing in the *d*- and *q*-brush pairs. Similarly (I_dX_d) and (I_qX_q) now denote the values of the steady rotational voltages produced by I_d and I_q on the other axis, in each case. We may rewrite them and also V_{q0}, which we can see represents the steady rotational voltage V_{q0} produced by the field current I_f.

The 2-axis diagram of Figure 8.16 is marked with the actual direction of voltages and currents appropriate to the generating conditions at a lagging power factor. In the diagram, phase a is shown in the position of the maximum value of terminal voltage (in firm lines) and the maximum terminal current (in dotted line depiction). Figure 8.16(b) shows, with the help of magnets the positions of the two component field patterns.

We can readily express the familiar 2-axis quantities in terms of the 3-phase quantities, through the 2-phase quantities.

Here,
$$\omega_r = -\omega_s$$

Also, the total self- and mutual-inductances of the pseudo-stationary windings are equal, so that:

$$L_d = M_d \quad \text{and} \quad L_q = M_q$$

Since we are primarily concerned with fundamental components only, we use suffix α to indicate 2-phase quantities and suffix a for 3-phase quantities. Using rms magnitudes for the poly-phase voltages and currents, we have:

$$V_d = I_q\,\omega_s\,M_q$$
$$= \sqrt{2}I_{q\alpha}\omega_s L_{q\alpha} = \sqrt{2}\,\frac{3}{2}\,I_{qa}x_{qa}$$

$$V_q = I_{q0} - I_d\,\omega_s\,M_d$$
$$= \sqrt{2}\,V_{q0\alpha} - \sqrt{2}I_{d\alpha}\omega_s L_{d\alpha}$$
$$= \sqrt{2}\left(\frac{3}{2}\right)V_{q0a} - \sqrt{2}\left(\frac{3}{2}\right)I_{da}X_{da}$$
$$I_d = \sqrt{2}\,I_{d\alpha} = \sqrt{2}I_{da}$$
$$I_q = \sqrt{2}\,I_{q\alpha} = \sqrt{2}I_{qa}$$

The total power as given for the 2-axis and the 3-phase systems is invariant, as seen by substitution:

$$P = V_d I_d + V_q I_q = 3V_{da}I_{da} + V_{qa}I_{qa}$$
$$= 3V_a I_a \cos\,\psi$$

The analysis shows that the phasor diagram representing the 3-phase rms phase quantities becomes a vector diagram denoting the steady 2-axis quantities if voltages are multiplied by $\sqrt{2}\left(\frac{3}{2}\right)$ and currents by $\sqrt{2}$.

8.5.4 Steady-state Operation of Salient-pole Synchronous Machines

In this section, the balanced steady-state operation of a 3-phase salient-pole synchronous machine running as both alternator and synchronous motor is considered.

Let V and I represent the rms values of the per-phase voltage and current in a 3-phase machine. The instantaneous values of these voltages and currents are given by:

$$V_a = \sqrt{2}V \sin \omega t \qquad i_a = \sqrt{2}I \sin(\omega t - \phi)$$

$$V_b = \sqrt{2}V \sin\left(\omega t - \frac{2\pi}{3}\right) \qquad i_b = \sqrt{2}I \sin\left(\omega t - \frac{2\pi}{3} - \phi\right)$$

$$V_c = \sqrt{2}V \sin\left(\omega t - \frac{4\pi}{3}\right) \qquad i_c = \sqrt{2}I \sin\left(\omega t - \frac{4\pi}{3} - \phi\right) \quad ...(8.98)$$

ϕ being the phase angle of lag. Also, the sine and *note the co-sine functions* so far chosen in the analysis have been adopted to enable the results of the general theory to be correlated with conventional theory.

The phase transformation $[C_1]$ replaced the balanced 3-phase system by a balanced 2-phase system which is given in Eq. (6.37) as:

$$\left.\begin{aligned} V_\alpha &= \sqrt{3}V \sin \omega t \qquad & i_\alpha &= \sqrt{3}I \sin(\omega t - \phi) \\ V_\beta &= -\sqrt{3}V \cos \omega t \qquad & i_\beta &= -\sqrt{3}I \cos(\omega t - \phi) \end{aligned}\right\} \quad ...(8.99)$$

In this case, the voltage and current in phase α lead those in phase β by $\left(\frac{\pi}{2}\right)$. In deriving the impedance matrix of 2-phase synchronous machine, we have observed in Section 7.4.1 that α and β are related to each other such that α lags β by $\left(\frac{\pi}{2}\right)$, and we should use β terms as $(-\beta)$ to conform to the above condition, so that the new values effectively become:

$$\left.\begin{aligned} V_\alpha &= \sqrt{3}V \sin \omega t \qquad & i_\alpha &= \sqrt{3}I \sin(\omega t - \phi) \\ V_\beta &= \sqrt{3}V \cos \omega t \qquad & i_\beta &= \sqrt{3}I \cos(\omega t - \phi) \end{aligned}\right\} \quad ...(8.100)$$

For the commutator transformation $[C_2]$ shown in Section 8.5.3, these two phase quantities are:

$$\left.\begin{aligned} V_q &= \sqrt{3}V \cos(\theta - \omega t) \qquad & i_q &= \sqrt{3}I \cos(\theta - \omega t + \phi) \\ V_d &= -\sqrt{3}V \sin(\theta - \omega t) \qquad & i_d &= -\sqrt{3}I \sin(\theta - \omega t + \phi) \end{aligned}\right\} \quad ...(8.101)$$

As the machine runs at all times under steady-state at synchronous speed, the angular frequency of rotation, ω_r in electrical radians per second equals the angular frequency of the supply, $\omega_s = 2\pi f$. And at time, $t = 0$, the terminal voltage of the a-phase

$$V_a = \sqrt{2}V \sin \omega t = 0$$

If at this instant the angle between the axis of the *a*-phase and the field axis is δ, then:

$$\theta = \omega t + \delta$$

so that:

$$\left.\begin{aligned} V_q &= \sqrt{3}V\cos\delta \qquad & I_q &= \sqrt{3}\,I\cos(\delta+\phi) \\ V_d &= -\sqrt{3}V\sin\delta \qquad & i_d &= -\sqrt{3}\,I\sin(\delta+\phi) \end{aligned}\right\} \qquad \text{...(8.102)}$$

As shown in Section 8.5.3, the equivalent direct- and quadrature-axis voltages and currents are independent of time, that is to say that they are direct voltages and currents. We can, therefore, introduce appropriate simplifications of general voltage equations. Thus, when all currents after transformation by $[C_2]$ have been executed, that is, I_F, I_d and I_q are all direct currents. Heaviside's differential operator, $p\left(=\dfrac{d}{dt}\right)$ acting upon these currents on the impedance matrix automatically disappears. Also, V_Q, the external voltage notionally applied to the *Q*-axis damper winding is zero, since no voltage can be induced in this winding and hence I_Q becomes zero. The damper windings are, therefore, ineffective in balanced steady-state operation and may be safely ignored.

By replacing V_q, V_d, I_q and I_d by the values derived in Eq. (8.102), writing $p = 0$, $R_2 = R$, so that the entities are expressed without subscripts—*V*, *I* and *R*—respresent the new armature parameters and quantities and omitting the row and the column pertaining to *Q* from the impedance matrix indicated in Eq. (8.59), along with suitable replacement of inductances L_d and L_q by equivalent synchronous impedances,

$$X_d = \omega_s L_d \quad \text{and} \quad X_q = \omega_s L_q$$

We obtain the simplified form of the resulting voltage equations:

$$\begin{bmatrix} V_F \\ \sqrt{3}V\cos\delta \\ \sqrt{3}V\sin\delta \end{bmatrix} = \begin{bmatrix} R_F & 0 & 0 \\ -\omega_s M_F & R & -X_d \\ 0 & X_q & R \end{bmatrix} \begin{bmatrix} I_F \\ \sqrt{3}\,I\cos(\delta+\phi) \\ \sqrt{3}\,I\sin(\delta+\phi) \end{bmatrix} \qquad \text{...(8.103)}$$

The first of the clustered equations:

$$V_F = R_F I_F$$

merely indicates that the field equation is a representation of Ohm's law under balanced steady-state condition. Using the remaining two equations and dividing both sides by $\sqrt{3}$, we get:

$$V\cos\delta = -\frac{\omega_s M_F}{\sqrt{3}} I_F + RI\cos(\delta+\phi) + X_d I\sin(\delta+\phi)$$

and

$$-V\sin\delta = X_q I\cos(\delta+\phi) - RI\sin(\delta+\phi) \qquad \text{...(8.104)}$$

A further simplification is in order. When the synchronous machine functions as an alternator on open circuit, the value of the load current $I = 0$. The second equation of Eq. (8.104) suggests that under this condition, δ must also be zero. Hence, δ is nomenclatured as the *load angle*. If, under no-load conditions of the machine functioning as a generator, V_0 is the open-circuit terminal voltage, from the first of Eq. (8.104),

$$V_0 = -\frac{\omega_s M_F}{\sqrt{3}} I_F$$

In our original sign convention followed in the book, we have consistently assumed the input voltage across the terminals of windings to be positive for motor action such that the applied voltage at the instantaneous positive terminals causes positive input currents into the windings. In other words, instead of saying that the generator produces negative voltage, we could also indicate that the generator current, which comes out of the instantaneous positive terminal is negative for a generator. Making this change in Eq. (8.104), we may therefore, write:

$$\left.\begin{aligned} V\cos\delta &= V_0 - RI\cos(\delta+\phi) - X_d I\sin(\delta+\phi) \\ V\sin\delta &= X_q I\cos(\delta+\phi) - RI\sin(\delta+\phi) \end{aligned}\right\} \quad \text{...(8.105)}$$

These are the operating equations of a salient-pole synchronous generator at balanced steady-state conditions.

We can construct the phasor diagram of the machine functioning as a generator. This is shown in Figure 8.17 for a lagging power factor. This can be drawn to a chosen scale of voltage and current, as follows:

Beginning at an arbitrary point O, let us draw the terminal voltage phasor V to the chosen scale of voltage along the positive direction horizontally. This is shown as OA. Also, let us draw the lagging current to the chosen current scale at a lagging angle ϕ from the terminal voltage phasor OA. This is denoted by OK in the diagram. Let us draw AH, parallel to OK to represent the resistance drop per phase as per the voltage scale. From H, we draw a perpendicular, HB, leading the current by $\left(\frac{\pi}{2}\right)$ and workout its length to the voltage scale equal to IX_q per phase. We join OB and extend its length to a point D. We also extend HB to a new point C and mark out HC = IX_d. By dropping a perpendicular CD on the line OD, we fix up the length OD. Hence, the length of OD is obtained by marking CD to equal IX_d and dropping a perpendicular CD on to line OB extended beyond B. We shall prove deductively that OD gives the magnitude and direction of the excitation voltage or the no-load voltage of the alternator at the specified speed and field excitation of the machine. We take the help of a geometrical proof.

We draw perpendiculars on line OB from points A and H, and let these normals meet line OB (which in extended form is also line OD) at points F and G respectively as shown. We also draw a line parallel to OB at A, so that line AJ meets line GH at J. We observe that angle BOK equals $(\phi + \delta)$. Also angles JAH = angle GHB = angle BCD and geometrically have values of $(\phi + \delta)$.

Since AJ = FG = IR cos $(\phi + \delta)$ to the chosen scale of voltages, and additionally OA is the terminal voltage V, length AF = $V \cos \delta$. We also observe that CH, to the chosen scale of voltage denotes IX_d. Hence, length GD equals $IX_d \sin(\phi + \delta)$.

Thus, geometrically referring to the phasor diagram drawn in Figure 8.17,

$$OD = OF + FG + GD$$

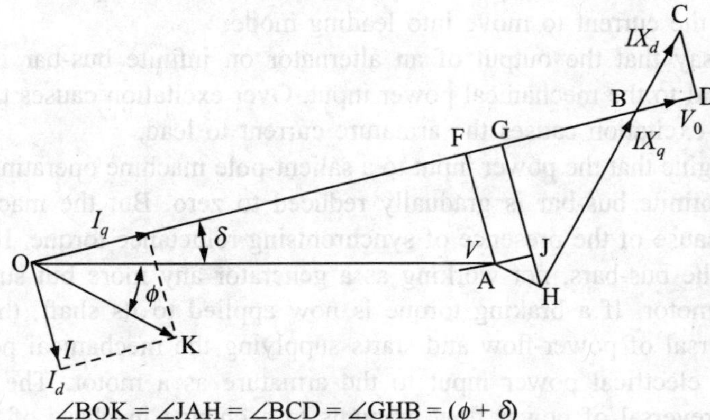

FIGURE 8.17 Phasor diagram of a salient-pole synchronous generator at lagging load.

can be written, in terms of the computed values of these components as follows:

$$OD = V \cos \delta + IR \cos(\phi + \delta) + IX_d \sin(\phi + \delta)$$

Comparing the above expression with the first of Eqs. (8.105), we immediately see that the length OD is the same as the excitation voltage V_0, and for the operating condition specified, positionally, it is ahead of the terminal voltage V by an angle δ or the load angle. Hence, the phasor diagram in Figure 8.17 has been proved deductively as correct and comprehensive.

We also observe from the diagram of Figure 8.17 that:

$$AF = HG - HJ \quad \text{and} \quad HG = IX_q \cos(\phi + \delta)$$

whereas HJ = $RI \sin(\phi + \delta)$.

Therefore, both equations in Eq. (8.105), specifying the functioning of the salient-pole synchronous machine working in the generator mode have been embodied in the phasor diagram of Figure. 8.17.

8.5.5 Salient-pole Synchronous Machine Functioning as a Synchronous Motor

It is pointed out is Section 8.5.2 that when a synchronous generator is connected to an infinite bus bar, it operates with many other synchronous machines in parallel, so that its effect on the terminal voltage and frequency becomes negligibly small. The power delivered by the machine to the mains is determined only by the input from the prime movers. Changing the excitation without changing the mechanical power input only brings about a change in the machine power factor without changing its power output. As the terminal voltage V is constant, the power output given by $3VI \cos \phi$, leads to changes in ($I \cos \phi$), while a change in excitation leads to simultaneous changes of I and ϕ, such that ($I \cos \phi$) is constant. If the excitation is increased so that the excitation voltage, V_0 of the machine is augmented, the armature reaction develops a demagnetizing component, and thus current lags. A reduction of excitation current in the machine must, therefore, be accompanied by the armature current producing a magnetizing component leading the current to move into leading mode.

Thus, we can say that the output of an alternator on infinite bus-bar is, neglecting the machine losses, equal to the mechanical power input. Over excitation causes the output current to lag, while under-excitation causes the armature current to lead.

Let us now imagine that the power input to a salient-pole machine operating as an alternator and connected to infinite bus-bar is gradually reduced to zero. But the machine speed must remain constant because of the presence of synchronising reluctance torque. It follows that the machine floats on the bus-bars, not working as a generator any more but supplying only the power losses as a motor. If a braking torque is now applied to its shaft, the machine starts experiencing a reversal of power-flow and starts supplying the mechanical power demand by drawing equivalent electrical power input to the armature as a motor. The voltage remains unaffected but the reversal of power flow amounts to a change in phase of the current. This is shown by the phasor diagram in Figure 8.18(a). This has been depicted for a synchronous motor operating at unity power factor, the diagram being drawn from the generator point of view. It shows that with the voltage phase drawn as reference to the right, the current in the portion right of *aa* plane corresponds to the generator operation, while to the left for motoring operation. [However, in considering synchronous motor, we should assume as positive currents for cases when the electric power flow into the machine, as per convention adopted and followed so far as positive power flow when power is fed into the electric circuit.] Hence, we should reverse the direction of current flow from left to right. This change is shown in Figure 8.18(b).

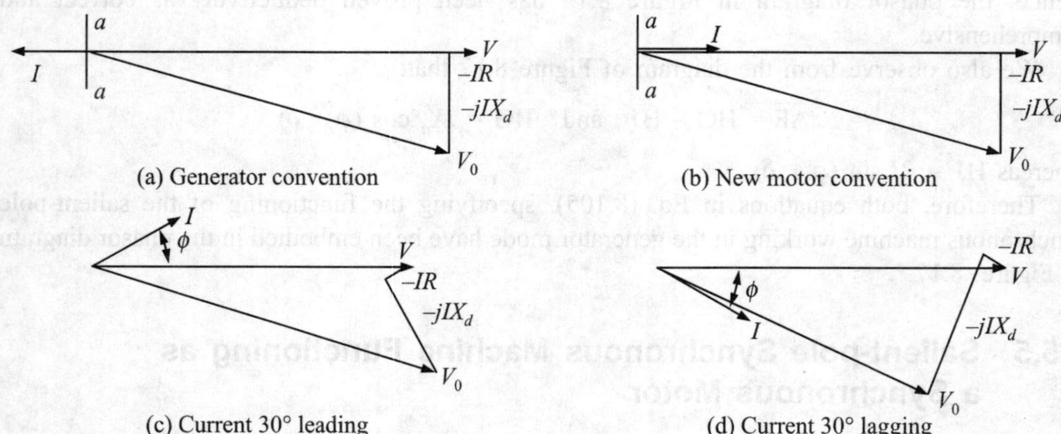

FIGURE 8.18 Phase diagram of salient-pole synchronous motor.

It merely indicates a change of convention as to the positive direction of current. ***We have to especially note that only the direction of current I and not IR and IX_d drops have changed in the relative directions shown in Figure 8.18(a) and 8.18(b). It must be remembered that when the motor diagram is used in Figure 8.18, then both IR and IX_d, drops are opposite to directions to what it would be for generator operation.***

Conceptualization of reversal of the terminal voltage, as adopted in many books, is neither easy to conveive physically nor is strictly correct. So, this convention of reversing the current direction is adopted in this book with clear emphasis that the direction of *IR* and IX_d drops are retained as per generator direction even while direction of *I* is reversed.

Thus, when working as a synchronous motor, the motor armature reaction presents a lagging power factor for under-excited operation while as a generator its armature reaction causes leading power factor operation, while under over-excited condition as a motor, the power factor is leading, although as a generator it is lagging.

These are illustrated by a set of *V*-curves at constant power delivery for synchronous machines when functioning as motors as well as generators under different excitations.

8.6 GENERALIZED EQUATIONS FOR 3-PHASE SALIENT-POLE SYNCHRONOUS MACHINES

A 3-phase alternator or synchronous motor has atleast four windings—three on stator carrying alternating current and one on the rotor having a d.c. excitation. In small machines, the stator sometimes carries the poles, and the d.c. windings are placed on the rotor member and are connected to slip-rings.

We undertake to drive equations of performance of these four windings. When a coil has a voltage *V* applied across its terminals and carries a current *i* from the positive terminal into the coil, the governing equation is:

$$V = ri + \frac{d}{dt}(\psi) = ri + p\psi$$

where

r represents the resistance of the coil

ψ denotes the flux-linkage in the coil

p stands for the Heavyside's operator $\left(\frac{d}{dt}\right)$

The above situation is a depiction for a synchronous motor.

A part of the flux-linkage is in leakage paths as far as the other coils are concerned. Hence, ψ includes the leakage flux. There are four simultaneous equations involving the phase and field currents, voltages, flux-linkages and winding resistances. These are shown in matrix form.

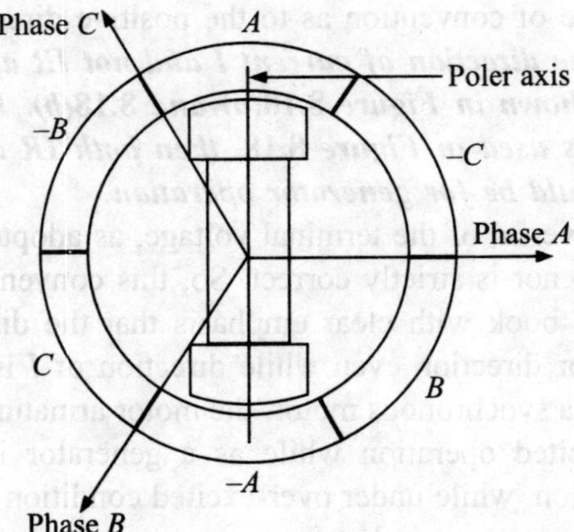

FIGURE 8.19 Phase-distribution in a synchronous machine and axes transformation for obtaining generalised equations.

Voltage equations

$$\begin{bmatrix} V_f \\ V_a \\ V_b \\ V_c \end{bmatrix} = \begin{bmatrix} R_f i_f + p\psi_f \\ R_a i_a + p\psi_a \\ R_b i_b + p\psi_b \\ R_c i_c + p\psi_c \end{bmatrix} \qquad \text{...(8.106)}$$

Flux-linkage equation

$$\begin{bmatrix} \psi_f \\ \psi_a \\ \psi_b \\ \psi_c \end{bmatrix} = \begin{matrix} \\ f \\ a \\ b \\ c \end{matrix} \begin{matrix} f & a & b & c \\ \end{matrix} \begin{bmatrix} l_{ff} & M_{af} & M_{bf} & M_{cf} \\ M_{af} & l_{aa} & M_{ab} & M_{ac} \\ M_{bf} & M_{ab} & l_{bb} & M_{bc} \\ M_{cf} & M_{ac} & M_{bc} & l_{cc} \end{bmatrix} \begin{bmatrix} i_f \\ i_a \\ i_b \\ i_c \end{bmatrix} \qquad \text{...(8.107)}$$

In general,

$$[\psi_{\text{phase}}] = [L_{ph}][i_{\text{phase}}] \qquad \text{...(8.108)}$$

8.6.1 Inductances

All inductances, except l_{FF} are functions of rotor position θ, and their values are computed as given in Section 8.2.1 and also below:

Self inductance of armature phases

Consider a current i amp in phase A, generating an mmf of fundamental frequency having amplitude F_1 along the axis of phase A. The permeance of air-gap at an angle θ from the polar axis, as given from Figure 8.20 is:

$$\lambda(\theta) = \lambda_0 + \lambda_2 \cos 2\theta$$

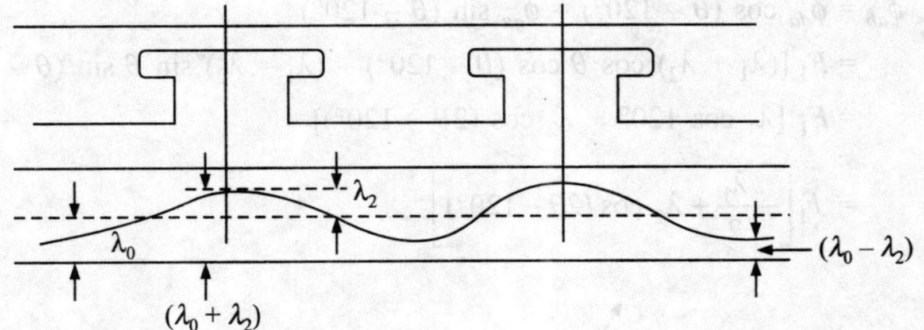

FIGURE 8.20 MMFs and permeances of salient-pole machines.

Consequently, the flux produced by F_1 along the polar (i.e., direct) axis is:

$$\phi_{da} = F_1(\lambda_1 + \lambda_2) \cos \theta \qquad ...(8.109)$$

There is also an additional flux produced by F_1 along the interpoler axis (i.e., the quadrature axis), which is given by:

$$\begin{aligned} \phi_{qa} &= F_1(\lambda_1 - \lambda_2) \cos (\theta + 90°) \\ &= - F_1(\lambda_1 - \lambda_2) \sin \theta \qquad ...(8.110) \end{aligned}$$

These two flux components may be resolved along the axis of phase A to yield the total flux linking phase A. Its value,

$$\begin{aligned} \phi_{aa} &= \phi_{da} \cos \theta - \phi_{qa} \sin \theta \\ &= F_1 [(\lambda_1 + \lambda_2) \cos^2 \theta + (\lambda_1 - \lambda_2) \sin^2 \theta] \\ &= F_1 [\lambda_1 + \lambda_2 \cos 2\theta] \qquad ...(8.111) \end{aligned}$$

The self-inductance of phase A is, therefore,

$$L_{aa} = L_l + L_1 + L_2 \cos 2\theta \qquad ...(8.112)$$

In the above expression, L_l is the leakage reactance component,

$$L_1 = N_{\text{eff}} F_1 \lambda_0 \qquad ...(8.113)$$

$$L_2 = N_{\text{eff}} F_1 \lambda_2 \qquad ...(8.114)$$

and, N_{eff} is the number of effective turns per phase.

Similarly, $$L_{bb} = L_l + L_1 + L_2 \cos 2(\theta - 120°) \qquad ...(8.115)$$

and $$L_{cc} = L_l + L_1 + L_2 \cos 2(\theta + 120°) \qquad ...(8.116)$$

Mutual inductance between armature phases

If the fluxes ϕ_{da} and ϕ_{qa} are generated by 1 amp current in phase A, and are resolved along the axis of phase B, we obtain the mutual flux between phase A and phase B. From Figure 8.21. This can be seen to be:

$$\phi_{ab} = \phi_{da} \cos(\theta - 120°) - \phi_{qa} \sin(\theta - 120°)$$
$$= F_1 [(\lambda_1 + \lambda_2) \cos\theta \cos(\theta - 120°) - (\lambda_1 - \lambda_2) \sin\theta \sin(\theta - 120°)]$$
$$= F_1 [\lambda_1 \cos 120° + \lambda_2 \cos(2\theta - 120°)]$$
$$= F_1\left[-\frac{\lambda_1}{2} + \lambda_2 \cos(2\theta - 120°)\right] \quad ...(8.117)$$

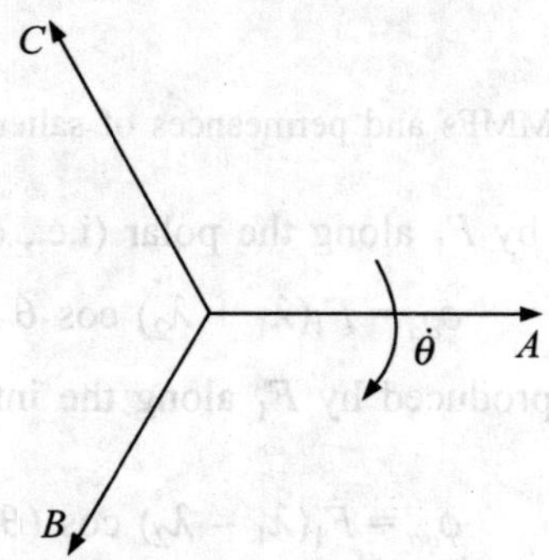

FIGURE 8.21 Location of *d*- and *q*-axis in relation to phase axes.

The resulting flux-linkage of phase *B* and its mutual inductance with phase *A* is given by:

$$\psi_{ab} = N_{\text{eff}}\, \phi_{ab} = N_{\text{eff}} F_1\left[-\frac{\lambda_1}{2} + \lambda_2 \cos(2\theta - 120°)\right]$$

and
$$m_{ab} = -\,0.5L_1 + L_2 \cos(2\theta - 120°) \quad ...(8.118)$$

Similarly,
$$m_{bc} = -\,0.5L_1 + L_2 \cos 2\theta \quad ...(8.119)$$

and
$$m_{ca} = -\,0.5L_1 + L_2 \cos(2\theta + 120°) \quad ...(8.120)$$

Mutual inductance between field winding and the armature phases

$$\left.\begin{aligned} m_{aF} &= M_{aF} \cos\theta \\ m_{bF} &= M_{aF} \cos(\theta - 120°) \\ m_{cF} &= M_{aF} \cos(\theta + 120°) \end{aligned}\right\} \quad ...(8.121)$$

M_{aF} is the maximum value of mutual inductance, when the polar axis (or the direct axis) is aligned to the axis of *A*-phase.

8.6.2 Consolidated Inductance Values

Self inductance	*Mutual inductance*
$l_{ff} = L_{FF}$	$m_{aF} = M_{aF} \cos \theta$
$l_{aa} = L_l + L_1 + L_2 \cos 2\theta$	$m_{bF} = M_{aF} \cos (\theta - 120°)$
$l_{bb} = L_l + L_1 + L_2 \cos 2(\theta - 120°)$	$m_{cF} = M_{aF} \cos (\theta + 120°)$
$l_{cc} = L_l + L_1 + L_2 \cos 2(\theta + 120°)$	
	$m_{ab} = -0.5L_1 + L_2 \cos (2\theta - 120°)$
	$m_{bc} = -0.5L_1 + L_2 \cos 2\theta$
	$m_{ca} = -0.5L_1 + L_2 \cos (2\theta + 120°)$

Power delivered

Instantaneous power delivered:

$$P_d = v_a i_a + v_b i_b + v_c i_c \quad \text{...(8.122)}$$

and, torque developed in the machine is:

$$T = \frac{\text{Total power associated with rotational voltages}}{\text{Mechanical angular velocity of the rotor}} \quad \text{...(8.123)}$$

General observations on the performance equations

The relations derived under Eq. (8.106) to Eq. (8.123) are the performance equations of a salient-pole synchronous machine at any state of operation. These are simultaneous differential equations in form with four unknown quantities, whose coefficients are periodic, containing rotor angles. They are, therefore, functions of time and pose special problems in their solutions unless more simplifying assumptions are applied in their solutions. The difficulty associated with coefficients being functions of rotor position, θ, is caused by the rotation of the field coils relative to the fixed armature phase windings. If, therefore, three coils rotating at the same speed relative to the rotor are used in a new reference frame, on which a transformation of Eq. (8.106) to Eq. (8.123) is carried out, the resulting relationships are expected to have coefficients independent of θ. Consequently, it is easy to solve the equations. This brings up the idea of choosing new equivalent coils properly, and the best choice is to have a coil centred on the polar axis (or the direct axis), and the second coil centred along the interpolar or quadrature axis. (A third coil, connected between the neutral of the machine and ground accounts for possible flow of zero-sequence or neutral currents under unbalanced operation). The use of two coils at right angles, leads to the concept that there is no mutually induced voltage due to transformer action in their respective coils. This system of axes is called the *f-d-q-0* axes.

A second type of reference frame is to choose two coils at right angles to each other, but fixed on to the stator, and a third coil connected between the neutral and the ground. These are known as α-β-γ coils or axes. The resulting differential equations in the fixed axes frame on the stator still contain coefficients that are time-varying, but they have the advantage that the number of unknown is reduced significantly, especially when the unbalanced operation of the machine is considered.

We now, transform Eq. (8.106) to Eq. (8.123) derived earlier, by using phase quantities, into equivalent relations using *f-d-q-0* axes as well as the f-α-β-γ axes.

Performance relations using f-d-q-0 axes

Figure 8.22 shows the two coils in the stator rotating at the rotor speed and centred along the polar and interpolar axes. *Q*-axis is shown ahead of *D*-axis, in conformity with the derivations used for the inductance expressions. In order to transform quantities in *f-a-b-c* axes into the corresponding *f-d-q-0* axes, the constraints of the system forming the basis of the transformation may be obtained by viewing the currents, mmfs, voltages and flux-linkages in the two axes sets. These are:

$$i_f = i_f \quad \text{...(8.124)}$$

$$i_d = k_d\,[i_a \cos\theta + i_b \cos(\theta - 120°) + i_c \cos(\theta + 120°)] \quad \text{...(8.125)}$$

$$i_q = k_q\,[i_a \cos(\theta + 90°) + i_b \cos(\theta - 120° + 90°) + i_c \cos(\theta + 120° + 90°)]$$

$$= k_q\,[i_a \sin\theta + i_b \sin(\theta - 120°) + i_c \sin(\theta + 120°)] \quad \text{...(8.126)}$$

and $$i_0 = k_0\,[i_a + i_b + i_c] \quad \text{...(8.127)}$$

where, k_d, k_q and k_0 are arbitrary constants.

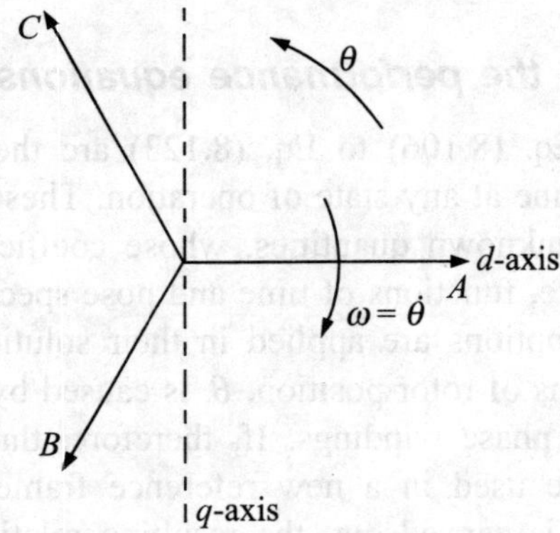

FIGURE 8.22 Resolving phase quantities into of and *q*-axes quantities.

These arbitrary constants are yet to be determined, but their values are very critical in the development of the theoretical equations and to determine the conditions of power invariance between the original and transformed axes systems.

A set of equations similar to Eq. (8.124) to Eq. (8.127), for voltages, mmfs and flux-linkages between the two axes are detailed below. The constraint or transformation matrix is:

$$[C] = \begin{array}{c|c|c|c|c|} & f & a & b & c \\ \hline f & 1 & 0 & 0 & 0 \\ \hline d & 0 & k_d \cos\theta & k_d \cos(\theta - 120°) & k_d \cos(\theta + 120°) \\ q & 0 & -k_q \sin\theta & -k_q \sin(\theta - 120°) & -k_q \sin(\theta + 120°) \\ \hline 0 & 0 & k_0 & k_0 & k_0 \\ \hline \end{array} \quad \text{...(8.128)}$$

The inverse transformation that transforms *f-d-q*-0 quantities in the phase quantities, has the following form:

$$
[C]^{-1} = \begin{array}{c} \\ f \\ a \\ b \\ c \end{array}
\begin{array}{c|c|c|c|}
f & d & q & 0 \\
\hline
1 & 0 & 0 & 0 \\
\hline
0 & \dfrac{2}{3}\dfrac{1}{k_d}\cos\theta & -\dfrac{2}{3}\dfrac{1}{k_q}\sin\theta & \dfrac{1}{3}k_0 \\
\hline
0 & \dfrac{2}{3}\dfrac{1}{k_d}\cos(\theta-120^\circ) & -\dfrac{2}{3}\dfrac{1}{k_q}\sin(\theta-120^\circ) & \dfrac{1}{3}k_0 \\
\hline
0 & \dfrac{2}{3}\dfrac{1}{k_d}\cos(\theta+120^\circ) & -\dfrac{2}{3}\dfrac{1}{k_q}\sin(\theta+120^\circ) & \dfrac{1}{3}k_0 \\
\hline
\end{array} \qquad \text{...(8.129)}
$$

Flux-linkage equations: The flux-linkage of the *f-d-q*-0 coils in terms of four currents is obtained from Eq. (8.107), upon transforming both sides using transformation matrix $[C]$ and its inverse $[C]^{-1}$. Thus,

$$
\begin{bmatrix} \psi_f \\ \psi_d \\ \psi_q \\ \psi_0 \end{bmatrix} = [C] \begin{bmatrix} \psi_f \\ \psi_a \\ \psi_b \\ \psi_c \end{bmatrix} = [C][L_{\text{phase}}][i_{\text{phase}}] = [C]\,[L_{\text{phase}}]\,[C]^{-1}\,[i_{f\text{-}d\text{-}q\text{-}0}]
$$

$$
= \begin{array}{c} \\ f \\ d \\ q \\ 0 \end{array}
\begin{array}{|c|c|c|c|}
f & d & q & 0 \\
\hline
L_{ff} & \dfrac{1}{k_d}M_{af} & & \\
\hline
\dfrac{3}{2}k_d M_{af} & L_d & & \\
\hline
 & & L_q & \\
\hline
 & & & L_l \\
\hline
\end{array}
\begin{array}{c} \\ f \\ d \\ q \\ 0 \end{array}
\begin{bmatrix} i_f \\ i_d \\ i_q \\ i_0 \end{bmatrix} \qquad \text{...(8.130)}
$$

where,

$$
L_d = L_l + \frac{3}{2}(L_1 + L_2) = L_l + \frac{3}{2}F_1 N_{\text{eff}}(\lambda_1 + \lambda_2) \qquad \text{...(8.131)}
$$

and

$$
L_q = L_l + \frac{3}{2}(L_1 - L_2) = L_l + \frac{3}{2}F_1 N_{\text{eff}}(\lambda_1 - \lambda_2) \qquad \text{...(8.132)}
$$

The above inductances are due to the permeances offered by the gap along the direct- and the quadrature-axis respectively. It should be noted from Eq. (8.130) that the impedances are no-longer functions of rotor position θ. This is the special advantage of transforming phase quantities into suitable set of *f-d-q*-0 axis quantities.

Voltage equations: The voltages V_f, V_d, V_q and V_0 in the new axes can be derived in terms of flux-linkages ψ_f, ψ_d, ψ_q, ψ_0 by transforming Eqs. (8.106), with the aid of $[C]$ and $[C]^{-1}$, as follows:

$$\begin{bmatrix} V_f \\ V_d \\ V_q \\ V_0 \end{bmatrix} = [C]\begin{bmatrix} V_f \\ V_a \\ V_b \\ V_c \end{bmatrix} = [C]\begin{bmatrix} r_f i_f + p\psi_f \\ ri_a + p\psi_a \\ ri_b + p\psi_b \\ ri_c + p\psi_c \end{bmatrix} = [C]\begin{bmatrix} ri_f \\ ri_a \\ ri_b \\ ri_c \end{bmatrix} + [C]p\begin{bmatrix} \psi_f \\ \psi_a \\ \psi_b \\ \psi_c \end{bmatrix}$$

$$= \begin{bmatrix} r_f i_f \\ ri_a \\ ri_b \\ ri_c \end{bmatrix} + [C]\,p[C]^{-1}\begin{bmatrix} \psi_f \\ \psi_d \\ \psi_q \\ \psi_0 \end{bmatrix} \qquad \text{...(8.133)}$$

As the elements of $[C]^{-1}$ are functions of time, the time derivatives of $[C]^{-1}$ $[\psi_{fdq0}]$ must be obtained by partial derivatives, which has the form:

$$\left\{\frac{d}{dt}[C]^{-1}\right\}[\psi_{fdq0}] + [C]^{-1}\left\{\frac{d}{dt}[\psi_{fdq0}]\right\}$$

The final result works out to be:

$$\begin{bmatrix} V_f \\ V_d \\ V_q \\ V_0 \end{bmatrix} = \begin{bmatrix} r_f i_f + p\psi_f \\ ri_d + p\psi_d - \left(\dfrac{k_d}{k_q}\right)\psi_q p\theta \\ ri_q + \left(\dfrac{k_q}{k_d}\right)\psi_d p\theta + p\psi_q \\ r_0 i_0 + p\psi_0 \end{bmatrix} \qquad \text{...(8.134)}$$

Impedance relations: When relation of flux-linkages [i.e., Eq. 8.130] is used, the voltages can be expressed in terms of currents, and the resulting voltage–current relation yields the generalized impedance of the synchronous machine. The equation can be written in the form:

$$[V_{fdq0}] = [Z_{fdq0}]\ [i_{fdq0}] \qquad \text{...(8.135)}$$

The generalized impedance matrix is shown in Section 8.6.3, for the condition of power-invariance, after the values of k_d, k_q and k_0 are computed. This is recorded under expression (8.143).

Power relations: The total instantaneous power delivered is given by:

$$P_d = V_a i_a + V_b i_b + V_c i_c$$

When *f-d-q-*0 axis quantities are substituted for the phase quantities using expression (8.129) and further multiplication the results of the delivered power in terms of the transformed quantities, it becomes:

$$P_d = \frac{2}{3k_d^2}V_d i_d + \frac{2}{3k_q^2}V_q i_q + \frac{1}{3k_0^2}V_0 i_0 \qquad ...(8.136)$$

Power invariant transformation in *f-d-q-*0 axes: So far, we did not determine the values of k_d, k_q and k_0 in Eq. (8.125) to Eq. (8.136). In order to determine their values, we compare the expressions (8.122) with Eq. (8.136). The former gives the instantaneous power in the phase reference axis, while the latter depicts the actual power in the *f-d-q-*0 axis frame. For power invariance between the two reference frames, the following condition must be satisfied:

$$\frac{2}{3k_d^2} = 1, \quad \frac{2}{3k_q^2} = 1 \quad \text{and} \quad \frac{1}{3k_0^2} = 1$$

This leads to the conclusion that, for power invariance, between phase systems, i.e. (*f-a-b-c*) and (*f-d-q-*0) system, we must select:

$$k_d = k_q = \sqrt{\frac{2}{3}} \quad \text{and} \quad k_0 = \sqrt{\frac{1}{3}}$$

Thus, to values of transformation matrix [*C*] and its inverse $[C]^{-1}$, assume the form:

$[C] =$ ...(8.137)

	f	*a*	*b*	*c*
f	1	0	0	0
d	0	$\sqrt{\frac{2}{3}}\cos\theta$	$\sqrt{\frac{2}{3}}\cos(\theta - 120°)$	$\sqrt{\frac{2}{3}}\cos(\theta + 120°)$
q	0	$-\sqrt{\frac{2}{3}}\sin\theta$	$-\sqrt{\frac{2}{3}}\sin(\theta - 120°)$	$-\sqrt{\frac{2}{3}}\sin(\theta + 120°)$
0	0	$\sqrt{\frac{1}{3}}$	$\sqrt{\frac{1}{3}}$	$\sqrt{\frac{1}{3}}$

and

$[C]^{-1} =$ ...(8.138)

	f	*d*	*q*	0
f	1	0	0	0
a	0	$\sqrt{\frac{2}{3}}\cos\theta$	$-\sqrt{\frac{2}{3}}\sin\theta$	$\sqrt{\frac{1}{3}}$
b	0	$\sqrt{\frac{2}{3}}\cos(\theta - 120°)$	$-\sqrt{\frac{2}{3}}\sin(\theta - 120°)$	$\sqrt{\frac{1}{3}}$
c	0	$\sqrt{\frac{2}{3}}\cos(\theta + 120°)$	$-\sqrt{\frac{2}{3}}\sin(\theta + 120°)$	$\sqrt{\frac{1}{3}}$

The transformation is called 'orthogonal' or 'unitary' transformation and is typified by the fact that the inverse of the transformation matrix is the transpose of the original matrix in Eq. (8.137). In general, as has already been elaborated in Section 6.2.3, the general form for orthogonal transformation is:

$$[A]^{-1} = [A^*]$$

8.6.3 Flux-linkage Current Relationship for Orthogonal Transformation

The final result has the form:

$$\begin{bmatrix} \psi_f \\ \psi_d \\ \psi_q \\ \psi_0 \end{bmatrix} = \begin{array}{c} \\ f \\ d \\ q \\ 0 \end{array} \begin{array}{c} \begin{array}{cccc} f & d & q & 0 \end{array} \\ \begin{bmatrix} L_{ff} & \sqrt{\frac{3}{2}}M_{af} & & \\ \sqrt{\frac{3}{2}}M_{af} & L_d & & \\ & & L_q & \\ & & & L_l \end{bmatrix} \end{array} \begin{array}{c} f \\ d \\ q \\ 0 \end{array} \begin{bmatrix} i_f \\ i_d \\ i_q \\ i_0 \end{bmatrix} \qquad \text{...(8.139)}$$

where

L_{ff} is the self-inductance of field

$$L_d = L_l + \frac{3}{2}N_{\text{eff}}(\lambda_1 + \lambda_2)F_1$$

$$L_q = L_l + \frac{3}{2}N_{\text{eff}}(\lambda_1 - \lambda_2)F_1$$

L_l = Leakage inductance of armature

Voltage equation for orthogonal transformation

$$\begin{bmatrix} V_f \\ V_d \\ V_q \\ V_0 \end{bmatrix} = \begin{bmatrix} r_f i_f + p\psi_f \\ ri_d + p\psi_d - \psi_q p\theta \\ ri_q + \psi_d p\theta + p\psi_q \\ ri_0 + p\psi_0 \end{bmatrix} \qquad \text{...(8.140)}$$

Voltage–current relation for orthogonal transformation

Combining Eq. (8.139) and Eq. (8.140), we get:

$$\begin{bmatrix} V_f \\ V_d \\ V_q \\ V_0 \end{bmatrix} = \begin{array}{c|cccc} & f & d & q & 0 \\ \hline f & r_f + L_{ff}p & \sqrt{\frac{3}{2}}M_{af}p & 0 & 0 \\ d & \sqrt{\frac{3}{2}}M_{af} & r + L_d p & -L_q p\theta & 0 \\ q & M_{af}p\theta & L_d p\theta & r + L_q p & 0 \\ 0 & 0 & 0 & 0 & r + L_l p \end{array} \begin{bmatrix} i_f \\ i_d \\ i_q \\ i_0 \end{bmatrix} \quad \text{...(8.141)}$$

which is the same as:

$$[V_{fdq0}] = [Z_{fdq0}]\,[i_{fdq0}] \quad \text{...(8.142)}$$

Generalized impedance matrix for orthogonal transformation

Elaborating relation (8.142), we observe that the impedance matrix is symmetrical in f- and d-axis.

It consists of two types of terms, one relating to transformer voltages and the second associated with $p\theta$ terms related to speed voltages. If the two types are separated, it will have the form:

$$[Z] = \begin{array}{c|cccc} & f & d & q & 0 \\ \hline f & r_f + L_{ff}p & \sqrt{\frac{3}{2}}M_{af}p & 0 & 0 \\ d & \sqrt{\frac{3}{2}}M_{af}p & r + L_d p & 0 & 0 \\ q & 0 & 0 & r + L_q p & 0 \\ 0 & 0 & 0 & 0 & r + L_l p \end{array} + \begin{array}{c|cccc} & f & d & q & 0 \\ \hline f & 0 & 0 & 0 & 0 \\ d & 0 & 0 & -L_q p\theta & 0 \\ q & M_{af}p\theta & L_d p\theta & 0 & 0 \\ 0 & 0 & 0 & 0 & 0 \end{array} \quad \text{...(8.143)}$$

Torque relation in orthogonal transformation

It is pointed out that the torque equals the power associated with speed induced voltages divided by the mechanical angular velocity, as given in Eq. (8.123). The speed voltages are:

(i) in the d-axis given by $-\psi_q\, p\theta$, and
(ii) in the q-axis associated with $\psi_d p\theta$

Also, the mechanical angular velocity

$$= \frac{2}{\text{No. of poles}} \times \text{Electrical angular velocity}$$

$$= \frac{2}{P} p\theta \qquad \text{...(8.144)}$$

where, P is the actual number of poles of the machine.

Hence,

$$T = \frac{P}{2}(\psi_d i_q - \psi_q i_d) \qquad \text{...(8.145)}$$

which when expanded, has final form:

$$T = \frac{P}{2}\left[\sqrt{\frac{3}{2}} M_{af} i_f i_q + L_d i_d i_q - L_q i_d i_q\right] \qquad \text{...(8.146)}$$

In a cylindrical rotor machine, since L_d and L_q are equal, the machine torque is given by:

$$T_d = \frac{P}{2}\sqrt{\frac{3}{2}}\, M_{af} i_f i_q \qquad \text{...(8.147)}$$

But, in salient-pole machines, a saliency torque, equalling

$$\frac{P}{2}(L_d - L_q) i_d i_q \qquad \text{...(8.148)}$$

is produced in addition. This is the reluctance torque and it exists even if the field excitation of the machine does not exist.

OBJECTIVE TYPE QUESTIONS

8.1 When the stator of an induction motor is fed from a non-sinusoidal source of voltage, and the spatial distribution of mmf is also non-sinusoidal, the resulting torques of fundamental and harmonics are all:

(a) super-synchronous corresponding to pole-numbers
(b) synchronous corresponding to pole-numbers
(c) sub-synchronous relating to pole-numbers
(d) all the three above alternatives

8.2 The rotor of an induction motor has a stand-still resistance of $r_2 = 0.10$ ohm. At a slip of 5%, its effective resistance will be:

(a) 0.5 ohm (b) 2.0 ohm
(c) 1.9 ohm (d) 0.19 ohm

8.3 The air-gap power of an induction motor is 12 kW at a slip of 6%. The rotor copper loss is:

(a) 1 kW (b) 0.72 kW
(c) 7.2 kW (d) .0720 kW

8.4 A 4-pole, 50 Hz induction motor is operating at 5% slip. The power input to the motor is 1000 kW. The torque developed is:

(a) 1000 N-m (b) 6048 N-m
(c) 6366 N-m (d) 5000 N-m

8.5 If we assume $r_1 = r_2$, the slip at which the maximum torque is developed for such a motor is in the region of:

(a) 10% (b) 20%
(c) 28% (d) 35%

8.6 The direction of rotation of a 3-phase induction motor can be reversed by reconnection of the line terminals *A-B-C* to:

(a) *C-A-B* (b) *A-C-B*
(c) *B-C-A* (d) None of these

if the motor was originally running with line terminals connected in sequence *A-B-C*.

8.7 The air-gap permeance of a generator/motor is given by:

$$P = 0.7 + 0.3 \cos 2\theta$$

The ratio of *q*-axis armature reactance, x_{aq}, to *d*-axis armature reactance, x_{ad}, is:

(a) 0.7 (b) 0.3
(c) 11/17 (d) 0.4

8.8 For a synchronous generator, when the value of the field current is higher, it operates at:

(a) with a leading power factor load (b) unity power factor load
(c) lagging power factor load (d) there is no way to predict

8.9 For a poly-phase synchronous motor, under excited operation corresponds to:

(a) leading p.f load (b) unity p.f load
(c) lagging power factor load (d) no way to predict

8.10 Saliency torque in a water wheel alternator exists:

(a) even when no field excitation is provided
(b) when the field in excited
(c) only when loaded
(d) both (a) and (b) are correct

8.11 Two non-salient pole alternators are supplying an electrical load when working in parallel. One is rated at 20 MVA at 13.2 kV and the other 15 MVA at 13 kV. Both have $X_d = 1$ pu. at their base value respectively.

At 13.2 kV terminal voltage, the currents shared are:

(a) equal (b) in the ratio of 20/15
(c) in the ratio of 2/1 (d) none of these

Answers

8.1 (d)	**8.2** (b)	**8.3** (b)	**8.4** (c)	**8.5** (a)	**8.6** (b)
8.7 (c)	**8.8** (c)	**8.9** (c)	**8.10** (d)	**8.11** (d)	

NUMERICAL PROBLEMS

8.1. A 15 kW 400 volt, 3-phase, 50 Hz wound rotor induction motor develops the rated output at the rated voltage and frequency and with the slip rings short-circuited. The maximum torque, equalling twice the full load torque occurs at a slip of 10%, with no external resistance included in the rotor circuit. If the rotational losses and the stator circuit resistance are assumed as negligibly small, determine:

(a) (i) slip and rotor speed at full-load torque
(ii) the rotor ohmic loss at full-load torque
(iii) starting torque at rated voltage and frequency
(iv) full-load efficiency

(b) The rotor resistance is, now, doubled by inserting external resistance in the rotor circuit. Estimate:
(i) the slip at maximum torque
(ii) full-load slip and the rotor speed
(iii) starting torque
(iv) rotor I^2R loss at full-load torque
(v) developed power at full-load torque and corresponding motor efficiency

[**Ans.** (a) (i) s_{fl} (full-load slip) = 0.0268; N_r (rotor speed) = 973 rpm, (ii) Rotor ohmic loss at full-load torque = 413.07 watt, (iii) Test = 58.291 N-m and (iv) efficiency at full-load torque = 97.32%; (b) (i) 0.054; 946.4 rpm, (ii) T_{efl} = 151.352 Nm, (iii) T_{est} = 116.425 Nm, (iv) Rotor ohmic loss = 849.54 watt (v) 94.64%]

8.2 A 3-phase, star-connected, 440 volt, 6 poles, 50 Hz induction motor has the following circuit parameters per phase:
$r = 0.8$ ohm; $r_2 = 0.3$ ohm,
$x_1 = x_2 = 2.0$ ohm, $x_m = 48$ ohm.
The motor is applied with half the applied voltage at half the original frequency.

(a) Obtain the current and power factor at the instant of starting and maximum torque conditions.

(b) Compare the results with the normal voltage and frequency conditions and comment on the effect.

[**Ans.** I_N, at normal voltage and frequency = 60.025 amp
$\cos \theta_N = 0.263$ (lagging)
I_R at reduced voltage and frequency = 54.629 amp
$\cos \theta_R = 0.4785$ (lagging)]

8.3 A synchronous machine is synchronized to an infinite bus-bar. Without changing the field current of the machine, it is made to operate as:

(a) a motor (b) a generator

Explain, in each case, whether the machine is delivering or absorbing reactive power.

[**Ans.** For (a) and also for (b), the machine absorbs reactive power from mains.]

8.4 Synchronizing power P_s, is defined as the rate at which synchronous power P varies with load angle, δ.

(a) A non-salient pole synchronous machine connected to an infinite bus-bar system is operating at unity power factor. Show that the synchronizing power per phase

$$P_s = \frac{V_t}{X_s}$$

where V_t is the per phase terminal voltage of bus-bar and X_s is the synchronous reactance.

(b) Using the phasor diagram of a cylinderical rotor synchronous motor, show that its per phase power output is given by:

$$P_{\text{output}} = \frac{E_f V_t}{Z_s} \sin(\delta + \alpha_a) - \frac{E_f^2}{Z_s^2} r_a$$

and $$\alpha_a = \tan^{-1} \frac{r_a}{X_s}$$

where all parameters have their usual meanings.

(c) Find an expression for the maximum per phase power output for the synchronous motor. [**Ans.** (c) $\frac{E_f V_t}{Z_s} - \frac{E_f^2}{Z_s^2} r_a$]

8.5 A star-connected, 3-phase, non-salient pole machine has a synchronous reactance of 4 ohm per phase with negligible armature resistance. It functions as an alternator on an infinite bus-bar and supplies 200 ampere at unity power factor to the bus-bars at 11 kV.

(a) If the excitation emf of the machine is increased by 15 per cent, without changing its prime-mover driving power, obtain the new values of current, power factor and load angle. Will there be any change in the value of the power supplied by the machine to the infinite bus-bars?

(b) If the increased excitation of part (a) above is held fixed at what power output would the alternator break away from synchronism? Find the corresponding values of current and power factor.

[**Ans.** (a) I_{a1} = 313.712 amp; $\cos \phi_1$ = 0.63753 (lag)
δ_1 = 6.24°; No change
(b) I_{a2} = 2430.6 amp; $\cos \phi$ = 0.7572 (lead)]

8.6 A turbo-alternator is connected to an infinite bus-bar through two transmission lines connected in parallel. The following details are known about the system:

Number of phases = 3

Alternator connection: Star

Bus-bar voltage: 11 kV

Impedance of transmission lines: 4 ohm per phase with negligible resistance

Impedance of machine per phase with negligible armature resistance: 3 ohm

(a) The mechanical input to the alternator and its excitation are adjusted so that the operating power factor is unity at the alternator terminals when a rated current of 500 amp is delivered to the bus-bar. Obtain the alternator terminal and excitation voltages and the active and reactive power delivered by the alternator.

(b) With the excitation fixed at its value as in part (a) above, the mechanical input to the alternator is manipulated so that no power transfer occurs between the bus-bar and the generator sets. Estimate under this condition:

(i) the lagging reactive power delivered to the bus-bar.

(ii) the alternator terminal voltage.

[**Ans.** (a) Terminal voltage, V_a = 6271.8 volt/phase
Alternator excitation voltage, E_f = 6448.7 volt/phase
Active power delivered to bus-bar = 3135.9 kW/phase
Reactive power delivered to bus-bar = 500 kVAr/phase
(b) (i) Reactive power delivered to bus-bar = 124.1 kVAr/phase
(ii) Alternator terminal voltage = 6390.8 volt/phase]

CHAPTER 9

Transients and Dynamics in Electrical Machines

9.1 INTRODUCTION

We have, so far, studied the steady-state operating conditions of dc and ac machines. But when disturbance is incident on these machines, the machine behaviour becomes quite different. A transient period of re-adjustment occurs between the initial condition following the disturbance and the final steady-state conditions becomes necessary because we wish to know in advance the machine response to the disturbance during this transient period. This chapter wishes to high-light this aspect of the behaviour of the machine both for electrical transients and the mechanical transients—the latter being identified as the dynamics—of dc and ac machines.

9.2 TRANSIENTS IN DC MACHINES

DC machines are used in applications demanding control of speed over a wide range with precision, and this is achieved by controlling the terminal voltage input to the field and the armature circuits or both.

It is common to introduce the following assumptions to reduce the complexity of the analysis:

1. Magnetic saturation is neglected. This assumption implies that the values of circuit inductances are independent of current magnitudes flowing through them.
2. The field mmf is taken as acting along the direct axis and there is no mutual inductance between the field and the armature and there is no demagnetizing effect due to armature reaction.

We shall study electrical transients in a dc generator and after this we shall consider the dynamics behaviour of a dc motor.

9.2.1 Separately Excited dc Generator

A schematic depiction of a separately excited dc generator is presented in Figure 9.1.

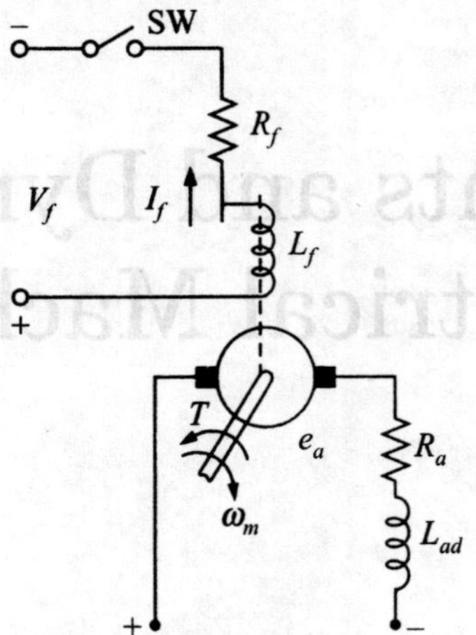

FIGURE 9.1 Schematic depiction of a separately excited dc generator.

The armature inductance is denoted as effective along the direct axis as the armature mmf acts along the direct axis. The basic equations for the dc machine, therefore, assume the form:

$$e = K_a \phi \omega_m \qquad ...(9.1)$$

and

$$T = K_a \phi i_a \qquad ...(9.2)$$

Since, magnetic linearly is assumed,

$$e = K_f i_f \omega_m \qquad ...(9.3)$$

and

$$T = K_f i_f i_a \qquad ...(9.4)$$

Field circuit transient

We first take up electrical transients resulting from changes in the excitation, that is the field circuit voltage. We assume that the armature circuit is open and the generator is running at a constant speed ω_m. The voltage equation for the field circuit after closure of the field circuit switch, has the form:

$$V_f = R_f i_f + L_f \frac{di_f}{dt} \qquad ...(9.5)$$

The Laplace transform of Eq. (9.5), with zero initial condition is as follows:

$$V_f(s) = R_f\, i_f(s) + L_f(s) I_f(s)$$
$$= I_f(s)\,[R_f + (s)L_f] \qquad ...(9.6)$$

And the transfer function relating the field current to the field voltage is:

$$\frac{I_f(s)}{V_f(s)} = \frac{1}{(R_f + sL_f)} = \frac{1}{R_f(1+s\tau_f)} \qquad ...(9.7)$$

where, $\tau_f = \dfrac{L_f}{R_f}$, denotes the time-constant of the field-circuit.

From Eq. (9.3), the generated voltage in the armature circuit is:

$$e = K_f\, i_f\, \omega_m = K_g\, i_f \qquad ...(9.8)$$

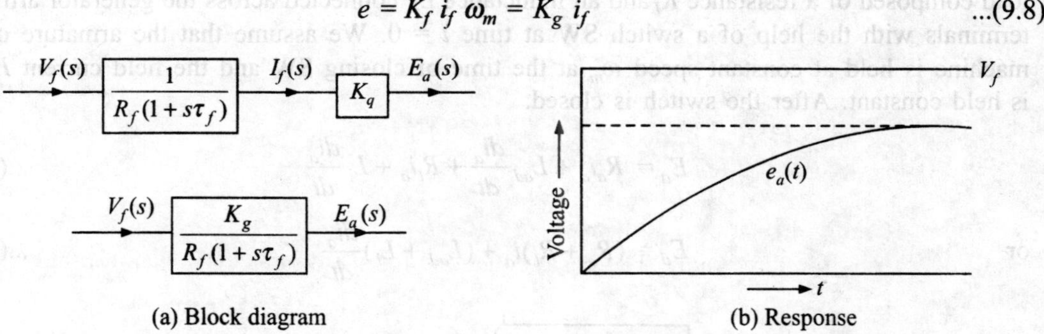

FIGURE 9.2 The field circuit transient (a) Block diagram (b) Response.

In Eq. (9.8), $K_g = (K_f \omega_m)$ is the slope of the straightline portion of the magnetization characteristic at speed ω_m, denoting e_a plotted against i_f. The Laplace transform of Eq. (9.8) is:

$$E_a(s) = K_g\, I_f(s) \qquad ...(9.9)$$

From Eqs. (9.7) to (9.9), the transfer function relating the generated armature voltage to the field current is, thus:

$$\frac{E_a(s)}{V_f(s)} = \frac{E_a(s)}{I_f(s)} \times \frac{I_f(s)}{V_f(s)} = \frac{K_g}{R_f(1+s\tau_f)} \qquad ...(9.10)$$

Equations (9.7), (9.9) and (9.10) are indicated in the block diagram form in Figure 9.2. The time domain response from in Figure 9.2. The time domain response due to the transfer function given in Eq. (9.10), due to a step change in V_f, is:

$$e_a(t) = \frac{K_g V_f}{R_f}(1 - e^{-t/\tau_f}) \qquad ...(9.11)$$
$$= E\,(1 - e^{-t/\tau_f}) \qquad ...(9.12)$$

where, $E_a = e_a(\infty) = \dfrac{K_g K_f}{R_f} = K_g I_f$, is the steady-state generated voltage and

$I_f = \dfrac{V_f}{R_f}$, is the steady-state field current.

The response, as already indicated, is depicted in Figure 9.2(b). It is a first-order response with a time constant of τ_f. The field circuit time constant, $\tau_f = \dfrac{L_f}{R_f}$, is quite large and its typical values are 0.1 to 2 second.

Armature circuit transient

We now discuss about an electrical transient in the armature circuit. Figure 9.3(a) shows a load composed of a resistance R_l and an inductance L_l, connected across the generator armature terminals with the help of a switch SW at time $t = 0$. We assume that the armature of the machine is held at constant speed ω_m at the time of closing SW and the field current I_f also is held constant. After the switch is closed:

$$E_a = R_a i_a + L_{ad}\frac{di_a}{dt} + R_l i_a + L_l\frac{di_a}{dt} \qquad ...(9.13)$$

or

$$E_a = (R_a + R_l)i_a + (L_{ad} + L_l)\frac{di_a}{dt} \qquad ...(9.14)$$

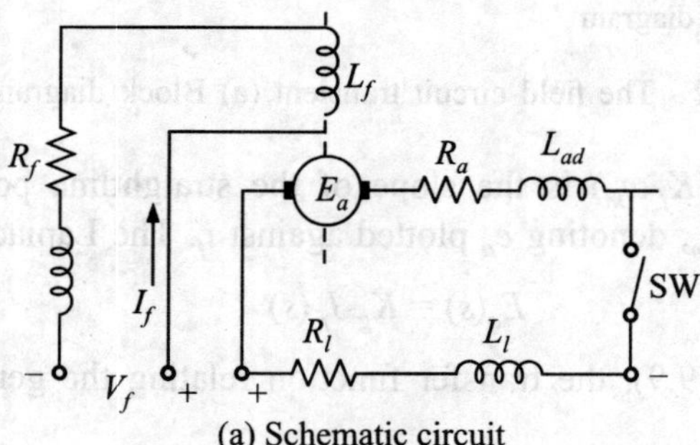

(a) Schematic circuit

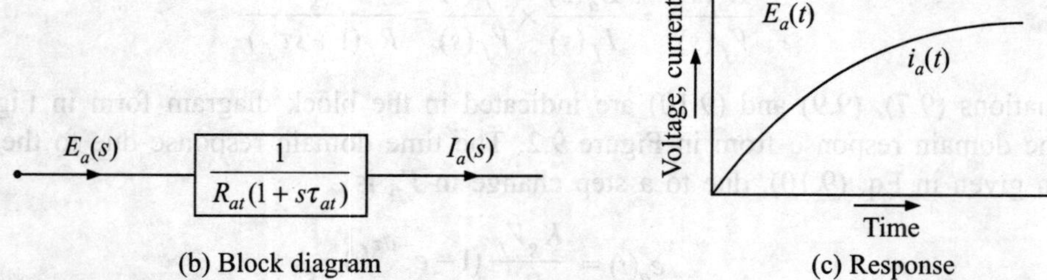

(b) Block diagram (c) Response

FIGURE 9.3 Armature circuit transient (a) Schematic diagram (b) Block diagram and (c) Response.

or
$$E_a = R_{at} i_a + L_{at} \frac{di_a}{dt} \quad \text{...(9.15)}$$

where $R_{at} = R_a + R_L$, the total resistance of the armature circuit
and $L_{at} = L_{aq} + L_L$, the total inductance of the armature circuit
The Laplace transform resulting from Eq. (9.15) is:

$$E_a(s) = R_{at}\, I_a(s) + L_{at}\, sI_a(s) \quad \text{...(9.16)}$$

and the transfer function is:

$$\frac{I_a(s)}{E_a(s)} = \frac{1}{R_{at}(1 + s\tau_{at})} \quad \text{...(9.17)}$$

where

$\tau_{at} = \dfrac{L_{at}}{R_{at}}$, is the armature circuit time-constant.

A block diagram representation of the transfer function of Eq. (9.17) is shown in Figure 9.3(b). The time domain response is

$$i_a(t) = \frac{E_a}{R_{at}}(1 - e^{-t/\tau_{at}}) \quad \text{...(9.18)}$$

The response is shown in Figure 9.3(c). This is also a first-order response with time constant τ_{at}. Usually, the value of τ_{at} is low and the armature current i_a builds up quickly.

Now, from Eqs. (9.10) and (9.17), the total transfer function relating the armature current to the field circuit voltage is:

$$\left.\begin{aligned} \frac{I_a(s)}{V_f(s)} &= \frac{I_a(s)}{E_a(s)} \times \frac{E_a(s)}{V_f(s)} \\ &= \frac{K_g}{R_f R_{at}(1 + s\tau_f)(1 + s\tau_{at})} \end{aligned}\right\} \quad \text{...(9.19)}$$

The corresponding block diagram representation is shown in Figure 9.4(a) and the response is denoted in Figure 9.4(b).

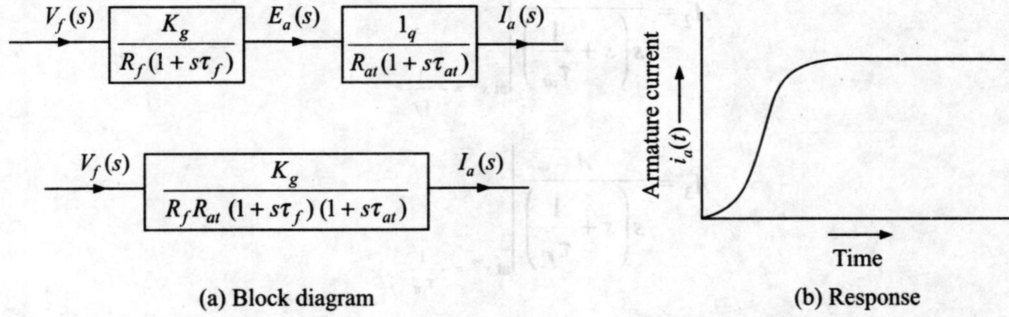

(a) Block diagram (b) Response

FIGURE 9.4 Field and armature current transient (a) Block diagram and (b) Response.

Now, for a step change of voltage in the field circuit,

$$V_f(s) = \frac{V_f}{s} \qquad \text{...(9.20)}$$

And, from Eqs. (9.19) and (9.20)

$$I_a(s) = \frac{K_g V_f}{R_f R_{at} s(1 + s\tau_f)(1 + s\tau_{at})} \qquad \text{...(9.21)}$$

or

$$I_a(s) = \frac{K_g V_f}{R_f R_{at} \tau_f \tau_{at} s \left(s + \dfrac{1}{\tau_f}\right)\left(s + \dfrac{1}{\tau_{at}}\right)} \qquad \text{...(9.22)}$$

that is,

$$I_a(s) = \frac{A}{s\left(s + \dfrac{1}{\tau_f}\right)\left(s + \dfrac{1}{\tau_{at}}\right)} \qquad \text{...(9.23)}$$

where

$$A = \frac{K_g K_f}{R_f R_{at} \tau_f \tau_{at}}$$

or

$$I_a(s) = \frac{A_1}{s} + \frac{A_2}{\left(s + \dfrac{1}{\tau_f}\right)} + \frac{A_3}{\left(s + \dfrac{1}{\tau_{at}}\right)} \qquad \text{...(9.24)}$$

where

$$A_1 = \left.\frac{A}{\left(s + \dfrac{1}{\tau_f}\right)\left(s + \dfrac{1}{\tau_{at}}\right)}\right|_{\text{at } s = 0} = A\tau_f \tau_{at}$$

$$A_2 = \left.\frac{A}{s\left(s + \dfrac{1}{\tau_{at}}\right)}\right|_{\text{at } s^{\infty} = -\frac{1}{\tau_f}}$$

$$A_3 = \left.\frac{A}{s\left(s + \dfrac{1}{\tau_f}\right)}\right|_{\text{at } s^{\infty} = -\frac{1}{\tau_{at}}}$$

The resulting time domain response of i_a is given by:

$$i_a(t) = A_1 + A_2 e^{-t/\tau_f} + A_3 e^{-t/\tau_{at}} \qquad ...(9.25)$$

From this expression, we note that A_1 gives the steady-state value of the armature current. This is given by:

$$A_1 = i_a(\infty) = \frac{K_g K_f}{R_f R_{at}} = \frac{K_g I_f}{R_{at}} = \frac{E_q}{R_{at}}$$

Hence, Figure 9.4(b) shows the response of the armature current i_a for a step change in the field circuit voltage.

EXAMPLE 9.1 The following parameters are given for a separately excited dc generator operating at its rated speed:

R_f = 100 ohm; L_f = 25 henry; R_a = 0.25 ohm;
L_{aq} = 0.02 henry; K_g = 100 volt per unit field current at the rated speed.

(a) The generator functions at the rated speed and a field voltage V_f is suddenly applied to the field winding.
Determine:
   (i) the armature voltage generated as a function of time.
   (ii) value of the steady-state armature voltage.

(b) If the generator continues to be deriver at its rated speed and a load constituted by a series connected combination of R_l = 1 ohm and L_l = 0.15 henry is connected to the armature terminals, while a field voltage of V_f = 200 Volts is suddenly applied to the field of the machine, determine the armature current as a function of time.

Solution:

(a) The field circuit time constant

$$\tau_f = \frac{L_f}{R_f} = \frac{25}{100} = 0.25 \text{ sec}$$

(i) Using Eq. (9.11), we get

$$e_a(t) = \frac{100 \times 200}{100}(1 - e^{-t/0.25})$$

$$= 200\,(1 - e^{-t/0.25})$$

(ii) $e_a(\infty) = 200$ volt

(b) We have, as before:

$$\tau_f = 0.25 \text{ sec}$$

and

$$\tau_{at} = \frac{L_{aq} + L_l}{R_a + R_l}$$

$$L_{aq} + L_l = 0.02 + 0.15 = 0.17 \text{ henry}$$
$$R_a + R_l = 0.25 + 1.0 = 1.25 \text{ ohm}$$

$$\therefore \qquad \tau_{at} = \frac{0.17}{1.25} = 0.136 \text{ sec}$$

Using, Eq. (9.22), we get

$$I_a(s) = \frac{100 \times 200}{100 \times 1.25 \times 0.25 \times 0.136\, s(s+4)\,(s+7.35)}$$

$$= \frac{4705.88}{s(s+4)(s+7.35)} = \frac{A_1}{s} + \frac{A_2}{(s+4)} + \frac{A_3}{(s+7.35)}$$

where,

$$A_1 = \left.\frac{4705.88}{(s+4)(s+7.35)}\right|_{s=0} = 160$$

$$A_2 = \left.\frac{4705.88}{s(s+7.35)}\right|_{s=-4} = -351$$

$$A_3 = \left.\frac{4705.88}{s(s+4)}\right|_{s=-7.35} = 191$$

Hence, from Eq. (9.25), we get:

$$i_a(t) = 160 - 351e^{-4t} + 191e^{-7.35t}$$

9.3 TRANSFER FUNCTION

A transfer function expresses a relationship between two quantities in a system. When applied to transient and dynamic problems, it describes, in particular, the relationship existing between the quantity controlled and the quantity which controls it.

9.3.1 Transfer Functions of a dc Machine under Open and Closed-loop Systems

In Figure 9.1, we have shown a schematic diagram a separately excited dc generator. Its output voltage is the armature voltage e_a, while the input quantities, which control e_a are (i) its field excitation, i_f and (ii) its armature speed ω_m. We can express this by following expression:

$$e_a(\omega_m, i_f) = \frac{PZ}{2\pi a}\phi_f(i_f)\omega_m$$

$$= K_a\, \phi_f\, \omega_m \qquad \text{...(9.26)}$$

where

ω_m = angular velocity of armature in radians per second.

P, Z, a and K_a are constants pertaining to the generator

[P = Number of poles, Z = Total number of armature conductors, a = Number of parallel paths formed in armature, K_a = Generator constant] and

ϕ_f = Flux per pole, which is usually a non-linear function of the field current.

In the linear region,

$$\phi_f = K_f\, i_f;\ (K_f = \text{A field constant})$$

and Eq. (9.26) can be modified to:

$$e_a = K_a\, K_f\, i_f\, \omega_m \qquad ...(9.27)$$

Therefore, there are two transfer functions for a dc generator:

(i) At constant field current:

$$G(\omega_m) = \frac{e_a}{\omega} = K_a K_f\, i_f \qquad ...(9.28)$$

(ii) At constant speed,

$$G(i_f) = \frac{e_a}{i_f} = K_a K_f\, \omega_m \qquad ...(9.29)$$

If the input to the field is taken as the voltage V_f, the resulting field current is:

$$i_f = \frac{V_f}{\left(R_f + \dfrac{L_f}{p}\right)} \qquad ...(9.30)$$

where, p is the Heaviside's differential operator $\left(\dfrac{d}{dt}\right)$, which is the total time derivative, i.e.,

$$p = \frac{d}{dt} \qquad ...(9.31)$$

Thus, the transfer function at constant speed modifies to:

$$G(p) = \frac{e_a}{V_f} = \frac{K_a K_f \omega_m}{R_f + L_f p} \qquad ...(9.32)$$

and its Laplace transform, with initial value of $i_f = 0$, is:

$$G(s) = \frac{K_a K_f \omega_m}{R_f + L_f s} = \frac{K_a K_f \omega_m}{R_f} \times \frac{1}{1 + \tau_f s} \qquad ...(9.33)$$

where

$$\tau_f = \frac{L_f}{R_f} = \text{Field time constant} \qquad ...(9.34)$$

$G(s)$ is also the gain factor of the open-loop control of the armature voltage applied to the field circuit.

If the transfer function has a general form:

$$G(s) = M \times \frac{1}{\tau s + \alpha} \qquad ...(9.35)$$

when a step input is applied to the field, the response of the armature voltage is obtained as follows:

$$L^{-1}\left\{\frac{M}{\alpha}\left(\frac{1}{s} - \frac{1}{s + \dfrac{\alpha}{\tau}}\right)\right\} = \frac{M}{\alpha}(1 - e^{-\alpha t/\tau}) \qquad ...(9.36)$$

The steady-state value of the response is $\left(\dfrac{M}{\alpha}\right)$ and the time-constant is $\left(\dfrac{\tau}{\alpha}\right)$.

The steady-state gain of the separately excited dc generator, with $M = \dfrac{K_a K_f \omega_m}{R_f}$ and $\alpha = 1$, as obtained from Eqs. (9.33) and (9.36), is:

$$G_{ss} = \frac{K_a K_f \omega_m}{R_f} \qquad ...(9.37)$$

When the field circuit has a low value of resistance, the steady-state gain is seen to be high. But this increases the time-constant τ_f of the field circuit and the dynamic response of the armature voltage to changes in field voltage becomes slower. The speed of the dynamic response and the steady-state gain act in counter-productive directions. A good design demands a compromise between the high speed of response and high gain. It is needless to mention that for this simple machine:

$$\frac{G_{ss}}{\tau_f} = \frac{K_a K_f \omega_m}{L_f}, \text{ is a constant} \qquad ...(9.38)$$

9.3.2 Closed-loop Gain and Time-constant

The build-up of armature voltage may be made faster by a feedback from the armature to the field circuit. Figure 9.5(a) depicts a block-diagram of a general closed-loop system with one feedback branch, and Figure 9.5(b), the actual circuit diagram pertaining to a separately excited dc generator.

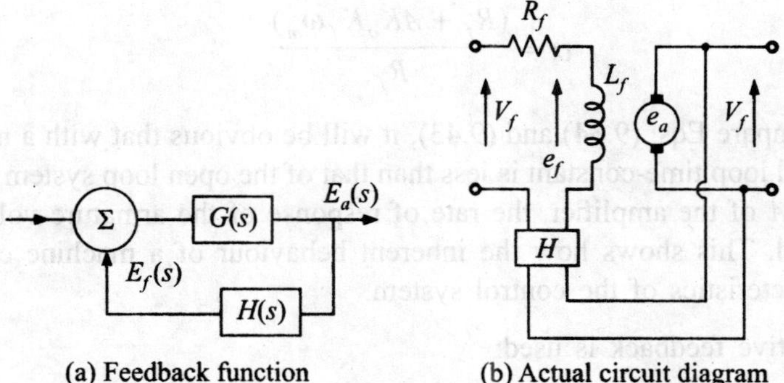

(a) Feedback function (b) Actual circuit diagram

FIGURE 9.5 Closed-loop system with feedback for a separately excited dc generator.

The input to the field winding is:

$$e_f = V_f \pm He_a \qquad \text{...(9.39)}$$

where, H is the transfer function of the feedback circuit.

The resulting armature voltage is given by:

$$e_a = Ge_f = G(V_f \pm He_a) \qquad \text{...(9.40)}$$

Here G is given by the relation (9.33) and the closed-loop transfer function assumes the form:

$$T = \frac{e_a}{V_f} = \frac{G}{1 \mp GH} \qquad \text{...(9.41)}$$

If the feedback loop has an amplifier of gain A, the result can have two cases:

Case I Let a negative feedback be used, so that $H = -A$, then:

$$T(s) = \frac{G(s)}{1 + AG(s)} = \frac{K_a K_f \omega_m}{R_f} \times \frac{1}{\tau_f s + 1 + \left(\dfrac{AK_a K_f \omega_m}{R_f}\right)} \qquad \text{...(9.42)}$$

and the steady-state gain and the time-constant are reduced. From Eqs. (9.35), (9.36) and (9.42), in the steady-state:

$$\left.\begin{aligned} G_{ss} &= \frac{M}{\alpha} = \frac{K_a K_f \omega_m}{R_f + AK_a K_f \omega_m} \\ \text{and} \qquad \tau_{c1} &= \frac{\tau}{\alpha} = \frac{L_f}{R_f + AK_a K_f \omega_m} \end{aligned}\right\} \qquad \text{...(9.43)}$$

where

$$M = \frac{K_a K_f \omega_m}{R_f} \qquad \text{...(9.44)}$$

and
$$\alpha = \frac{(R_f + AK_aK_f\omega_m)}{R_f} \qquad ...(9.45)$$

If we compare Eqs. (9.34) and (9.43), it will be obvious that with a negative feedback the closed loop time-constant is less than that of the open loop system, and by adjusting the gain A of the amplifier, the rate of response of the armature voltage can be duly controlled. This shows how the inherent behaviour of a machine can be altered by the characteristics of the control system.

Case II If a positive feedback is used:

$$T(s) = \frac{K_aK_f\omega_m}{R_f} \times \frac{1}{T_f(s) + \left(1 - \frac{AK_aK_f\omega_m}{R_f}\right)} \qquad ...(9.46)$$

and the time-constant is affected to:

$$T_{c1+} = \frac{L_f}{R_f - AK_aK_f\omega_m} \qquad ...(9.47)$$

In this case, the system can be stable only if the time-constant is positive. Thus, with a positive feedback (or $H = +A$), a stable system results, if

$$R_f > AK_a K_f \omega_m \qquad ...(9.48)$$

This can be achieved by controlling K_f, A and ω_m. K_f may be reduced using large air-gap in the machine, thereby yielding reduced flux per pole per ampere of field current, that is low values of ϕ_f. Therefore, slow-speed machines with large air-gaps could be made to have stable open-circuit characteristics than machines with short air-gaps and high speeds.

The system considered is a first-order one because it contains only one energy storage element—the inductance of the field winding. A separately excited dc generator normally forms a part of the source of excitation of a large synchronous machine, or the source of dc supply in a Ward-Leonard system of speed control of dc motors. Relation between gain, speed of response and stability of second or higher order systems may also be obtained by a similar procedure.

9.3.3 Dynamic Response of First and Second Order Linear Systems

Step response of open-loop system

The transfer function of a separately excited dc generator is observed to be:

$$G(s) = \frac{E_a(s)}{V_f(s)} = \frac{K_aK_f\omega_m}{R_f} \times \frac{1}{\tau_f s + 1} \qquad ...(9.49)$$

If the generator is operating at steady-state with the armature open-circuited, and with a field current,

$$I_{f0} = \frac{V_{f0}}{R_f} \quad ...(9.50)$$

The steady-state open-circuit voltage is given by:

$$E_{a0} = \frac{K_a K_f \omega_m}{R_f} \times V_{f0} \quad ...(9.51)$$

Since, the system is taken as linear, the change in armature voltage for a step change in field voltage experiences a sudden change from V_{f0} to V_{ft}, so that:

$$\Delta V_f = V_{f1} - V_{f0}$$

and the resulting armature voltage is given by:

$$e_a(t) = e_{a0} + L^{-1}\left[\frac{K_a K_f \omega_m}{R_f} \times \frac{1}{\tau_f s + 1} \times \frac{\Delta V_f}{s}\right] \quad ...(9.52)$$

But,

$$L^{-1}\left[\frac{1}{s} \times \frac{1}{(\tau_f s + 1)}\right] = L^{-1}\left[\frac{1}{s} - \frac{1}{s + \dfrac{1}{\tau_f}}\right]$$

$$= 1 - e^{-(t/\tau_f)} \quad ...(9.53)$$

Therefore,

$$e_a(t) = E_{a0} + \frac{K_a K_f \omega_m}{R_f}(1 - e^{-(t/\tau_f)})(V_{f1} - V_{f0}) \quad ...(9.54)$$

The final expression for armature voltage then becomes

$$E_a = E_{a0} + \frac{K_a K_f \omega_m}{R_f}(V_{f1} - V_{f0})$$

$$= \frac{K_a K_f \omega_m}{R_f} V_{f1} \quad ...(9.55)$$

Compare the expression with relation obtained in Eq. (9.12) in conjunction with Eq. (9.3).

First-order systems in cascade in open-loop

When the armature has a load supplied by its terminals, a step change in field voltage causes the terminal voltage across the armature terminals as well as the armature current to change. The system now denotes two first-order systems controlled in cascade in the open-loop configuration.

If we refer to the circuit diagram shown in Figure 9.6, the following equations in Laplace transform are arrived at:

$$I_f(s) = \frac{V_f(s)}{R_f + \dfrac{L_f}{s}} \quad ...(9.56)$$

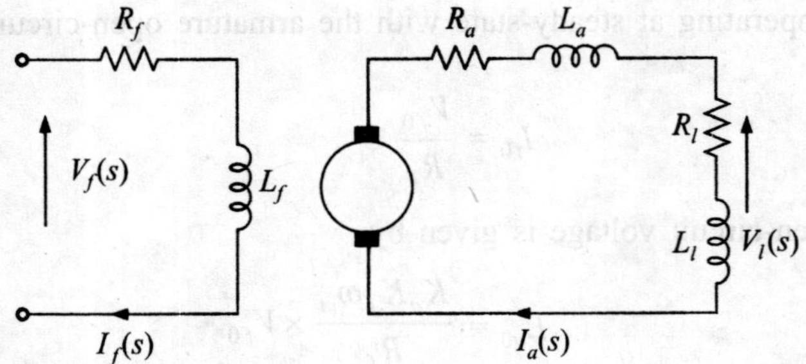

FIGURE 9.6 Two first order systems in cascade in open-loop.

$$E_a(s) = K_a K_f \omega_m I_f(s)$$

$$= \frac{K_a K_f \omega_m}{R_f} \frac{1}{\tau_f s + 1} \times \frac{1}{\tau_a s + 1} V_f(s) \qquad \text{...(9.57)}$$

$$V_l(s) = (R_l + L_l s) I_a(s)$$

$$= \frac{K_a K_f \omega_m R_l}{R_f + R} \frac{\tau_L s + 1}{(V_f s + 1)(\tau s + 1)} V_f(s) \qquad \text{...(9.58)}$$

where

R_l and L_l = Resistance and inductance of load

R_a and L_a = Resistance and inductance of armature

R and L = Total resistance and inductance of armature and load

$[R = R_l + R_a;\ L = L_l + L_a]$

τ_l = Time constant of load = $\dfrac{L_l}{R_l}$

τ = Time constant of armature circuit, that is, $\dfrac{L}{R} = \dfrac{L_l + L_a}{R_l + R_a}$

The resulting transfer function of the load voltage as a function of the field voltage therefore, becomes:

$$G(s) = \frac{V_l(s)}{V_f(s)} = \frac{K_a K_f \omega_m R_l}{R_f \cdot R} \times \frac{\tau_L s + 1}{(\tau_f + 1)(\tau s + 1)} \qquad \text{...(9.59)}$$

When a step change in V_f is imposed, $V_f(s) = \dfrac{V_f}{s}$, and the load voltage is:

$$V_L(s) = K \frac{\tau_1 s + 1}{s(\tau_f s + 1)(\tau s + 1)} V_f \qquad \text{...(9.60)}$$

where

$$K = \frac{K_a K_f \omega_m R_l}{R_f R} \qquad \text{...(9.61)}$$

Also,
$$\frac{\tau_L s+1}{s(\tau_f s+1)(\tau s+1)}=\frac{1}{s}-\frac{\tau_f-\tau_L}{\tau_f-\tau}\times\frac{1}{s+\frac{1}{\tau_f}}-\frac{\tau_l-\tau}{\tau_f-\tau}\times\frac{1}{s+\frac{1}{\tau}} \quad ...(9.62)$$

And, the resulting time variation of load voltage assumes the form:

$$V_L(t)=\frac{K_a K_f \omega_m R_l V_f}{R_f R}\left[1-\frac{\tau_f-\tau_L}{\tau_f-\tau}e^{-(1/\tau_f)}-\frac{\tau_l-\tau}{\tau_f-\tau}e^{-(1/\tau)}\right] \quad ...(9.63)$$

with a steady-state value of:

$$V_L = K_a K_f \omega_m \frac{V_f}{R_f}\times\frac{R_l}{R} \quad ...(9.64)$$

The response voltage $V_L(t)$ given in Eq. (9.63) contains two exponentially decaying transient terms. One of these terms decays with the field-time-constant τ_f, whereas the second term decays with time constant τ of the armature circuit. When two non-interacting first order systems are connected in cascade in the open-loop configuration, the response contains transient components whose time constants equal those of the individual first-order systems.

9.3.4 Step-response of Closed-loop Systems

First order systems

If a negative feedback loop is incorporated as shown in Figure 9.5, the transfer function for the armature voltage in terms of the field-winding voltage is given as follows:

$$T(s)=\frac{E_a(s)}{V_f(s)}=\frac{G(s)}{1+H(s)G(s)} \quad ...(9.65)$$

where

$$G(s)=\frac{K_a K_f \omega_m}{R_f}\times\frac{1}{\tau_f s+1} \quad ...(9.66)$$

and $H(s)$ = Feedback function

If an amplifier is incorporated, the feedback function is $H(s) = A$, the amplifier gain. Then,

$$T(s)=\frac{K_0}{1+AK_0}\times\frac{1}{1+\frac{\tau_f}{1+AK_0}s} \quad ...(9.67)$$

where

$$K_0=\frac{K_a K_f \omega_m}{R_f} \quad ...(9.68)$$

The armature voltage, for a step change is V_f, the field applied voltage, is given by:

$$E_a(s) = \frac{K_0 V_f}{1 + AK_0} \times \frac{1}{1 + \frac{\tau_f}{1 + AK_0} s} \times \frac{1}{s} \quad ...(9.69)$$

and its time variation, referring to Eq. (9.53) is obtained as follows:

$$e_a(t) = L^{-1} E_a(s)$$

$$= \frac{K_0 V_f}{1 + AK_0} \left[1 - e^{-(1 + AK_0)\frac{L}{\tau_f}} \right] \quad ...(9.70)$$

Thus, the steady-state closed-loop gain is:

$$G_{ss} = \frac{K_0}{1 + AK_0} = \frac{K_a K_f \omega_m}{R_f + K_a K_f \omega_m A} \quad ...(9.71)$$

whereas, the corresponding open-loop gain is observed in relation to (9.43) to be $\frac{K_a K_f \omega_m}{R_f}$.

The time-constant of the closed-loop system, with negative feedback is:

$$\tau_{cL-} = \frac{\tau_f}{1 + AK_0} = \frac{R_f}{R_f + K_a K_f \omega_m A} \times \tau_f \quad ...(9.72)$$

The resulting dynamic response is quicker than in open-loop configuration—the dynamic response has a time constant of τ_f.

Two first order systems in cascade

Let a feedback loop be introduced from the load voltage to the field circuit, as shown in Figure 9.7. The resulting transfer function from the load voltage to the field voltage is given by:

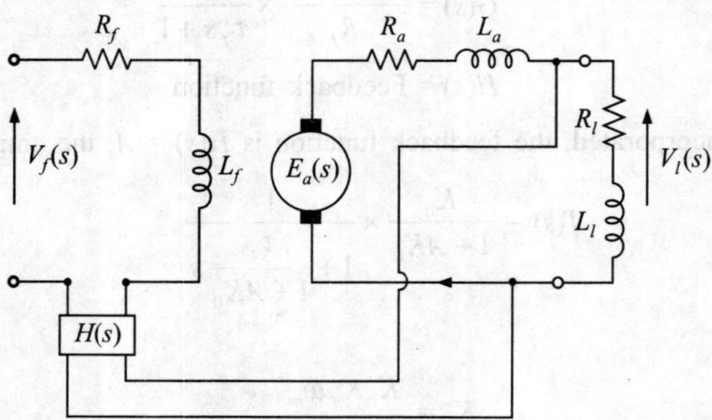

FIGURE 9.7 Tow first order systems of Figure 9.6 with close-loop feedback.

$$T(s) = \frac{V_l(s)}{V_f(s)} = \frac{G(s)}{1+[H(s)+G(s)]} \quad ...(9.73)$$

and with

$$G(s) = \frac{K_a K_f \omega_m R_l}{R_f R} \times \frac{(\tau_L s+1)}{(\tau_f s+1)(\tau s+1)}$$

where, $H(s)$ is the feedback function.

With a proportional control applied as before, $H(s) = A$–a number. Then the load voltage may be expressed as follows:

$$V_L(s) = T(s)V_f(s) = \frac{K\tau_L}{\tau_f \tau} \times V_f(s) \times \frac{s+\frac{1}{\tau_L}}{s^2+2\zeta\omega_n s+\omega_n^2} \quad ...(9.74)$$

where,

$$K = \frac{K_a K_f \omega_m R_l}{R_f R} \quad ...(9.75)$$

$$\omega_n^2 = \frac{1+AK}{\tau_f \tau} \quad ...(9.76)$$

$$2\zeta\omega_n = \frac{(\tau_f + \tau + AK\tau_L)}{\tau_f \tau} \quad ...(9.77)$$

Let $\omega_d^2 = \omega_a^2 (1-\zeta^2)$ ...(9.78)

Then, for a step change in V_f, from Eq. (9.67), with $V_f(s) = \frac{V_f}{s}$:

$$V_l(s) = \frac{KV_f}{1+AK}\left[\frac{1}{s} - \frac{s+\zeta\omega_n}{(s+\zeta\omega_n)^2+\omega_d^2} - \frac{\zeta-\omega_n\tau_l}{\sqrt{1-\zeta^2}} \times \frac{\omega_d}{(s+\zeta\omega_n)^2+\omega_d^2}\right] \quad ...(9.79)$$

and, the resulting time variation of load voltage is given by:

$$V_l(t) = \frac{KV_f}{1+AK}\left[1 - \frac{\sqrt{1+\tau_l^2\omega_n^2\left(1-\frac{2\zeta}{\tau_l\omega_n}\right)}}{1-\zeta^2} e^{(-\zeta\omega_n t)} \sin(\omega_d t+\phi)\right] \quad ...(9.80)$$

for $0 \le \zeta \le 1$

where,

$$\tan\phi = \frac{\sqrt{1-\zeta^2}}{(\zeta-\omega_n\tau_l)} \quad ...(9.81)$$

The closed-loop time constant with a loaded armature, from Eq. (9.80), is:

$$\tau_{\text{load}} = \frac{1}{\zeta\omega_n} = \frac{2\tau}{\tau_f + \tau + AK\tau_L}\tau_{fs} \quad \text{over } 0 \le \zeta \le 1 \qquad \text{...(9.82)}$$

and the corresponding value of open-circuit time constant from Eq. (9.72) is:

$$\tau_{0c} = \frac{1}{1 + AK_0}\tau_f \qquad \text{...(9.83)}$$

Thus, when two first-order systems are connected in cascade and are interconnected with a closed loop negative feedback, the response is described in the same manner as that of a second order system as presented by Eqs. (9.74) to (9.81). The response voltage given in Eq. (9.80) is in a damped oscillatory form and is valid when ζ lies between 0 and 1. The frequency of oscillation is given by:

$$f_d = \frac{1}{2\pi}\omega_d = \frac{\omega_n}{2\pi}\sqrt{1-\zeta^2} \qquad \text{...(9.84)}$$

When, on the other hand ζ is greater than 1, the response is over-damped, consisting of two exponentially decaying terms with different time constants, τ_1 and τ_2. These are negative reciprocals of two roots of the denominator of expression (9.74) in the relation:

$$s^2 + 2\zeta\omega_n s + \omega_n^2 = 0 \quad \text{for} \quad \zeta > 1$$

or

$$\left(s + \frac{1}{\tau_1}\right)\left(s + \frac{1}{\tau_2}\right) = 0 \qquad \text{...(9.85)}$$

The values of the time constants are:

$$\tau_1 = \left[\zeta\omega_n + \omega_n\sqrt{(\zeta^2 - 1)}\right]^{-1} \qquad \text{...(9.86)}$$

and

$$\tau_2 = \left[\zeta\omega_n - \omega_n\sqrt{(\zeta^2 - 1)}\right]^{-1} \qquad \text{...(9.87)}$$

thereby, resulting a response as in Eq. (9.74):

$$V_l(t) = \frac{K_a K_f \omega_m R_l V_f}{R_f R}\left[1 - \frac{\tau_1 - \tau_l}{\tau_1 - \tau_2}e^{-t/\tau_1} - \frac{\tau_2 - \tau_l}{\tau_2 - \tau_1}e^{-t/\tau_2}\right] \qquad (9.88)$$

The critical value of ζ is unity, and for this case, Eq. (9.74) assumes the following form for a step change in field voltage:

$$V_l(s) = \frac{K\tau_L V_f}{\tau_f \tau}\frac{\left(s + \frac{1}{\tau_L}\right)}{s(s+\omega_n)^2} \quad \text{for} \quad \zeta = 1$$

$$= \frac{KV_f}{1 + AK}\left[\frac{1}{s} - \frac{1}{s+\omega_n} - (1 - \tau_L\omega_n)\frac{\omega_n}{(s+\omega_n)^2}\right] \qquad \text{...(9.89)}$$

The voltage across the load terminals is therefore,

$$V_l(t) = \frac{KV_f}{1+AK}\left[1-\{1+(1-\tau_l\omega_n)\omega_n t\}e^{-\omega_n t}\right] \quad ...(9.90)$$

Neglecting inductance of the armature circuit, the load voltage for a step change in field voltage becomes:

$$V_l(t) = \frac{KV_t}{1+AK}\left[1-e^{-\left\{\frac{t(1+AK)}{\tau_f}\right\}}\right] \quad ...(9.91)$$

Since the time constant τ_l vanishes when the inductance of the armature and the load are neglected, and it has the same form as in Eq. (9.70) for the closed loop response of a first order system with negative feedback. The only difference is in the value of K, which, for the two cases are:

(a) For the open circuit condition:

$$K = K_0 = \frac{K_a K_f \omega_m}{R_f} \quad ...(9.92a)$$

(b) For resistive load:

$$K = \frac{K_a K_f \omega_m}{R_f}\cdot\frac{R_l}{(R_a + R_l)} \quad ...(9.92b)$$

If the load resistance is much larger than the armature resistance, that is $R_l >> R_a$, the two expressions shown in Eqs. (9.70) and (9.91) have identical values.

9.4 VOLTAGE REGULATION OF SEPARATELY-EXCITED DC GENERATOR

A dc generator normally operates from its open circuit condition with no armature current to its rated load condition. For this entire range of armature current, the terminal voltage across load should ideally remain constant, and we wish to record the general design requirements for achieving such a voltage profile. If the voltage applied to the field circuit is maintained constant and the field rheostat value is adjusted according to the armature current, we can observe in Figure 9.8(a) and (b), that the generated voltage equals the terminal voltage at no-load for which the field current has a value of I_{f0}. This leads to the value of the field resistance to be:

$$R_f + R_{r0} = \frac{V_f}{I_{f0}} \quad ...(9.93)$$

so that,

$$R_{f0} = \frac{V_f}{I_{f0}} - R_f \quad ...(9.94)$$

yields the maximum value of external resistance required in the field circuit.

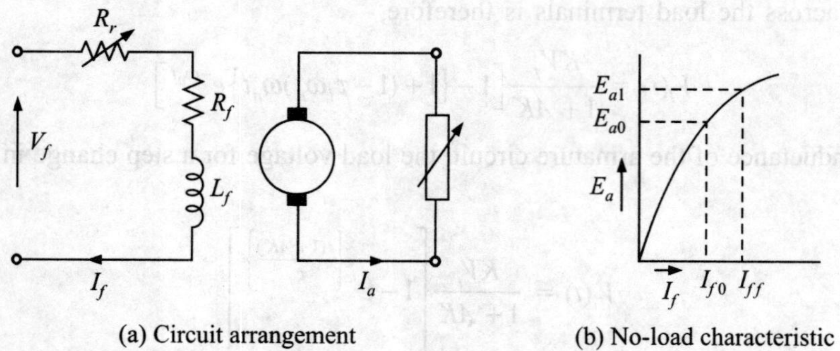

(a) Circuit arrangement (b) No-load characteristic

FIGURE 9.8 Voltage regulation of separately excited dc generator.

At full-load, the generated voltage naturally exceeds the no-load voltage in accordance with the voltage drop in the armature resistance and the brush voltage drop. This is indicated as follows:

$$E_{a1} = V_l + I_a R_a + I_a R_b \qquad ...(9.95)$$

where, R_b is the effective brush resistance. From the open-circuit characteristic shown in Figure 9.8(b), the required field current is observed to be I_{ff}. The minimum value of resistance required in the field rheostat is, therefore,

$$R_{rf} = \frac{V_f}{R_{ff}} - R_a \qquad ...(9.96)$$

The armature reaction modifies the equation as explained in the following illustration:

EXAMPLE 9.2 A 620 kW, 750 rpm, 600 volt dc generator has an open-circuit characteristic given below, at 750 rpm:

Field current, I_f	1.5	2.4	3.6	4.9	7.2	11.5	14.4	1.8
Generated voltage, E_a	200	300	400	500	600	700	750	800

The armature has a resistance of 0.020 ohm and the brush voltage drop of 2 volts. Further the armature reaction demagnetizing effect of full-load current in the armature amounts to 10% of the main field flux.

Determine the range of the resistance required in the field rheostat, if the field is excited by a constant voltage of 600 volts. The field winding has a resistance of 25 ohm. The terminal voltage across the load has to be maintained constant at 600 V.

Solution: At no load,

E_{a0} = 600 volt and I_{f0} = 7.2 amp

$$\therefore \qquad R_{r0} = \frac{600}{7.2} - 25 = 58.3 \text{ ohm}$$

At full-load,

$$I_a = 1000 \text{ amp}$$

∴ Required generated voltage to be produced

$$E_{al} = 600 + 1000 \times 0.002 + 2$$
$$= 622 \text{ volt}$$

As this voltage is generated by the resultant of the main field flux and the armature-reaction flux, the net ampere-turns acting on the armature corresponds to 8.1 amp of field current, as read off from the no-load characteristic of the machine (given) for a generated voltage of 622 volt. Therefore, the actual value of the main field current required is $I_{ff} = 9$ amp, obtained by

$$I_{ff} = \frac{8.1}{9} = 9 \text{ amp}$$

[Net I_{ff} = Actual I_f /demagnetizing factor]

$$R_f = \frac{600}{9} - 25 = 41.71 \text{ ohm}$$

Values of resistance required in the field rheostat at other load currents could be estimated under assumption that the armature reaction demagnetizing effect is proportional to the load current as shown:

I_a	0	200	400	600	800	1000	amp
E_a	600	606	610	614	618	622	volt
Net I_f	7.2	7.5	7.7	7.8	7.95	8.1	amp
Demagnetizing factor	1	0.98	0.96	0.94	0.92	0.9	Demagnetizing factor
Actual I_f	7.2	7.65	8.03	8.3	8.64	9.0	amp
$R_f + R_r$	83.3	78.5	74.8	72.3	69.5	66.7	ohm
R_r	58.3	53.5	49.8	47.3	44.5	41.7	ohm

$$R_{ff} + R_r = \frac{600}{\text{Actual } I_f}; \quad R_r = \frac{600}{I_f} - R_f \quad [R_f = 25\ \Omega]$$

In principle, the hardware for the voltage regulation scheme needs to be arranged by using a motor-operated rheostat controlled and actuated by the error voltage obtained between the terminal voltage and a reference voltage. The scheme is presented in Figure 9.9.

A constant reference voltage is applied across a potentiometer P_1, and its arm position selected suitably for the value at which the armature voltage has to be held. A second potentiometer P_2 is also connected across the load terminals of the generator. The variable point A of P_2 is set such that, with a constant voltage (of 600 volt in the example) across the load terminals, there is a zero voltage across the points A–B. As the load increases across the generator armature, the generator terminal voltage drops and the voltage across the potentiometer decreases. This will make point A to be at a lower voltage than B. The error voltage, positive at B is therefore developed. This error voltage could be supplied to a small dc motor actuating the contacts of the potentiometer rheostat P_2. This actuator motor rotates so as to decrease the total resistance of the field circuit, thereby, increasing the field current and also the generated voltage. As the terminal voltage across the load is restored to its original value, the potential difference between A and B is, once again, zero and the actuator motor stops automatically.

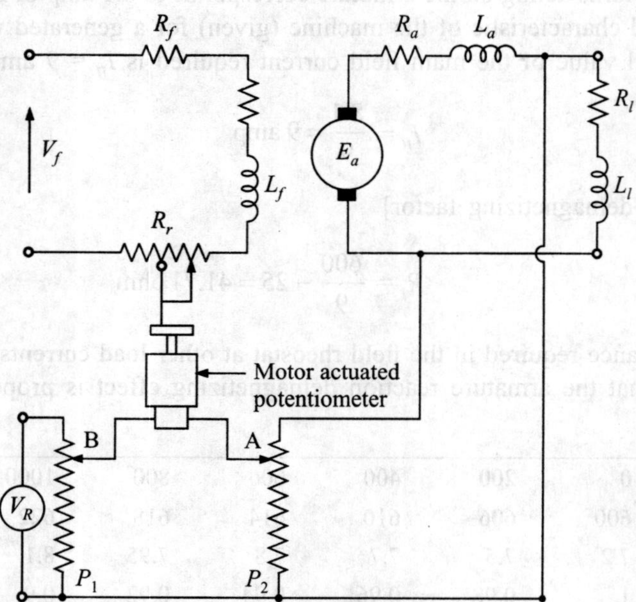

FIGURE 9.9 Schematic diagram of arrangement of component for voltage regulation.

With reduced load on the generator terminals, point A is at a higher potential than B, and the rheostat-motor rotates in the reverse direction. The generator field current, now, reduces till a balance is achieved and the terminal voltage equals the reference value.

9.5 SHORT-CIRCUIT IMPOSED ON ARMATURE TERMINALS OF SEPARATELY EXCITED GENERATOR

If the load terminals of a dc generator are suddenly short-circuited, a dangerously high current tends to flow in the armature of the generator. Although the circuit breaker operates to protect the armature and the machine, it is also necessary and desirable to have arrangement which would reduce the field current during the period of transition so that the generator armature voltage automatically reduces. To illustrate the short-circuit protection scheme, let us consider Example 9.2. Let us assume that the generator is delivering a full-load current of 1000 amp and developes a voltage of 622 volt in its armature prior to experiencing the short-circuit.

The field current prior to the short-circuit is 9.0 amp, and the total resistance of the field circuit is 66.7 ohm. If we assume that the field winding inductance and the armature inductance have values of L_a = 20 henry and L = 0.01 henry respectively, when the sudden short-circuit is imposed, the resulting short-circuit current can be obtained from the differential equation:

$$0.01\frac{di}{dt} + R_a i = E_a - \text{Brush drop} \qquad ...(9.97)$$

And, at i = 1000 amp,

$$0.01\frac{di}{dt} + 0.02i = 622 \quad 2 = 620$$

With an initial value of i = 1000 amp, the armature current is given by the relation:

$$i = 1000 + \frac{600}{0.02}\left(1 - e^{\frac{0.02}{0.01}t}\right)$$

and the final value of short-circuit current would be:

$$i_{sc} = 31000 \text{ amp}$$

If we assume that the safe short-circuit current should be limited to twice the full-load current, that is, 2000 amp in the instant case, the circuit breaker has to open the armature circuit in a time t_c, according the relation:

$$2000 = 1000 + 30000\,(1 - e^{-2t_c})$$

which yields a value of t_c = 17 milli-secs. With a sustained short-circuit, the generated voltage should not be more than:

$$E_a = 2000 \times 0.02 + 2 = 42 \text{ volt}$$

From the magnetization characteristic at 750 rpm (provided in Example 9.2), the field current needed should be brought down to 0.3 amp from the original value of 9.0 amp, within 17 milli-second.

It is a common practice to incorporate a high resistance R in the field circuit which is kept initially short-circuited with the help of a normally close-circuited contactor. This is inserted in the field circuit immediately following the occurrence of the short-circuit across the load terminals by the operation of an under-voltage relay.

In Example 9.2, the field circuit resistance is 66.7 ohm, the field inductance is 20 henry, the current is 9 amp, and the voltage is 600 volt, introduction of the resistance of R is suddenly done, the resulting current in the field circuit is given by the differential equation:

$$20\frac{di_f}{dt} + (66.7 + R)i_f = 600$$

And, with an initial value of $i_f = 9$ amp, we can solve the differential equation for i_f, as follows:

$$i_f = \frac{600}{66.7 + R}\left[1 + 0.015Re^{\left\{(66.7+R)\times\frac{0.017}{20}\right\}}\right]$$

The trial and error solution yields:

$$R = 4575 \text{ ohm}$$

It will be prudent to point out that the above simplified solution has ignored the influence of the following factors:

(a) The armature reaction during the period following the short-circuit is heavy, and therefore it will reduce the generated voltage from the constant value of 600 volt, assumed in estimating the growth of short-circuit current. A longer duration than 17 milli-secs is thus permissible.
(b) The collapse of the terminal voltage tends to operate the rheostat motor such that it increases the field current. A limit to the motor operation must be introduced so that it stops at a pre-set spot on the rheostat. However, this factor is not serious since the value of R is close to hundred times the resistance of the rheostat and the motor response is usually small.
(c) Due to the presence of armature reaction, the field current required for developing 42 volt is longer than 0.3 sec. Hence, the field current would be reduced to a safe value even before 17 milli-second.
(d) In a practical situation, the circuit breaker clearing time must be predetermined before selecting proper value of R. The preceding value of 4575 ohm has been found for a sustained short-circuit, which is safer.

9.6 DYNAMICS OF SEPARATELY EXCITED DC MOTORS

The mechanical transient behaviour (or the dynamics) of machines, when studied are grouped under dynamic behaviour of machines during motoring operation.
D.C. motors are extensively used in applications where precise control of speed and torque are envisaged over a wide range. A common method of control is to utilize a separately excited dc motor with constant field excitation. The basic equations relating the motor speed ω and the developed torque T_d for dc motor are:

$$\omega = \frac{V_t - I_a R_a}{K_a \phi} \qquad \text{...(9.98)}$$

and

$$T_d = K_a\, \phi\, I_a \qquad \text{...(9.99)}$$

where

V_t = Applied voltage to the motor armature
I_a = Armature current

ϕ = Flux per pole, interacting with armature

R_a = Resistance of armature circuit

$$K_a = \text{Constant for the motor} = \frac{PZ}{2\pi a}$$

where P denotes the number of poles and a stands for the number of parallel paths into which the armature winding is divided by the brushes, while Z represents the total number of armature conductors.

For a given shaft torque, the speed ω may be varied by controlling:

(i) the applied voltage, V_t
(ii) the armature circuit resistance R_a
(iii) the flux per pole, ϕ

The flux per pole, ϕ, may be varied by controlling the field current, I_f, of the motor. In the linear region, the flux per pole may be expressed as a constant times I_f, that is,

$$\phi = K_f I_f$$

The relationships listed under Eqs. (9.98) and (9.99) pertain to steady-state conditions, and may be of relevance when the variation of all quantities is very slow. When changes are made, it is assumed that they are varied from one steady-state value to another without disturbing the system otherwise.

The complete set of dynamic equations is important from the point of view of stability and operation of control equipment that need to be adopted to yield the desirable operating characteristics. The governing equations, written down with reference to Figure 9.10, and Figure 9.11 show (a) the relationship (b) the corresponding block diagrams and (c) the schematic diagrams which help in establishing these expressions on analogue computer. The detailed analogue representation is depicted in Figure 9.12.

There are three inputs to the systems:

1. V_t, the armature terminal voltage
2. T_L, the applied load torque (or braking torque) which may be constant, and
3. V_f, the voltage applied to the motor-field.

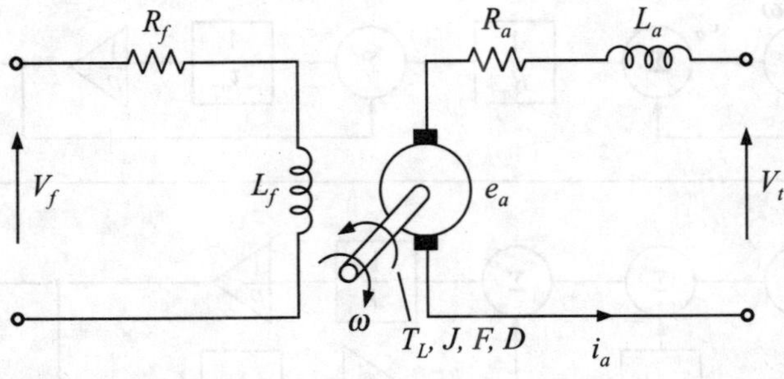

FIGURE 9.10 Speed control of dc motor—a schematic diagram.

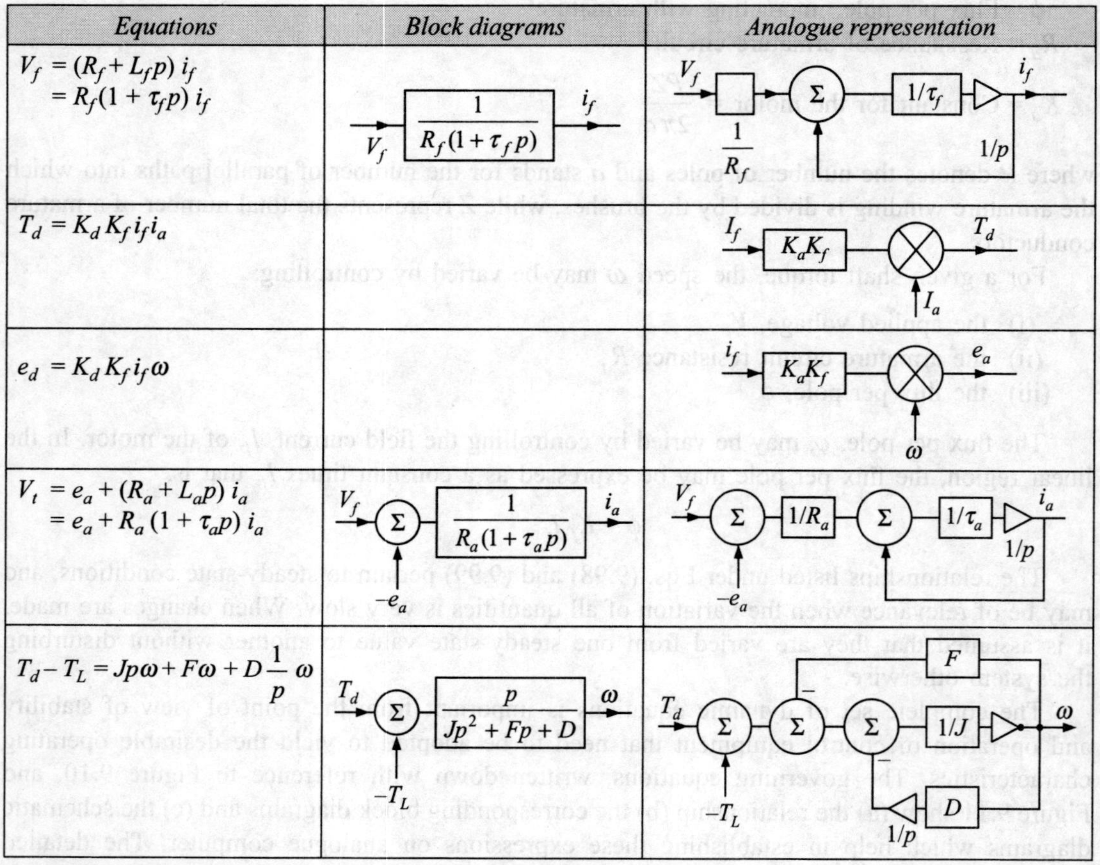

Equations	*Block diagrams*	*Analogue representation*
$V_f = (R_f + L_f p)\, i_f$ $= R_f(1 + \tau_f p)\, i_f$	V_f → $\dfrac{1}{R_f(1+\tau_f p)}$ → i_f	V_f, $\dfrac{1}{R_f}$, Σ, $1/\tau_f$, $1/p$, i_f
$T_d = K_d K_f i_f i_a$		I_f, $K_a K_f$, I_a, T_d
$e_d = K_d K_f i_f \omega$		i_f, $K_a K_f$, ω, e_a
$V_t = e_a + (R_a + L_a p)\, i_a$ $= e_a + R_a(1 + \tau_a p)\, i_a$	V_f → Σ → $\dfrac{1}{R_a(1+\tau_a p)}$ → i_a; $-e_a$	V_f, Σ, $1/R_a$, Σ, $1/\tau_a$, $1/p$, i_a, $-e_a$
$T_d - T_L = Jp\omega + F\omega + D\dfrac{1}{p}\omega$	T_d → Σ → $\dfrac{p}{Jp^2 + Fp + D}$ → ω; $-T_L$	T_d, Σ, Σ, Σ, F, $1/J$, ω, D, $1/p$, $-T_L$

FIGURE 9.11 Equations, block diagrams and analogue depiction of speed control of separately controlled dc motor.

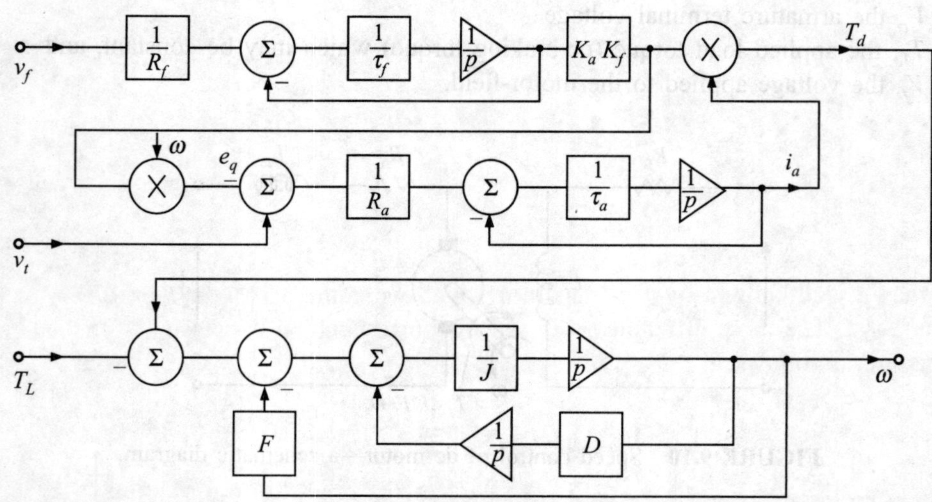

FIGURE 9.12 Complete analogue representation for the separately excited dc motor speed-control scheme.

9.6.1 V_f and T_L Held Constant, Step Change in V_t

In the linear region of the magnetizing characteristic,

$$\phi = K_f I_f \quad ...(9.100)$$

$$e_a = K_a K_f I_f \omega \quad ...(9.101)$$

$$V_t = K_a K_f I_f \omega + (R_a + L_a p) i_a \quad ...(9.102)$$

$$T_d = K_a K_f I_f i_a \quad ...(9.103)$$

and

$$T_d - T_l = Jp\omega + F\omega + D\frac{1}{p}\omega \quad ...(9.104)$$

In Eq. (9.104), the parameters J, F and D stand for the moment of inertia, damping and spring constants of the rotor.

If

$$K = K_a K_f i_f \quad ...(9.105)$$

eliminating i_a, the Laplace transform of the shaft speed is:

$$\omega(s) = \frac{KV_t(s) - T_L R_a - T_L L_a s}{L_a J s^3 + (R_a J + L_a F)s^2 + (R_a F + L_a D + K^2)s + RD} \quad ...(9.106)$$

In general, the shaft stiffness, D, has a very small influence, and if we neglect it, a simple expression for $\omega(s)$ is obtained, whose inverse transform may be obtained for a step change in V_t, as indicated below:

$$\omega(s) = -\frac{T_L}{J}\frac{s+\alpha}{s(s^2 + 2\zeta\omega_n s + \omega_n^2)} \quad ...(9.107)$$

where,

$$\alpha = \frac{T_L R_a - KV_t}{L_a T_L} \quad ...(9.108)$$

$$2\zeta\omega_n = \frac{R_a J + L_a F}{L_a J} \quad ...(9.109)$$

and

$$\omega_n^2 = \frac{R_a F + K^2}{L_a J} \quad ...(9.110)$$

This has a well-known inverse transform which we came across in Eqs. (9.74) to (9.90) in Section 9.3.4. The three different cases and the corresponding expressions for their inverse transforms are as follows:

Case (i) $0 < \zeta < 1$:

$$L^{-1}\left[\frac{s+\alpha}{s(s^2 + 2\zeta\omega_n s + \omega_n^2)}\right] = \frac{\alpha}{\omega_n^2}\left[1 - \frac{\sqrt{\alpha^2 - 2\alpha\zeta\omega_n + \omega_n^2}}{\alpha\sqrt{1-\zeta^2}} \times e^{(-\zeta\omega_n t)\sin(\omega_d t + \phi)}\right] \quad ...(9.111)$$

where

$$\omega_d^2 = \omega_n^2 (1-\zeta^2) \quad \text{...(9.112)}$$

and

$$\phi = \text{arc tan}\left[\frac{\sqrt{1-\zeta^2}}{s-\dfrac{\omega_n}{\alpha}}\right] \quad \text{...(9.113)}$$

Case (ii) $\zeta = 1$:

$$L^{-1}\left[\frac{s+\alpha}{s(s+\omega_n)^2}\right] = \frac{\alpha}{\omega_n^2}\left[1-\left\{1+\left(1-\frac{\omega_n}{\alpha}\right)\omega_n t\right\}e^{-\omega_n t}\right] \quad \text{...(9.114)}$$

Case (iii) $\zeta > 1$:

$$L^{-1}\left[\frac{s+\alpha}{s(s^2+2\zeta\omega_n s+\omega_n^2)}\right] = L^{-1}\left[\frac{s+\alpha}{s\left(s+\dfrac{1}{\tau_1}\right)\left(s+\dfrac{1}{\tau_2}\right)}\right]$$

$$= \frac{\alpha}{\omega_n^2}\left[1-\frac{\omega_n}{2\alpha\sqrt{\zeta^2-1}}\{(1-\alpha\tau_1)\,e^{-t/\tau_1}-(1-\alpha\tau_2)\,e^{-t/\tau_2}\}\right] \quad \text{...(9.115)}$$

where

$$\frac{1}{\tau_1} = \omega_n\left(\zeta+\sqrt{\zeta^2-1}\right) \quad \text{...(9.116)}$$

and

$$\frac{1}{\tau_2} = \omega_n\left(\zeta-\sqrt{\zeta^2-1}\right) \quad \text{...(9.117)}$$

The expressions for speed obtained from Eqs. (9.107) to (9.117), assume values of:

Case (i) $0 < \zeta < 1$:

$$\omega(t) = \frac{KV_l - R_aT_l}{R_aF+K^2}\left[1-\sqrt{\frac{1-2\zeta\dfrac{\omega_n}{\alpha}+\left(\dfrac{\omega_n}{\alpha}\right)^2}{1-\zeta^2}}\;e^{-\zeta\omega_n t}\sin(\omega_d t+\phi)\right] \quad \text{...(9.118)}$$

Case (ii) $\zeta = 1$:

$$\omega(t) = \frac{KV_t - R_aT_L}{R_aF+K^2}\left[1-\left\{1+\left(1-\frac{\omega_n}{\alpha}\right)\omega_n t\right\}e^{-\omega_n t}\right] \quad \text{...(9.119)}$$

Case (iii) $\zeta > 1$:

$$\omega(t) = \frac{KV_t - R_aT_L}{R_aF+K^2} + \frac{T_L}{J}\frac{1}{2\omega_n\sqrt{\zeta^2-1}}\left[(1-\alpha\tau_1)e^{-t/\tau_1}-(1-\alpha\tau_2)e^{-t/\tau_2}\right] \quad \text{...(9.120)}$$

The steady-state angular velocity in all three cases (cases (i), (ii) and (iii)), becomes:

$$\omega(t) = \frac{KV_t - R_a T_L}{R_a F + K^2} \qquad ...(9.121)$$

Neglecting armature time constant, τ_a: In case of dc motors, the electrical time constants are usually much smaller than the mechanical time constants. Further, the inductance of the armature is a very small quantity and contributes only marginally on speed changes. It is usual to analyse the speed variation for a preliminary estimate, by neglecting the armature inductance, L_a. The resulting expression of Eq. (9.107), then assumes the form:

$$\omega(s) = \frac{KV_t - R_a T_l}{R_a J} \frac{1}{s\left(s + \dfrac{K^2 + R_a F}{R_a J}\right)} \qquad ...(9.122)$$

and thus the speed is:

$$\omega(t) = \frac{KV_t - R_a T_L}{R_a F + K^2}\left[1 - e^{\left\{\frac{-t(R_a F + K^2)}{R_a J}\right\}}\right] \qquad ...(9.123)$$

The armature current is given [from Eq. (9.102)] as:

$$i_a(t) = \frac{V_t - K\omega}{R_a} = \frac{V_t}{R_a} - \frac{K}{R_a}\frac{KV_l - R_a T_l}{R_a F + K^2}\left[1 - e^{\left\{\frac{-t(R_a F + K^2)}{R_a J}\right\}}\right] \qquad ...(9.124)$$

And, their steady-state values become:

$$\omega_{ss} = \frac{KV_t - R_a T_L}{R_a F + K^2} \qquad ...(9.125)$$

and

$$I_a = \frac{V_t F + KT_L}{R_a F + K^2} \qquad ...(9.126)$$

It may readily be seen from Eqs. (9.125) and (9.126) that:

$$V_t - I_a R_a = K\omega_{ss}$$

as per the original operation relating to and governing the motor operation.

From the preceeding analysis, we can readily conclude that:

(a) When the field current has a high operating value, the machine constant:

$$K = K_a K_f \, i_f$$

is high in value and the time-constant τ is less, since

$$\tau = \frac{R_a J}{K^2 + R_a F}$$

Hence, the motor accelerates rapidly.

(b) With a low value of field current, however, τ is high and the motor takes longer time to attain the final speed.

EXAMPLE 9.3 A westing house generalised machine set is known to have the following parameters:

$$J = 0.0236 \text{ kg m}^2$$
$$F = 0.2 \times 10^{-2} \text{ Newton-metre/radian/per sec}$$
$$K_a K_f = 1.4 \text{ ohm/rad/sec}$$
$$= \text{Volt in ampere per ampere of field current per rad/sec of armature angular velocity.}$$

When the driving motor is initially at result and is unloaded, a step voltage of 115 volts is suddenly applied to the motor armature while the field is already excited. Estimate the variation in speed, under the following conditions:

(a) $R_a = 27$ ohm; $I_f = 0.4$ amp
(b) $R_a = 112$ ohm; $I_f = 0.4$ amp
(c) $R_a = 112$ ohm; $I_f = 0.5$ amp

Solution:

(a)

$$K = K_a K_f I_f = 0.56$$
$$R_a F + K^2 = 27 \times 0.2 \times 10^{-2} + (0.56)^2 = 0.367$$
$$\text{Time constant, } \tau = \frac{R_a J}{(R_a F + K^2)} = 1.75 \text{ sec}$$
$$\text{Steady-state speed} = \frac{K V_t}{(R_a F + K^2)}$$
$$= \frac{0.56 \times 115}{0.367} = 176 \text{ rad/sec}$$
$$= 1680 \text{ rpm.}$$

(b)

$$K = 0.56 \quad \text{[as in (a)]}$$
$$R_a F + K^2 = 112 \times 0.2 \times 10^{-2} + (0.56)^2$$
$$= 0.54$$
$$\text{Time constant, } \tau = \frac{112 \times 0.0236}{0.54} = 4.96 \text{ sec}$$
$$\text{Steady-state speed, } \omega_{ss} = \frac{0.56 \times 115}{0.54} = 120 \text{ rad/sec}$$
$$= 1150 \text{ rpm}$$

(c)

$$K = 1.4 \times 0.5 = 0.7$$
$$R_a F + K^2 = 112 \times 0.2 \times 10^{-2} + (0.7)^2$$
$$= 0.714$$

$$\text{Time constant, } \tau = \frac{R_a J}{(R_a F + K^2)}$$

$$= \frac{112 \times 0.236}{0.714}$$

$$= 3.75 \text{ sec}$$

The steady-state speed,

$$\omega_{ss} = \frac{K V_t}{R_a F + K^2} = \frac{0.75 \times 115}{0.714} = 113.5 \text{ rad/sec}$$

$$= 1080 \text{ rpm}$$

When the inductance of the armature L_a, and the stiffness of the shaft D, are neglected, the equations further simply to:

$$V_f = R_f(1 + \tau_f p)\, i_f \qquad \text{...(9.127)}$$

$$T_d = K_a K_f\, i_f\, i_a \qquad \text{...(9.128)}$$

$$e_a = K_a K_f\, i_f\, \omega \qquad \text{...(9.129)}$$

$$V_t = e_a + R_a\, i_a \qquad \text{...(9.130)}$$

and

$$Jp\omega + F\omega = T_d - T_l$$

$$= K_a K_f\, i_f\, i_a - T_L \qquad \text{...(9.131)}$$

and the resulting differential equation for ω in terms of V_f is non-linear because of the term V_f^2, and has the form:

$$Jp\omega + F\omega + \frac{K_a^2 K_f^2}{R_a} \frac{V_f^2 \omega}{(R_f + L_f p)^2} = \frac{K_a K_f V_t}{R_a} \frac{V_f}{(R_f + L_f p)} - T_L \qquad \text{...(9.132)}$$

The above equation is difficult to solve unless a simplistic procedure, shown below, is adopted.

A step change in V_f (the voltage applied across the field winding) causes a change in the field current given by:

$$i_f(t) = \frac{V_f}{R_f}\left\{1 - e^{-t/\tau_f}\right\} \qquad \text{...(9.133)}$$

The speed at which the armature is rotating depends on the accelerating torque $(T_d - T_L)$ and so, on the growth of field current, which determines the value of T_d as per Eq. (9.128). The field current also governs e_a—and so the armature current i_a as decided by Eqs. (9.129) and (9.130). If the variation of ω for a step change in i_f is calculated then for the actual variation of i_f given by Eq. (9.133), the resulting change in ω may be computed by using Duhamel's integral formula.

Let $\omega_s(t)$ be the variation of ω brought about by a step change in i_f. Then, the actual variation of ω, when the field current changes in accordance with Eq. (9.133) is:

$$\omega(t) = \omega_0 i_f(t) + \int_0^t \omega_s'(\tau)\, i_f(t - \tau)\, d\tau \qquad \text{...(9.134)}$$

where

$$i_f(t) = \frac{V_f}{R_f}\left\{1 - e^{-t/\tau_f}\right\}$$

ω_0 = Initial speed and

$$\omega_s(t) = \frac{\partial}{\partial \tau}\omega_s(\tau)$$

The response of ω for a step change in i_f is obtained by combining Eqs. (9.128) and (9.131), thus leading to:

$$Jp\omega + F\omega + \frac{K_a^2 K_f^2}{R_a} i_f^2 \omega = \frac{K_a K_f V_l}{R_a} i_f - T_L \qquad \text{...(9.135)}$$

Equation (9.135) is still non-linear because of the product $i_f^2\omega$. Analytical methods exist for solving such equations provided that the change in field current is very small and the perturbation method may be used. A linearization technique could also be adopted. A step-by-step procedure is, however, the most suitable and can be computerized.

The problem becomes more complex, if the load torque T_L is not constant but depends on the shaft speed.

9.7 WARD-LEONARD CONTROL OF DC MOTORS

Ward-Leonard control is extensively applied in dc motors exclusively for industrial uses involving speed variation and reversal of large motors. The control scheme consists of a separately-excited motor and a separately-excited generator, whose armatures are connected to each other. The motor field is supplied from a dc source and the field current is not adjusted during normal running. The generator field is connected to a separate source of dc supply and it is provided with a provision for adjustment. The scheme is shown in Figure 9.13.

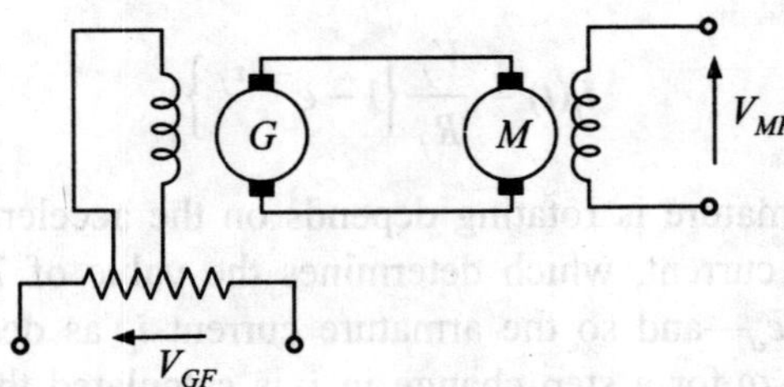

FIGURE 9.13 Schematic arrangement of Ward-Leonard system of speed control and reversal of direction of rotation of dc motor.

Ward-Leonard system is also used for position control of the shaft of a dc motor and for maintaining constant speed under variable loads on the motor shaft. This is done by mounting a tachometer with a feedback loop to the generator field. The block diagrams and analogous representations of these cases are appropriately detailed.

9.7.1 Speed Control

The complete circuit diagram of the physical model of Ward-Leonard system used for speed control of motor shaft under a load torque, T_L is presented in Figure 9.14. The inductances of armatures are neglected and the magnetic circuits are assumed to be operating in their linear regions. The generator shaft speed is also taken as constant during the development of the mechanical model.

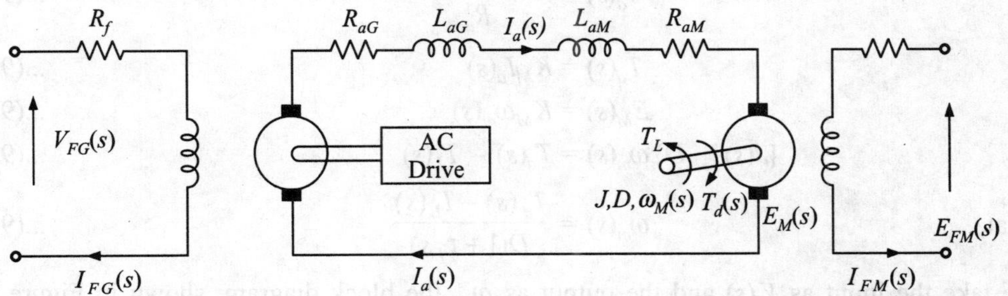

FIGURE 9.14 Electrical circuit diagram of Ward-Leonard speed and position control.

Let,

$V_{fG}(s)$ = The voltage applied to the generator field.

$I_{fG}(s)$ = Generator field current.

R_{fG}, L_{fG} = The resistance and inductance of the generator field.

$E_{aG}(s)$ = Voltage generated in generator armature.

$K_a = \dfrac{E_{aG}(s)}{I_{fG}(s)}$ = Constant at linear magnetic circuit with generator speed held constant.

$E_M(s)$ = Generator voltage at the motor armature, that is constant = $K_M\omega_M(s)$.

K_M = A constant of the motor when the field current is held constant.

$\omega_M(s)$ = Motor speed which is controlled by changing V_{fG}.

$R_a = R_{aM} + R_{aG}$ = Aggregate resistance of the firmly connected armature circuit of the generator and motor including all other windings like compensating and interpole windings and brush drop.

$I_a(s)$ = Current flowing in the firmly connected armature circuit.

$T_d(s)$ = Torque developed by motor armature.

$T_L(s)$ = Load torque, which may either be constant or dependent on the motor speed, $\omega_M(s)$.

J = Inertia of rotating parts of the motor shaft and the load.

D = Viscous damping torque in rads/sec.

$\tau_f = \dfrac{L_{fG}}{R_{fG}}$ = Generator field time constant.

$\tau_L = \dfrac{J}{D}$ = Mechanical time-constant of the load.

From the physical model, we can write down the following relations:

$$V_{fG}(s) = (R_{fG} + L_{fG}(s)\ I_{fG}(s)$$
$$= R_{fG}(1 + \tau_f s)\ I_{fG}(s) \qquad ...(9.136)$$

$$E_a(s) = K_a I_{fG}(s) \qquad ...(9.137)$$

$$I_a(s) = \frac{E_a - E_M(s)}{R_a} \qquad ...(9.138)$$

$$T_d(s) = K_M I_a(s) \qquad ...(9.139)$$

$$E_M(s) = K_M \omega_m(s) \qquad ...(9.140)$$

$$\{J(s) + D\}\ \omega_m(s) = T_d(s) - T_L(s) \qquad ...(9.141)$$

so that,

$$\omega_m(s) = \frac{T_d(s) - T_L(s)}{D(1 + \tau_L s)} \qquad ...(9.142)$$

If we take the input as $V_f(s)$ and the output as ω_m, the block diagram, shown in Figure 9.15, can be developed on the basis of Eqs. (9.136) to (9.142), as shown:

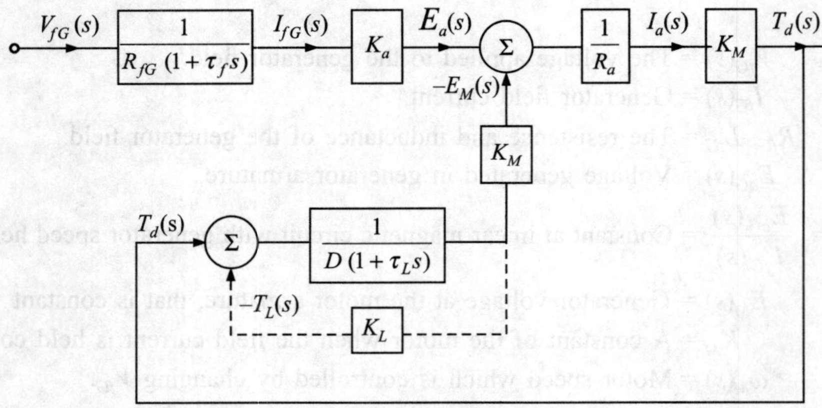

FIGURE 9.15 Block diagram of Figure 9.14.

9.7.2 Constant Speed Regulation of a dc Motor Using Ward-Leonard System

Ward-Leonard arrangement can also effectively maintain the speed of a dc motor that it uses for driving to a constant speed under condition of different load supplied by it. The arrangement uses a tacho-generator on the motor shaft. The output voltage, which is proportional to the speed of the motor shaft, is fedback to the generator field circuit of Ward-Leonard set as indicated in Figure 9.16. The resulting applied voltage to the generator field is given by:

$$V_{fG}(s) = A[V_R - K_t \omega_m(s)] \qquad ...(9.143)$$

where

A = The amplifier gain

K_t = The tacho-generator constant in volt/rad/sec

V_R = Reference voltage.

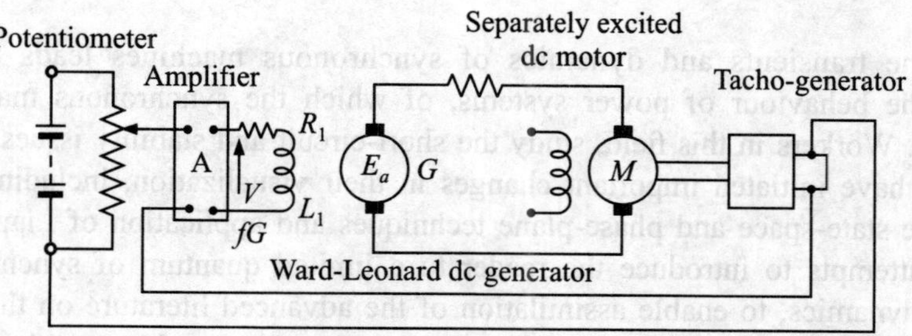

FIGURE 9.16 Constant speed regulation of a dc motor using Ward-Leonard control.

The potentiometer in this figure is adjusted for the desired constant speed of the motor so that, under steady state, the voltage applied to the generator field:

$$V_{fG} = A(V_R - K_t\omega_{ss}) \qquad ...(9.144)$$

If, now, the motor speed drops due to increase in shaft load, this causes a drop in the tacho-generator voltage output. This, in turn, develops an error voltage and also the voltage effective across the generator field is increased, and causes a higher value of field current, thus augmenting the generator voltage across the Ward-Leonard set. As a result the motor speeds up until the desired constant speed is attained by the motor as per earlier setting of the potentiometer.

Similarly, if the load decreases, on the motor shaft, it speeds up, causing the error voltage to decrease, thus resulting in a decrease in field-current to the Ward-Leonard generator, and a drop in generator output voltage. The motor slows-down up to a point when the selected constant speed is reached.

The resulting block diagram in Figure 9.17 now uses a negative feedback introduced by the tacho-generator output in Figure 9.16.

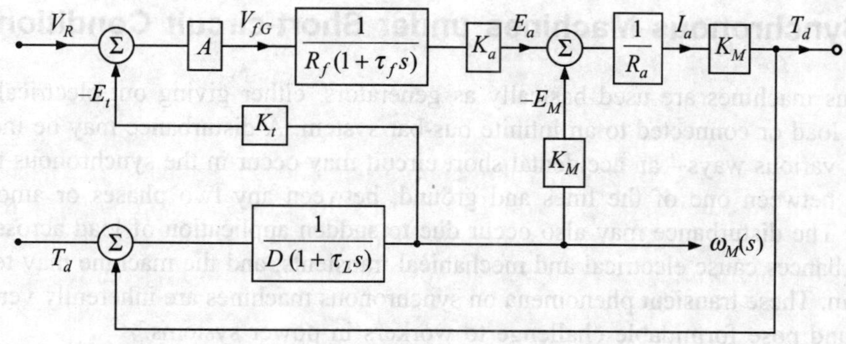

FIGURE 9.17 Block diagram for Figure 9.16.

9.8 TRANSIENT AND DYNAMICS OF SYNCHRONOUS MACHINES

9.8.1 Introductory Remarks

A study of the transients and dynamics of synchronous machines leads to the basis of appreciating the behaviour of power systems, of which the synchronous machines form an important part. Workers in this field, study the short-circuit and stability issues of synchronous machines and have initiated important changes in their visualization, including the analytical techniques like state-space and phase-plane techniques and application of Liapunov functions. This section attempts to introduce the reader to a limited quantum of synchronous machine transient and dynamics, to enable assimilation of the advanced literature on the area. Some of the problems faced in the area with reference to synchronous machines include:

(a) machine response under short-circuit operation.
(b) parallel operation, synchronizing and load sharing of alternators.
(c) voltage regulation, excitation system and control of reactive power.
(d) operation of a single machine on infinite bus-bar under load changes and changes in system reactances due to system faults.
(e) Alternator and synchronous motor stability during short circuit.
(f) shaft torque resulting from resonance conditions of connected electrical system, natural frequency of shafts, leading to shaft failure in turbo-alternators connected to series capacitance compensated long transmission lines—popularly known as sub-synchronous resonance conditions.

Problems associated with multi-machine systems, considering interconnection through reactances due to long transmission lines under fault condition assume special importance. In such cases, the stability considerations become very intricate. We introduce the case of a single alternator connected to an infinite bus-bar through reactances as an introduction to the associated problem in order to present techniques for solving problem using methods available for solution of non-linear differential equations like the Equal Area Criterion of Stability.

9.8.2 Synchronous Machines under Short-circuit Conditions

Synchronous machines are used basically as generators, either giving out electrical power to an isolated load or connected to an infinite bus-bar system. A disturbance may be incident in a machine in various ways—an accidental short circuit may occur in the synchronous machine—may occur between one of the lines and ground, between any two phases or among all the three lines. The disturbance may also occur due to sudden application of load across the lines. Such disturbances cause electrical and mechanical transients, and the machine may tend to lose synchronism. These transient phenomena on synchronous machines are inherently very complex in nature, and pose formidable challenge to workers in power systems.

Short-circuit imposed on an open-circuited synchronous generator

Figure 9.18 presents a schematic depiction of a 3-phase synchronous machine, for which the rotor is taken as rotating at some speed ω_m, and the field current I_f develops an open-circuit voltage of E_{fG} in each phase. If the stator terminals are now short-circuited, a large transient current flows through them. This transient eventually dies down and a steady state short-circuit current starts flowing whose value is given by:

$$I_{sc} = \frac{E_{fG}}{X_s} \qquad ...(9.145)$$

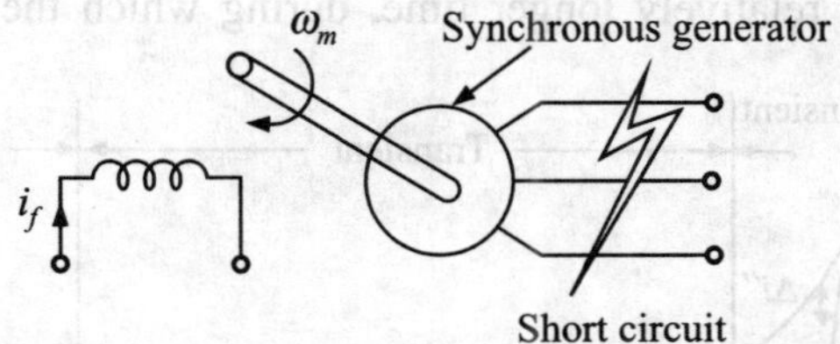

FIGURE 9.18 Sudden 3-phase short-circuit imposed on synchronous generator.

If E_{fG} = 1 pu and X_s = 1 pu (this is the typical value), the steady-state short-circuit current is 1 pu. This is a good feature of the synchronous machine. If the short-circuit is sustained, it will not damage the armature. However, at the instant the short-circuit is imposed, the armature current may assume very high values—as high as 5 to 10 pu. In order to determine the circuit breaker rating or relay setting of the protective system, a prior knowledge of the armature current during this initial period assumes special important.

Now, before imposition of the short-circuit, the flux leakages of the field winding and the damper winding were constant. No current in the damper winding was present and the only current in the field winding was a dc current. However, with the incidence of short-circuit, armature current starts flowing, whose mmf directly opposes the mmf of the field winding. The flux linkages of both field and damper windings are, therefore, affected. In order to maintain the component fluxes constant at their initial values, the induced components of current, now, flows both in field and damper windings. The induced currents in the field and damper windings decay because of the winding resistances in these circuits.

The change in field current is shown in Figure 9.19, and it is clear that the induced currents are equivalent to an augmentation of the field excitation and therefore, a large current flows in the armature circuit immediately following the incidence of short-circuit on the armature terminals.

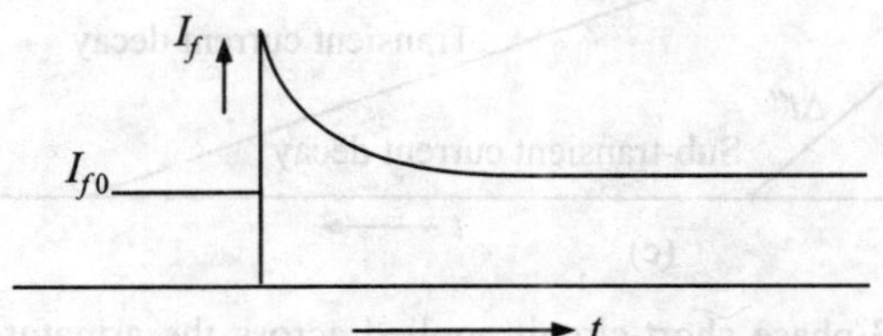

FIGURE 9.19 Change in field current immediately following a three-phase short-circuit across armature.

In salient-pole synchronous machines, the damper windings are located on the rotor pole faces. For cylindrical-rotor machines, the damper windings are usually not there. However, during transient periods, currents that are induced directly in the rotor body, produce the same effect as the damper currents in a salient-pole machine.

A typical trace of short-circuit current in a stator phase of the synchronous machine is shown in Figure 9.20(a), under a three-phase short-circuit, suddenly applied to the armature terminals.

These symmetrical trace of short-circuit currents due to sudden short-circuit on armature terminals can be obtained using oscillogram. The envelop of the wave is detailed in Figure 9.20(b), and it shows three distinct periods lasting only a few cycles following incident of short-circuit—the sub-transient period during which the current decreases very rapidly, the transient period lasting relatively longer time, during which the current decrease is more

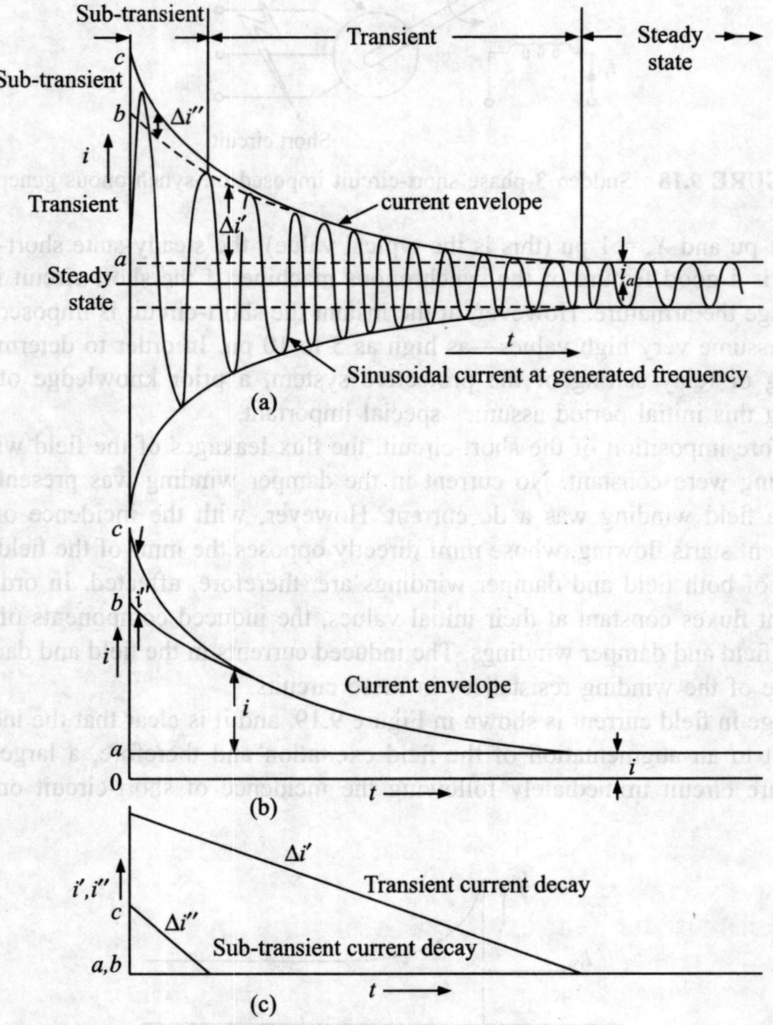

FIGURE 9.20 A sudden 3-phase short-circuit applied across the armature terminals: (a) symmetrical armature current (b) envelop of current (c) decay of current difference.

moderate and finally the steady-state period, during which the current stabilizes to a value given by Eq. (9.145). These three successive periods merge through nearly exponential decays.

In Figure 9.20(b), the current difference $\Delta i'$—the difference between the transient envelope and the steady-state amplitude—and similarly $\Delta i''$—the difference between the sub-transient envelope and an extrapolation of the transient envelope—when plotted on semi-log coordinates, they decay linearly, as shown in Figure 9.20(c), thus indicating that they are exponential decays.

During the sub-transient period, because of the demagnetizing effect of armature current, currents are induced in the field winding as well as the damper winding, to maintain flux-constancy of the pre-fault conditions. The damper current decays rapidly because of the small time constant of the damper circuit. The behaviour of the stator current during the period is controlled by the damper winding.

During the transient period, the damper winding current has decayed to zero. The behaviour of the armature current during this period is governed by the field winding current. It is decayed with a large time constant.

A critical scrutiny of the short-circuit oscillogram shows that the symmetrical wave-form of current in Figure 9.20(a) is a special case and not a general case. The more usual short-circuit currents are not symmetrical about the zero current axis but definitely show dc components responsible for offset waves.

The dc component in the short-circuited armature currents is due to the flux-linkage of a phase at the instant the short-circuit is imposed. If the flux-linkage of a phase is zero at the instant the short-circuit has been applied, no dc component is required to maintain the flux-linkage at that zero value, and the short-circuit current wave for that phase is symmetrical. On the other hand, if the flux-linkage of a phase is non-zero at the instant of short-circuit, the dc component must appear in the current of that phase to keep the flux linkage constant. The dc component decays with the armature time constant. We may note that in a three-line system the sum of the dc components in the three phases at all times adds up to zero. The largest dc component occurs in a phase current when the flux-linkage of that phase is maximum at the instant of application of short-circuit.

Short-circuit on a loaded synchronous generator

Let us assume that a 3-phase generator is delivering power to a load or to an infinite bus-bar. If a short-circuit is imposed at the machine terminals, the short-circuit current passes through a sub-transient period and a transient period, and finally settles down to a steady-state period of operation. When the short-circuit is imposed, the machine reactance changes from X_d to X_d''. The excitation voltages also change to satisfy the initial condition of flux-linkage constancy.

A detailed analysis the phenomenon following short-circuit of a loaded synchronous machine is quite complex, because it involves many coupled circuits with initial currents.

9.8.3 Synchronous Generators Operating in Parallel

When number of alternators function in parallel, the following functional requirements assume special significance:

(a) The alternators should preferably be of similar ratings and should, as far as possible, share the total load in proportion to their ratings;

(b) When a change of load comes about, all the concerned alternators should continue to function in synchronism and share the total load;
(c) Their frequencies must be the same and the prime-movers must operate at synchronous speeds;
(d) The machines should not lose stability when under fault conditions on the system, and should have sufficient restoring force to take care of the resulting shocks and disturbances.

The above observations lead to the idea of synchronizing power, requirement of damper winding, swing curves, steady-state and transient stability limits, and solution of non-linear differential equation for Equal-Area Criterion for stability, state-space and phase-plane technique, Routh-Harowitz criterion for stability and also Liapunov stability using Liapunov function.

Load-staring under Steady-state Conditions

We now take up the case of two alternators functioning in parallel:

Let the combination supply a total load of:

$$W_L = P_L + jQ_L$$

at a bus-bar voltage, V, so that

$$I_L = \left[\frac{W_L}{V}\right]^* = A + jB \qquad ...(9.146)$$

* *In this discussion, the above quantities are assumed to be in their per-unit values on a chosen base quantity. Whenever actual numbers are used, they denote the values for one phase.*

If the generators are identical, each contributes to one-half of the load current. In general, we refer to Figure 9.21, and the load is supplied by two alternators—generator I and generator II:

I_{I} and I_{II} : pu currents supplied by the two machines
E_{I} and E_{II} : Corresponding excitation voltages
δ_{I} and δ_{II} : Respective power angles
X_{I} and X_{II} : Corresponding synchronous reactances
$m/(1 - m)$: Ratio of their ratings
β : Reactive component of current supplied by machine I
and $V\angle 0°$: the bus-bar voltage of the combination.

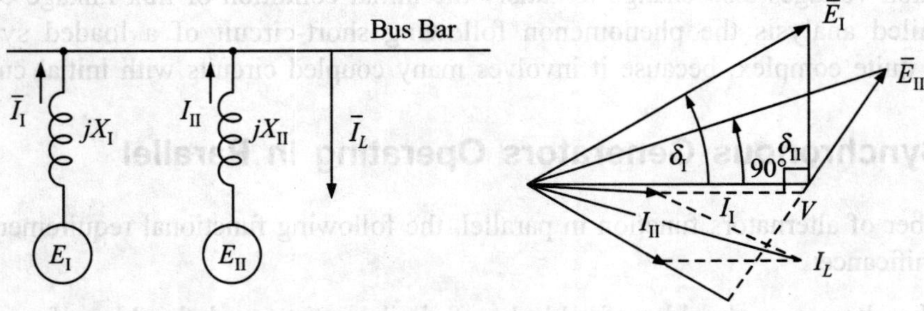

FIGURE 9.21 Two generators in parallel, sharing load.

The component of real power, as shared by the two generators, is the ratio of their ratings. The in-phase component of current of the machines is, therefore, mA for the machine I and $(1 - m)A$ for machine II. The reactive components of current supplied by the machines is β (for machine I). In other words:

$$I_{\text{I}} = mA + j\beta \quad \text{...(9.147)}$$

and

$$I_{\text{II}} = (1 - m)A + j(B - \beta) \quad \text{...(9.148)}$$

Considering cylindrical rotor configuration, the excitation voltage expressions assume the form:

$$E_{\text{I}}\angle\delta_1 = V + jX_{\text{I}}\,(mA + j\beta) \quad \text{...(9.149)}$$

and

$$E_{\text{II}}\angle\delta_2 = V + jX_{\text{II}}\,[(1 - m)A + j(B - \beta)] \quad \text{...(9.150)}$$

From the above relations, we get:

$$E_{\text{II}}^2 = (V - \beta X_{\text{I}})^2 + (mAX_{\text{I}})^2 \quad \text{...(9.151)}$$

and

$$E_{\text{II}}^2 = [V - (B - \beta)X_{\text{II}}]^2 + [(1 - m)AX_{\text{II}}]^2 \quad \text{...(9.152)}$$

Also,

$$\tan\delta_{\text{I}} = \frac{mA\,X_{\text{I}}}{(V - \beta X_{\text{I}})} \quad \text{...(9.153)}$$

and

$$\tan\delta_{\text{II}} = \frac{(1-m)A\,X_{\text{II}}}{[V - (B-\beta)X_{\text{II}}]} \quad \text{...(9.154)}$$

At this load condition, if the excitation voltage in machine I is varied, the load component of current remains maintained at their original value, but the remaining values of $\bar{E}_{\text{II}}$, δ_{I}, δ_{II} and β undergo changes. These variations are given by the following relations, based on the original relations recorded under Eqs. (9.149) to (9.154):

$$\beta = \left[V - \sqrt{E_{\text{I}}^2 - (mA\,X_{\text{I}})^2}\right]\frac{1}{X_{\text{I}}} \quad \text{...(9.155)}$$

$$\delta_{\text{I}} = \text{arc tan}\left[\frac{mA\,X_{\text{I}}}{V - \beta X_{\text{I}}}\right] \quad \text{...(9.156)}$$

$$E_2 = \left[\{V - (B-\beta)X_{\text{II}}\}^2 + \{(1-m)A\,X_{\text{II}}\}^2\right]^{1/2} \quad \text{...(9.157)}$$

$$\delta_{\text{II}} = \text{arc tan}\left[\frac{(1-m)A\,X_{\text{II}}}{V - (B-\beta)X_{\text{II}}}\right] \quad \text{...(9.158)}$$

Circulating current: When the reactive component of current, β, delivered by machine I has the same sign as the total reactive current B of the load, the remaining component of reactive current delivered by machine II, $(B - \beta)$ also has the same sign at B. However, when the excitation of machine I is increased to a value such that β exceeds the total reactive current B, that is—$\{|\beta| > |B|\}$ or $|B - \beta|$ is negative. This leads to production of circulating current caused by $\bar{E}_{\text{I}}$ against the excitation voltage E_{II} of machine II. The limiting values of E_{I}, δ_{I} and δ_{II} and also E_{II}, for the initiation of circulating current, can be computed for the condition $\beta = B$. The correspond condition can be arrived from Eq. (9.149) to Eq. (9.154) as:

$$E_{\text{I}} = \sqrt{(V - B X_{\text{I}})^2 + (mA X_{\text{I}})^2} \quad ...(9.159)$$

$$\delta_{\text{I}} = \text{arc tan}\left[\frac{mA X_{\text{I}}}{V - B X_{\text{I}}}\right] \quad ...(9.160)$$

$$E_{\text{II}} = \sqrt{V^2 + [(1-m)A X_{\text{II}}]^2} \quad ...(9.161)$$

$$\delta_{\text{II}} = \text{arc tan}\,\frac{[(1-m)A X_{\text{II}}]}{V} \quad ...(9.162)$$

And, if E_{I} is increased beyond the value given by Eq. (9.159), a circulating current, given below, is initiated, as given from Eq. (9.155):

$$I_c = |\beta - B|$$

$$= \left|\frac{1}{X_{\text{I}}}\left[V - \sqrt{E_{\text{I}}^2 - (mA X_{\text{I}})^2}\right] - B\right| \quad ...(9.163)$$

The machine that has more excitation voltage than necessary for the proportionate share of load current runs hot due to consequent increased losses due to its stator resistance.

EXAMPLE 9.4 Two star-connected alternators, rated at 13.8 kV (line-to-line), having the following ratings:

Machine 1 3 phase, 50 Hz, 19200 kVA, 800 amp and X_1 = 10 ohm/phase (1 pu);

Machine 2 3 phase, 50 Hz, 9600 kVA, 400 amp and X_2 = 20 ohm/phase (1 pu);

They supply a constant load of 23.04 MW at 0.8 pf lagging, at a bus-bar voltage of 13.8 kV (line-to-line).

Determine:

(i) The limiting excitation voltages E_1 and E_2 and power angles δ_1 and δ_2, along with the currents I_1 and I_2 supplied by the machines while sharing the load;

(ii) The limiting value of excitation voltage E_1 of machine 1 for causing flow of circulating current (that is, machine 1 carries all the reactive current of the total load). Also obtain values of E_1, E_2, δ_1 and δ_2, and also I_1 and I_2, under the above condition of functioning.

(iii) If the excitation voltage E_1 is increased to 120% of part (i), find the operating conditions of the two machines.

Solution: Voltage per phase at line terminals:

$$= \frac{13.8 \times 10^3}{\sqrt{3}} = 8\angle 0° \text{ kV}$$

Bus-bar load current,

$$I_L = \frac{23040}{\sqrt{3} \times 13.8} \frac{0.8 - j0.6}{0.8}$$

$$= (960 - j720) \text{ amp}$$

$$= 1200 \angle -36.87 \text{ amp}$$

Ratio of ratings of the two machines:

$$= \frac{19200}{9600} = \frac{2}{1}$$

Hence, $m = \frac{2}{3}$ and $(1 - m) = \frac{1}{3}$

(i) Current supplied by machine 1:

$$I_1 = mI_L = \frac{2}{3} \times 1200 \angle -36.87°$$

$$= 800\angle -36.87°$$

$$= (640 - j480) \text{ amp}$$

$$= (0.64 + j0.48) \text{ k-amp}$$

$$I_2 = (1 - m)I_L = 400\angle -36.87°$$

$$= (320 - j240) \text{ amp}$$

$$= (0.32 - j0.24) \text{ k-amp}$$

Excitation voltages:

$$E_1 = \overline{V} + jX_1I_1 = 8\angle 0° + \text{j}10 \times 0.8\angle -36.87°$$

$$= 14.3\angle 26.5° \text{ kV per phase}$$

and $\quad E_2 = \overline{V} + jX_2I_2 = 14.3\angle 26.5° \text{ kV line-to-line}$

or $\quad = 1.8 \text{ pu}$

$E_1 = E_2$ and $\delta_1 = \delta_2 = 26.5$

(ii) At the condition of commencement of circulating current:

$$\beta = B = -j720 \text{ amp} = 0.72 \text{ k-amp}$$

$\therefore \quad E_1 = \{(8 + 0.72)^2 + (6.4)^2\}^{1/2}$

$$= 16.5 \text{ kV per phase}$$

or $\quad E_1 = 28.6 \text{ kV (line-to-line)}$

$$= 2.06 \text{ pu}$$

and $\quad \delta_1 = \text{arc tan} \frac{6.4}{15.2} = 23°$

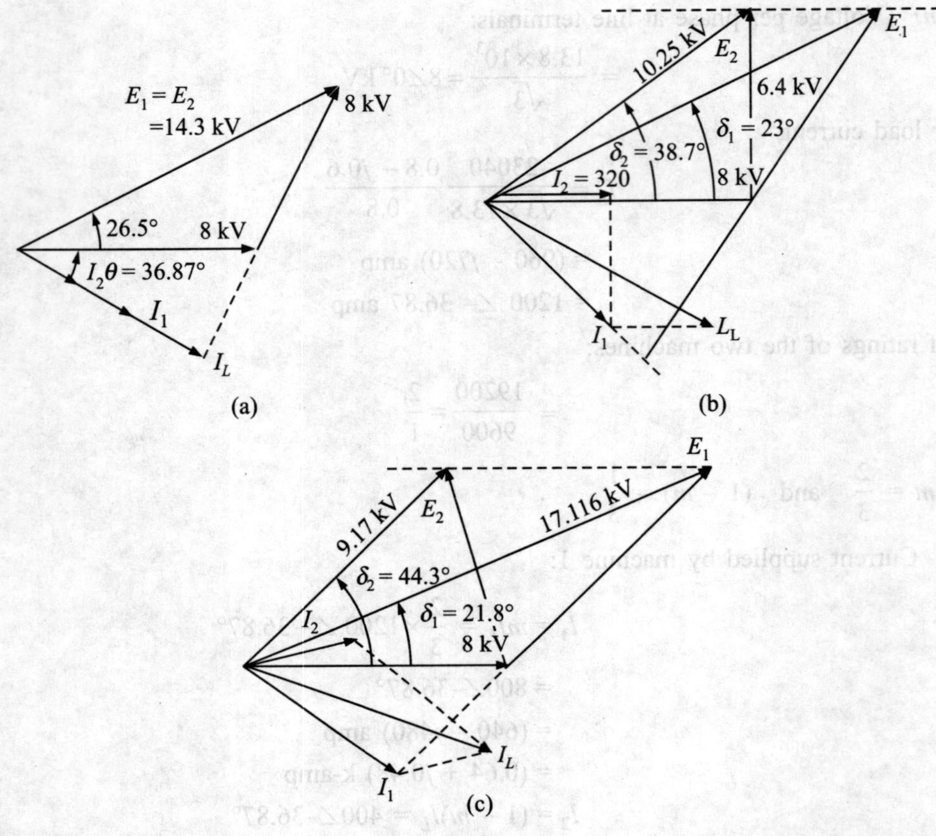

FIGURE 9.22 Phasor diagrams (a), (b) and (c) for parts (i), (ii) and (iii) respectively.

Also, $E_2 = (8^2 + 6.4^2)^{1/2} = 10.25$ kV per phase

$= 17.75$ kV (line-to-line)

$= 1.28$ pu

and $$\delta_2 = \text{arc tan}\,\frac{6.4}{8} = 38.7°$$

$$I_1 = (0.64 - j0.72) \text{ k-amp}$$
$$I_2 = (0.32 - j0) \text{ k-amp}$$

Hence,

$$W_1 = (15.36 + j17.28) \text{ MVA}$$
$$W_2 = (7.68 + j0) \text{ MVA}$$

(iii) When E_1 = 120% of 14.3 kV, that is, 17.16 kV (which is equivalent to $E_1 = 2.15$ pu)

$$\beta = \frac{1}{10}\left\{8 - \sqrt{(17.16)^2 - (6.4)^2}\right\}$$
$$= 0.792 \text{ k-amp}$$

which is 10% in excess of load requirements of $B = -\ 0.72$ k-amp.

Hence, the circulating current is 72 amp.

Thus,

$$I_1 = 0.64 - j0.792 \text{ k-amp}$$

$$I_2 = 0.32 + j0.072 \text{ k-amp}$$

and

$$I_L = 0.96 - j0.072 \text{ k-amp}$$

Hence,

$$E_2 = 8 + j20(0.32 + j0.072)$$
$$= 9.17\angle 44.3° \text{ kV/phase}$$
$$= 15.9 \text{ kV (line-to-line)}$$
$$= 1.15 \text{ pu}$$

Also,

$$\delta_1 = \text{arc tan}\left[\frac{6.4}{8 + 7.92}\right] = 21.8°$$

$$\delta_2 = 44.3°$$

$W_1 = (15.96 + j19)$ MVA

$W_2 = (7.68 - j1.73)$ MVA

Under this condition, the machine 1 carries 27% of overload.

Synchronizing of alternators

It is a very common practice in power system applications to switch synchronous machines out of service and to reconnect them to the bus-bars. Such reconnection involves synchronization of a machine to the bus-bars of the power systems.

The process of reconnection through switching must ensure avoidance of shocks and jerks on the on-coming machine because it sets up vibration of machine foundation, mechanical forces on machine conductors and induced currents on machine pole faces. In the worst possible condition, the excessive forces assume values to those under sudden short-circuit across the machine terminals by voltages equalling twice the voltage for which the machine is rated. An improper or faulty synchronization of an alternator to a power system configuration, with a number of machines functioning in parallel—as is encountered in an infinite bus-bar system—may pose very severe conditions, as the inductance across such an on-coming machine terminals are much lower than that of a single machine (as an infinite bus-bar has theoretically zero impedance), leading to development of very large forces.

We can closely examine the necessary and sufficient conditions that need to be achieved to enable satisfactory synchronizing.

1. *The phase-sequence of the on-coming machine and that of the bus-bar across the circuit breaker should be identical*

The phase-sequence of the machine and the bus-bar are usually tested during installation of the machine in the first place. It is not likely that it gets modified and it is not required to be checked before every synchronizing action. If the phase sequences do not match, there will be swings in alternator current as the periodic phase difference between the voltages persists. This current swing occurs at the rated frequency and at twice the rated voltage of the machine.

2. *The voltage magnitude of the on-coming machine and the bus-bar must be equal across the circuit breaker poles*

If the voltage magnitudes are unequal, there is a circulating current in the machine winding, as soon as the circuit breaker is closed. A voltmeter connected through a potential transformer is kept connected across two poles of the circuit breaker along with a change-over switch to ensure equity of voltages of the on-coming machine and the corresponding terminals of the bus-bar.

3. *The frequency of the two voltages must be equal*

This condition is practically very hard to be satisfied in exact terms, nor is it strictly aimed at. This is because of the inertia of extremely heavy rotating parts of the on-coming machine and its prime-mover. If a strict equity of frequency is not satisfied, the two sets of 3-phase voltages rotate relative to each other at a slip frequency. Even when the magnitudes of voltages (condition 2) and the phase sequence (condition 1) are met, the current swings initially occur due to nearly twice the system voltage but at a low slip frequency with the synchronizing circuit breaker closed. In normal synchronizing practice, the machine is kept slower than that of the bus-bar, as it is better to pull up rotating mass to the expected speed. When the on-coming machine is at a slower speed, the system supplies synchronizing power. Immediately after the circuit breaker is closed, the on-coming machine receives electric power from the bus-bar and acts as a motor such that the machine accelerates the rotor and the exact synchronous speed is attained. This is called the phenomenon of pulling into step. A very small difference in speed at the time of closing the circuit breaker at the switching instant actually becomes necessary when using synchroscopes, as has been indicated later.

4. *At the instant of closing the circuit breaker, the on-coming machine voltage phasors and those of the bus-bar system must be aligned and co-phasal*

When conditions (1), (2) and (3) are satisfied, the circuit breaker is closed at an instant, when the machine and bus-bar voltages are aligned. Else, the current that flows in the on-coming alternator has the same characteristic as the current during a sudden 3-phase short-circuit across the machine terminals. The value of the current is governed by the sub-transient reactance, the transient reactance and then by the synchronous reactance.

Synchronization of an alternator to a bus-bar system can be brought about after the conditions elaborated under (1), (2) and (3), have been ensured and the connection brought about at the instant of (4) above, by using Dark and Bright Lamp Method. But this practice, although used earlier in power system applications in the past, is currently replaced by using synchroscopes in place of the lamp method in view of the inbuilt accuracy of the method and the growth of ratings of the involved machines.

In one form of the synchroscope, a small induction motor is used whose direction of rotation denotes the direction of the rotating field set up by two windings on its stator. One of the phases is energised by one phase of this small induction motor, while the other is fed from the corresponding phase of the bus-bar. When the two voltages are aligned and are of equal frequency, there is no phase displacement between the magnetic fields produced by the winding and the rotor is in stand-still position. This is indicated by a pointer on a dial, normally arranged to point in 12 O'clock position. In actual practice, the pointer is allowed to rotate slowly with the on-coming machine running marginally slower than the bus-bar. It is usually arranged to rotate clockwise under this condition. To provide for time-lag in operating mechanisms, the circuit breaker solenoid switch is closed when the pointer is in the 11 O'clock position on the dial.

It is not a good practice for the on-coming machine to start sharing load immediately after it is synchronized to the bus-bar system, as the machine experiences shock on the prime mover and the armature. It should be loaded gradually through proper control of the prime-mover governor. We should allow the on-coming machine to float on the bus-bar at its initial stages following synchronization.

Current during synchronizing

Cylindrical rotor machine: Figure 9.23 shows a uniform air-gap alternator whose excitation voltage E has the same amplitude as the bus-bar voltage V. Let:

X = the general value of reactance of machine per phase
X'' = the sub-transient value of reactance per phase
X_s = the synchronous reactance per phase
δ = the power angle between V and E

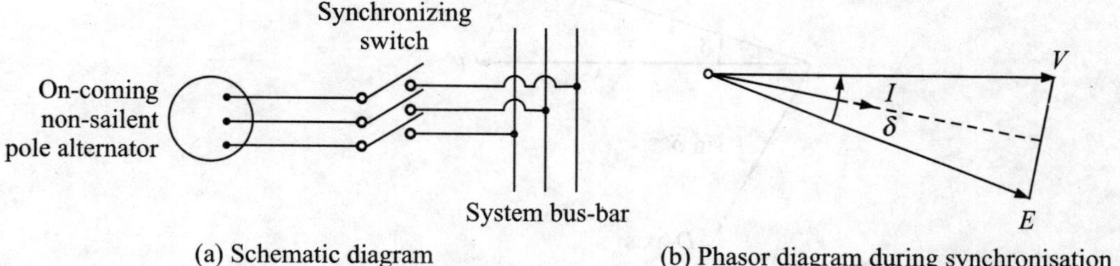

FIGURE 9.23 Synchronizing current when a cylindrical rotor machine is connected to infinite bus-bar.

In general,

$$V_b = (E - V) = (E \cos \delta - V) + jE \sin \delta \qquad ...(9.164)$$

Closing the circuit breaker contacts is equivalent to impressing a voltage equal but opposite to the voltage existing across the breaker contacts, so that the resulting current flowing into the machine is given by:

$$I_{\text{synchronizing}} = \frac{V - E}{jX}$$

During the first few cycles, the machine offers only the sub-transient value of reactance, $X = X''$. Hence,

$$I''_{\text{syn}} = \frac{V - E}{jX''} \qquad ...(9.165)$$

which assumes a value under steady-state equal to:

$$I_s = \frac{V - E}{jX_s} \qquad ...(9.166)$$

This current is nearly in phase with the bus-bar voltage V, and when the machine voltage lags the bus-bar voltage, a power given by:

$$W_{syn} = VI_s^* \quad ...(9.167)$$

flows into the on-coming alternator. This is termed as the synchronizing power and it accelerates the rotor of the on-coming machine, thus reducing the power-angle δ to zero.

Salient-pole machine: The phasor diagram shown in Figure 9.24, indicates that the quadrature-axis is ahead of the bus-bar voltage by an angle δ. The machine voltage is along the quadrature axis and is equal in magnitude to the bus-bar voltage. Then, the resulting voltage equations along the quadrature- and direct-axis are:

$$V_q = E - V \cos \delta \quad ...(9.168)$$

$$V_d = -V \sin \delta \quad ...(9.169)$$

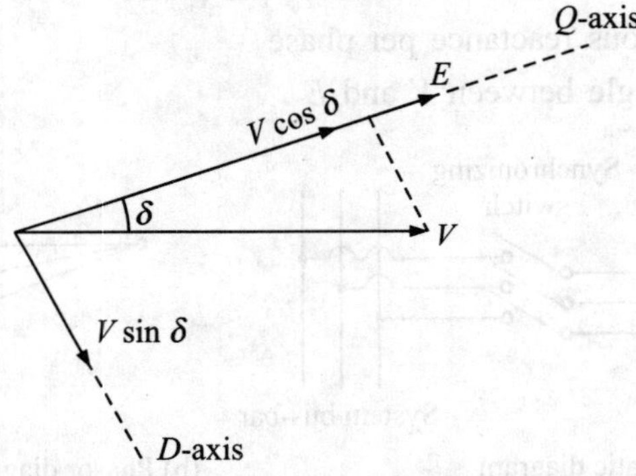

FIGURE 9.24 Synchronizing current and power for a salient pole alternator when synchronizing.

But, we recall that:

$$V_d = p(\psi_d) - \psi_q(p\theta) \quad ...(9.170)$$

and

$$V_q = \psi_d(p\theta) + p(\psi_q) \quad ...(9.171)$$

and

$$\psi_d = M_{af}\, i_f + L_d i_d \quad ...(9.172)$$

$$\psi_q = L_q i_q \quad ...(9.173)$$

Taking up only the sub-transient range, the field current does not influence changes taking place in the armature. Hence, the above equations yield:

$$L_d \frac{di_d}{dt} - \omega L_q i_q = V_d \quad ...(9.174)$$

and

$$\omega L_d\, i_d + L_q \frac{di_q}{dt} = V_q \quad ...(9.175)$$

together with initial values of $i_q(0) = i_d(0) = 0$ at $t = 0$. If we take Laplace transforms of equations (9.174) and (9.175), and keeping in mind that V_d and V_q are constant voltages, then:

$$L_d sI_d - \omega L_q I_q = \frac{V_d}{s} \quad ...(9.176)$$

and

$$\omega L_d I_d + L_q sI_q = \frac{V_q}{s} \quad ...(9.177)$$

where, $I_d = L[i_d]$ and $I_q = L[i_q]$ are the Laplace transforms.

Therefore, we obtain from the above simultaneous equations:

$$I_d = \frac{sV_d + \omega V_q}{L_d s(s^2 + \omega^2)} = \frac{V_d}{X_d} \frac{\omega}{s^2 + \omega^2} + \frac{V_q}{X_d}\left(\frac{1}{s} - \frac{s}{s^2 + \omega^2}\right) \quad ...(9.178)$$

and

$$I_q = \frac{V_q - \omega V_d}{L_q s(s^2 + \omega^2)} = \frac{V_q}{X_q} \frac{\omega}{\omega^2 + s^2} - \frac{V_d}{X_q}\left(\frac{1}{s} - \frac{s}{s^2 + \omega^2}\right) \quad ...(9.179)$$

In the sub-transient range, $X_d = X_d''$ and $X_q = X_q''$. Then by taking the inverse Laplace transforms of equations (9.178) and (9.179) and using Eqs. (9.168) and (9.169) for V_d and V_q, the direct- and quadrature-axis currents are given by:

$$i_d(t) = -\frac{V \sin \delta}{X_d''} \sin \omega t + \frac{E - V}{X_d''}(1 - \cos \omega t) \quad ...(9.180)$$

and

$$i_q(t) = \frac{E - V \cos \delta}{X_q''} \sin \omega t + \frac{V \sin \delta}{X_q''}(1 - \cos \omega t) \quad ...(9.181)$$

The resulting current in A-phase can be shown to be:

$$i_A = i_d'' \cos \theta - i_q'' \sin \theta$$

where, θ is the angle between the axis of phase A and the d-axis of the rotor at the instant of closing the circuit-breaker, and the fundamental component of i_A is due to sudden short-circuit occurring with twice the rated voltage.

9.8.4 Swing-equation of a Single Machine

In power systems, a sudden change in operating conditions is a common feature. Such situations occur when:

(a) Electrical loads connected to alternators change.

(b) Faults occur in systems, bringing about altered system impedances in the machine circuit.

(c) Mechanical loads on the synchronous motor shaft change.

In such situations, the balance between the mechanical power and the developed electrical power gets disturbed and the rotor of the machine starts oscillating in its attempt to adjust itself to the new situation, in order to attain steady-state operation in the altered state. During the oscillation, it is mandatory that the machine maintains stability. This calls for developing sufficiently large electrical and mechanical restoring forces to counter-act the disturbance imposed on it without losing synchronism and enable restoration of its synchronous speed.

The following issues are needed to be critically examined:

1. The resulting differential equations that consider the dynamic performance of machines under a disturbance.
2. Solution of the differential equations and their application in designing simple systems.
3. Developing the restoring forces that the machine is capable of generating.
4. Techniques available for examining stability conditions of systems.

Derivation of swing equation

The dynamic equation of a rotating machine in general, during mechanical disturbance is qualified by:

$$\text{Torque of imbalance, } T_b = \text{Initial torque, } T_l + \text{Damping torque, } T_d + \text{Stiffness torque of shaft, } T_{st} \quad ...(9.182)$$

Of these component torques, the shaft stiffness, T_{st}, in large machines is comparatively less when compared to other terms and, for this derivation, has been neglected all through. We develop expressions for the remaining terms:

(a) Internal torque, T_l: We consider an alternator first. It receives power mechanically amounting to P_m watt from the prime mover and delivers a power P_e electrically, which amounts to:

$$P_e = VI \cos \phi \text{ watt/phase} \quad ...(9.183)$$

When expressed in terms of the terminal voltage, V, the excitation voltage E, the power angle between E and V phases, δ, and the machine reactance, X_d (and X_q), it assume the following form:

(i) For *cylindrical rotor machines:*

$$P_e = \frac{VE}{X_s} \sin \delta = P_{\max} \sin \delta \quad ...(9.184)$$

(ii) For *salient-pole machines:*

$$\left.\begin{aligned} P_e &= \frac{VE}{X_d} \sin \delta + \frac{V^2}{2}\left(\frac{1}{X_d} - \frac{1}{X_q}\right) \sin 2\delta \\ &= P_{\max} \sin \delta + P_r \sin 2\delta \end{aligned}\right\} \quad ...(9.185)$$

In the steady-stage, the developed electrical power nearly equals the input mechanical power. But when a change in electrical load is imposed, the following conditions occur:

(a) The terminal voltage V is held constant for an infinite bus-bar system;
(b) The excitation voltage E also remains constant till the voltage regulator is brought into action;
(c) The shaft input power P_n is also held constant until the governor of the mechanical power system initiates change of shaft power input system.

The way the machine can react to meet the demand for the change is, therefore, to fall back on its stored kinetic energy, through a change in the value of the power angle δ. δ denotes the angle between the inter-polar axis (q-axis), which is ahead of the terminal voltage phasor. As the terminal voltage phasor V, which makes an angle ϕ with the load power factor angle and hence the phasor I, the maximum value of the rotating mmf of armature reaction becomes the agent for meeting the balance. Hence:

$$P_b = P_m - P_e \text{ in watt}$$

which has the form:

$$P_b = P_m - P_{\max} \sin \delta, \text{ for cylindrical rotor machines} \qquad ...(9.186)$$

and

$$P_b = P_m - \{P_{\max} \sin \delta + P_r \sin 2\delta\} \text{ for salient pole machines} \qquad ...(9.187)$$

which can be expressed in Newton-metre units as follows:

$$T_m = \frac{P_b}{\omega_m} = \frac{P}{2}\frac{1}{\omega} P_b \text{ Nm} \qquad ...(9.188)$$

$$\omega_m = \frac{2\pi n_s}{60} \qquad ...(9.189)$$

where

ω_m = mechanical angular velocity

n_s = synchronous speed in rps
ω = electrical angular velocity
P = number of poles of the machine

The synchronous speed is given by:

$$\omega = 2\pi f = \frac{P}{2}\omega_{ms} = \frac{P}{2}\frac{2\pi\omega_s}{60} \qquad ...(9.190)$$

(b) The intertial torque, T_i: The change in speed under the influence of balance power causes an acceleration or retardation of the the rotating mass. This leads to development in inertial torque:

$$T_I = J\frac{d\omega_m}{dt} = \frac{2}{P}J\frac{d\omega}{dt} \text{ in Nm} \qquad ...(9.191)$$

where, J is the polar mass moment of inertia in kg/m^2.

However,

Instantaneous angular velocity = Angular velocity at synchronous speed
+ Change in angular velocity from synchronous speed ...(9.192)

that is,

$$\omega = \omega_s + \frac{d\delta}{dt} \qquad ...(9.193)$$

so that,

$$\frac{d\omega}{dt} = \frac{d^2\delta}{dt^2} \qquad ...(9.194)$$

and

$$T_l = \frac{2}{P} J \frac{d^2\delta}{dt^2} \qquad ...(9.195)$$

(c) Damping torque, T_D: The damping torque in synchronous machines is produced due to induction in rotor pole faces and the damper windings when the speed of the rotor deviates from synchronous speed. At very low values of slip, this torque is directly proportional to the change in speed. Its value has the form:

$$T_D = D\frac{d\delta}{dt} \text{ in Nm} \qquad ...(9.196)$$

where, D is the damping constant.

If we substitute in Eq. (9.182) the relations shown in Eqs. (9.184) or (9.185) as the case may be, Eqs. (9.195) or (9.196), the dynamic swing equation for the machine, neglecting shaft stiffness, has the form:

$$\frac{2}{P} J \frac{d^2\delta}{dt^2} + D\frac{d\delta}{dt} = \frac{P}{2}\frac{1}{\omega} P_b(\delta) \qquad ...(9.197)$$

which is the same as writing:

$$\frac{4\omega}{P^2} J \frac{d^2\delta}{dt^2} + \frac{2\omega}{P} D \frac{d\delta}{dt} = P_b(\delta) \qquad ...(9.198)$$

Equation (9.198) takes into account the actual angular velocity ω. Its deviation from the actual angular velocity is small. The dynamic behaviour of synchronous machines is of interest only when small perturbations occurs near synchronous speed. It is possible, therefore, to approximate ω by introducing ω_s instead, so that the resulting swing equation has the form:

$$\frac{4\omega}{P^2} J \frac{d^2\delta}{dt^2} + \frac{2\omega_s}{P} D \frac{d\delta}{dt} = P_b(\delta) \qquad ...(9.199)$$

which may assume a further simplified form:

$$J' \frac{d^2\delta}{dt^2} + D' \frac{d\delta}{dt} = P_b(\delta) \qquad ...(9.200)$$

In the above expression,

$$J' = \frac{4\omega_s}{P^2} J \qquad ...(9.201)$$

and

$$D' = \frac{2\omega_s}{P} D$$

In the above relations, the initial conditions are:

$$\delta = \delta_0 \text{ and } \frac{d\delta}{dt} = \delta_0 \text{ at } t = 0$$

Thus, for a cylindrical rotor alternator, the swing equation has the form:

$$J'\frac{d^2\delta}{dt^2} + D'\frac{d\delta}{dt} = P_m - P_{\max} \sin \delta \qquad ...(9.202)$$

and the swing equation for a salient-pole alternator is:

$$J'\frac{d^2\delta}{dt^2} + D'\frac{d\delta}{dt} = P_m - \{P_{\max} \sin \delta + P_r \sin 2\delta\} \qquad ...(9.203)$$

where, $P_{\max}$ and P_r have values recorded in expressions (9.184) and (9.185) respectively. The quantity J' can be conveniently indicated in terms of kinetic energy of the rotor at synchronous speed.

If μ denotes the angular momentum of the rotor at synchronous speed in J/rad/sec units and D_K is the kinetic energy of the rotor at synchronous speed, we get:

$$D_K = \frac{1}{2}\omega_{ms}\mu = \frac{1}{2}\omega_{ms}^2 J = \frac{2}{P^2}\omega_{ms}^2 J \qquad ...(9.204)$$

where, P is the number of poles of the machine.

Therefore,

$$J' = \frac{4\omega_s}{P^2} J = \frac{2}{\omega_s} D_K$$

$$= \frac{D_k}{\pi f} \qquad ...(9.205)$$

In Eqs. (9.202) and (9.203), the power angle δ is expressed in electrical radian. In general practice, it is customary to denote δ in electrical degrees. Hence, the right hand side of the swing equation assumes a new form:

$$\frac{D_K}{180f}\frac{d^2(\delta^\circ)}{dt^2} + \frac{4\pi^2 f}{180P}\frac{d(\delta^\circ)}{dt} = P_b(\delta^\circ) \qquad ...(9.206)$$

where (δ°) denotes the value of δ in electrical degrees, P stands for the number of poles and P_b denotes the balance of power $[P_m - P_e]$, on the shaft.

The swing equation presented in expressions (9.200) and (9.206) are both non-linear differential equations and are not easily solvable. We use the expression derived in Eq. (9.200) for future analysis only and not expression (9.206).

When the load on the alternator is less, the power angle δ is also small. If a sudden change in mechanical power input occurs, and it is not very large in magnitude, the swing equation may be approximated by introducing

$$\sin \delta = \delta$$

It then converts the swing equation to a linear, second order, ordinary differential equation with constant coefficients, whose solution is easily obtainable. Under linearization, the governing equation assumes the form:

$$J' \frac{d^2\delta}{dt^2} + D' \frac{d\delta}{dt} = P_m - P_e'(\delta)$$

where

$P_e' = P_{\max}$ for a cylinderical rotor machine

$P_e' = P_{\max} + 2P_r$ for a salient-pole machine

Although the solution of the equation contains the initial values of both δ and $\left(\dfrac{d\delta}{dt}\right)$ at $t = 0$, the angular velocity cannot change instantaneously with heavy rotating masses. Thus, for all practical purposes, δ, at $t = 0$, that is, $\delta_0 = 0$, and this further simplifies the solution for the power angle swing equation.

Further, we note a very important characteristic of the solution due to the presence of the term $\left[\dfrac{P_m}{P_e'} - \delta_0\right]$—the final power angle minus the initial power angle δ_0. This is the total change in the value of δ, caused by the changed mechanical power input to the generator.

If the initial power angle δ_0 is subtracted from both sides of the expression of $\delta(t)$, the resulting expression is obtained for the change in power angle from its initial value, or $[\delta(t) - \delta_0]$. This quantity is of greater interest than the detailed solution for the value of the power angle itself. In the original differential Eq. (9.200), we can write for the change in power angle, from its initial value of δ_0, as:

$$\delta = \delta_0 + \Delta\delta \qquad \text{...(9.207)}$$

Then,

$$\frac{d\delta}{dt} = \frac{d}{dt}(\Delta\delta) \text{ and } \frac{d^2\delta}{dt^2} = \frac{d^2}{dt^2}(\Delta\delta) \qquad \text{...(9.208)}$$

Then, the new form of equation (9.200) becomes:

$$J' \frac{d^2(\Delta\delta)}{dt^2} + D' \frac{d(\Delta\delta)}{dt} + P_e'(\Delta\delta) = (P_m - P_e'\delta_0) \qquad \text{...(9.209)}$$

with new initial conditions:

$$\Delta\delta = 0 \text{ and } \frac{d}{dt}(\Delta\delta) = 0 \text{ at } t = 0 \qquad ...(9.210)$$

Equation (9.209) could also be written in the revised form:

$$\frac{d}{dt^2}(\Delta\delta) + 2\zeta\omega_0\frac{d(\Delta\delta)}{dt} + \omega_n^2(\Delta\delta) = (K - \omega_n^2\delta_0) \qquad ...(9.211)$$

where,

$$K = \frac{P_m}{J'}$$

$$2\zeta\omega_0 = \frac{D'}{J'}$$

$$\omega_n^2 = \frac{P'_e}{J'}$$

so that, $\dfrac{K}{\omega_n^2} = \dfrac{P_m}{P'_e}$, the final value of the power angle.

Now, taking Laplace transform:

$$L(\Delta\delta) = (K - \omega_n^2\delta_0)\frac{1}{s(s^2 + 2\zeta\omega_n s + \omega_n^2)} \qquad ...(9.212)$$

together with initial condition shown in expressions (9.210), that:

$$\Delta\delta = 0 \text{ and } \frac{d}{dt}(\Delta\delta) \text{ at } t = 0$$

so that, for $\zeta = 1$:

$$\Delta\delta = \left(\frac{K}{\omega_n^2} - \delta_0\right)\left[1 - (1 + \omega_n t)\, e^{-\omega_n t}\right] \qquad ...(9.213)$$

Similar expression for $(\Delta\delta)$ could also be obtained for $0 < \zeta < 1$ and for $\zeta > 1$. The solutions are also valid for synchronous motor, when a sudden change in mechanical load occurs on the shaft. In this case, the balance power, $(P_b = P_e - P_m)$ causes the swing. It is only required to reverse the sign attached to δ.

Behaviour under sudden change of load

For a synchronous generator, severe disturbance arises if a short-circuit is accidentally incident across the machine terminals, as discussed earlier. For a synchronous motor, the disturbance may befall due to sudden application of load torque to its shaft. It is very important to predict the ability of the machine to remain in synchronism even after disturbance is incident. We study the dynamic behaviour and discuss the techniques for predicting whether or not a synchronous machine remains in synchronism after the disturbance is incident.

Steady-state stability limit: The power and torque developed by a poly-phase synchronous machine, when operating on a power system, are given by the following expressions:

$$P = \frac{mVE}{X_s} \sin \delta \qquad ...(9.214)$$

and

$$T_e = \frac{mVE}{\omega_s X_s} \sin \delta \qquad ...(9.215)$$

or

$$T_e = T_{\max} \sin \delta \qquad ...(9.216)$$

If the load torque is slowly increased on the shaft, the torque angle increases. The load torque can be increased to the value $T_{\max}$, for $\delta = 90°$. If the load torque is further increased, the machine loses synchronism because the torque developed is less than the load torque. The load torque, can thus be, increased to its maximum value provided it is increased slowly. This torque is identified as the steady-state stability limit or the static stability limit.

Dynamic stability limit: If the load torque is applied suddenly the machine may lose synchronism even if the load torque is less than the maximum torque that the machine is capable of producing for the specified operating conditions. The maximum value of the load torque that can be applied suddenly without losing synchronism can be estimated from the dynamic behaviour of the machine.

Let us consider an unloaded synchronous motor supplied from a power system. If losses are neglected, the power angle δ is zero. If a sudden load torque is applied to its shaft, the motor slows down, thereby, increasing the torque angle δ. And as δ increases, the motor develops torque to meet the load torque, and as the value of δ reaches the value at which the load torque is balanced by the motor torque, the machine has already slowed down because the kinetic energy of the moving rotor has already provided the load torque. However, because of inertia δ increases beyond the value at which it produced the load torque, so that the machine develops more torque than required for the load. The deceleration now decreases and δ reaches a maximum value $\delta_{\max}$ and then swings back. The torque angle thus oscillates around the load torque required for the applied load. Because of damping in the system, the torque angle settles down to the required value of δ eventually.

The sub-transient period lasts for a very short-time immediately following the imposition of the change. But the oscillation of δ lasts for several cycles. It can be assumed that the transient equivalent circuit provides a satisfactory depiction of the electrical system, and we can denote the torque-δ relationship as follows:

$$T_d = \frac{mVE'}{\omega X'_d} \sin \delta \qquad ...(9.217)$$

where, E' is the voltage behind the transient reactance before the imposition of the disturbance, and its value is determined from the relation:

$$E' = V - jI_a X'_d \qquad ....(9.218)$$

The equation could also be expressed as:

$$T_d = T_{\max} \sin \delta \qquad ...(9.219)$$

The oscillation in δ as a function of time was indicated in the previous section by solving the differential equation that describes the dynamic system. The torque balance equation for a synchronous motor could be written as:

$$T_d = T_a + T_D + T_m \quad \text{...(9.220)}$$

where

T_d = torque developed by the motor

$T_a = K_j \dfrac{d^2\delta}{dt^2}$ is the acceleration torque

$T_D = K_D \dfrac{d\delta}{dt}$ is the damping torque

T_m = load (or mechanical) torque

We have seen that it leads to an equation (valid for the motor) as:

$$K_j \frac{d^2\delta}{dt^2} + K_D \frac{d\delta}{dt} = T_{\max} \sin \delta - T_m \quad \text{...(9.221)}$$

The above equation, termed as swing equation, is non-linear.

9.8.5 Stability of Synchronous Machine Operating on Infinite Bus-bar

In order to find out whether or not a synchronous machine operates in its stable region and maintain its stability under change in power input (or electrical power input in case of synchronous motors or mechanical power input in case of synchronous generators), it is not necessary to obtain a solution of the swing equation shown in Eq. (9.200), which gives the final value of the power angle δ as a function of t.

For small oscillations and hunting, we approximated the solution by linearizing the swing equation. But we noted that the swing equation was essentially non-linear and a solution from a particular initial condition could not be arrived at easily.

If the system maintains stability, the new value of δ reaches a steady-state as the time elapses following the disturbance, and $\left(\dfrac{d\delta}{dt}\right)$ attains a value of zero. The stability or instability of a machine is totally governed by initial operating conditions and the nature and magnitude of the disturbance imposed on it. The power–angle curve of the machine, in turn, is controlled by the excitation voltage E and the armature reactance X. When these quantities are known, the stability can be determined from the Swing Equation itself, by a number of techniques listed below:

1. Equal Area Criterion evolved by Crary
2. Phase Plane Technique
3. Lipunov Technique

We take up the Equal Area Criterion first, followed by a brief introduction to the other two methods. These two methods are some what advanced in nature. A complete coverage of these methods are considered beyond the scope of this book.

Equal area criterion for stability

if we introduce a slight modification to the swing equation (9.200) by eliminating the damping torque component as small, it assumes the new form:

$$J'\frac{d^2\delta}{dt^2} = P_b(\delta) = P_m - P_e(\delta) \qquad \text{...(9.222)}$$

It is easy to see that if the system is stable without the damping, the damping component makes stability even more certain. Figure 9.25 presents the power–angle curve indicating electromagnetic power as a sign curve of the power angle δ as given for a cylindrical rotor machine in Eq. (9.184). However, the criterion is equally applicable to any function of $P_e(\delta)$— such as that of salient-pole machine shown in Eq. (9.185).

The input mechanical power P_m is now considered suddenly changed from a value of P_{m1} to P_{m2}. With this change in the operating condition, the rotor experiences a sudden accelerating torque (or a net balance power input $P_b(\delta) = P_{m2} - P_{m1}$, denoted by the ordinate Aa in Figure 9.25.

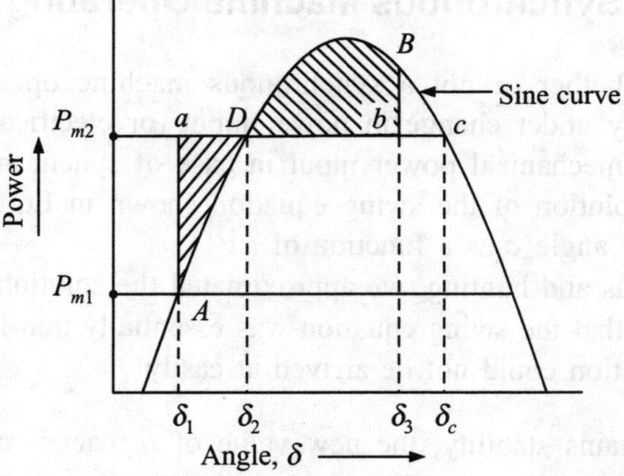

FIGURE 9.25 Equal area criterion for stability.

During the consequent acceleration, the rotor swings past the equilibrium position D, at a power angle δ_2.

The equal area criterion states that the rotor swings to the maximum power angle δ_3 (denoted by point B), such that area $ADaA$ = area $DBbD$.

The maximum permissible area above the mechanical power angle line P_{m2} is $CBDC$, which is permissible for the above swing. If the (area $ADaA$) exceeds the area $cBDc$, the machine loses stability. Thus for stable operation, the maximum permissible power angle δ_3 must be below the critical value of the permissible power angle δ_c determined by point c in Figure 9.25. This is the point of intersection of the mechanical power input line and the power angle curve $P_e(\delta)$. Otherwise the generator swings beyond δ_c and the mechanical power input exceeds the electromagnetic power, and the machine speeds up, losing synchronism and land itself in the unstable region.

When area ADaA becomes equal to area DcBD, the machine is said to have attained the limit of dynamic stability. Under this condition, the value of Aa is said to be the value of dynamic stability limit.

$\delta_c = (\pi - \delta_2)$ for a non-salient-pole machine. For a salient-pole machine, δ_c has a lesser value.

This criterion is also valid for a synchronous motor, where P_{m1} and P_{m2} indicate the load power including the stray power losses.

The above criterion may be proved by taking simplistic Eq. (9.222) as follows:

$$\frac{d^2\delta}{dt^2} = \frac{1}{J'}[P_m - P_e(\delta)]$$

In Figure 9.25, P_{m2} is the new value of the input mechanical power at the start of oscillation, and the initial value of $\delta_0 = \delta_1$. The above equation could, therefore, be written as:

$$\frac{d^2\delta}{dt^2} = \frac{1}{J'}[P_{m2} - P_e(\delta)] \qquad ...(9.223)$$

or

$$\frac{d\delta}{dt}\left(\frac{d^2\delta}{dt^2}\right) = \frac{1}{J'}[P_{m2} - P_e(\delta)]\frac{d\delta}{dt}$$

or

$$\frac{1}{2}\frac{d}{dt}\left(\frac{d^2\delta}{dt^2}\right) = \frac{1}{J'}[P_{m3} - P_e(\delta)]\frac{d\delta}{dt}$$

or

$$\left(\frac{d\delta}{dt}\right)^2 = \int_{\delta_1}^{\delta}\frac{2}{J'}[P_{m3} - P_e(\delta)]\,d\delta \qquad ...(9.224)$$

The initial value of power angle being δ_1.

If the system is stable, $\frac{d\delta}{dt} = 0$ at $\delta = \delta_2$, the position of maximum swing,

or

$$\int_{\delta_2}^{\delta_1}[P_{m2} - P_e(\delta)]\,d\delta = 0 \qquad ...(9.225)$$

Now, referring to Figure 9.25, the integral in Eq. (9.225) is the area *DBbD* which is the same as the area *ADaA*.

Geometrically, it means that the area below the line P_{m2} must equal the area above the line P_{m2}, in both cases the upper limit being the power angle curve.

For a known value of the initial angle, δ_1 can be obtained analytically by determining the integral of Eq. (9.224) or Eq. (9.225), and then equate it to zero. Thus, for a cylindrical rotor machine:

$$\left(\frac{d\delta}{dt}\right)^2 = \frac{2}{J'}\int_{\delta_1}^{\delta_2}[P_{m2} - P_{\max}(\sin\delta)]\,d\delta$$

$$= \frac{2}{J'}[P_{m2}(\delta_3 - \delta_1) + P_{\max}(\cos\delta_2 - \cos\delta_1)]$$

$$= 0 \qquad ...(9.226)$$

Therefore,

$$P_{\max}\cos\delta_3 + P_{m2}\delta_3 = P_{\max}\cos\delta_1 + P_{m2}\delta_1 \qquad ...(9.227)$$

An iterative (or graphical) method yields the value of δ_3 from the values of P_{m2}, $P_{\max}$ and δ_1. The maximum value of δ_1, so that the machine does not lose stability, can also be determined using Eq. (9.227). In this case, $\delta_3 = \delta_c$ in Eq. (9.227) and with specified values of P_{m2} and $P_{\max}$, the iterative (or graphical) method gives the minimum value of δ_1. It may be noted that the left hand side of Eq. (9.227) is minimum when $\delta_3 = \delta_c$.

We now, consider a synchronous generator operating under faulted conditions and investigate about its stability. Figure 9.26 shows a synchronous generator G connected to an infinite bus-bar through two transmission lines, each with a series reactance X_t. When both lines are functional, the series reactance between the voltage E and V is

$$X_1 = X_s + 0.5X_t$$

for a cylindrical rotor machine. For a salient-pole machine, each of the synchronous reactances—X_d and X_q—has to be increased by $0.5X_t$.

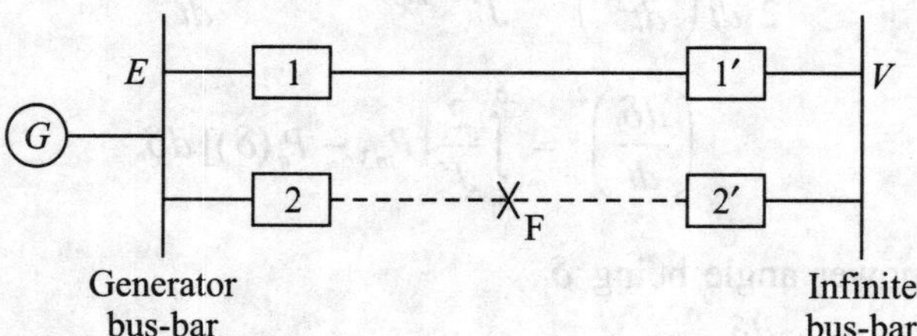

FIGURE 9.26 Synchronous generator G, connected to infinite bus-bar through two lines, to illustrate the change in reactance when a fault occurs on one line and is cleared.

Let a fault occur on the second line as shown at a point F.

This alters the total reactance scenario between E and V voltages to a new value X_2. The power angle curve is, therefore, altered. A few cycles later, the circuit breakers 2 and 2′ are considered as opened to clear the fault, so that the post-fault reactance becomes X_3.

During these three stages of operation—identified as the pre-fault, fault and the post-fault conditions—the mechanical power input, P_m, to the rotor of the generator and the excitation voltage E remain substantially constant.

Therefore, the three power angle curves (we only consider non-salient pole case for the fault) are:

(a) Pre-fault:
$$P_{e1}(\delta) = \frac{EV}{X_1} \sin \delta \qquad ...(9.228)$$

(b) Fault:
$$P_{e2}(\delta) = \frac{EV}{X_2} \sin \delta \qquad ...(9.229)$$

(c) Post-fault:
$$P_{e3}(\delta) = \frac{EV}{X_3} \sin \delta \qquad ...(9.230)$$

In general, we assume $X_1 > X_3 > X_2$, so that the power angle curves are as shown in Figure 9.27. For the sake of discussion, we consider the maximum values of $P_{e1} > P_{e3} > P_{e2}$.

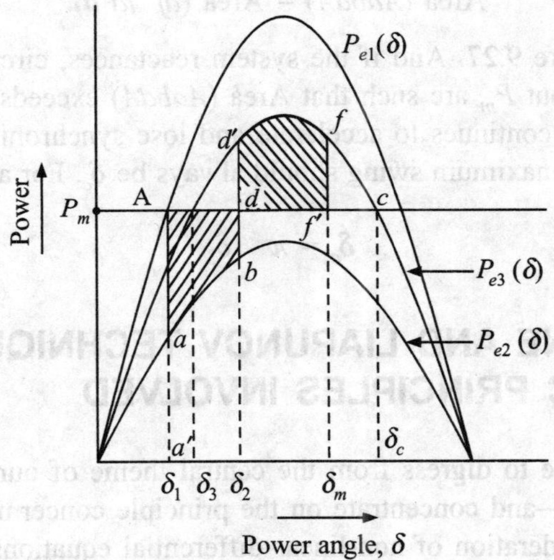

FIGURE 9.27 Power-angle curves of the pre-fault, fault and post-fault conditions.

Before the fault occurred, the machine was operating with a steady-state power angle of δ_1, whose value was determined by the intersection of the constant mechanical input power line P_m and the pre-fault power-angle curve, $P_{e1}(\delta)$.

Immediately after occurrence of the fault, the electrical power delivered falls to *Aa* (see Figure 9.27) at the same power angle δ_1, but on the power angle curve $P_{e2}(\delta)$. There is, therefore, a resulting balance power or accelerating power *Aa* and the rotor retards with a resulting increase in δ. If $P_{e2\ max}$—the maximum power delivered with a load reactance of X_2 during the fault condition is less than P_m and the circuit breakers do not open, then evidently there is an accelerating power always, and the rotor loses synchronism, thus leading to instability.

However, when the rotor reaches angle δ_2, let the circuit breakers 2 and 2′ open to clear the fault. This suddenly increases the developed power to d' on the post-fault power angle curve $P_{e3}(\delta)$. We would observe that the rotor now begins to experience a decelerating power when the power angle swings occur on the portion ($d'fc$). The problem now reduces to determining:

(a) The maximum angle of swing δ_m of the rotor, if the system is stable.

(b) The minimum value of δ_1 of the pre-fault power angle which results in the power angle swing to reach δ_c, the critical angle required for stability.

(c) With a given initial value of angle δ_1, to determine the point b, at which the circuit breakers must open such that the machine does not lose synchronism, or in other words, the maximum swing does not exceed δ_c.

We observe that during the post-fault condition, the machine settles down the power angle δ_3, determined by the point of intersection of the constant mechanical input power P_m with the post-fault power angle curve $P_{e3}(\delta)$. If the circuit breaker clearing time and therefore, the angle δ_2 is known, the maximum swing of δ is such that

$$\text{Area }(AabdA) = \text{Area }(df'fd'd)$$

as in the diagram in Figure 9.27. And if the system reactances, circuit breaker clearing times and mechanical power input P_m are such that Area $(AabdA)$ exceeds Area $(df'fd'd)$ above the line of P_m, then the rotor continues to accelerate and lose synchronism.

The original value of maximum swing should always be δ_c. For a non-salient pole machine its value is:

$$\delta_c = \pi - \delta_3$$

9.9 PHASE PLANE AND LIAPUNOV TECHNIQUE—SOME BASIC PRINCIPLES INVOLVED

In this section, we propose to digress from the central theme of our discussion—the stability of synchronous machines—and concentrate on the principle concerning the general aspects of stability. It involves consideration of non-linear differential equations.

We choose to discuss a simple problem in this realm—the classical problem of motion of a simple pendulum and observe the development of the phase plane and Liapunov techniques.

Let us consider a simple pendulum shown in Figure 9.28. For this model, let:

M = The mass of the pendulum bob
L = Length of suspension
θ = Displacement of the bob from its natural equilibrium position at a general instant of time, t.

Then, $(\dot{\theta})$, is the angular velocity at the general instant of time t. Also let: θ_0 and $(\dot{\theta}_0)$ represent their respective initial values at time $t = 0$.

If g denotes the acceleration due to gravity, and $\omega^2 = \dfrac{g}{L}$, ω being the angular velocity at a general time t, then v is the linear velocity of the pendulum at time t, given by:

$$V = L(\dot{\theta})$$

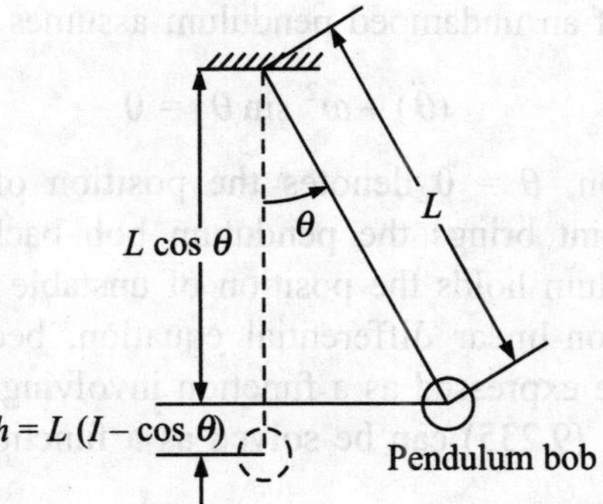

FIGURE 9.28 A simple pendulum considered to clarify the stability criterion.

If we use the Lagrangian formulation of the governing differential equation of motion, we develop expressions for the potential energy D_P and the kinetic energy D_K functions of the pendulum bob as:

$$\text{Potential energy, } D_P = Mgh$$

$$= MgL(1 - \cos \theta) \qquad \text{...(9.231)}$$

and the kinetic energy,

$$D_K = \frac{1}{2} M v^2$$

$$= \frac{1}{2} ML^2 (\dot{\theta})^2 \qquad \text{...(9.232)}$$

Hence, the value of the Lagrangian

$$\alpha = D_K - D_P = ML^2 \left[\frac{1}{2} (\dot{\theta})^2 - \omega^2 (1 - \cos \theta) \right] \qquad \text{...(9.233)}$$

This leads to:

$$\left.\begin{aligned} \frac{\partial \alpha}{\partial (\dot{\theta})} &= ML^2 (\dot{\theta}) \\ \frac{d}{dt} \left\{ \frac{\partial \alpha}{\partial (\dot{\theta})} \right\} &= ML^2 (\ddot{\theta}) \\ \text{and} \qquad \frac{\partial \alpha}{\partial \theta} &= - MgL \sin \theta \end{aligned}\right\} \qquad \text{...(9.234)}$$

Substituting these values in Lagrangian equation:

$$\frac{d}{dt} \left\{ \frac{\partial \alpha}{\partial (\dot{\theta})} \right\} - \left\{ \frac{\partial \alpha}{\partial \theta} \right\} = 0$$

And the dynamic equation of an undamped pendulum assumes the form:

$$(\ddot{\theta}) + \omega^2 \sin\theta = 0 \tag{9.235}$$

From the above equation, $\theta = 0$ denotes the position of stable equilibrium. A small displacement around this point brings the pendulum bob back to the above position. It is obvious that if $\theta = \pi$, pendulum holds the position of unstable equilibrium.

Equation (9.235) is a non-linear differential equation, because the dependent varriable occurs as sin θ, which can be expressed as a function involving θ and its higher powers using Taylor series. In general, Eq. (9.235) can be solved as a function of t and it results in elliptic functions.

For small displacements about the equilibrium position, $\theta = 0$, Eq. (9.235) could be linearized by expressing the (sin θ) function in expanded form by using Taylor series:

$$\sin\theta = \theta \frac{\theta^3}{3!} + \frac{\theta^5}{5!} \qquad \text{...(9.236)}$$

and if we use only the first term, the new form of the differential equation becomes:

$$\frac{d^2\theta}{dt^2} + \omega^2\theta = 0 \qquad \text{...(9.237)}$$

which can be solved as:

$$\theta(t) = A\cos(\omega t + \phi) \qquad \text{...(9.238)}$$

in which,

A = Amplitude of the swing, and
ϕ = A phase angle.

Now, at $t = 0$,

$$A\cos\phi = \theta_0 \qquad \text{...(9.239)}$$

and

$$\frac{d\theta}{dt} = (\dot{\theta}_0) = \omega A \sin\phi \qquad \text{...(9.240)}$$

Thus, expanding the co-sine term in Eq. (9.238) and using the initial values of Eqs. (9.239) and (9.240), lead to:

$$\theta(t) = \mathrm{A}\cos\phi\cos\omega t - A\sin\phi\sin\omega t$$

$$= \theta_0\cos\omega t + \frac{1}{\omega}(\dot{\theta}_0)\sin\omega t \qquad \text{...(2.241)}$$

Also,

$$\{(\dot{\theta})(t)\} = -\omega\sin\omega t\,\theta_0 + \cos\omega t(\dot{\theta}_0) \qquad \text{...(2.242)}$$

We observe that the solution for displacement θ and angular velocity $(\dot{\theta})$ clearly depend on the initial conditions. The linearized form of the differential equation is derived very easily with the displacement and velocity as functions of time. If the time parameter is eliminated from Eqs. (9.241) and (9.242), we get the important concept of dynamics, namely, *the phase plane trajectory of the pendulum* describing the velocity $(\dot{\theta})$ as a function of the displacement θ.

and

$$\left.\begin{aligned} \theta &= A \cos(\omega t + \phi), \text{ and } \theta_0 = A \cos\phi \\ (\dot{\theta}) &= -\omega A \sin(\omega t + \phi); \quad (\dot{\theta}_0) = -\omega A \sin\phi \end{aligned}\right\} \qquad ...(9.243)$$

Therefore,

$$\theta^2 + \frac{1}{\omega^2}(\dot{\theta})^2 = A^2 = (\theta_0)^2 + \frac{1}{\omega^2}(\dot{\theta}_0)^2 \qquad ...(9.244)$$

which leads to:

$$\frac{\theta^2}{(\theta_0)^2 + \dfrac{(\dot{\theta}_0)^2}{\omega^2}} + \frac{(\dot{\theta})^2}{\omega^2(\theta_0)^2 + (\dot{\theta}_0)^2} = 1 \qquad ...(9.245)$$

When the value of ω is specified, that is, when the length of suspension, L, is given, Eq. (9.245) denotes a family of ellipses for different sets of initial values of (θ_0) and $(\dot{\theta}_0)$. A coordinate plane in which the abscissa stands for displacement (θ) and the ordinate, the velocity $(\dot{\theta})$, is called *phase plane.*

The phase-plane Eq. (9.245) can be directly and quickly obtained by the following procedure, and apply equally well when the time solutions are not possible to obtain.

Let us represent:

$$x_1 = \theta \quad \text{and} \quad x_2 = (\dot{\theta}) = \frac{d\theta}{dt} \qquad ...(9.246)$$

Without going through the round about way obtained in the foregoing discussion above as recorded in Eq. (9.231) through Eq. (9.245), it may be assumed that x_1 and x_2 in Eq. (9.246) are ordinary mathematical variables and not reactances, hence,

$$\dot{x}_1 = (\dot{\theta}) = x_2 \qquad ...(9.247)$$

Also, from the original differential Eq. (9.237):

$$\dot{x}_2 = (\ddot{\theta}) = -\omega^2\theta = -\omega^2 x_1 \qquad ...(9.248)$$

Then, using Eqs. (9.247) and (9.248), the ratio of differentials is given by:

$$\frac{dx_2}{dx_1} = \frac{\dot{x}_2}{\dot{x}_1} = \frac{-\omega^2 x_1}{x_2} \qquad ...(9.249)$$

which is also the slope of the phase plane trajectory. Equation (9.249) is now separable in x_1 and x_2, and are integrable. Thus,

$$x_2 dx_2 = -\omega^2 x_1 dx_1$$

or

$$x_2^2 = -\omega^2 x_1^2 + c$$

which has the form:

$$(\dot{\theta})^2 = -\omega^2(\theta)^2 + c \qquad ...(9.250)$$

In this relation, c denotes a constant of integration with respect to x_1.

At $t = 0$, we may write:

$$(\dot{\theta}_0)^2 = -\omega^2(\theta_0)^2 + c$$

and the velocity $(\dot{\theta})$ is obtained in terms of the displacement as follows:

$$(\dot{\theta})^2 = -\omega^2(\theta)^2 + (\dot{\theta}_0)^2 + \omega^2(\theta_0)^2 \quad ...(9.251)$$

The above expression immediately reduces to the phase plane equation recorded under Eq. (2.45). The phase plane technique for the non-linear form of the simple pendulum—Eq. (9.235) follows exactly the some steps.

Before discussing the derivation of this case, it may be pointed out that as the terms in the Taylor series for sin θ in Eq. (9.236) are increased in the pendulum Eq. (9.235), the resulting solution tends to approach the exact phase plane relation of Eq. (9.235), as is indicated in the tabular form given below:

S. No.	*No. of terms in Series*	*Phase Plane Equation*	
1.	θ	$(\dot{\theta})^2 - (\dot{\theta}_0)^2 = \omega^2(\theta_0^2 - \theta^2)$	...(9.252)
2.	θ, θ^3	$(\dot{\theta})^2 - (\dot{\theta}_0)^2 = \omega^2\{(\theta_0^2 - \theta^2)\} - \frac{\omega^2}{12}(\theta_0^4 - \theta^4)$	...(9.253)
3.	....	....	
4.	Exact sin θ solution	$(\dot{\theta})^2 - (\dot{\theta}_0)^2 = 2\omega^2(\cos\theta - \cos\theta_0)$	...(9.254)

It may be noted from the above sequence that by taking progressively more terms in the Taylor series, the phase plane equation finally leads to the exact solution, because:

$$\omega^2(\theta_0^2 - \theta^2) - \frac{\omega^2}{12}(\theta_0^4 - \theta^4) + ... = 2\omega^2(\cos\theta - \cos\theta_0) \quad ...(9.255)$$

Now, for the original non-linear differential equation (9.235), the exact phase plane Eq. (9.255) is obtained directly as follows:

Let $x_1 = \theta$ and $x_2 = (\dot{\theta})$

Then $\dot{x}_1 = x_2$ and $\dot{x}_2 = -\omega^2 \sin x_1$.

From Eq. (9.235),

$$\frac{dx_2}{dx_1} = \frac{\dot{x}_2}{\dot{x}_1} = \frac{-\omega^2 \sin x_1}{x_2} \quad ...(9.256)$$

and

$$(\dot{\theta})d(\dot{\theta}) = -\omega^2 \sin\theta \, d\theta \quad ...(9.257)$$

Integrating Eq. (9.256) and using the initial conditions $\theta = \theta_0$ and $(\dot{\theta}) = (\dot{\theta}_0)$ at time $t = 0$, there results:

$$(\dot{\theta})^2 - (\dot{\theta}_0)^2 = 2\omega^2(\cos\theta - \cos\theta_0)$$

which is the same as Eq. (9.255).

In the phase plane trajectory of variation of θ and $(\dot{\theta})$, for known values of ω and the given initial conditions, may now be plotted.

9.9.1 Liapunov Criterion of Stability and the Liapunov Function for a Simple Pendulum

The total energy associated with a simple pendulum is:

$$U(\theta, \dot{\theta}) = \text{Kinetic energy} + \text{Potential energy}$$

or

$$U(\theta, \dot{\theta}) = \frac{1}{2} ML^2 (\dot{\theta})^2 + MgL\,(1 - \cos\theta) \qquad \text{...(9.258)}$$

If we use the phase plane Eq. (9.254), we get

$$U(\theta, \dot{\theta}) = \frac{1}{2} ML^2 \left[(\dot{\theta}_0)^2 + 2\omega^2 (1 - \cos\theta_0) \right]$$

and since,

$$(\dot{\theta})^2 - 2\omega^2 \cos\theta = (\dot{\theta}_0)^2 - 2\omega^2 \cos\theta_0 \qquad \text{...(9.259)}$$

as well as, $\omega^2 = \dfrac{g}{L}$.

The total energy is constant and its value depends only on the initial displacement and velocity—θ_0 and $(\dot{\theta}_0)$. Its time rate of change is consequently also zero. Also from relation (9.258), we get

$$\frac{dU}{dt} = ML^2 (\dot{\theta})(\ddot{\theta}) + MgL \sin\theta$$

$$= ML^2 (\dot{\theta}) \{ (\ddot{\theta}) + \omega^2 \sin\theta \}$$

$$= 0$$

from the original equation recorded in Eq. (9.235).

For this section, therefore, there exists a function $U(\theta, \dot{\theta})$ whose time rate of change is zero. This indicates in very simple terms, the Liapunov criterion for stability.

The basis of the direct method of Liapunov may be stated as follows:

'If for a dynamic system, there exists a positive function in total energy $U(\theta, \dot{\theta})$, whose time rate of change near the equilibrium position is zero or negative, then the original system is stable.'

The function is called *Liapunov Function* of the system.

For the problem of a simple pendulum, the total energy itself is the Liapunov function. But for more complicated problems, involving non-linear equations, examining the stability of the system involves determination of the Liapunov function of the system. If the existence of such a function, and the region in which the Liapunov function is valid, are established, then the stability of the system is assured within the region.

9.10 STABILITY CONSIDERATION FOR A SIMPLE ALTERNATOR OPERATING ON INFINITE BUS-BAR SYSTEM

Having been introduced to principles of phase plane and Liqpunov techniques applied to the study of stability problems of a simple pendulum, we can now revert back to the original problem of stability considerations for a single alternator functioning in conjunction with an infinite bus-bar system, as already considered in Section 9.8.5. The swing equation for such a system is derived in Eq. (9.200), and has the form:

$$\frac{d^2\delta}{dt^2} + \frac{D'}{J'}\frac{d\delta}{dt} = \frac{1}{J'}[P_m - P_e(\delta)] \qquad ...(9.260)$$

Here, $P_e(\delta) = P_{\max}\sin\delta$, and $P_{\max}\sin\delta + P_r\sin 2\delta$ for cylindrical rotor and salient pole configurations, as were shown in Eqs. (9.184) and (9.185) respectively.

The swing equation is a second-order non-linear differential equation. Its form recorded in Eq. (9.260) is similar to that of a simple pendulum—recorded in Eq. (9.235), with the exception of additional terms involving dumping and a constant term P_m corresponding to mechanical power input. Also, the term $P_e(\delta)$ for the salient-pole configuration contains a possible term with (2δ).

We restrict our consideration to phase plane technique applied to non-salient pole machines only.

9.10.1 Phase Plane Technique Applied to Cylindrical Rotor Machines

The swing equations to be considered in this section, have the form ignoring damping, given by:

$$\frac{d^2\delta}{dt^2} = (\ddot{\delta}) = \frac{P_m}{J'} - \frac{P_{\max}}{J'}\sin\delta \qquad ...(9.261)$$

As before, if we assume:

$$\left.\begin{aligned} x_1 &= \delta \text{ and } x_2 = (\dot{\delta}) \\ \dot{x}_1 &= x_2 \end{aligned}\right\} \qquad ...(9.262)$$

so that

Equation (9.262) could be rewritten in the new form incorporating the assumptions of Eq. (9.262) as:

$$(\dot{x}_2) = (\ddot{\delta}) = \frac{P_m}{J'} - \frac{P_m}{J'}\sin x_1 \qquad ...(9.263)$$

Therefore,

$$\frac{dx_2}{dx_1} = \frac{\dot{x}_2}{\dot{x}_1} = \frac{P_m - P_{\max}\sin x_1}{J'x_2} \qquad ...(9.264)$$

or

$$J'x_2dx_2 = (P_m - P_{\max}\sin x_1)\,dx_1 \qquad ...(9.265)$$

Integrating Eq. (9.265), and using the initial conditions:

$$x_1 = \delta = \delta_0 \text{ and } x_2 = (\dot{\delta}) = 0 \quad \text{at } t = 0$$

the phase plane equation has the form in (δ) and (δ_0) as follows:

$$\left(\frac{d\delta}{dt}\right)^2 = \omega^2 = \frac{2}{J'}[P_m(\delta - \delta_0) + P_{\max}(\cos\delta - \cos\delta_0)] \quad ...(9.266)$$

The above relation is identical to Eq. (9.224). For the value of $\delta = \delta_m$ considered in Figure 9.27 while considering the equal area criterion for stability, we observe that the two techniques are equivalent. For known values of J', P_m, $P_{\max}$ and δ_0, the trajectory of the angular velocity, $\omega = (\dot{\delta})$, when the power angle changes, is qualified by Eq. (9.266).

In obtaining the phase plane trajectory for the swing equation [Eq. (9.261)], we ignored the damping term, in keeping with our earlier assumption made under equal area criterion. If we had not made this assumption, then the resulting equation containing the terms $x_1 = \delta$ and $x_2 = (\dot{\delta})$ could not be separated into terms containing x_1 only, and x_2 only as we had simplified in Eq. (9.265).

If we still insisted with assumptions:

$$x_1 = \delta \text{ and } x_2 = \dot{x}_1 = (\dot{\delta})$$

we would have ended up, while deriving the phase plane trajectory expression by obtaining

$$\frac{dx_2}{dx_1} = \frac{\dot{x}_2}{\dot{x}_1} = \frac{P_m - P_{\max}\sin x_1 - D'x_2}{J'x_2} \quad ...(9.267)$$

Hence, we would have ended up with a product term of $D'x_2dx_1$, thus leading to problems with integration.

Such difficulties require advanced techniques, which are beyond the scope of the present discussion.

9.10.2 Liapunov Stability and Liapunov Function for Alternator

The stability of a dynamic system of a non-linear differential equation by which it is governed, can be examined using the theorem of Liapunov, which states that:

'The equilibrium of a system is stable, if there exists a positive definite function of the form $U(\theta, \dot{\theta}, t)$, such that the total time derivative $\left(\frac{dU}{dt}\right)$ is not positive'.

The function $U(\theta, \dot{\theta}, t)$ is identified as the Liapunov function of the state variables, that is, the variables that describe the dynamic motion of the system in terms of the displacement, velocity and time, in this case. The function is positive definite, if it satisfies the following properties:

1. It is a continuous scalar function of the state variables $\theta, \dot{\theta}$ and t;
2. The partial derivatives with respect to the state variables exist and are continuous;

3. It is a positive function with the region of stability and is zero at the equilibrium point;
4. Its time derivative is negative within the region of stability and zero at the equilibrium point.

The Liapunov theorem stated above does not provide a practical method whereby the Liapunov function may be determined. While there are sophisticated analytical methods developed for obtaining Liapunov functions for the dynamical problems of various degrees of complexity, experience still forms the best guide.

We had discussed the problem concerning a simple pendulum in Section 9.9.1, and showed that the total energy is constant near the equilibrium point. This also served as the Liapunov function for the system in case of the simple pendulum.

If we follow the above basis, a Liapunov function is constructed in case of an alternator for its stability consideration when connected to an infinite bus-bar. The starting point is, therefore, to determine the total energy of the system near the equilibrium position and stretch the region further till the properties required of the Liapunov function fail to hold.

This region, in general, consists of the range δ from the stable equilibrium value of δ_0 to the unstable equilibrium point (but excluding it) δ_c, as indicated in Figure 9.25. The original differential equation or the swing equation may, therefore, be used to obtain the required function.

The total energy of the system is the sum of the potential energy, D_P and the kinetic energy, D_K, of the system. The kinetic energy has a quadratic term in one of the state variables—$(\dot{\theta})$—the angular velocity. The potential energy, on the other hand, is governed by the displacement—δ—which appeared in some of the terms of the original non-linear differential equation listed in Eq. (9.235) in the form of $[P_m - P_{\max} \sin \delta]$.

Let us examine the process of reasoning in connection with the simple pendulum and its stability near the equilibrium point at $\theta = 0$. The governing differential equation was:

$$\ddot{\theta} + \omega^2 \sin\theta = 0$$

The estimated kinetic energy was given by:

$$D_K = \frac{1}{2} ML^2 (\dot{\theta})^2 = ML^2 \left\{ \frac{1}{2} (\dot{\theta})^2 \right\}$$

The value of potential energy was obtained using the term containing the displacement θ in the original differential equation, which had the form in Eq. (9.231) as:

$$D_P = MgL(1 - \cos\theta)$$

This could have been expressed in the form:

$$U = \int_0^{\theta} MgL \sin\theta \, d\theta = ML^2 \int_0^{\theta} \omega^2 \sin\theta \, d\theta \qquad \text{...(9.268)}$$

The resulting Liapunov function for this problem given in expression (9.258) could therefore be constructed from the original differential equation by writing the total energy in the modified form:

$$U(\theta, \dot{\theta}) = ML^2\left[\frac{1}{2}(\dot{\theta})^2 + \omega^2\int_0^{\theta}\sin\theta\, d\theta\right] \qquad ...(9.269)$$

The function satisfies the following properties:

(a) It is continuous in the state variable θ and $\dot{\theta}$

(b) Its partial derivatives with respect to the state variables are:

(i) $\dfrac{dU}{d\theta} = ML^2\omega^2 \sin\theta$, and

(ii) $\dfrac{dU}{d\dot{\theta}} = ML^2(\dot{\theta})$

They exist and are continuous.

(c) The function is positive over the range $0 < \theta < \pi$

(d) At the equilibrium point, when $\theta = 0$ and $(\dot{\theta}) = 0$, the value of function equals zero.

(e) Its total derivative is:
$$\frac{\partial U}{\partial t} = \frac{\partial U}{\partial \theta}(\dot{\theta}) + \frac{\partial U}{\partial(\dot{\theta})}(\ddot{\theta})$$
$$= (\dot{\theta})\left[\omega^2 \sin\theta + (\ddot{\theta})\right] ML^2$$
$$= \text{zero}$$

We also observe that the system with a damping term in the original differential equation would have resulted in a negative value for $\left(\dfrac{dU}{dt}\right)$ in the region of stability.

(f) The point $\theta = 0$ is the position of minimum potential energy while the point $\theta = \pi$ is the point of maximum potential energy. The system has a point of stable equilibrium at $\theta = 0$, and unstable equilibrium at $\theta = \pi$. The region of stability of this system is $-\pi < \theta < \pi$, excluding the point $\theta = \pi$, which is a point of singularity.

Similar steps when undertaken to investigate the stability of an alternator connected to an infinite bus-bar, yield the following results:

(a) Clearly δ_0 and δ_c are equilibrium points in Figure 9.25. The former is the stable and the latter is unstable. The kinetic energy of the rotor corresponds to the term $\dfrac{1}{2}\left(\dfrac{d\delta}{dt}\right)^2$.

The potential energy term may be obtained from the original differential equation:

$$J'\frac{d^2\delta}{dt^2} - P_m + P_{max}\sin\delta = 0 \qquad ...(9.270)$$

Let $\theta = \delta - \delta_0$ = Angle of deviation from the position of stable equilibrium.

The Liapunov function could, now, be rewritten as:

$$U(\delta, \dot{\delta}) = \frac{1}{2} J'(\dot{\delta})^2 + \int_0^{\theta} [-P_m + P_{\max} \sin(\alpha + \delta_0)]\, d\alpha \qquad \text{...(9.271)}$$

The region of validity of this function is given by:

$$0 < \theta < (\delta_c - \delta_0)$$

The existence of the function ensures stability provided that $\left(\frac{dU}{dt}\right)$ is negative or zero.

Since damping has been ignored, $\frac{dU}{dt} = 0$

This may be verified from Eq. (9.271) as given below:

$$\frac{dU}{dt} = (\dot{\delta}) J' \frac{d(\dot{\delta})}{dt} - P_m \frac{d\theta}{dt} + P_{\max} \sin(\theta + \delta_0) \frac{d\theta}{dt} \qquad \text{...(9.272)}$$

But,

$$\frac{d\theta}{dt} = \frac{d}{dt}(\delta - \delta_0) = \frac{d\delta}{dt} = \dot{\delta} \qquad \text{...(9.273)}$$

Therefore, from Eqs. (9.272) and (9.273)

$$\frac{dU}{dt} = (\dot{\delta}) \left[J'(\ddot{\delta}) - P_m + P_{\max} \sin \delta \right] = 0 \qquad \text{...(9.274)}$$

OBJECTIVE TYPE QUESTIONS

9.1 In a separately dc generator, used to control the excitation of a synchronous machine or as a source of dc supply in a Ward–Leonard speed control of dc motors, when the field circuit has a low value of resistance:

(i) the steady-state gain yielded by the machine is high.
(ii) the dynamic response of the armature voltage is slow.
(iii) the speed of dynamic response and the steady-state gain act in counter-productive directions.
(iv) dynamic response of the machine is not dependent on the field resistance.
(v) steady-state gain and dynamic response act in supportive direction

Choose the correct statements:

(a) statement (i) is true
(b) statement (ii) is true
(c) statements (i), (ii) and (iii) are true
(d) statements (i), (ii) and (v) are true

9.2 A feedback from the armature to the field circuit in a dc separately-excited generator, actually:

(i) makes the system faster

(ii) makes the system slower and complicated
(iii) makes the system desirable from control point of view
(iv) makes the system undesirable

Choose the correct statements:

(a) statements (i) and (iii) are valid
(b) statement (i) is invalid
(c) statements (ii) and (iv) are valid
(d) statement (ii) is invalid.

9.3 Choose the correct alternative:

A feedback of armature circuit terminal voltage to the field circuit voltage, through an amplifier in case of a dc separately excited generator results in an effective improvement in the closed loop time constant compared to that of the open loop system:

(a) with a negative feedback
(b) with a positive feedback
(c) with either positive or negative feedback

Answers

9.1 (c) **9.2** (a) **9.3** (a)

NUMERICAL PROBLEMS

9.1 (i) Develop suitable relationships for the transient analysis of a separately excited dc generator.

(ii) A separately excited dc generator has the following circuit and operational parameters:

Resistance of the field circuit, $R_f = 120$ ohm
Self-inductance of the field, $L_f = 40$ henry
Resistance of the armature, $r_a = 0.15$ ohm
Self-inductance of armature, $l_a = 0.3$ mH
Effective motional inductance between q-axis armature winding and d-axis armature flux, $M_d = 0.8$ H
Steady-state operational speed, $n_q = 1500$ rpm

The machine is considered as operating at its rated speed with its field winding unexcited and with the armature terminals open circuited.

(a) Obtain the armature voltage as a function of time, if a constant voltage of 150 volt is suddenly applied across the field terminals. Plot the voltage developed as a function of time qualitatively.

(b) If the armature terminals were initially short-circuited and the machine operated as described in part (a) sketch the short-circuit current curve qualitatively as a function of time.

(c) Allow the armature voltage in part (a) to attain a steady-state value. If the armature terminals are now connected to a load consists in a resistance of 2.85 ohm and an inductance of 1.7 mH, in series, obtain the armature current and the armature terminal voltage as functions of time with load parameters included.

(d) Estimate the electromagnetic torque as a function of time when the machine was operating as specified in part (a) and also its mean value.

(e) After the armature current in part (c) attains steady-state value, an additional resistance of 0.4 ohm is inserted in the armature circuit. Find the new value of armature current in the time-domain, after the change.

[**Ans.** (a) $e_a = 157.08\ (1 - e^{-3t})$ volt (b) $i_a(t) = 1570.8\ (1 - e^{-3t})$ amp
(c) Armature terminal voltage across load
$v_a(t) = 149.23 - 15.7\ e^{-667t}$ volt
$i_a(t) = 5.236(1 - e^{-667t})$ amp
(d) $T_e = -54.83(1 - e^{-667t})$ Nm, Average $T_e = -54.83$ Nm
(e) $i_{\text{final}} = (46.2 + 6.16e^{-588.24t})$ amp]

9.2 Express the speed of a separately excited dc motor in the time domain and analysis the nature of speed variation under transient conditions, when

(a) The motor is subjected to a sudden application of rated voltage across the armature terminals under no-load but with a steady excitation current.

(b) The motor was running at no-load speed, while a load torque is suddenly applied to its shaft.

(c) The motor was running under a torque at steady-state with the rated applied voltage, while an additional voltage was applied across its terminals.

(d) The motor was operating under steady-speed with a constant load-torque and with a constant applied voltage, when load torque was suddenly changed.

[**Ans.** (a) $\omega_r(t) = \dfrac{K_m V_t}{r_a(D + D_e)}(1 - e^{-t/\tau_m})$

(b) $\omega_r(t) = \dfrac{K_m V_t}{r_a(D + D_e)} - \dfrac{T_L}{(D + D_e)}(1 - e^{-t/\tau_m})$

(c) $\omega_r(t) = \dfrac{K_m V_t}{r_a(D + D_e)}(V_t + \delta V_t) - \dfrac{T_l}{D + D_e} - \dfrac{K_m \delta V_t}{r_a(D + D_e)} e^{-t/\tau'_m}$

(d) $\omega_r(t) = \dfrac{K_m V_t}{r_a(D + D_e)} - \dfrac{T_l}{D + D_e} + \dfrac{\delta T_l}{D + D_e} e^{-t/\tau'_m}$]

9.3 A separately excited dc motor has the following circuit parameters:

Motor torque constant, $K_m = 2$ Nm/arm. current
Armature resistance, $r_a = 0.4$ ohm
Self-inductance of armature, $l_a = 0.1$ henry
Field circuit resistance, $R_f = 115$ ohm
Polar moment of inertia of the motor and the load, $J = 12$ kg m^2
Rated voltage of the motor, $V_a = 230$ volt

If the armature carries a constant armature current of 50.0 amp when the rated voltage is applied across the armature and the field circuit, obtain

(a) The value of the constant load torque.

(b) The armature voltage is suddenly reduced to 210 volt without any change in the excitation voltage. Determine the new speed of the motor as a function of time. Also find the final steady-state speed of the motor, after the above change is imposed on the motor. [**Ans.** (a) 100 Nm (b) $\omega_r(t) = 95 + 10e^{-0.86t}$]

9.4 Write down the general voltage equations of an inter-connected system of two separately excited dc machines leading to Ward–Leonard method of speed control of dc motors and obtain the transient impedance of the system. From this relation, derive the behaviour of the Ward-Leonard system under all operating conditions with constant value of motor field excitation current.

9.5 Two separately excited dc machines are connected electrically to form a Ward–Leonard system. The generator unit is run at a constant speed but it is kept unexcited, while the motor field is excited but kept at stand-still conditions:

(a) A constant voltage of 40 volt is considered as suddenly applied to the generator field at $t = 0$. Determine the expression for the motor speed as a function of time.

(b) After the motor speed has reached steady-state value, when the motor was operating as in (a) above, consider a mechanical torque equivalent of 20 Nm is suddenly applied on the motor shaft. Obtain the variation of speed as a function of time, counting the instant of time, $t = 0$, from the instant of application of torque to the motor shaft.

The following parameters are known about the generator:

Armature resistance, $r_{ag} = 0.4$ ohm
Field resistance, $r_{fg} = 100$ ohm
Generator voltage constant, $K_g = 6000$ volt/amp

For the separately excited machine functioning as a motor:

Armature resistance, $r_{am} = 0.6$ ohm
Motor constant, $K_m = 2$ Nm/arm. ampere
Total inertia of motor armature and load = 10 kg-m^2
and mechanical damping constant = 1 Nm sec/radian.

Ignore the effect of inductance in the generator armature and field, and the self-inductance of motor armature.

[**Ans.** (a) $\omega_m(t) = [96 + 24e^{-2.5t} - 120e^{-0.5t}]$ rad/sec
(b) $\omega_m(t) = 96 - 4(1 - e^{-0.5t})$
$= 92 + 4e^{-0.5t}$]

9.6 A cumulatively compounded dc generator has its shunt field connected separately excited and has the following circuit parameters:

Self-inductance of the shunt-field, $L_f = 50$ henry
Resistance of the shunt field, $R_f = 100$ ohm

Considering that the corresponding value of the armature circuit are:

$l_a = 0.03$ henry and $r_a = 0.6$ ohm

For the series field winding, the circuit parameters are:

$l_s = 0.02$ henry; $r_s = 0.4$ ohm

and the mutual inductance between the shunt and series field is 1 henry.

If generator emf constant at the rated speed of 100 rads/sec, $K_{da} = 100$ volt/per 100 rads/sec and $K_{ds} = 0.5$ volt per 100 rads/sec, and the generator is operating at steady-state speed of rated value with its armature terminals open-circuited, while the separately excited shunt field is suddenly switched with a dc voltage of 200 volt, find the nature of increase in the generator voltage. [**Ans.** $v_t(t) = 200 - 204e^{-2t}$]

CHAPTER 10

Analysis of Three-phase Induction Motor under Unbalanced Voltage and Asymmetrical Conditions

10.1 INTRODUCTION

Application of symmetrical components to the analysis of induction motor under unbalanced conditions, is taken up in this chapter. We assume symmetrically wound machine, without any reaction between the different sequence quantities, and application of a balanced set of voltages of a given phase sequence to the stator or the rotor member. This yields balanced currents of the same sequence at any value of slip. We wish to undertake the analysis of the following five conditions of operation:

1. Unbalanced voltage fed to the stator of a symmetrical machine.
2. Stator fed with one phase (say phase A), open circuited.
3. Stator having two of its phases (say phases B and C) shorted and fed from a single-phase supply between the healthy phase and the shorted phases (B and C).
4. Unbalanced impedance connected to the rotor terminals of an otherwise symmetrical machine.
5. Operation with an open-circuited rotor phase of a wound-rotor induction motor.

10.2 REVIEW OF OPERATIONAL CHARACTERISTIC WITH APPLICATION OF POSITIVE-SEQUENCE VOLTAGE TO THE STATOR MEMBER

When a balanced set of a positive-sequence voltages is applied to the stator terminals of the machine, the rotor rotates along the positive-sequence, that is, along the direction of the rotating magnetic field at a fractional slip. The operation can be specified by characteristics derived from the equivalent circuit shown in Figure 10.1, and elaborated in Section 8.4.1.

In Figure 10.1, the machine parameters R_s represents resistance per phase of the stator winding:

x_s stands for the leakage reactance per phase of the stator winding, at supply frequency.

R_r' signifies the resistance per phase of the rotor, referred to stator.

x_r' denotes the leakage reactance per phase of the rotor referred to the stator at supply frequency.

i_{s1} is the positive-sequence current in the stator referred to phase A.

i_{r1}' is the positive-sequence current in the rotor referred to phase A.

e_{s1} signifies the positive-sequence line-to-neutral voltage in the stator referred to phase A.

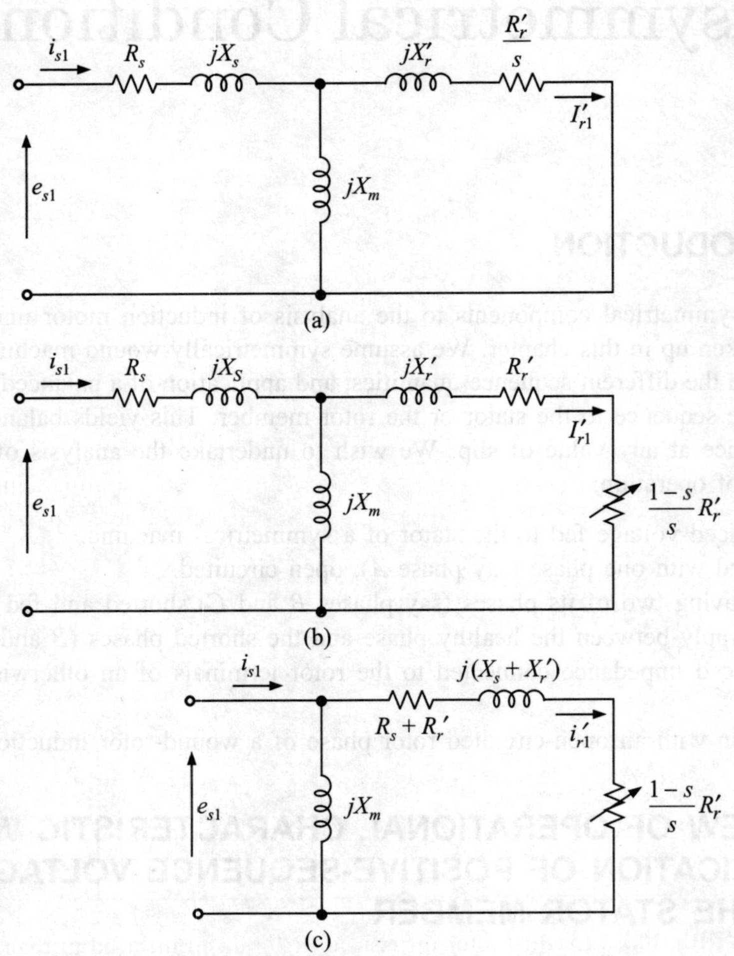

FIGURE 10.1 Equivalent circuit of an induction motor.

As is well known, the application of voltage e_{s1} to the stator produces a rotating magnetic field, rotating at synchronous speed in the air-gap and cuts across the conductors on the rotor member. It, therefore, produces a circulating current in the closed circuited rotor winding at a slip frequency s. This circulating current produces a second field that rotates at a slip velocity with respect to the rotor. Since, the rotor speed is only $(1 - s)$ times the synchronous velocity, the rotor field set up by current i_{r1} traverses at synchronous velocity relative to the stator, which is the same as s times the synchronous speed with respect to the rotor in the direction of the rotor.

The above leads to the development of the equivalent circuit for the positive-sequence as shown in Figure 10.1(a), and characteristics of the induction motor can be obtained from the solution of the network. It is a common practice to recast the equivalent circuit by splitting the resistance $\left(\dfrac{R_r'}{s}\right)$ into two components (i) R_r' and (ii) $\dfrac{1-s}{s}R_r'$ as indicated in Figure 10.1(b), and to simplify it further by placing the shunt magnetizing reactance jX_m directly across the terminals as indicated in Figure 10.1(c), as is normally done for transformers.

Since the power absorbed by R_r' denotes the copper loss and since this is the only loss considered, the power absorbed in the component resistance equivalent $\left(\dfrac{1-s}{s}\right)R_r'$ denotes the shaft power developed. Therefore, calling P_1 as the total shaft power,

$$P_1 = 3\left(\frac{1-s}{s}\right)R_r'(I_{r1}')^2 \qquad ...(10.1)$$

This is the total shaft torque T_1 (in synchronous watt units)—see Section 8.4.1—and

$$T_1 = K3\frac{R_r'}{s}(I_{r1}')^2 \qquad ...(10.2)$$

where, K is an arbitrary constant dependent upon the units.

It must be indicated that no effort was made to include the friction, windage and core losses in this equivalent circuit. They need to be considered in addition separately. If external impedance is introduced in the rotor circuit, its effect should be included by inserting an equivalent value across the rotor terminals of the equivalent circuit as shown in Figure 10.2.

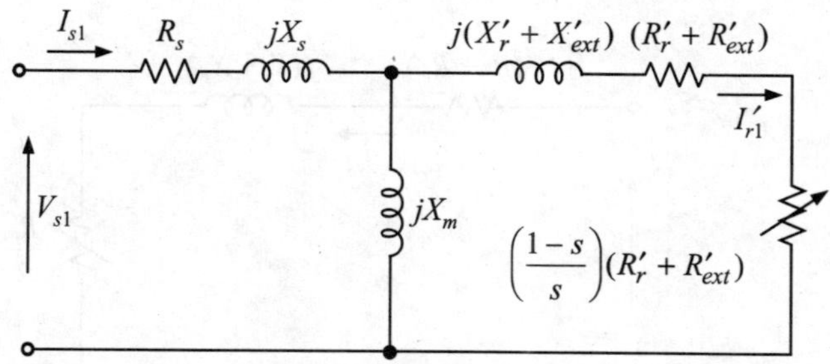

FIGURE 10.2 External rotor impedance $(R_{ext} + jX_{ext})$ included in rotor.

10.3 CHARACTERISTICS WITH APPLICATION OF NEGATIVE-SEQUENCE VOLTAGES TO THE STATOR

Application of negative-sequence voltage is effectively the same as interchanging the supply leads of two of the three supply voltage terminals. In such a case, a field is set up which rotates at synchronous speed in a direction opposite to that set up by the positive-sequence voltage.

The characteristics for the negative-sequence may be obtained by selecting an appropriate value of slip in association with the positive-sequence equivalent circuit of Figure 10.1. Figure 10.3 depicts the relation between the positive-sequence slip s and the negative-sequence slip s_2.

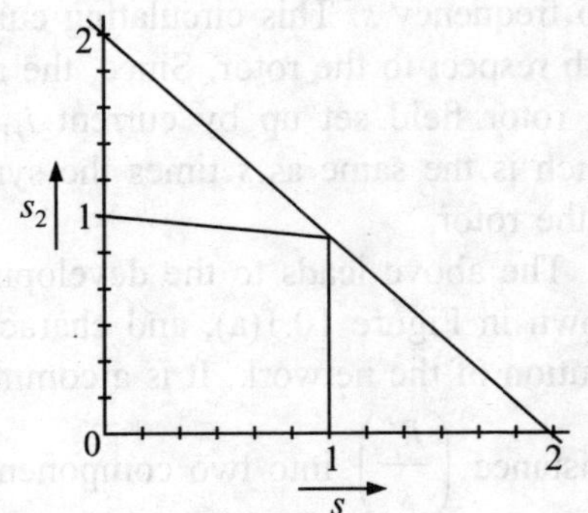

FIGURE 10.3 Relation between the positive-sequence slip s and the negative-sequence slip s_2.

As can be readily seen, for small values of s, s_2 is very nearly equal to 2, and at stand-still conditions, $s = s_2 = 1$, so that the relation between s and s_2 is given by:

$$s_2 = 2 - s \qquad ...(10.3)$$

The equivalent circuit for the negative-sequence can be derived by replacing s in Figure 10.1(a), (b) and (c) by $(2 - s)$, as indicated in Figure 10.4 for the approximate equivalent circuit only in Figure 10.1(c).

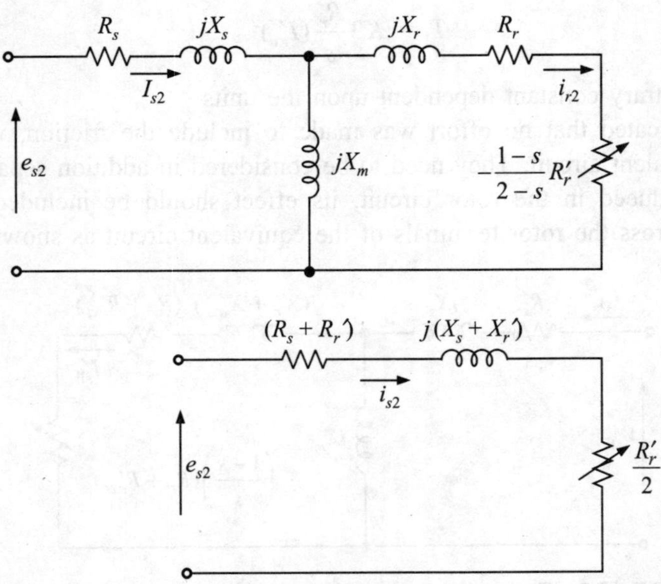

FIGURE 10.4 Negative-sequence equivalent circuit of an induction motor including the approximate equivalent circuit.

The power for the negative-sequence is obtained by estimating the power absorbed in $\left\{-\frac{1-s}{2-s}R_r'\right\}$ which, for the three phases works out to be:

$$-3\left\{\frac{1-s}{2-s}\right\}R_r'(I_{r2}')^2$$

The negative-sequence torque,

$$T_2 = K\frac{3R_r'(I_{r2}')^2}{(2-s)} \quad ...(10.4)$$

Since $(R_r' + jX_r')$ seldom exceeds 15 to 20 per cent and jX_m is usually of the order of 400 to 500 per cent, the effect of jX_m on the rotor current may be ignored. Further, in the approximate equivalent circuit for negative-sequence, the load component of the rotor resistance $\left\{-\frac{1-s}{2-s}\right\}R_r'$ is replaced by $-\left(\frac{R_r'}{2}\right)$ as shown in the later part of Figure 10.4.

10.4 CHARACTERISTIC WHEN BOTH POSITIVE AND NEGATIVE-SEQUENCE VOLTAGES ARE APPLIED TO THE STATOR

When an unbalanced set of voltages is applied to the stator of an induction motor, the voltages can be transformed into an equivalent positive and negative-sequence voltages. The motor, therefore, functions with both the voltage components applied to it simultaneously. The currents of the two sequences do not react with each other. Their effects can, therefore, be considered by computing separately by considering circuits elaborated in Sections 10.2 and 10.3, and taking their aggregate effects.

The total shaft power available from the motor works out to be:

$$P_{\text{total}} = 3\left[\frac{1-s}{s}R_r'(\dot{I}_{r1}')^2 - \frac{1-s}{2-s}R_s'(I_{r2}')^2\right] \quad ...(10.5)$$

and the total shaft torque works out to be:

$$T_{\text{total}} = 3KR_r'\left[\frac{\dot{I}_{r1}^2}{s} - \frac{(i_{r2}')^2}{2-s}\right] \quad ...(10.6)$$

We take a numerical example to indicate the significance power, torque and loss relations for operation with unbalanced voltage applied to machine terminals.

EXAMPLE 10.1 A 3-phase, star-connected, 400 volt, 50 Hz, induction motor has the following circuit parameters in ohm per phase, referred to stator:

R_s, the stator resistance per phase = 0.15 Ω

X_s, the stator reactance per phase when referred to stator side = 0.45 Ω

R_s', the rotor resistance per phase, when referred to stator side = 0.12 Ω

X'_r, the rotor reactance per phase when referred to stator side = 0.45 Ω

X_m, the magnetizing reactance of the machine as seen from the stator side = 28.5 Ω

Fixed losses (core, friction and windage losses) of the machine = 400 watt

Stator to rotor effective turns ratio = 1.2

Compute:

(a) Complete the torque-speed characteristic of the machine for 100% positive- and negative-sequence voltage across the stator terminals. Hence, indicate power output, torque developed and machine losses when the voltage unbalance is such that the negative-sequence voltage is 10% of the positive-sequence voltage.

(b) The stator current, output torque and efficiency at a slip of 4%.

Solution:

(a) Characteristics of the machine for 100% positive-sequence and 100% negative-sequence voltages applied to the stator are shown in Figure10.5(a) and (b) respectively. The output powers at the shaft and losses are also shown for the two cases, and the torques are displayed. The negative-sequence cases have the same shape as the positive-sequence cases, but are inverted above the axis at $s = 0$ for power and loss

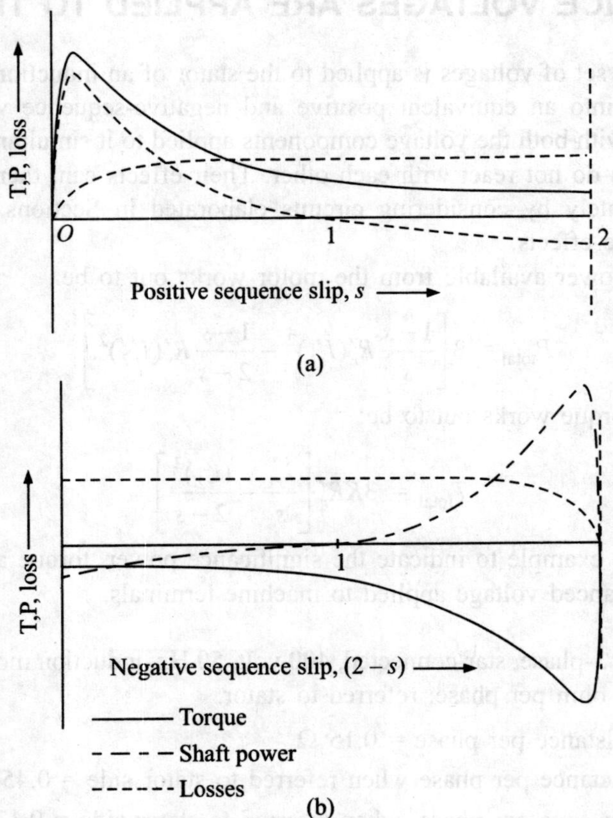

FIGURE 10.5 Induction motor characteristics for 100 per cent applied voltage in stator.

and double inverted about $s = 0$ and slip axis in case of torque. If we assume 100 per cent positive-sequence and 10% negative-sequence voltages to be impressed simultaneously, with the load such that the slip is 4 per cent, then the torque due to positive-sequence is about 98% of normal in a positive sense.

Since, for the negative-sequence, the torque varies as the square of the applied voltage, a 10% contribution contributes to a negative torque of a very small quantity compared to positive-sequence torque. Unbalance in voltages therefore contributes to negligible reduction in torque characteristics. But as far as the losses are concerned, it is considerable. The unbalance in voltage, therefore, causes heating and loss in machine rating if negative-sequence component is large.

(b) Using the equivalent circuit of the induction motor shown in Figure 10.6, we may solve the network for different values of s. The result is plotted as indicated in Figure 10.5. A sample calculation for 100% positive-sequence applied voltage with the motor operating at slip $s = 0.04$ is shown:

$$Z_{\text{MN}} = (0.15 + j0.45) + \frac{\left\{\dfrac{0.12}{0.04} + j0.45\right\}\{j28.50\}}{\left\{\dfrac{0.12}{0.04} + j0.45 + j28.50\right\}}$$

$$= (0.15 + j0.45) + \frac{\{3.0 + j0.45\}\{j28.5\}}{(3.0 + j28.95)}$$

$$= 3.02 + j1.188$$

$$= 3.25\angle 21.47° \text{ ohm}$$

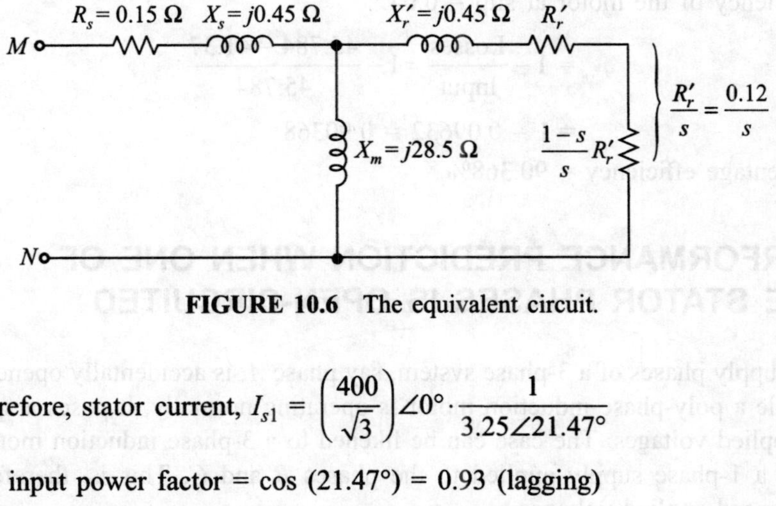

FIGURE 10.6 The equivalent circuit.

Therefore, stator current, $I_{s1} = \left(\dfrac{400}{\sqrt{3}}\right)\angle 0° \cdot \dfrac{1}{3.25\angle 21.47°}$

The input power factor = cos (21.47°) = 0.93 (lagging)

Total stator copper loss = $3I_{s1}^2 R_s$

$$I_{s1} = \frac{400}{\sqrt{3} \times 3.25} = 71.06 \text{ amp}$$

Hence, stator copper loss = $3 \times (71.06)^2 \times 0.15 = 2272.3$ watt

$$\text{Total stator power input} = \sqrt{3} V_{\text{line}} I_{s1} \cos\theta$$

$$= \sqrt{3} \times (400) \times (71.06) \times (0.93) = 45.784 \text{ kW}$$

Total power input to rotor across the air-gap,

$$P_r = \text{stator power input} - \text{stator copper losses}$$

$$P_{\text{rotor}} = 45.784 - 2.272 = 43.512 \text{ kW}$$

Mechanical power developed by machine at slip $s = 0.04$, P_{mech}:

$$P_{\text{mech}} = (1 - s)\, P_{\text{rotor}}$$

$$= (0.96) \times (43.512) = 41.77 \text{ kW}$$

Shaft power output, P_{shaft}

$$-P_{\text{shaft}} = P_{\text{mech}} - \text{Fixed losses}$$

$$= 41.77 - 0.400 \text{ kW} = 41.37 \text{ kW}$$

$$\text{Output torque} = \frac{P_{sh}}{\omega_r} = \frac{41.37}{(1-s)\omega_s}$$

where, $s = 0.04$ and $\omega_s = 50\pi$

$$T_{\text{output}} = \frac{41370}{(0.96) \times (50\pi)} = 274.34 \text{ Nm}$$

Efficiency of the motor at slip = 0.04:

$$= 1 - \frac{\text{Losses}}{\text{Input}} = 1 - \frac{45.784 - 41.37}{45.784}$$

$$= 1 - 0.09632 = 0.90368$$

Percentage efficiency = 90.368%

10.5 PERFORMANCE PREDICTION WHEN ONE OF THE STATOR PHASES IS OPEN-CIRCUITED

If one of the supply phases of a 3-phase system, say phase A, is accidentally opened by blowing of a fuse, while a poly-phase induction motor is operating normally, it poses a special case of unbalanced applied voltages. The case can be likened to a 3-phase induction motor effectively supplied with a 1-phase supply applied to the phases B and C. This is, therefore, a special case of unbalanced applied voltages.

It is relatively easy to treat the machine originally connected in star, although consideration of stator connected in delta, can be handled by applying conversion of the original delta connection by an equivalent star connection.

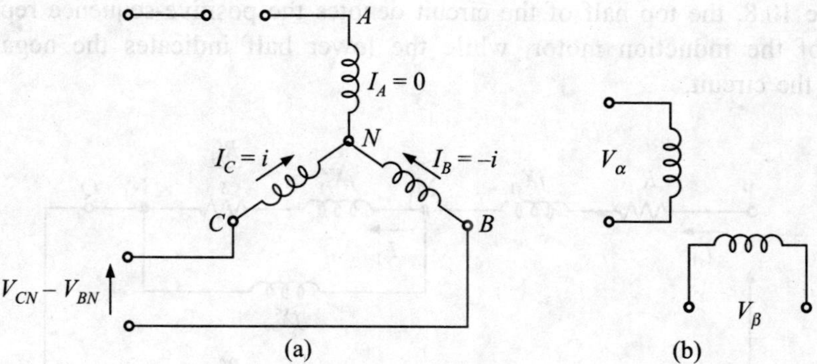

FIGURE 10.7 Depiction of a 3-phase induction motor with one of its phase (say, phase *A*) open circuited (a) 3-phase motor depiction (b) equivalent 2-phase depiction with 0-α-β transformation.

Referring to Figure 10.7, the phase currents assume the relations:

$$\left.\begin{aligned} I_a &= 0 \\ I_b &= -I_c = -i \end{aligned}\right\} \qquad \text{...(10.7)}$$

which leads to the relations:

$$\begin{aligned} i_{s1} &= \frac{1}{3}(I_a + aI_b + a^2 I_c) \\ &= \frac{1}{3}(0 + ai - a^2 i) = \left(\frac{a - a^2}{3}\right) i \\ &= +j\frac{i}{\sqrt{3}} \end{aligned} \qquad \text{...(10.8)}$$

and

$$\begin{aligned} i_{s2} &= \frac{1}{3}(I_a + a^2 I_b + aI_c) \\ &= \frac{1}{3}(0 + a^2 i - ai) = -\left(\frac{a - a^2}{3}\right) i \\ &= -j\frac{i}{\sqrt{3}} \end{aligned} \qquad \text{...(10.9)}$$

The only voltage effective between the terminals *B* and *C* is E_{s1}. Its value can be reconstructed as follows:

Since,

$$\left.\begin{aligned} E_{sA} &= E_{s0} + E_{s1} + E_{s2} \\ E_{sB} &= E_{s0} + a^2 E_{s1} + aE_{s2} \\ E_{sC} &= E_{s0} + aE_{s1} + a^2 E_{s2} \end{aligned}\right\} \qquad \text{...(10.10)}$$

$$\begin{aligned} E_{sA} &= E_{sC} - E_{sB} \\ &= (a - a^2)(E_{s1} - E_{s2}) = j\sqrt{3}(E_{s1} - E_{s2}) \end{aligned}$$

or

$$E_{s1} = -E_{s2} = \frac{E_{sA}}{j\sqrt{3}} \qquad \text{...(10.11)}$$

In Figure 10.8, the top half of the circuit denotes the positive-sequence representation of the circuit of the induction motor, while the lower half indicates the negative-sequence depiction of the circuit.

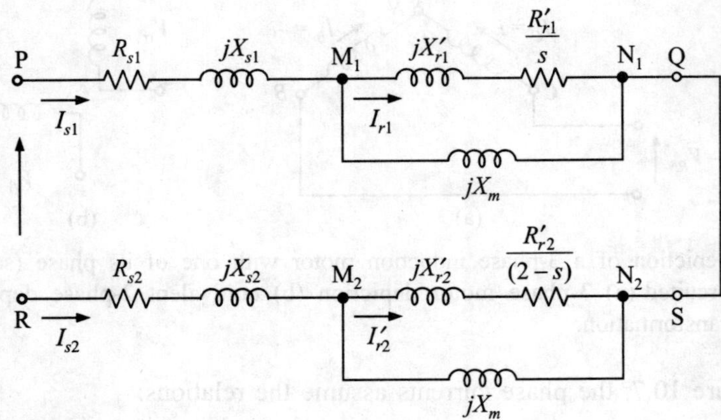

FIGURE 10.8 Equivalent circuit of the induction motor for condition $I_{s1} = - I_{s2}$ to be satisfied, with $E_{s1} = E_{s2}$ as the voltage applied across points P-R equalling $\left(E_s A / j\sqrt{3}\right)$.

The total shaft power output and the net torque output of the motor can be obtained using Eqs. (10.5) and (10.6) respectively.

It is a common practice to introduce approximation in the equivalent circuit of the induction motor, by shifting the magnetizing branch jX_m to the input terminals from its actual position across the rotor input terminals M_1N_1, in order to bring about simplification in computation. However, since all the impedances in the negative circuit component, except for the impedance jX_m across terminals $M_2 - N_2$ in this branch are of the order of leakage reactance, the magnetic branch jX_m—connected between $M_2 - N_2$ in the negative-sequence branch component can be ignored as its net effect is not much of consequence.

This approximation, when applied in the equivalent circuit, gives the new shape as shown in Figure 10.9, with the magnetization branch jX_m due to the positive-sequence component (originally connected across terminal $M_1 - N_1$) brought across the terminals P-R in Figure 10.8.

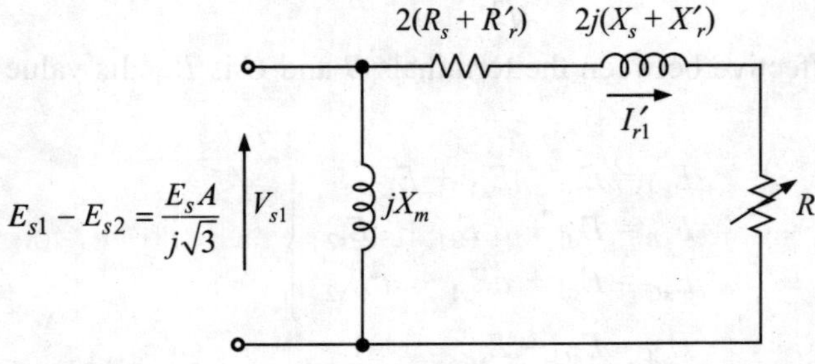

FIGURE 10.9 Equivalent circuit of Figure 10.8 simplified by placing magnetizing branch across input terminals.

The variable component of the approximated equivalent circuit, shown in Figure 4.15, now becomes:

$$R = \left\{\frac{1-s}{s} - \frac{1-s}{(2-s)}\right\} R_r'$$

$$= \frac{2(1-s)^2}{s(2-s)} R_r' \qquad \text{...(10.12)}$$

for the direction of current flow taken into account.

The total shaft power output works out to be:

$$P_{\text{shaft}} = \frac{6(1-s)^2}{s(2-s)} (R_r')(I_r')^2 \text{ watt} \qquad \text{...(10.13)}$$

and the net torque available at the shaft has the form:

$$T_{\text{shaft}} = \frac{6(1-s)}{s(2-s)} (R_r')(I_r')^2 \text{ Syn. watt} \qquad \text{...(10.14)}$$

A scrutiny of Eq. (10.14) shows that, for small values of slip, the torque produced by the motor continues to be positive. This indicates that the machine continues to run as a motor when the fuse of one of the supply lines blows, if the load is not high. But at $s = 1$, corresponding to stand-still conditions, the machine does not develop any torque and will not start.

Incidentally, for the condition:

$I_a = 0$; $I_b = -I_c$, as is obtained in the case under review, we find by referring to relations (6.40), that:

$$I_0 = 0; \quad I_\alpha = 0 \quad \text{and} \quad I_\beta = \sqrt{2} I_\beta$$

The type of operation of a 3-phase induction motor is therefore similar to operation of an equivalent 2-phase induction motor with the above current constraints. We may refer to Figure 10.3 for clarifications.

10.6 STATOR WINDING WITH TWO PHASES SHORTED WHILE VOLTAGE APPLIED ACROSS HEALTHY AND SHORTED PHASES

The operating condition is depicted in Figure 10.10. The single phase supply is effective between terminals A and B–C in parallel.

The constraints are now applied to the voltages, which take the form:

$$V_{cN} = V_{bN} \quad \text{and} \quad V = V_{cN} - V_{bN}$$

where, V is the applied voltage.

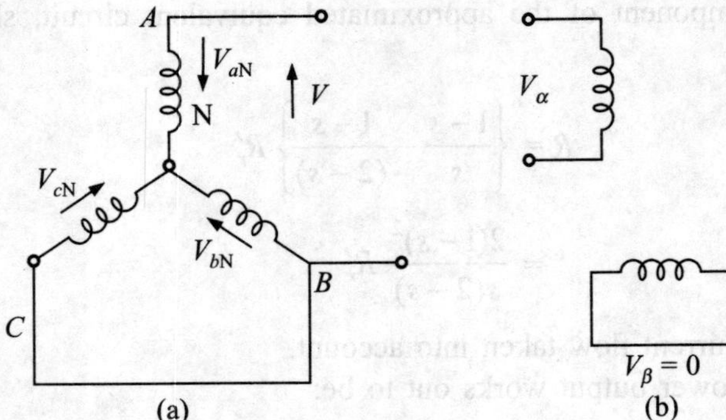

FIGURE 10.10 Depiction of a 3-phase induction motor with two of its phases (say phase B and C) as shorted and single phase voltage effective between terminals A and B–C. (a) 3-phase motor depiction and (b) Equivalent 2-phase presentation with 0-α-β transformation.

Since, as per our convention,

$$V_{AN} + V_{BN} + V_{CN} = 0$$

(An assumption that depends on the transferred impedance matrix Z'), then

$$V_{CN} = V_{BN} = -\frac{1}{2} V_{AN}$$

so that we can write

$$V_{AN} = \frac{2}{2} V; V_{BN} = -\frac{1}{2} V \quad \text{and} \quad V_{CN} = -\frac{1}{2} V$$

and using the voltage equations of Eq. (6.40), we get:

$$V_{CB} = 0 \quad \text{and} \quad V_\alpha = \sqrt{\frac{2}{3}} V, \; V_\beta = 0$$

This type of operation is, therefore, equivalent to that shown in Figure 10.10(b). The single phase voltage is now effective across the α-phase of the equivalent 2-phase machine and its value is, $V_\alpha = \sqrt{\frac{2}{3}} V$, while the phase β is now effectively short-circuited.

The distinctions between the imposed constraints indicated under Sections 10.5 and 10.6 present interesting studies.

It is left to the readers to work out the details for the machine operation for the present case in the form of equivalent circuit representation, total shaft power output and net shaft torque production.

10.7 ROTOR CIRCUIT CLOSED WITH UNBALANCED IMPEDANCES

A symmetrical and balanced 3-phase induction motor leads to unbalanced operation when its secondary rotor circuit is closed with unbalanced impedances.

For the analysis of the situation, positive-sequence voltage is taken as applied to the stator terminals, and currents of fundamental frequency in the stator produce slip frequency currents in the rotor member, which are unbalanced. These unbalanced currents at slip frequency in the rotor are resolved into positive- and negative-sequence currents at slip frequency. The negative-sequence components set up fields that rotate negatively with respect to the rotor at slip frequency and, in turn, induce currents in the stator having frequency equal to fundamental frequency minus twice the slip frequency.

We attack the problem by analysing the reactions on the machine into two groups:

1. Those associated with fundamental frequency in the stator and slip frequency of positive-sequence in the rotor, and
2. Those linked to slip frequency of negative-sequence in the rotor and fundamental minus twice slip frequency in the stator.

These two groups of reactions are tied together in the end using the nature and characteristics of the unbalanced impedances through which the rotor circuit is completed.

Figure 10.11 schematically depicts the induction motor circuit under review.

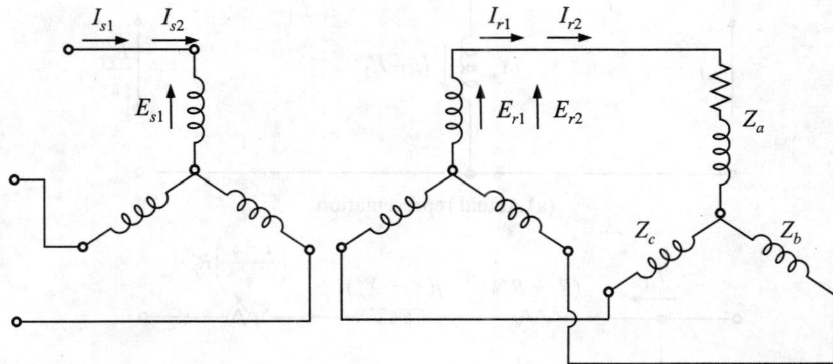

FIGURE 10.11 A symmetrically balanced induction motor with unbalanced impedances connected in rotor circuit.

We undertake the analysis by taking Group 1 first.

With the positive-sequence voltage E_{s1} applied to the stator, the resultant air-gap flux rotating at synchronous angular velocity ω_s induces in the stator a voltage E_{g1} per phase and in the rotor an effective voltage sE_{g1} referred to the stator (where s is the fractional slip) is produced. The impedance drop in the stator is governed by the relationship:

$$E_{s1} = E_{g1} + (R_s + jX_s)\, I_{s1} \qquad \text{...(10.15)}$$

In the rotor circuit, the voltage equation assumes the form:

$$sE_{g1} = (R'_r + jsX'_r)I'_{r1} + E'_{r1} \qquad \text{...(10.16)}$$

where, E'_{r1} is the rotor terminal voltage due to positive-sequence. It is clearly at slip frequency and can be measured at the terminals of the rotor.

Equation (10.16) leads to:

$$E_{g1} = \left(jX'_r I'_{r1} + \frac{R'_r}{s} I'_{r1} \right) + \frac{E'_{r1}}{s} \qquad \text{...(10.17)}$$

The air-gap flux is produced by the resultant of the currents I_{s1} and I_{r1}. Therefore, considering I_{s1} as positive when magnetizing and I'_{r1} as positive when demagnetizing, the proportionality between these currents and the voltage produced by the air-gap flux may be expressed by the relation:

$$E_{g1} = jX_m(I_{s1} - I'_{r1}) \qquad \text{...(10.18)}$$

The corresponding equivalent circuit diagram can now be obtained based on the Eqs. (10.15), (10.17) and (10.18).

This is shown in Figure 10.12(a) and the approximate diagram is presented in Figure 10.12(b).

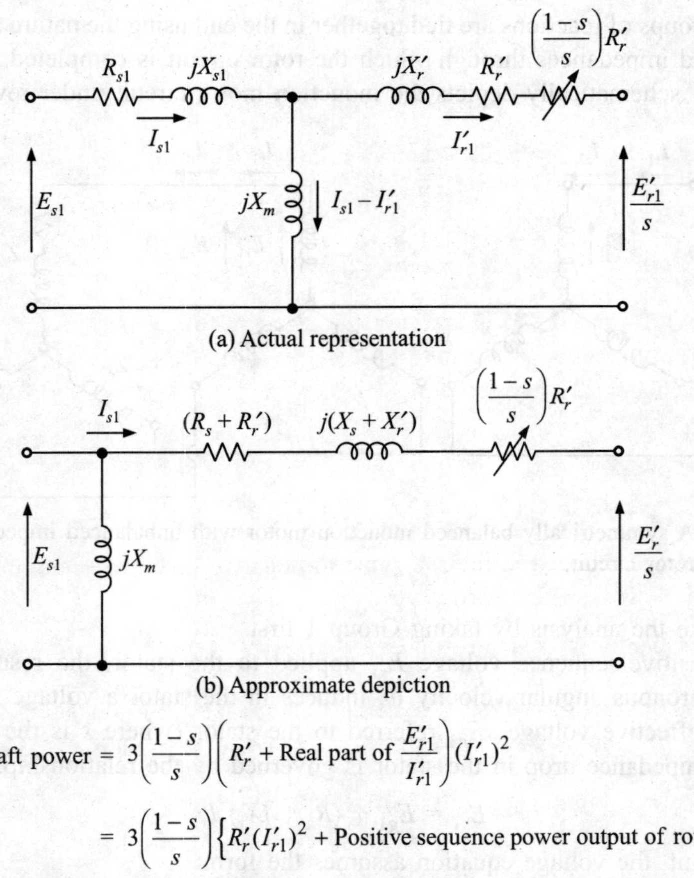

$$\text{Shaft power} = 3\left(\frac{1-s}{s}\right)\left(R'_r + \text{Real part of } \frac{E'_{r1}}{I'_{r1}}\right)(I'_{r1})^2$$

$$= 3\left(\frac{1-s}{s}\right)\left\{R'_r(I'_{r1})^2 + \text{Positive sequence power output of rotor}\right\}$$

FIGURE 10.12 Equivalent positive-sequence network for the induction motor with voltage E'_{r1} appearing across the rotor (a) Actual representation (b) Approximate.

From the above diagram, we observe that:

$$\frac{E'_{r1}}{s} = E_{s1} - \left[R_s + \frac{R'_r}{s} + j(X_s + X'_r)\right]I'_{r1}$$

which leads to:

$$E'_{r1} = sE_{s1} - \left[sR_s + R'_r + js(X_s + X'_r)\right] I'_{r1} \qquad \text{...(10.19)}$$

We now take up the second group. The development of the equivalent circuit for the negative-sequence is carried out by assuming a negative-sequence voltage E'_{r2} of frequency *sf* applied across the rotor terminals and assuming the rotor to be stationary, while the stator is rotating with a relative speed of $(1 - s)\omega_s$ in the negative direction. This produces the same relative condition as if the stator is stationary and the rotor is rotating at a speed of $(1 - s)\omega_s$ in the positive direction.

The applied voltage develops a flux rotating at a velocity $(-s\omega_s)$, and since the assumed relative velocity of the stator is $-(1 - s)\omega_s$, the velocities assumes the same value at $s = 0.5$. Hence, for this value of the slip, no negative-sequence voltage is induced in the stator and the characteristic corresponds to that of a normal induction motor operating at synchronous speed.

For $s < 0.5$, the negative-sequence velocity of the stator is greater than the negative velocity of the flux produced by the negative-sequence voltage. The machine, therefore, functions as an induction generator.

However, for $s > 0.5$, the effective negative velocity of the stator is less than the negative velocity of the flux produced by negative-sequence voltage, and the machine functions like an induction motor.

With voltage V'_{r2} considered as applied to the rotor, and with this operation, the current flowing out of the rotor taken as positive (as has been indicated in Figure 10.11), the voltage equation for the rotor can be written as:

$$E'_{r2} = E'_{g2} - (R'_r + jsX'_r)\, I'_{r2} \qquad \text{...(10.20)}$$

where

I'_{r2} is the negative-sequence rotor current.

E'_{g2} is the voltage induced in the rotor due to negative flux corresponding to the negative-sequence reactions inside the machine.

The angular velocity of the flux that produced E_{g2} is, relative to the rotor, $(-s\omega_s)$ and with respect to the stator $(1 - 2s)\omega_s$. The same flux induces the voltage $\left\{-\left(\frac{1-2s}{s}\right)E_{g2}\right\}$ in the stator winding. When the system impedance to the flow of the stator current of frequency $(1 - 2s)f$ is zero, this voltage must equal the impedance drop of the stator at the frequency $(1 - 2s)f$, so that the voltage in the stator has the form:

$$-\left(\frac{1-2s}{s}\right)E_{g2} = R_s(-I_{s2}) - j(1 - 2s)X_s\,(-I_{s2}) \qquad \text{...(10.21)}$$

where, I_{s2} is the stator current associated with the negative-sequence current in the rotor. In order to be in line with the previous assumptions, the direction of the flow is considered positive as shown in Figure 10.13. The sign used in convention with the reactance drop term is negative because at small slips the machine operates as an induction generator. We can arrive at the expression for E_{g2} (by manipulation of Eq. 10.21), as follows:

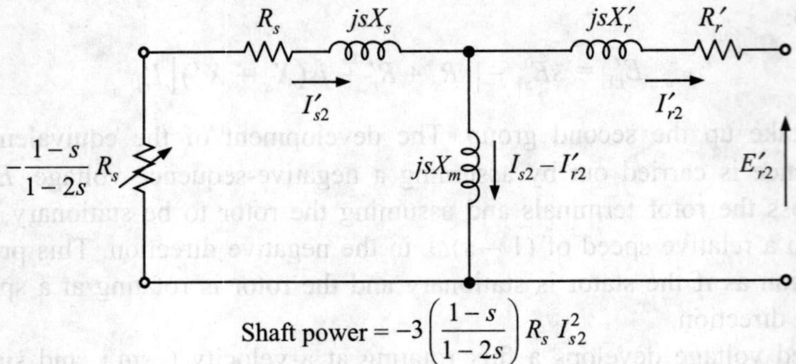

FIGURE 10.13 Equivalent circuit diagram for the negative-sequence currents in the rotor, with slip-frequency voltage of frequency *sf* assumed applied to the rotor (with rotor effectively stationary while the stator is effectively rotating at $(1-s)\omega_s$ in the reverse direction with its terminals shorted through supply).

$$E_{g2} = \left(\frac{s}{1-2s}\right)R_s I_{s2} - jsX_s I_{s2} \quad ...(10.22)$$

$$= -\left\{\frac{(1-s)}{(1-2s)}R_s I_{s2} + R_s I_{s2}\right\} - jsX_s I_{s2} \quad ...(10.23)$$

We take the positive value of I_{s2} to be magnetizing and positive value of I'_{r2} as de-magnetizing, as in the case of positive-sequence counterparts I_{s1} and I'_{r1}, as indicated soon after relations in Eq. (10.17). Hence, E_{g2} is proportional to $(I_{s2} - I'_{r2})$. We use the same constant of proportionality as was used for the positive-sequence case of currents. Since E_{g2} is measured at slip-frequency, the equation governing E_{g2} is as follows:

$$E_{g2} = jsX_m(I_{s2} - I'_{r2}) \quad ...(10.24)$$

We have used Eqs. (10.20), (10.23) and (10.24) in evolving the equivalent circuit of Figure 10.13 for the negative-sequence network of the machine.

The power absorbed by the resistance $\left\{-\frac{(1-s)}{(1-2s)}R_s\right\}$ stands for the shaft power output as the total power input from the rotor can indicate only copper loss and the shaft power.

At small values of slip *s*, the quantity is negative, indicating that the shaft power output for the negative-sequence is in opposition to that of the positive-sequence.

For $s > 0.5$, the opposite is valid, and the negative- and positive-sequence powers are additive.

The voltage and current relations for the negative-sequence circuit can be written down using Figure 10.13 as follows:

$$E'_{r2} = I'_{r2}\left[R'_r + jsX'_r + \frac{jsX_m[-R_s + jX_s(1-2s)]}{-R_s + j(1-2s)(X_s + X_m)}\right] \quad ...(10.25)$$

After having obtained the two sequence networks, we could consider the tie-up between them and the externally connected unsymmetrical impedances connected to the rotor circuit. For the impedances connected to the rotor circuit—Z_a, Z_b and Z_c as shown in Figure 10.11, the phase voltages and currents at the terminals can be written down as:

$$\left.\begin{aligned} E_a - E_c &= Z_a I_a - Z_c I_c \\ E_b - E_a &= Z_b I_b - Z_a I_a \\ E_c - E_b &= Z_c I_c - Z_b I_b \end{aligned}\right\} \qquad ...(10.26)$$

The phase voltages and currents could now be expressed in terms of the sequence voltages and currents as follows:

$$E'_{r1} = \frac{1}{3}(Z_a + Z_b + Z_c)I'_{r1} + \frac{1}{3}(Z_a + a^2 Z_b + aZ_c)I'_{r2} \qquad ...(10.27)$$

and

$$E'_{r2} = \frac{1}{3}(Z_a + aZ_b + a^2 Z_c)I'_{r1} + \frac{1}{3}(Z_a + Z_b + Z_c)I'_{r2} \qquad ...(10.28)$$

The sequence voltages at the rotor terminals could be eliminated between Eqs. (10.27) and (10.28) and Eqs. (10.19) and (10.25). The remaining equations could then be solved for the sequence currents and the values of sequence voltages could be obtained.

The actual analytical expressions are quite cumbersome. From the sequence currents and voltages, the motor performance regarding torque, power loss, phase voltages and currents can be computed, for different values of external impedances connected in rotor circuit, using numerical values.

10.8 OPERATION OF INDUCTION MOTOR WITH ONE OPEN-CIRCUITED ROTOR PHASE

The operation of an induction motor with an open-circuited rotor phase is considered as a special case. We use the material developed in Section 10.7 for the analysis.

We assume that the stator of the induction motor is fed from a stable and balanced 3-phase supply system. The phase *a* of its rotor is open-circuited while the phases *b* and *c* of the rotor terminals are short-circuited without any external impedance connected. Thus:

$$Z_a = \infty, \text{ while } Z_b = Z_c = 0$$

The circuit conditions indicate that the situation can be likened to a 3-phase transmission line with line-to-line fault imposed on it. Application of symmetrical components to the analysis indicates that the positive- and negative-sequence equivalent circuits are to be connected back-to-back in parallel opposition.

If we use the equivalent circuits developed in Figure 10.12 and Figure 10.13 for the analysis of the induction motor operating with the fault, the condition satisfies:

$$I'_{r1} = -I'_{r2} \text{ and } E'_{r1} = E'_{r2}$$

If the constants of the equivalent circuit for the positive-sequence currents is multiplied by $\left(\frac{1}{s}\right)$, then the same current flows when $\left(\frac{E'_{r2}}{s}\right)$ is impressed across the rotor terminals of the negative-sequence network, as if E'_{r2} were impressed across the unchanged equivalent circuit diagram for the negative-sequence (see Figure 10.13).

It, therefore, follows that if this negative-sequence network is connected across $\left(\frac{E'_{r1}}{s}\right)$ in the positive-sequence diagram [see Figure 10.12(b)] as show in Figure 10.14(a), both conditions representing the fault are satisfied.

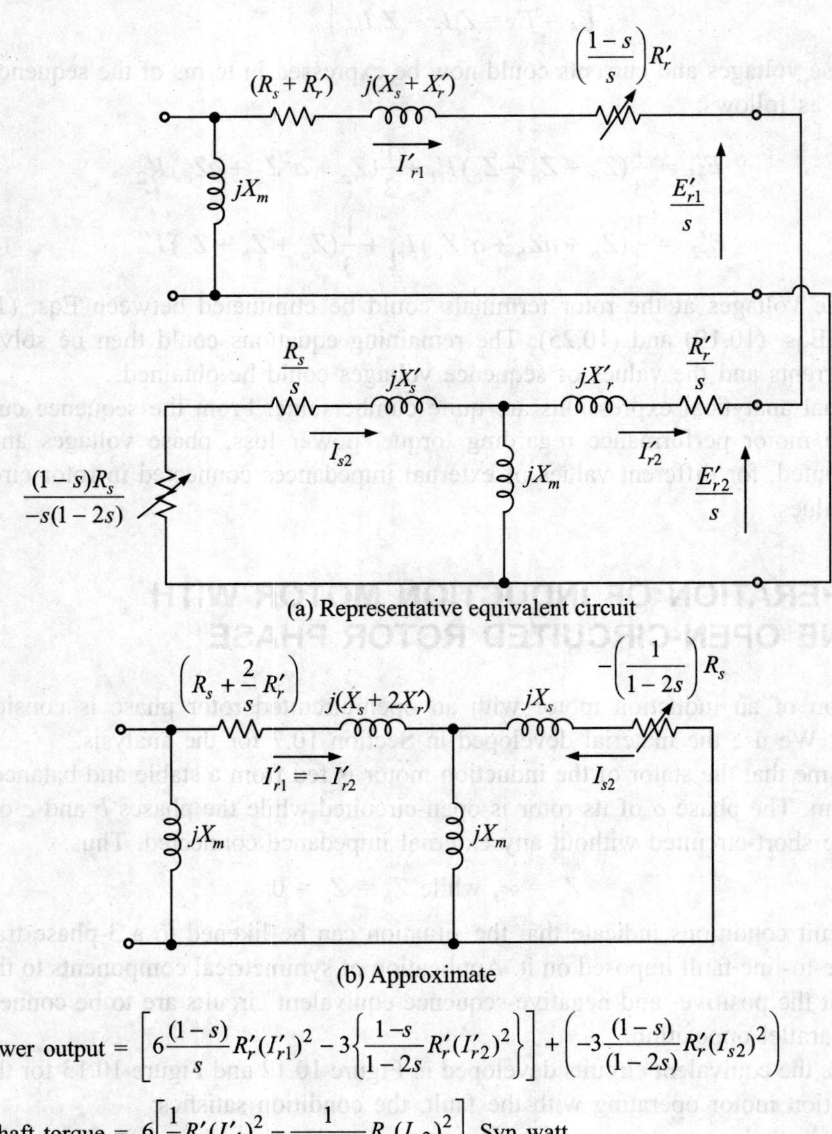

$$\text{Shaft power output} = \left[6\frac{(1-s)}{s}R'_r(I'_{r1})^2 - 3\left\{\frac{1-s}{1-2s}R'_r(I'_{r2})^2\right\}\right] + \left(-3\frac{(1-s)}{(1-2s)}R'_r(I_{s2})^2\right)$$

$$\text{Shaft torque} = 6\left[\frac{1}{s}R'_r(I'_{r1})^2 - \frac{1}{(1-2s)}R_s(I_{s2})^2\right] \text{ Syn watt}$$

$$\text{Loss} = 3[R_s + 2R'_r](I'_{r1})^2 + 3R_s(I_{s2})^2$$

FIGURE 10.14 Equivalent circuit-diagrams for open-circuited rotor phase of the induction motor.

A critical scrutiny of Figure 10.14(a) shows that the currents indicated are the actual currents flowing in the machine. We observe that the power loss does not represent the actual value of power loss in the negative sequence part of the diagram. Hence, the advantage of retaining the two networks in the diagram is lost. The only advantage of the diagram is that it enables determination of the terminal voltage of the rotor. Hence, the diagram may be simplified to the shape as shown in Figure 10.14(b).

Since the voltage applied to the stator is wholly of positive-sequence, the total power input into the terminals of the network in Figure 10.14(b) indicates that this input is composed of the sum of the losses in the indicated resistances. If we subtract the actual I^2R losses due to the positive- and negative-sequence currents in the stator and rotor, this yields the total shaft power output. The shaft torque can now be obtained by multiplying the term by $\dfrac{1}{(1-s)}$, in synchronous watt.

The characteristics are shown schematically in Figure 10.15. The motor torque and shaft power are plotted as functions of speed. The curves present the case for an open-circuited rotor phase, and also for the normal balanced short-circuited rotor.

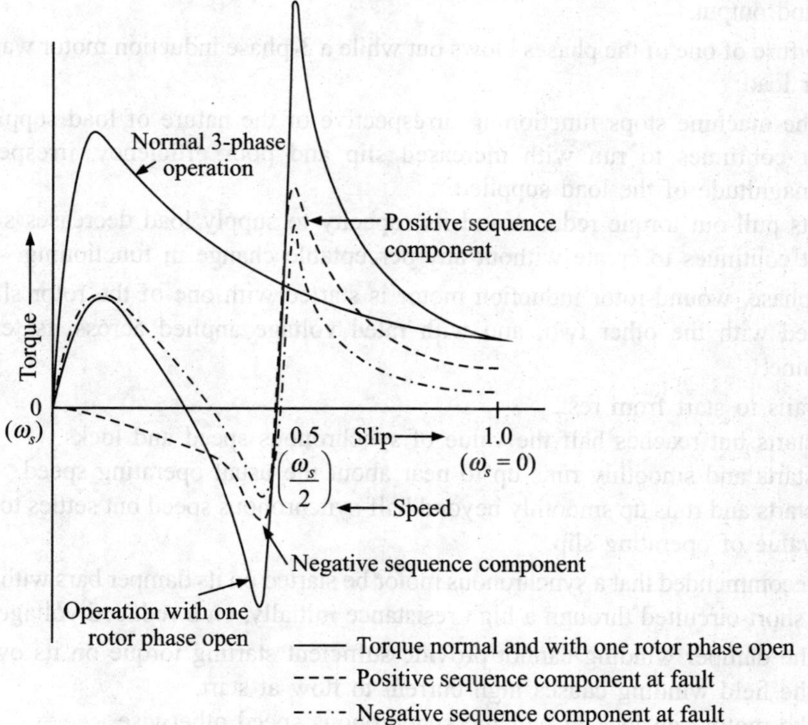

Figure 10.15 Torque-speed characteristic of a 3-phase induction motor with one rotor phase open-circuited, and its comparison with torque for normal operation.

It may be noted that at half-speed the motor develops a significantly small value of the shaft power output. While the motor is capable of starting with a relatively high starting torque, it is incapable of accelerating beyond a speed approximately half synchronous speed, at which it gets locked. ***This phenomenon is commonly observed by students when they attempt to start a synchronous motor using its damper bars, with the rotor field winding short-circuited.***

Furthermore, the pull-out characteristic of the induction motor in the vicinity of half speed is usually considerably more than the normal pull-out value near its synchronous speed.

OBJECTIVE TYPE QUESTIONS

10.1 A 3-phase induction motor operates on unbalanced voltage supply. If the unbalance is such that, with the positive-sequence component voltage taken as 100 per cent, the negative-sequence voltage component is only 10 per cent, then:

(a) the machine performance suffers perceptably by overall deterioration.
(b) the machine losses are not affected significantly, but the torque and power rating are significantly reduced.
(c) the machine displays increased losses with perceptable deterioration in rating, but the torque is not reduced visibly.
(d) the efficiency is adversely affected as the losses increase, with less effect on torque and output.

10.2 If the fuse of one of the phases blows out while a 3-phase induction motor was functioning under load:

(a) the machine stops functioning, irrespective of the nature of load supplied by it.
(b) it continues to run with increased slip and poor efficiency, irrespective of the magnitude of the load supplied.
(c) its pull-out torque reduces and its capacity to supply load decreases significantly.
(d) it continues to create without any perceptable change in functioning.

10.3 A 3-phase, wound-rotor induction motor is started with one of the rotor slip-rings non-shorted with the other two, and with rated voltage applied across its terminals. The machine:

(a) fails to start from rest.
(b) starts but reaches half the value of synchronous speed and locks.
(c) starts and smoothly runs up to near about the usual operating speed.
(d) starts and runs up smoothly beyond half-synchronous speed but settles to an increased value of operating slip.

10.4 It is recommended that a synchronous motor be started on its damper bars with its excitation field short-circuited through a high resistance initially, on a reduced voltage, because:

(a) the damper winding cannot provide sufficient starting torque on its own.
(b) the field winding causes high current to flow at start.
(c) the motor can lock at half the synchronous speed otherwise.
(d) the field winding can get damaged due to high inductance voltage being induced in it during starting.

Answers

10.1 (c) **10.2** (b) and (c) **10.3** (b) **10.4** (d)

CHAPTER 11

Concept of Space Phasor and its Application

11.1 INTRODUCTION

In past practice, dc machines were extensively employed in industrial applications especially where variable speeds were called for. This was because they offered special advantages in that, in such machines the flux and the torque could be readily controlled using control of the field and armature currents. A separately excited dc motor configuration was especially preferred where fast response and four-quadrant operation at near-zero speeds were required.

DC motors are identified with a number of inherent shortcomings. They require periodic maintenance and call for industrial interruptions and also present restrictions of their possible use in corrosive and explosive environments because of the presence of commutators and brushes. They also suffer limitations in capacity under high speed functioning.

AC machines do not suffer from the above limitations. They present a relatively simple and rugged construction with accompanying freedom from maintainability requirements and economy in functioning. They are also suitable to be designed with higher values of power density.

With the development of high speed switching and power electronic technology the above trend is inexorably shifting towards cheaper and more effective power conditioners and single-motor ac drives in industrial applications.

In conventionally controlled ac drives, control of torque is achieved by controlling the armature current, but in this case, both the phase angle and the current magnitude get affected. In dc drive systems, the orientation of the field flux and the armature mmf are fixed by the commutator and brushes. They are naturally decoupled, while in ac drive systems, the field flux and the spatial angle of the armature mmf require external control. In its absence, the spatial angle among the various fields changes with load, giving rise to avoidable oscillatory dynamic response.

With vector control of ac drives, the torque and flux producing current components can be effectively decoupled. The transient response characteristics yielded by vector controlled ac drives have similarity with those of separately excited dc machines. The space phasor theory is very closely related to the two-axis theory of electrical machines. The simplicity and compactness of the space phasor equations and the very clear physical pictures obtained by its applications, add further to the advantages.

Space phasor quantities–voltages, currents, mmfs, flux-densities, flux-linkages etc. are introduced by using both physical and mathematical considerations.

To start with, we first consider a smooth air-gap ac machine with symmetrical two-pole, 3-phase winding. Figure 11.1 depicts such a schematic.

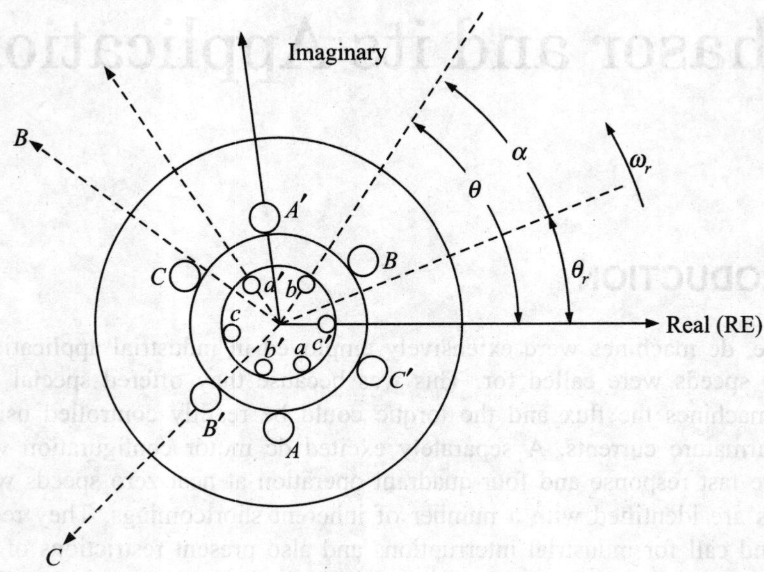

FIGURE 11.1 Schematic representation of a 2-pole, 3-phase smooth air-gap ac machine.

The slotting effects is ignored at this stage, and we assume that the permeability of the metallic part of the magnetic circuit as infinitely large. Also the flux-density distribution at the air-gap is assumed as radial and the iron-losses and end-effects are negligible.

In the figure, the stator and rotor windings have been depicted by multi-turn full-pitched coils, both on the stator member and the rotor member, on the two sides of the air-gap. They represent distributed windings, which at every instant of time, produce sinusoidally distributed mmf waves centred along the magnetic axis of the respective representative phase windings. The phase windings are displaced by $(2\pi/3)$ electrical degrees from each other. θ_r is the rotor angle, the angle between the magnetic axes of the stator winding (denoted by upper case letters A, B, C) and the rotor winding (depicted by lower case letters a, b, c). The speed of the rotor is $\left(\omega_r = \dfrac{d\theta}{dt}\right)$, and is assumed as positive in the counter-clockwise direction, as indicated in Figure 11.1, deliberately, although it was taken as clockwise in earlier part.

11.1.1 Space Phasor of Stator mmfs and Stator Currents

If the stator winding is supplied by a system of 3-phase currents $i_A(t)$, $i_B(t)$ and $i_C(t)$, which vary arbitrarily in time, but the neutral point is kept isolated so that zero-sequence stator currents cannot flow, then

$$i_0(t) = i_A(t) + i_B(t) + i_C(t) = 0 \quad \text{...(11.1)}$$

If we assume that the stator phase windings have equal number of effective turns per phase,

$$N_{SE} = N_S \, K_{WS}$$

where N_S are the number of turns and K_{WS} is the winding factor of the stator winding, then the resultant mmf distribution $F_S(\theta, t)$ is given by Eq. (11.2) if θ is the angle around the periphery with reference to the axis of phase A depicted along the real axis indicated as RE in Figure 11.1.

$$F_S(\theta, t) = N_{SE}\left[i_A(t)\cos\theta + i_B\cos\left(\theta - \frac{2\pi}{3}\right) + i_C(t)\cos\left(\theta - \frac{4\pi}{3}\right)\right] \quad \text{...(11.2)}$$

The above relation could also be expressed as:

$$F_S(\theta, t) = \frac{3}{2} N_{SE}\text{RE}\left[\frac{2}{3}\{i_A(t) + a i_B(t) + a^2 i_C(t)\}e^{-j\theta}\right] \quad \text{...(11.3)}$$

The component within the third brackets marked as real and multiplied by $e^{-j\theta}$, that is:

$$i_S(t) = \frac{2}{3}\{1 i_A(t) + a i_B(t) + a^2 i_C(t)\} = \left|\overline{i}_S\right| e^{j\alpha_s} \quad \text{...(11.4)}$$

is the complex space phasor of 3-phase stator currents in the complex plane, with reference to the stationary reference frame fixed to the stator member. In the above expression, the operators, 1, a and a^2 are spatial operators. Although these spatial operators are the same as the time operators when complexors are considered during the steady-state analysis with sinusoidal voltages and currents, their usage is notionally different.

In Eq. (11.4), $\left|\overline{i}_S\right|$ is the modulus of the stator space phasor and α_s is the phase angle with reference to the real axis of the stationary reference frame fixed to the stator. We specify the real axis of the stator by SD. Since all currents i_A, i_B and i_C vary with time—both in magnitude and phase angle—physically, the space phasor of the currents decides the instantaneous magnitude and spatial displacement of the peak of the sinusoidal stator mmf distribution produced by the three stator winding components.

Therefore, the space phasor of stator mmfs can be defined as:

$$F_S(\theta, t) = N_{SE}\, i_S(t)$$

$$= \overline{F}_{SA}(t) + \overline{F}_{SB}(t) + \overline{F}_{SC}(t) \quad \text{(11.5)}$$

In the above relation $F_{SA}(t)$, $F_{SB}(t)$ and $F_{SC}(t)$ denote the space phasors of the respective phase mmfs.

Thus the 3-phase space phasor of the stator currents can be obtained by aggregating the space phasors due to each phase current, $\left|\overline{I}_A\right| = 1 i_A(t)$; $\left|\overline{I}_B\right| = a i_B(t)$; $\left|\overline{i}_C\right| = a^2 i_C(t)$, and in the symmetrical steady-state (where the currents are symmetrical and sinusoidal, forming a balanced 3-phase system),

$$i_A = I_S \cos(\omega_1 t - \phi_s)$$

$$i_B = I_S \cos\left(\omega_1 t - \phi_s - \frac{2\pi}{3}\right)$$

and
$$i_C = I_S \cos\left(\omega_1 t - \phi_s - \frac{4\pi}{3}\right)$$

Thus, the space phasor of the stator currents is:

$$\bar{I}_S = I_S e^{j(\omega_1 t - \phi_s)}$$

where, I_S is constant. This is, therefore, indicated as a circle in the complex plane, and the tip of this space phasor rotates in the positive direction of space. The angular speed of this space phasor is constant and equals the synchronous speed.

The space phasor of the stator currents is identical with the positive sequence complexure of 3-phasor currents. Or in other words, in this case, the time-phasor diagrams are identical with the space phasor diagrams.

If, however, the 3-phase currents form an asymmetrical system, in the steady-state the locus of the stator-current space phasor can assume an elliptical form or even a straightline. In the transient state, the space-phasor can have arbitrary shapes.

The assumption of sinusoidal mmf distribution permits the space phasor mmfs of 3-phases to be added together. It is also possible to introduce the space phasor concept by utilizing 2-axis theory. The space phasor of the stator currents could be defined as phasors whose real part equals the instantaneous values of *d*-axis stator current components, $i_{SD}(t)$ and the imaginary parts equaling the quadrature-axis stator current components, $i_{SQ}(t)$. Thus, the stator current space-phasor in the stationary reference frame fixed to the stator takes the form:

$$\bar{i}_S = i_{SD}(t) + j i_{SQ}(t) \qquad \text{...(11.6)}$$

With symmetrical 3-phase machines, the direct and quadrature-axis stator currents are fictitious, two-phase current components, which may be related to the actual 3-phase currents, i_A, i_B and i_C, as shown

$$i_{SD} = K\left[i_A - \frac{1}{2} i_B - \frac{1}{2} i_C\right] \qquad \text{...(11.7)}$$

and
$$i_{SQ} = K\frac{\sqrt{3}}{2}(i_{SB} - i_{SC}) \qquad \text{...(11.8)}$$

where, K is an arbitrary constant. For the non-power invariant form of phase transformation from 3-phase to 2-phase components, $K = \frac{2}{3}$, while for the power-invariant form of transformation, $K = \sqrt{\frac{2}{3}}$.

Thus, from the definition of space phasors of the stator currents contained in Eq. (11.4), if $K = \frac{2}{3}$, the real-axis component becomes:

$$\text{RE}(\bar{i}_S) = \text{RE}\left\{\frac{2}{3}(1i_A + ai_B + a^2 i_C)\right\}$$

$$= \frac{2}{3}\left(i_A - \frac{1}{2}i_B - \frac{1}{2}i_C\right) = i_{SD} \qquad \text{...(11.9)}$$

which tallies with Eq. (11.7), for the non-power invariant form of phase transformation.

The imaginary-axis component, similarly, yields:

$$\text{IM}(\bar{i}_S) = \text{IM}\left\{\frac{2}{3}(1i_A + ai_B + a^2 i_C)\right\}$$

$$= (i_B - i_C)\frac{1}{\sqrt{3}} = i_{SQ} \qquad \text{...(11.10)}$$

which tallies with Eq. (11.8).

In a 2-phase machine, i_{SD} and i_{SQ} are actual (and not transformed) currents that flow in the two stator windings *SD* and *SQ* in space quadrature.

It may be noted that the space-phasor contains no zero-sequence components. If there is a zero-sequence component, we have to introduce an additional definition with this component. In general, the instantaneous zero-sequence current component of stator currents is specified by Eq. (11.1) as per relation

$$i_0(t) = \{i_A(t) + i_B(t) + i_C(t)\}K_1 \qquad \text{...(11.11)}$$

where, K_1 is an arbitrary constant such that $K_1 = \frac{1}{3}$ for the non-power invariant form and $K_1 = \frac{1}{\sqrt{3}}$ for the power invariant form of transformation.

If the non-power-invariant form of transformation is used, the projections of a space-phasor quantity on the corresponding phase axis, directly yield the instantaneous values of the phase variables having the same magnitude, if there are no zero-sequence components. This can be noted from Figure 11.2, for the stator currents.

Numerically, it signifies that by using $i_0(t) = 0$, and also using the non-power-invariant form of the zero-sequence current component with the space-phasor of the stator currents, the following relation is obtained:

$$\text{RE}(\bar{i}_S) = \text{RE}\left\{\frac{2}{3}(1i_A + ai_B + a^2 i_C)\right\}$$

$$= \frac{2}{3}\left\{i_A - \frac{1}{2}i_B - \frac{1}{2}i_C\right\} = i_A \qquad \text{...(11.12)}$$

and

$$\text{RE}(a^2\bar{i}_S) = \text{RE}\left\{\frac{2}{3}(a^2 i_A + i_B + ai_C)\right\} = i_B \qquad \text{...(11.13)}$$

Also
$$\text{RE}(a\bar{i}_S) = \text{RE}\left\{\frac{2}{3}(ai_A + a^2 i_B + i_C)\right\} = i_C \quad ...(11.14)$$

If, however, the zero sequence current is not zero, the phase-variable stator currents are given by:

$$i_A = \text{RE}(\bar{i}_S) + i_0 \quad ...(11.15)$$

$$i_B = \text{RE}(a^2\bar{i}_S) + i_0 \quad ...(11.16)$$

$$i_C = \text{RE}(a\bar{i}_S) + i_0 \quad ...(11.17)$$

and it follows that the zero-sequence current components are added to the corresponding projection of the space-phasor along the corresponding axes as shown in Figure 11.2.

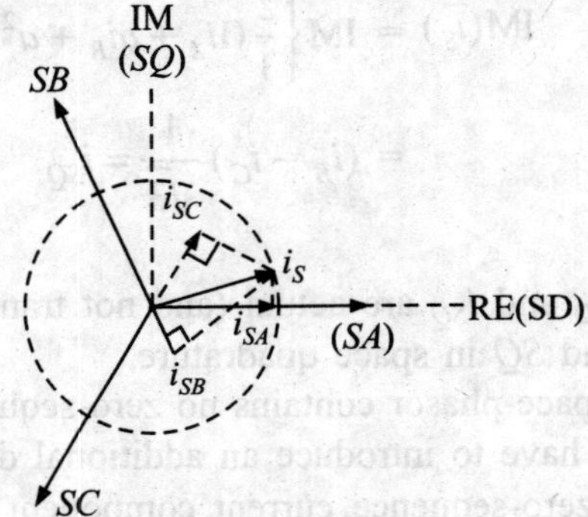

FIGURE 11.2 Projections of the stator current space phasor.

11.2 SPACE PHASOR OF ROTOR MMFS AND ROTOR CURRENTS

We can derive the resultant rotor mmf produced by a symmetrical 3-phase, sinusoidally distributed rotor winding with similar considerations to those of the resultant stator mmf. We assume that the rotor phase a is displaced from the stator phase A by an angle θ_r and an angle α around the periphery with respect to the axis of the rotor winding a as shown in Figure 11.1.

If the rotor winding has an effective equal number of turns per phase

$$N_{rE} = N_r K_{wr}$$

where, N_r is the number of turns per phase and K_{wr} the winding factor of the rotor winding, then by similar consideration as adopted in Eq. (2, 2), with no zero-sequence rotor current, the resultant rotor mmf distribution F_{rE} (θ, t), when carrying currents $i_a(t)$, $i_b(t)$ and $i_c(t)$ in the three rotor phases, is:

$$F_{rE}(\theta, t) = N_{rE}\left\{i_a(t)\cos\alpha + i_b(t)\cos\left(\alpha - \frac{2\pi}{3}\right) + i_c(t)\cos\left(\alpha - \frac{4\pi}{3}\right)\right\} \quad ...(11.18)$$

And, introducing complex notation, we express Eq. (11.18) as follows:

$$F_{rE}(\theta, t) = \frac{3}{2} N_{rE}\text{RE}\left[\frac{2}{3}\left\{1i_{ra}(t) + ai_{rb}(t) + a^2 i_{rc}(t)\right\} e^{-j\alpha}\right] \quad ...(11.19)$$

In Eq. (11.19), the complex quantity multiplied by $e^{-j\alpha}$, is the rotor current space phasor, $\bar{i}_r$.

$$\bar{i}_r = \frac{2}{3}\{i_{ra}(t) + ai_{rb}(t) + a^2 i_{rc}(t)\} = |\bar{i}_r| e^{j\alpha_r} \quad ...(11.20)$$

expressed in reference frame fixed on the rotor. The real axis of this reference frame has been denoted by α and its imaginary axis by β, as shown in Figure 11.3. The speed of this reference frame is $\omega_r = \frac{d\theta_r}{dt}$, where θ_r is the rotor angle (as shown in Figure 11.1). [This definition is similar to that of the stator current space phasor indicated with respect to stationary reference frame, given in Eq. (11.4).]

The rotor current space phasor gives the instantaneous magnitude and angular displacement of the peak of the sinusoidally distributed rotor mmf derived by the sinusoidally distributed rotor windings.

It follows that the definitions of the space-phasor quantities with respect to their respective reference frames are similar, and it is possible to introduce the definition of rotor-current space phasor by direct application of two-axis theory.

If $i_{r\alpha}$ and $i_{r\beta}$ denote the instantaneous values of the direct and the quadrature-axis rotor current components of the 3-phase rotor currents, they are related to Eqs. (11.7) and (11.8) respectively to the instantaneous values of actual 3-phase currents, in accordance with the following relations:

$$i_{r\alpha} = K\left\{i_a - \frac{1}{2}i_b - \frac{1}{2}i_c\right\} \quad ...(11.21)$$

and

$$i_{r\beta} = K\frac{\sqrt{3}}{2}\{i_b - i_c\} \quad ...(11.22)$$

where, $K = \frac{2}{3}$ for the non-invariant power form of transformation. The definition of space phasor for the rotor currents in the reference frame fixed to the rotor is, therefore:

$$\bar{i}_r = i_{r\alpha} + ji_{r\beta} \quad ...(11.23)$$

The real part of $\bar{i}_r$ provides the transformed d-axis rotor current and the imaginary part in Eq. (11.23) gives the transformed q-axis rotor current. These currents can vary arbitrarily in time. As before, a machine with 2-phase rotor winding, $i_{r\alpha}$ and $i_{r\beta}$ are non-transformed, actual rotor currents, which flow in the rotor windings α and β respectively.

Equations (11.19) and (11.20) are the instantaneous version of rotor mmf wave, that rotates at speed, ω_r. It can be expressed in terms of the rotor current space phasor and the angles θ and θ_r, as indicated, if it is considered that, with reference to Figure 11.1, $\alpha = \theta - \theta_r$:

$$F_{rE} = (\theta,\ \theta_r,\ t)$$

$$F_r(\theta,\ \theta_r,\ t) = \frac{3}{2}N_{rE}\text{RE}[\bar{i}_r e^{-j(\theta-\theta_r)}]$$

$$= \frac{3}{2}N_{rE}\text{RE}(\bar{i}'_r e^{-j\theta}) \quad ...(11.24)$$

where

$$\bar{i}'_r = \bar{i}_r e^{j\theta_r} = |\bar{i}_r| e^{j\alpha'_r} \quad ...(11.25)$$

is the space phasor of the rotor currents expressed in the stationary reference frame. Substitution of Eq. (11.20) into Eq. (11.25) yields:

$$\bar{i}'_r = |\bar{i}_r|\, e^{j\alpha'_r} = |\bar{i}_r|\, e^{j(\alpha_r + \theta_r)} \qquad ...(11.26)$$

The fact that $\bar{i}_r$ is the space phasor of the rotor currents expressed in the stationary reference frame could also be proved as shown below. The relationship between the reference frames fixed to stator (whose direct and quadrature-axis are depicted by D and Q respectively), and the reference frames fixed to the rotor that rotates at an angular velocity of ω_r, is shown in Figure 11.3.

Now, according to Eq. (11.20), the space phasor of rotor currents when expressed in the rotating reference frame is given by:

$$\bar{I}_r = |\bar{I}_r|\, e^{j\alpha_r}$$

where α_r is the angle of the current space phasor with respect to α-axis. This is stationary with respect to the moving rotor. The angle is shown in Figure 11.3.

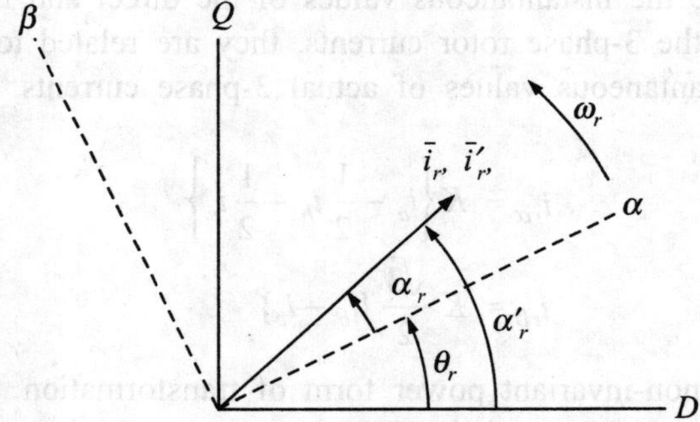

FIGURE 11.3 Relationship between the stationary and rotating reference frames of the stator and the rotor.

$\bar{i}_r$ is the space phasor of the rotor currents expressed with reference to the stationary stator frame. Figure 11.3 clearly indicates that:

$$\bar{i}'_r = |\bar{i}_r|\, e^{j\alpha'_r}$$

where

$$\alpha'_r = \alpha_r + \theta_r$$

Hence, the above indicates full agreement with Eq. (11.26).

11.3 MAGNETISING CURRENT SPACE PHASOR

The combined effect of the stator and rotor expression leads to the resultant mmf wave to be given by the sum of the stator and rotor mmf distributions:

$$F(\theta, \theta_r, t) = F_S(\theta, t) + F_r(\theta, \theta_r, t) \qquad ...(11.27)$$

Using Eqs. (11.3), (11.4) and (11.24), we get:

$$F(\theta, \theta_r, t) = \frac{3}{2} N_{SE} \left\{ \text{RE}\,(\bar{i}_S\, e^{-j\theta}) + \frac{N_{rE}}{N_{SE}}\, RE\,(\bar{i}'_r\, e^{-j\theta}) \right\} \qquad ...(11.28)$$

so that

$$F(\theta, \theta_r, t) = \frac{3}{2} N_{SE} \text{RE} \left\{ \left(\bar{i}_S + \frac{N_{rE}}{N_{SE}}\, \bar{i}'_r \right) e^{-j\theta} \right\} \qquad ...(11.29)$$

is obtained as given by Eq. (11.29). The complex current inside second brackets multiplied by $e^{-j\theta}$ is the magnetising current space phasor, expressed in the reference frame fixed to the stator. This could, therefore, also be denoted by:

$$\bar{i}_m = \bar{i}_S + \left(\frac{N_{rE}}{N_{SE}} \right) \bar{i}'_r \qquad ...(11.30)$$

This is the sum of the stator-current space phasor and the rotor-current space phasor—both expressed in the stationary reference frame fixed to the stator. However, if the rotor current space phasor is referred to the stator, the reference factor is

$$N_{SE}/N_{rE} = N_S K_{WS}/N_r K_{wr}$$

when $\bar{I}'_r$ is divided by the above factor, then $(N_{rE}\bar{i}_r)/N_{SE}$ is obtained. This is the rotor current space phasor referred to the stator and expressed in the stationary reference frame.

By a special selection of the turns ratio, that is, if the reference factor equals the ratio of the magnetising inductance and the rotor self reactance, Eq. (11.30) yields the rotor magnetising current space phasor expressed in the secondary reference frame $\bar{I}_{mr}$.

The rotor magnetising current space phasor is extensively utilized when the rotor-flux-oriented control of ac machine is employed.

11.4 FLUX-LINKAGE SPACE PHASOR

Now we, derive expressions for the stator and rotor flux linkages in stationary and rotating reference frames.

11.4.1 Stator Flux-linkage Space Phasor in the Stationary Reference Frame

Using definitions of the stator current and rotor current space phasors, it is possible to define the space phasor of the stator flux linkage $\bar{\psi}_S$, the terms of the instantaneous values of the flux-linkages of the three stator windings. Thus, in the stationary reference frame fixed to the stator, the total flux-linkage space phasor can be expressed as follows:

$$\bar{\psi}_S = \frac{2}{3}(\psi_{SA} + a\psi_{SB} + a^2\psi_{SC}) \qquad ...(11.31)$$

in which the instantaneous values of the phase variable flux-linkage components are as follows:

$$\psi_A = \bar{L}_S i_{SA} + \bar{M}_S i_{SB} + \bar{M}_S i_{SC} + \bar{M}_{Sr} \cos\theta_r i_{ra} + \bar{M}_{Sr} \cos\left(\theta_r + \frac{2\pi}{3}\right) i_{rb} + M_{Sr} \cos\left(\theta_r + \frac{4\pi}{3}\right) i_{rc} \quad ...(11.32)$$

$$\psi_B = \bar{M}_S i_{SA} + L_S i_{SB} + \bar{M}_S i_{SC} + \bar{M}_{Sr} \cos\left(\theta_r + \frac{4\pi}{3}\right) i_{ra} + \bar{M}_{Sr} \cos\theta_r i_{rb} + \bar{M}_{Sr} \cos\left(\theta_r + \frac{2\pi}{3}\right) i_{rc} \quad ...(11.33)$$

and

$$\psi_C = \bar{M}_S i_{SA} + \bar{M}_S i_{SB} + L_S i_{SC} + \bar{M}_{Sr} \cos\left(\theta_r + \frac{2\pi}{3}\right) i_{ra} + \bar{M}_{Sr} \cos\left(\theta_r + \frac{4\pi}{3}\right) i_{rb} + M_{Sr} \cos\theta_r i_{rc} \quad ...(11.34)$$

In the above expressions, $\bar{L}_S$ denotes the self-inductance of the stator phase windings, $\bar{M}_{Sr}$ is the mutual inductance between stator windings, and $\bar{M}_{Sr}$ stand for the maximum value of the stator-to-rotor mutual inductance between phases. The phase variable flux-linkages contain six-flux-linkage terms—a self-linkage component produced by stator current of the corresponding stator winding, two mutual stator flux-linkage components due to the other two stator currents, besides the three stator-to-rotor mutual flux-linkage components due to the three rotor phase currents.

Substitution of Eqs. (11.32), (11.33) and (11.34) in Eq. (11.31) results in the under-noted space-phasor equation for the stator flux-linkages, if Eqs. (11.4), (11.20) and (11.25) are also considered:

$$\begin{aligned}\bar{\psi}_S &= L_S \bar{i}_S + L_m \bar{i}'_r \\ &= L_S \bar{i}_S + L_m \bar{i}_r e^{j\theta_r}\end{aligned} \quad (11.35)$$

where, $L_S = \bar{L}_S - \bar{M}_S$, is the total 3-phase stator inductance and L_m is the magnetising three-phase inductance or in other words, $L_m = \left(\frac{3}{2}\right)\bar{M}_{Sr}$.

The stator-flux-linkage space phasor indicates the modulus and phase-angle of the peak of the sinusoidal stator flux distribution in the air-gap.

In Eq. (11.35), there are two flux-linkage space phasor components. $L_S \bar{i}_S$, is the self-flux-linkage space phasor of the stator phases, caused by the stator currents. The second component $L_m \bar{i}'_r$, is a mutual flux-linkage space phasor, due to rotor currents expressed in the stationary reference frame. Equation (11.35) is general in nature and holds as valid even under non-linear magnetic conditions when the leakage or main flux paths are saturated. In such cases, the values of L_S and L_m do not remain constant, but depend on the value of currents flowing in the machine.

As usual, it is possible to define the stator flux-linkages in terms of the direct and quadrature-flux linkages components—ψ_{SD} and ψ_{SQ}.

$$\bar{\psi}_S = \psi_{SD} + j\psi_{SQ} \quad ...(11.36)$$

We can define the direct-axis stator flux-linkage component as:

$$\psi_{SD} = L_S i_{SD} + L_m i_{rd} \qquad ...(11.37)$$

and, similarly, the quadrature-axis stator flux-linkage component by:

$$\psi_{SQ} = L_S i_{SQ} + L_m i_{rq} \qquad (11.38)$$

The relation connecting the instantaneous values of direct and quadrature-axis flux-linkage components ψ_{SD} and ψ_{SQ} and the instantaneous values of the corresponding 3-phase stator flux-linkage components ψ_{SA}, ψ_{SB} and ψ_{SC} are similar to the relation connecting 2-phase and 3-phase currents depicted in Eqs. (11.9) and (11.10). The currents i_{SD}, i_{SQ} and and i_{rd}, i_{rq}, in the above relations are of instantaneous values along *D*-axis and *Q*-axis of the stator (depicted with capital suffices) and rotor (indicated by lower-case suffices). All the four currents are defined in the stationary reference frame attached to the stator member and they could arbitrarily vary with time. The rotor currents i_{rd} and i_{rq} are, further related to the rotor currents $i_{r\alpha}$ and $i_{r\beta}$ through Eq. (11.25), and the latter current components, defined by Eqs. (11.21) and (11.22) are the two axis components of the rotor currents in the reference frame fixed to the rotor. Therefore, it follows from Eq. (11.25) that:

$$\bar{i}_r' = i_{rd} + j i_{rq} = \bar{i}_r e^{j\theta_r} \qquad ...(11.39)$$

The following transformational relationship between *d*, *q* and α, β components of the rotor currents is observed, if for convenience the matrix form is used. This matrix form of transformation between *d*, *q* and α, β are also encountered in the conventional unified theory applied to electrical machines.

$$\begin{bmatrix} i_{rd} \\ i_{rq} \end{bmatrix} = \begin{bmatrix} \cos\theta_r & -\sin\theta_r \\ \sin\theta_r & \cos\theta_r \end{bmatrix} \begin{bmatrix} i_{r\alpha} \\ i_{r\beta} \end{bmatrix}$$

$$= \begin{bmatrix} C_2^{-1} \end{bmatrix} \begin{bmatrix} i_{r\alpha} \\ i_{r\beta} \end{bmatrix} \qquad ...(11.40)$$

The inverse of the above matrix is termed as the commutator transformation matrix.

11.4.2 Rotor Flux-linkage Space Phasor in the Rotating Reference Frame Attached to the Rotor Member

The space phasor of the rotor flux-linkages referred to the rotor reference frame and rotating at a speed of ω_r, may be defined as given below:

$$\bar{\psi}_r = \frac{2}{3}\{\psi_{ra}(t) + a\psi_{rb}(t) + a^2\psi_{rc}(t)\} \qquad ...(11.41)$$

where, $\psi_{ra}(t)$, $\psi_{rb}(t)$ and $\psi_{rc}(t)$ are the instantaneous values of the rotor flux-linkages in the rotor phases *ra*, *rb* and *rc* respectively. In terms of the instantaneous values of the stator and rotor currents, they may be alternatively expressed as follows:

$$\psi_{ra} = \bar{L}_r i_{ra} + \bar{M}_r i_{rb} + \bar{M}_r i_{rc} + \bar{M}_{Sr} \cos\theta_r i_{SA} + \bar{M}_{Sr} \cos\left(\theta_r + \frac{4\pi}{3}\right) i_{SB} + \bar{M}_{Sr} \cos\left(\theta_r + \frac{2\pi}{3}\right) i_{SC} \qquad ...(11.42)$$

$$\psi_{rb} = \bar{M}_r i_{ra} + \bar{L}_r i_{rb} + \bar{M}_r i_{rc} + \bar{M}_{Sr} \cos\left(\theta_r + \frac{2\pi}{3}\right) i_{SA} + \bar{M}_{Sr} \cos\theta_r i_{SB} + \bar{M}_{Sr} \cos\left(\theta_r + \frac{4\pi}{3}\right) i_{SC} \qquad ...(11.43)$$

$$\psi_{rc} = \bar{M}_r \bar{i}_{ra} + \bar{M}_r i_{rb} + \bar{L}_r i_{rc} + M_{Sr} \cos\left(\theta_r + \frac{4\pi}{3}\right) i_{SA} + \bar{M}_{Sr} \cos\left(\theta_r + \frac{2\pi}{3}\right) i_{SB} + M_{Sr} \cos\theta_r i_{SC} \qquad ...(11.44)$$

In the above expressions, $\bar{L}_r$ denotes the self-inductance of the rotor phase windings, $\bar{M}_r$ represents per phase mutual inductance between two rotor windings. All three rotor flux-linkage components $\psi_{ra}(t)$, $\psi_{rb}(t)$ and $\psi_{rc}(t)$ contain three flux-linkage components produced by the rotor currents in addition to three mutual inductance flux-linkage components developed by the appropriate stator currents. We can make possible a significant simplification by substituting Eq. (11.42) through Eq. (11.44) into Eq. (11.41). Hence, the space phasor of rotor flux-linkages in the rotor reference frame can be obtained as:

$$\bar{\psi}_r = L_r \bar{i}_r + L_m \bar{i}'_S \qquad ...(11.45)$$

where, $L_r = \bar{L}_r - \bar{M}_r$, the total 3-phase rotor inductance and $\bar{i}'_S$ is the space phasor of the stator currents, expressed in terms of the reference frame fixed to the rotor.

As before, Eq. (11.45) consists of two terms:

(i) The space phasor $L_r \bar{i}_r$, which is produced by the rotor self-flux linkage space phasor expressed in the rotor reference frame due only to rotor currents, and;

(ii) The space phasor $L_m \bar{i}'_S$ — a mutual flux-linkage space phase, exclusively produced by stator currents and expressed in the rotor reference frame.

It is also possible to define the rotor flux-linkage space phasor in terms of 2-axis components ($\psi_{r\alpha}$ and $\psi_{r\beta}$). This has the form:

$$\bar{\psi}_r = \psi_{r\alpha} + j\psi_{r\beta} \qquad ...(11.46)$$

and Eq. (11.45), *d*-axis rotor flux-linkage component could be specified as

$$\psi_{r\alpha} = L_r i_{r\alpha} + L_m i_{SD} \qquad ...(11.47)$$

while *q*-axis rotor flux-linkage components becomes

$$\psi_{r\beta} = L_r i_{r\beta} + L_m i_{SQ} \qquad ...(11.48)$$

In Eqs. (11.47) and (11.48), currents $i_{r\alpha}$, $i_{r\beta}$ and i_{SD} and i_{SQ} are *d*-axis and *q*-axis rotor and stator current components respectively, which are all expressed in the reference frame fixed to the rotor.

11.4.3 Rotor Flux-linkage Space Phasor When Referred in the Stationary Reference Frame

The rotor flux-linkage components in the reference frame fixed to the rotor member: $(\psi_{r\alpha}, \psi_{r\beta})$—are related to the rotor flux-linkage components expressed in the stationary reference frame (ψ_{rd}, ψ_{rq}) by the same transformation $e^{j\theta_r}$, as indicated in Eq. (11.29) for the rotor currents. The following relations hold:

$$\overline{\psi}'_r = \psi_{rd} + j\psi_{rq} = \overline{\psi}_r e^{j\theta_r}$$
$$= (\psi_{r\alpha} + j\psi_{r\beta})\, e^{j\theta_r} \qquad \text{...(11.49)}$$

It can be put in matrix form as:

$$\begin{bmatrix} \psi_{rd} \\ \psi_{rq} \end{bmatrix} = \begin{bmatrix} \cos\theta_r & -\sin\theta_r \\ \sin\theta_r & \cos\theta_r \end{bmatrix} \begin{bmatrix} \psi_{r\alpha} \\ \psi_{r\beta} \end{bmatrix} \qquad \text{...(11.50)}$$

The transformation matrix in Eq. (11.50) is the inverse of the commutator matrix for transformation as shown in Eq. (11.40).

Now, by substitution of Eq. (11.45) in Eq. (11.49), the space phasor of the rotor flux-linkage in the stationary reference frame could be:

$$\overline{\psi}'_r = L_r \overline{i}'_r + L_m \overline{i}'_S e^{j\theta_r}$$
$$= L_r \overline{i}'_r + L_m \overline{i}_S \qquad \text{...(11.51)}$$

This contains two flux-linkage components—a self flux-linkage component produced by rotor currents but expressed in the stator reference frame $(L_r \overline{i}'_r)$ and a mutual flux-linkage component produced by the stator currents and also expressed in the stationary frame, $(L_m \overline{i}_S = L_m \overline{i}'_s e^{j\theta_r})$. Therefore, the stator currents in the stationary reference frame are related to stator current components in the rotating reference frame fixed to the rotor member by the following complex transformation:

$$\overline{i}_S = \overline{i}'_S e^{j\theta_r} \qquad \text{...(11.52)}$$

where $\overline{i}_S$ and $\overline{i}'_s$ denote, in terms of their two-axis components as:

$$\left.\begin{aligned} \overline{i}_S &= i_{SD} + j i_{SQ} \\ \overline{i}'_s &= i_{Sd} + j i_{Sq} \end{aligned}\right\} \qquad \text{...(11.53)}$$

Thus, from Eq. (11.52), the transformed stator current space phasor in the rotating reference frame attached to the rotor member, is obtained from the space phasor of the stator currents expressed with reference to the stationary reference frame as

$$\overline{i}'_S = \overline{i}_S e^{-j\theta_r} \qquad \text{...(11.54)}$$

The corresponding 2-axis form could also be arrived at by substitution of Eq. (11.53) in Eq. (11.54) shown by

$$\begin{bmatrix} i_{Sd} \\ i_{Sq} \end{bmatrix} = \begin{bmatrix} \cos\theta_r & \sin\theta_r \\ -\sin\theta_r & \cos\theta_r \end{bmatrix} \begin{bmatrix} i_{SD} \\ i_{SQ} \end{bmatrix} \quad ...(11.55)$$

These are the stator currents used in equations (11.47) and (11.48). The transformation in Eq. (11.54) is pictorially indicated in Figure 11.4.

In Eq. (11.4), the stator current space phasor indicated in the stator reference frame ($\overline{I}_S$) is expressed in terms of its modulus $\left(\left|\overline{I}_s\right|\right)$ and its phase angle (α_s). These are also shown in Figure 11.4 as $\overline{I}_s = \left|\overline{I}_s\right| e^{j\alpha_s}$. However, when the space phasor of the stator currents is expressed in the reference frame rotating at the rotor speed ω_r, it follows that

$$\overline{i}_S' = \left|\overline{i}_s\right| e^{j\alpha_s'}$$

where, α_s' is the angle between $\overline{i}_s'$ and the real-axis (*Sd*) of the rotating reference frame. But referring to Figure 11.4,

$$\alpha_s' = \alpha_s - \theta_r$$

and

$$\overline{i}_s' = \left|\overline{i}_s\right| e^{j(\alpha_s - \theta_r)}$$

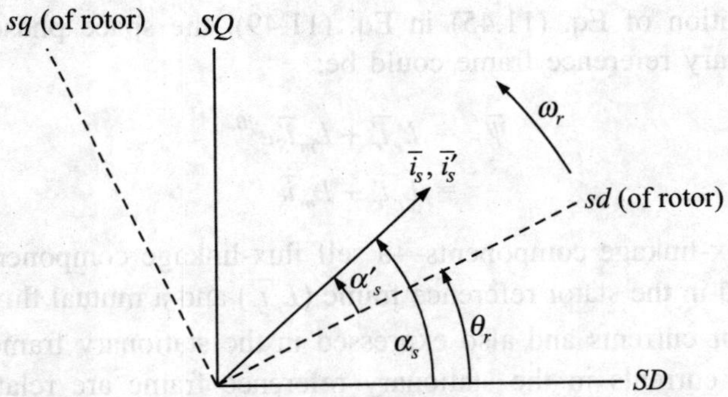

FIGURE 11.4 Transformation of stator current space phasor.

11.4.4 Stator Flux-linkage Space Phasor in the Reference Frame Attached to the Rotor

If it is necessary to know the space phasor of the stator flux-linkages referred to the rotating reference frame ($\overline{\psi}_S$) in terms of the stator flux-linkage space phasor expressed in the stationary reference frame ($\overline{\psi}_S$), then as per Eqs. (11.54),

$$\overline{\psi}_s' = \overline{\psi}_S e^{-j\theta_r} \quad ...(11.56)$$

In 2-axis form, by substituting

$$\overline{\psi}_S' = \psi_{sd} + j\psi_{sq}$$

and

$$\psi_S = \psi_{SD} + j\psi_{SQ}$$

we get:

$$\begin{bmatrix}\psi_{Sd}\\ \psi_{Sq}\end{bmatrix}=\begin{bmatrix}\cos\theta_r & \sin\theta_r\\ -\sin\theta_r & \cos\theta_r\end{bmatrix}\begin{bmatrix}\psi_{SD}\\ \psi_{SQ}\end{bmatrix} \quad ...(11.57)$$

where, the transformation matrix is the commutator transformation $[C_2]$. Incidentally, substitution of Eq. (11.35) in Eq. (11.56) yields ψ_s', if the Eqs. (11.39) and (11.54) are also considered

$$\begin{aligned}\psi_s' &= (L_S\overline{i}_S + L_m\overline{i}_r')\, e^{-j\theta_r}\\ &= L_S\overline{i}_S' + L_m\overline{i}_r\end{aligned} \quad ...(11.58)$$

where $\overline{i}_s'$ and i_r are the space phasors of the stator and rotor currents respectively and are expressed in the reference frame fixed to the rotor. By resolving into real and imaginary parts and using the notation introduced in equations (11.23) and (11.53) the d-axis and the q-axis stator flux-linkages in the rotor reference frames are obtained as follows:

$$\psi_d = L_S i_d + L_m i_\alpha \quad ...(11.59)$$

and

$$\psi_q = L_S i_q + L_m i_\beta \quad ...(11.60)$$

In the above relations, i_d, i_q and i_α, i_β are the 2-axis stator and rotor current respectively to the reference frame fixed to the rotor.

11.5 SPACE PHASORS OF THE STATOR AND ROTOR VOLTAGES

The space phasors of the stator and rotor voltages can be defined similarly to the space phasor quantities defined earlier. Thus, the stator voltage space phasor in the stationary reference frame is:

$$\begin{aligned}V_S &= \frac{2}{3}\{V_{SA}(t) + aV_{SB}(t) + a^2V_{SC}(t)\}\\ &= V_{SD} + jV_{SQ}\end{aligned} \quad ...(11.61)$$

and, the rotor voltage space phasor in the reference frame fixed to the moving rotor is given by:

$$\begin{aligned}V_r &= \frac{2}{3}\{V_{ra}(t) + aV_{rb}(t) + a^2V_{rc}(t)\}\\ &= V_{r\alpha} + jV_{r\beta}\end{aligned} \quad ...(11.62)$$

In Eqs. (11.61) and (11.62), $V_{SA}(t)$, $V_{SB}(t)$, $V_{SC}(t)$ and $V_{ra}(t)$, $V_{rb}(t)$ and $V_{rc}(t)$ are the instantaneous values of stator and rotor phase voltages respectively and V_{SD}, V_{SQ} and V_{rd}, V_{rq} are the corresponding d-axis and q-axis components. The relation between the 3-phase and 2-phase voltages are also shown. Hence,

$$\begin{aligned}V_{SD} &= \text{RE}\left[\frac{2}{3}\{V_{SA}(t) + aV_{SB}(t) + a^2V_{SC}(t)\}\right]\\ &= \frac{2}{3}\left\{V_{SA} - \frac{1}{2}V_{SB} - \frac{1}{2}V_{SC}\right\}\end{aligned} \quad ...(11.63)$$

and

$$V_{SQ} = \text{IM}\left[\frac{2}{3}\left\{V_{SA} + aV_{SB} + a^2V_{SC}\right\}\right]$$

$$= (V_{SB} - V_{SC})/\sqrt{3} \qquad ...(11.64)$$

Similarly, the under-notated relations hold good for rotor voltage components

$$V_{r\alpha} = \frac{2}{3}\left\{V_{ra} - \frac{1}{2}V_{rb} - \frac{1}{2}V_{rc}\right\} \qquad ...(11.65)$$

and

$$V_{r\beta} = (V_{rb} - V_{rc})/\sqrt{3} \qquad ...(11.66)$$

We have pointed out in Section 11.2 that the space phasors do not contain the zero-sequence components. If there are zero-sequence voltages, they need to be considered separately in terms of the under-noted stator and rotor zero-sequence components:

$$V_0 = \frac{1}{3}\{V_A(t) + V_B(t) + V_C(t)\} \qquad ...(11.67)$$

and

$$V_0 = \frac{1}{3}\{V_a(t) + V_b(t) + V_c(t)\} \qquad ...(11.68)$$

If we use Eqs. (11.63), (11.64) and (11.67) in the matrix format, we get

$$\begin{bmatrix} V_0 \\ V_D \\ V_Q \end{bmatrix} = \frac{2}{3}\begin{bmatrix} \frac{1}{2} & \frac{1}{2} & \frac{1}{2} \\ 1 & -\frac{1}{2} & -\frac{1}{2} \\ 0 & \frac{\sqrt{3}}{2} & -\frac{\sqrt{3}}{2} \end{bmatrix}\begin{bmatrix} V_{SA} \\ V_{SB} \\ V_{SC} \end{bmatrix} \qquad ...(11.69)$$

The transformation matrix in Eq. (11.69) is the inverse of the phase transformation matrix $[C_1]$, and a similar transformation matrix is valid for the rotor voltages. Similarly, Eqs. (11.12) to (11.14), without zero-sequence components, yield the space phasor of voltages, on the corresponding axes as shown:

$$V_{SA} = \text{RE}(\overline{V}_S) \qquad ...(11.70)$$

$$V_{SB} = \text{RE}(a^2\overline{V}_S) \qquad ...(11.71)$$

$$V_{SC} = \text{RE}(a\overline{V}_S) \qquad ...(11.72)$$

The stator voltage space phasor, expressed in the stationary reference frame (V_S) can be transformed in the stator space phasor with reference to frames attached to the rotor ($\overline{V}_s'$), by relation similar to expressions shown under (11.54), by

$$\overline{V}_S' = \overline{V}_S\, e^{-j\theta_r}$$

$$= V_{SD} + jV_{SQ} \qquad ...(11.73)$$

If this is resolved in the d-axis and the q-axis components, the same transformation matrix $[C_2]$ appears as shown in Eq. (11.57).

The rotor voltage space phasor in the reference frame fixed to the rotor ($\bar{V}_r$) can be, similarly, expressed in the reference frame fixed to the stator ($\bar{V}_r'$). In this case, complex transformation has to be used—it being the same as was used in Eq. (11.39). Hence,

$$\bar{V}_r' = \bar{V}_r e^{j\theta_r}$$

$$= V_{rd} + jV_{rq} \qquad \text{...(11.74)}$$

When it is in 2-axis form, the same transformation matrix $[C_2]^{-1}$ appears as in Eq. (11.40). Thus, for a relationship between V_A, V_B, V_C and V_d, V_q, we can use Eqs. (11.73) and (11.61), leading to:

$$V_{SD} + jV_{SQ} = \frac{2}{3}\{V_{SA}(t) + aV_{SB}(t) + a^2V_{SC}(t)\}\,(\cos\theta_r - j\sin\theta_r) \qquad \text{...(11.75)}$$

Resolving Eq. (11.75) into real and imaginary parts, we get

$$V_{SD} = \frac{2}{3}\left\{V_{SA}\cos\theta_r + V_{SB}\cos\left(\theta_r - \frac{2\pi}{3}\right) + V_{SC}\cos\left(\theta_r - \frac{4\pi}{3}\right)\right\} \qquad \text{...(11.76)}$$

and $$V_{SQ} = -\frac{2}{3}\left\{V_{SA}\sin\theta_r + V_{SB}\sin\left(\theta_r - \frac{2\pi}{3}\right) + V_{SC}\sin\left(\theta_r - \frac{4\pi}{3}\right)\right\} \qquad \text{...(11.77)}$$

These specify the so-called Park Transformation. Incidentally, if the zero-sequence stator voltage component defined by Eq. (11.67) is combined with Eqs. (11.76) and (11.77), the under-noted matrix form is obtained:

$$\begin{bmatrix} V_{S0} \\ V_{SD} \\ V_{SQ} \end{bmatrix} = \frac{2}{3}\begin{bmatrix} \frac{1}{2} & \frac{1}{2} & \frac{1}{2} \\ \cos\theta_r & \cos\left(\theta_r - \frac{2\pi}{3}\right) & \cos\left(\theta_r - \frac{4\pi}{3}\right) \\ -\sin\theta_r & -\sin\left(\theta_r - \frac{2\pi}{3}\right) & -\sin\left(\theta_r - \frac{4\pi}{3}\right) \end{bmatrix}\begin{bmatrix} V_{SA} \\ V_{SB} \\ V_{SC} \end{bmatrix} \qquad \text{...(11.78)}$$

11.6 MECHANISM OF ELECTROMAGNETIC TORQUE PRODUCTION

In this section, we discuss the mechanism of electromagnetic torque production in dc and ac machines. For simplicity, we neglect the possible effect of magnetic non-linearity due to saturation. The space phasor formulation of the electromagnetic torque is taken up for both dc and ac machines to establish analogy between their mechanism of torque production.

11.6.1 Electromagnetic Torque Production in dc Machines

Figure 11.5 depicts the schematic diagram of compensated dc machine with smooth air-gap and with slotting effects neglected.

On the stationary member, there are the field (f) and the compensating windings (c), while on the rotor member, we house the armature winding (a). Let us assume that the current in the field winding is i_f, and let it produce an excitation flux-linkage ψ_f.

If a current i_a flows in the armature winding, the interaction of the armature current and the excitation flux linkage produces a force (F) acting on the conductors as indicated in Figure 11.5. Since the excitation flux is in space quadrature current, maximum forces are applied to the shaft. Hence, the position of the armature winding is optimal for electromagnetic torque production.

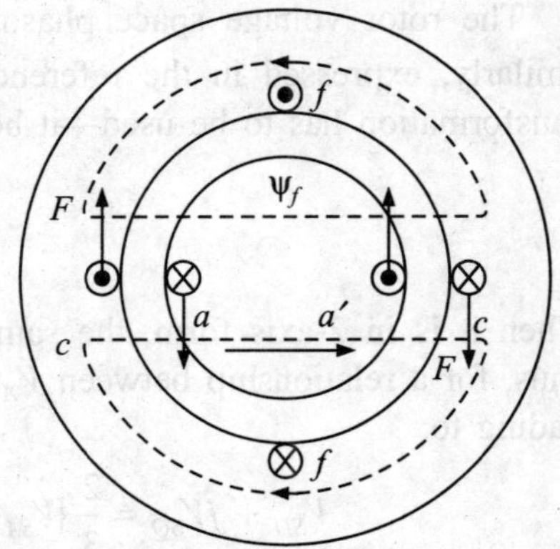

FIGURE 11.5 Electromagnetic torque produced in dc machines.

The armature winding also produces a field which is superposed on the field developed by the field winding. It is in space quadrature with respect to the excitation flux and the resultant field gets displaced from its optimal original position with the load armature current flowing. This effect is normally neutralized with the use of the compensating winding (c), which is made to carry a current i_c, equal to $(-i_a)$. The currents in the compensating winding and the field winding are, therefore responsible for the production of electromagnetic torque. This torque acts against the armature.

The space phasors of the currents in the field winding, the compensating winding and armature winding and also the excitation flux-linkage are shown in Figure 11.6. These are denoted respectively by $\bar{i}_f, \bar{i}_c, \bar{i}_a$ and $\bar{\psi}_f$. The arbitrary stationary reference frames D and Q have been used so that these space phasors are expressed in the stationary reference frames fixed to the stator of the machine.

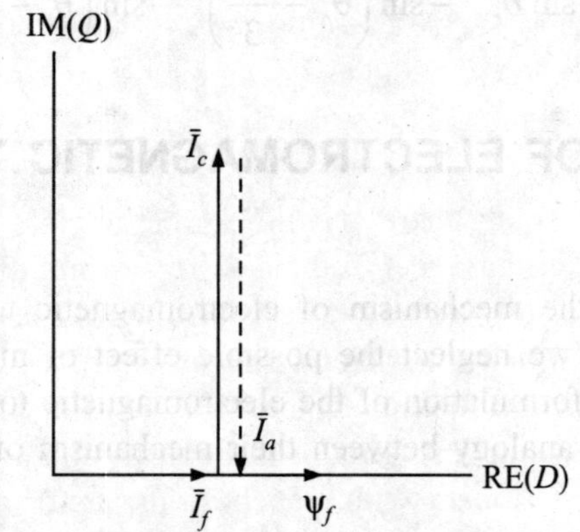

FIGURE 11.6 Space phasors in dc machines.

The current in the field winding develops the excitation flux-linkage, and the interaction of the excitation flux with the currents in the armature winding and the compensating winding produces the electromagnetic torque—both under transient and steady-state conditions. When the magnetic condition is linear, we express the instantaneous value of the electromagnetic torque (T_e) as a vector product of the flux-linkage and the armature current space vectors. This leads to the following expression:

$$T_e = K\overline{\psi}_f \times \overline{i}_d \qquad \text{...(11.79)}$$

In the above equation, K is a constant and the cross sign (×) represents the vector product. Since the two space phasors are in space quadrature, equation (11.79) can be simplified into the following form:

$$T_e = K\psi_f i_a \qquad \text{...(11.80)}$$

where ψ_f and i_a are the moduli of the respective space phasors, which are the instantaneous values of the excitation flux-linkage and armature currents.

11.6.2 Electromagnetic Torque developed in ac Machines

In general, for a symmetrical 3-phase or a symmetrical 2-phase ac machine with smooth air-gap, the developed instantaneous electromagnetic torque in phasorical form can be shown to have shapes similar to that of dc machines in Eqs. (11.79), that is:

$$T_e = K\overline{\psi}_S \times \overline{I}_r' \qquad \text{...(11.81)}$$

which could be expressed as

$$T_e = K\left|\overline{\psi}_S\right|\left|\overline{I}_r'\right| \sin\gamma \qquad \text{...(11.82)}$$

where under-linear magnetic conditions (that is at unsaturated operating conditions), K is a constant and $\overline{\psi}_S$ and $\overline{I}_r'$ are the space phasors of the stator flux-linkage and rotor currents expressed in stationary reference frames. The electromagnetic torque, which is a cross product of the flux-linkage and the rotor current phase phasor, as shown in Eq. (11.81), could also be expressed in reference frames other than the stationary reference frame. However, in such cases with other transformed space phasors shall be present in the expression for the electromagnetic torque—for example $\overline{\psi}_S$ and $\overline{I}_r$ corresponding to the space phasors of the stator flux-linkage and rotor currents, referred to the rotor reference frame.

In Eq. (11.82), is shown, $\gamma = 90°$ and $\left|\overline{\psi}_S\right|$ and $\left|\overline{i}_r\right|$ are the moduli of the stator flux-linkage and the rotor current space phasors, where γ is the torque angle.

While in dc machines, the armature current and the main flux distribution are fixed in space quadrature due to commutator, and where torque control could be achieved by independently controlling the excitation flux-linkage (ψ_f) and the armature current (I_a), and by virtue of the distribution of flux and current, maximum torque is developed for the field and armature current configurations, in ac machines it is more difficult to obtain the situation because these quantities are coupled and are stationary with respect to the stator and rotor respectively. They also depend on the modulus, frequency and phase angles of the stator currents.

In machines with squirrel-cage rotors, it is not possible to monitor the rotor current readily. But it is possible to obtain two current components—one of which is a flux-producing current component and the other, the torque producing current component. The instantaneous electromagnetic torque can be formulated as functions of these two components of current independent of each other in order to achieve decoupled control.

We present the proof of Eq. (11.81). This is based on energy considerations. An expression for the electromagnetic torque can be obtained as the rate of change of the mechanical output energy–$\left\{\frac{dW_{\text{mech}}}{dt}\right\}$, this being the mechanical power P_{mech}, which is represented by the instantaneous rotor speed and the electromagnetic torque. Thus,

$$P_{\text{mech}} = \frac{d}{dt}(W_{\text{mech}}) = T_e\omega_r \qquad \text{...(11.83)}$$

The input electrical energy, (W_e), has to account for the energy related to:

1. Stator and rotor losses (we identify this as W_{loss} component)
2. The stored magnetic energy in the field (W_{field}), and
3. The mechanical output energy (W_{mech}).

In other words:

$$W_e = W_{\text{loss}} + W_{\text{field}} + W_{\text{mech}} \qquad \text{...(11.84)}$$

and the differential mechanical output energy is derived from above as follows:

$$dW_{\text{mech}} = dW_e - dW_{\text{loss}} - dW_{\text{field}} \qquad \text{...(11.85)}$$

In Eq. (11.85), the differential input electrical energy for a doubly-fed poly-phase machine,

$$dW_e = \frac{3}{2}\,\text{RE}\left\{\overline{v}_S\ \overline{i}_S^{\,*} + \overline{v}'_r\overline{i}'^{\,*}_r\right\} dt \qquad \text{...(11.86)}$$

where, the stator voltage and current space phasors ($\overline{V}_S, \overline{I}_S$) and the rotor voltage and current space phasors ($\overline{V}'_r, \overline{I}'_r$) are expressed in the reference frame fixed to the stator. Relation (11.86) reveals that physically, the total instantaneous power is the sum of the instantaneous powers of the stator:

$$P_S = \frac{3}{2}\,\text{RE}(\overline{v}_S\overline{i}_S^{\,*}) \qquad \text{...(11.87)}$$

and the instantaneous power of the rotor:

$$P_r = \frac{3}{2}\,\text{RE}\,(\overline{v}'_r\ \overline{i}'^{\,*}_r)$$

$$= \frac{3}{2}\,\text{RE}\,(\overline{v}_r\ \overline{i}_r^{\,*}) \qquad \text{(11.88)}$$

The asterisk mark represents the complex conjugate. The rate of change of the electrical energy must equal the total power $P_S + P_r$. Equations (11.87) and (11.88) can be shown to be expressing the stator and rotor space phasor voltages and currents in terms of the instantaneous phase power components as specified by Eqs. (11.4), (11.20), (11.25), (11.61), (11.62) and (11.74), in which there are no zero-sequence stator and rotor voltages and current components.

The losses are constituted of I^2R losses in stator and rotor-winding resistances, the hysteresis and eddy current losses in magnetic materials, and friction losses between contact surfaces and windage losses contributed by the surrounding air, besides dielectric losses in the electric fields. If only the I^2R losses are accounted for due to simplicity, then

$$dW_{\text{loss}} = \frac{3}{2}\left\{R_S\left|\bar{i}_S\right|^2 + R_r\left|\bar{i}_r\right|^2\right\}dt \qquad \text{...(11.89)}$$

where, R_S and R_r indicate the stator and rotor resistances.

The third differential energy component, the differential field energy, could be obtained as the rate of change of the magnetic energy stored in the field, and its magnitude equals the input power less the sum of the stator and rotor losses and the mechanical power.

Thus
$$\frac{dW_{\text{field}}}{dt} = \frac{3}{2}\,\text{RE}\left\{\bar{v}_{si}\,\bar{i}_s^* + \bar{v}_{ri}\,\bar{i}_r'^*\right\} \qquad \text{...(11.90)}$$

Here, $\bar{v}_{si}$ and $\bar{v}_{ri}$ are the space phasors of the induced stator and rotor transformer emf's respectively, both expressed in terms of the stationary reference frame fixed to the stator. Thus,

$$\bar{v}_{si} = \frac{d\psi_S}{dt} \qquad \text{...(11.91)}$$

and similarly, the induced rotor emf due to rate of change of the rotor flux-linkage, as:

$$\bar{v}_{ri}' = \frac{d\bar{\psi}_r'}{dt} \qquad \text{...(11.92)}$$

Substituting Eqs. (11.91) and (11.92) in Eq. (11.90), we get the under-noted expression for the differential magnetic energy stored in the field:

$$dW_{\text{field}} = \frac{3}{2}\left\{\text{RE}\left(\bar{i}_S^*\,\frac{d\bar{\psi}_S}{dt} + i_r'^*\,\frac{d\bar{\psi}_r'}{dt}\right)\right\}dt \qquad \text{...(11.93)}$$

and, it follows that the stored magnetic energy has the form:

$$W_{\text{field}} = \frac{3}{2}\,\text{RE}\left\{\bar{i}_s^*\,\bar{\psi}_S + \bar{i}_r^*\,\bar{\psi}_r'\right\}$$

which is the physically expected result. It could also be put into the non-space-phasor form, if the space-phasor quantities are expressed in their phase-variable component terms.

By substitution of Eqs. (11.86), (11.89) and (11.93) into Eq. (11.85), the following expression for the differential mechanical energy could now be obtained:

$$dW_{\text{mech}} = dW_{\text{mech}\,S} + dW_{\text{mech}\,r} \qquad \text{...(11.94)}$$

where, $dW_{\text{mech}\,S}$ and $dW_{\text{mech}\,r}$ are the mechanical energies due to the stator and rotor respectively,

and
$$dW_{\text{mech}\,S} = \frac{3}{2}\left[\text{RE}\left(\bar{v}_S\,\bar{i}_S^*\right) - R_S\left|\bar{i}_S\right|^2 - \text{RE}\left(\bar{i}_S^*\,\frac{d\bar{\psi}_s}{dt}\right)\right]dt \qquad \text{...(11.95)}$$

and $$dW_{\text{mech } r} = \frac{3}{2}\left[\text{RE}(\overline{v}_r' \overline{i}_r'^*) - R_r \left|\overline{i}_r'\right|^2 - \text{RE}\left(\overline{i}_r^* \frac{d\overline{\psi}_r'}{dt} \right) \right] dt \qquad ...(11.96)$$

In the stationary reference frame, the stator voltage space phasor $\overline{v}_S$ could only be countered by the stator ohmic drop $(R_s\overline{i}_s)$ and the rate of change of the stator flux-linkage $\left(\frac{d\overline{\psi}_s}{dt}\right)$— as indicated later in Section 11.7. Thus, it follows from Eq. (11.95) that $dW_{\text{mech } S}$ must be zero. Also, in the same reference frame the rotor ohmic voltage drop, $(R_r\overline{i}_r)$ plus the rate of change of the rotor flux-linkage $\left(\frac{d\overline{\psi}_r'}{dt}\right)$, and the rotational voltage $(-j\omega_r\overline{\psi}_r')$—as indicated in Section 11.7—that in the stationary reference frame, the rotor voltage space-phasor relation must also contain this rotational voltage term, because the rotor is rotating in the direction of the stator phase *A* to stator phase B as shown in Figure 11.1 Therefore, by considering Eqs. (11.83), (11.94), (11.95) and (11.96), the mechanical power is obtained as follows:

$$\begin{aligned} P_{\text{mech}} &= \frac{3}{2}\text{RE}\,(-j\omega_r\overline{\psi}_r'\overline{i}_r'^*) \\ &= \frac{3}{2}\omega_r RE(-j\overline{\psi}_r'\,\overline{i}_r'^*) \\ &= -\frac{3}{2}\omega_r\overline{\psi}_r' \times \overline{i}_r \end{aligned} \qquad ...(11.97)$$

Examination of Eq. (11.97) reveals that the mechanical power is proportional to the instantaneous rotor speed and the vector product of the rotor-flux-linkage and the rotor current space phasor. It is also noted by considering Eq. (11.83) that the electromechanical torque is expressed as follows:

$$T_e = -\frac{3}{2}\overline{\psi}_r' \times \overline{i}_r' \qquad ...(11.98)$$

For a machine with 2*P* poles, we have to multiply the above expression by P. It should be emphasised that in Eq. (11.98), the rotor flux-linkage and the rotor current space-phasors— $\overline{\psi}_r'$ and $\overline{i}_r'$ —are expressed in stationary reference frames, but since torque is invariant to change in reference frames, the expression

$$T_e = \left(-\frac{3}{2}\overline{\psi}_r \times \overline{i}_r\right)$$

is also equally valid where $\overline{\psi}_r$ and $\overline{i}_r$ are the space-phasors of the rotor flux-linkage and rotor currents respectively, but are expressed in terms of the reference frame fixed to the rotor member.

Equation (11.98) assumes many different forms, if we obtain Eq. (11.81) as shown below.

From Eqs. (11.51) and (11.52), the rotor flux-linkage space phasors, when expressed with respect to the stationary reference frame, should contain two flux-linkage components—$L_r\overline{i}_r'$, produced by the rotor currents only where L_r is the self-inductance of a rotor winding, and the other part, $L_m\overline{i}_S$, the mutual flux-linkage component produced by the stator currents where L_m is the magnetising inductance and $\overline{i}_S$ is the space phasor of stator currents in the stationary reference frame. This leads to putting Eq. (11.98) into the following format:

$$T_e = -\frac{3}{2}(L_r\bar{i}'_r + L_m\bar{i}_S)\times\bar{i}'_r$$

$$= -\frac{3}{2}L_m(\bar{i}_S \times \bar{i}'_r) \quad ...(11.99)$$

In this relation we have used the well-known property that ***the vector product of a vector with itself is zero.***

We could expand Eq. (11.99) and obtain the following relation:

$$T_e = -\frac{3}{2}L_m\bar{i}_S \times \bar{i}'_r = -\frac{3}{2}\frac{L_m}{L_S}(L_S\bar{i}_S + L_m\bar{i}'_r)\times\bar{i}'_r \quad ...(11.100)$$

since the cross product $(\bar{i}'_r \times \bar{i}'_r)$ is zero. L_S in Eq. (11.100) is the self-inductance of a stator winding. As per Eq. (11.35), the term $(L_S\bar{i}_S + L_m\bar{I}'_r)$ equals the space phasor of the stator flux-linkage expressed in terms of the stationary reference frame $(\bar{\psi}_S)$, and contains a self flux-linkage component produced by the stator currents $(L_S\bar{i}_S)$ and a mutual flux-linkage produced by the rotor currents $(L_m\bar{i}'_r)$. Thus, the electromagnetic torque produced by a 2-pole machine could be expressed as:

$$T_e = -\frac{3}{2}\frac{L_m}{L_S}\bar{\psi}_S \times \bar{i}'_r \quad ...(11.101)$$

If the effects of satuation of the main flux and leakage flux are neglected, the maglising inductance L_m and the total self-inductance L_S are both constants. Hence Eqs. (11.81) and (11.101) become identical.

The general expressions obtained for the instantaneous value of the electromagnetic torque are valid for symmetrical 3-phase and 2-phase machines, and no restriction has been made on the time variation of the stator and rotor currents other than that there can be no zero-sequence stator or rotor currents. The space phasors have been derived on the assumptions that the spatial distribution of the flux density and current density are sinusoidal. This is a significant assumption frequently considered in the Unified Theory of Electrical Machines. We also ignore the non-sinusoidal flux and current densities in the present text, which could have been considered in the parameters of the machine.

11.7 VOLTAGE RELATIONS OF SMOOTH AIR-GAP MACHINES IN DIFFERENT REFERENCE FRAMES

We consider in this section, the stator and rotor voltage differential equations, valid for transient and steady-state operation of smooth air-gap machines, using 3-phase as well as 2-phase variables, along with space phasors. The concept underlying space phasor theory yields great simplification of the voltage equations when compared with their phase variable forms. It is also possible to yield both physical and mathematical analysis of the dynamics in terms of phase variable equations, but this is already covered in Chapter 9, and is not developed for space phasor considerations in order to avoid complications.

We briefly repeat the 3-phase voltage relations as well as the 2-phase relations, for ready comparison between the non-sapce phasor and space phasor forms.

11.7.1 Phase Variable Voltage Relations When Referred to Natural Reference Frame

The stator voltage relations are formulated in the stationary reference frame fixed with stator while the rotor relations are developed in the rotating reference frame fixed to the rotor, in their phase-variable forms.

3-phase Model

A symmetrical 3-phase, 2-pole, smooth air-gap machine with symmetrically distributed windings is considered as shown earlier in Figure 11.1, schematically. The effects of mmf space harmonics are ignored and it is assumed that the stator and rotor voltages and currents vary in time in an arbitrary manner. The phase variable form of voltage relations, as obtained in the configuration is as follows, for the stationary reference frame attached to the stator member:

$$v_{SA}(t) = R_S i_{SA}(t) + \frac{d\psi_{SA}(t)}{dt} \quad \text{...(11.102)}$$

$$v_{SB}(t) = R_S i_{SB}(t) + \frac{d\psi_{SB}(t)}{dt} \quad \text{...(11.103)}$$

$$v_{SC}(t) = R_S i_{SC}(t) + \frac{d\psi_{SC}(t)}{dt} \quad \text{...(11.104)}$$

In Eqs. (11.102) to (11.104), the entities $v_{SA}(t)$, $v_{SB}(t)$, $v_{SC}(t)$, $i_A(t)$, $i_B(t)$ and $i_C(t)$ are the instantaneous values of the stator voltages and currents, while R_S is the resistance of the stator phase winding, while $\psi_{SA}(t)$, $\psi_{SB}(t)$, and $\psi_{SC}(t)$ are the instantaneous values of the stator flux-linkage in A, B and C phases respectively, as already specified in Eqs. (11.32), (11.33) and (11.34). Similar relations also hold for the rotor voltage expressions when put down with reference frame fixed to the rotor:

$$v_{ra}(t) = R_r i_{ra}(t) + \frac{d\psi_{ra}(t)}{dt} \quad \text{...(11.105)}$$

$$v_{rb}(t) = R_r i_{rb}(t) + \frac{d\psi_{rb}(t)}{dt} \quad \text{...(11.106)}$$

$$v_{rc}(t) = R_r i_{rc}(t) + \frac{d\psi_{rc}(t)}{dt} \quad \text{...(11.107)}$$

where, $V_{ra}(t)$, $V_{rb}(t)$, $V_{rc}(t)$, $i_{ra}(t)$, $i_{rb}(t)$ and $i_{rc}(t)$ stand for the instantaneous values of rotor phase voltages and currents, R_r is the resistance of rotor per phase and $\psi_{ra}(t)$, $\psi_{rb}(t)$ and $\psi_{rc}(t)$ are the instantaneous values of rotor flux-linkages for the phases a, b and c respectively as was already specified by Eqs. (11.42), (11.43) and (11.44) earlier.

The stator and rotor voltage equations Eqs. (11.102) to (11.107), Eqs. (11.32) to (11.34) and Eqs. (11.42) to (11.44) may be combined and laid down in a single-matrix equation:

$$\begin{bmatrix} V_{SA} \\ V_{SB} \\ V_{SC} \\ V_{ra} \\ V_{rb} \\ V_{rc} \end{bmatrix} = \begin{bmatrix} R_S + pL_S^- & p\bar{M}_S & p\bar{M}_S & p\bar{M}_{Sr}\cos\theta & p\bar{M}_{Sr}\cos\theta_1 & p\bar{M}_{Sr}\cos\theta_2 \\ p\bar{M}_S & R_S + pL_S & p\bar{M}_S & p\bar{M}_{Sr}\cos\theta_2 & p\bar{M}_{Sr}\cos\theta & p\bar{M}_{Sr}\cos\theta_1 \\ p\bar{M}_S & p\bar{M}_S & R_S + pL_S^- & p\bar{M}_{Sr}\cos\theta_1 & p\bar{M}_{Sr}\cos\theta_2 & p\bar{M}_{Sr}\cos\theta \\ p\bar{M}_{Sr}\cos\theta & p\bar{M}_{Sr}\cos\theta_2 & p\bar{M}_{Sr}\cos\theta_1 & R_r + pL_r^- & p\bar{M}_r & p\bar{M}_r \\ p\bar{M}_{Sr}\cos\theta_1 & p\bar{M}_{Sr}\cos\theta & p\bar{M}_{Sr}\cos\theta_2 & p\bar{M}_r & R_r + pL_r & p\bar{M}_r \\ p\bar{M}_{Sr}\cos\theta_2 & p\bar{M}_{Sr}\cos\theta_1 & p\bar{M}_{Sr}\cos\theta & p\bar{M}_r & p\bar{M}_r & R_r + pL_r^- \end{bmatrix} \begin{bmatrix} i_A \\ i_B \\ i_C \\ i_a \\ i_b \\ i_c \end{bmatrix} \qquad ...(11.108)$$

In this sets of equations, $p = \left(\frac{d}{dt}\right)$, stands for Heavisides differential operator. The angles θ, θ_1 and θ_2 are defined as $\theta = \theta_r$, $\theta_1 = \left(\theta_r + \frac{2\pi}{3}\right)$ and $\theta_2 = \left(\theta_r + \frac{4\pi}{3}\right)$, and θ_r is the rotor angle as shown in Figure 11.1. In so far as the stator self-inductance of one phase winding, L_s^- is concerned, it can be expressed as the sum of the stator leakage inductance L_{Sl} and the stator magnetising inductance L_{Sm}, so that:

$$L_S = L_{Sl} + L_{Sm} \qquad ...(11.109)$$

The relationship for mutual inductance between two stator phases for the mmf to be sinusoidally distributed, has the form:

$$\bar{M}_S = L_{Sm}\cos\left(\frac{2\pi}{3}\right) = -L_{Sm}/2$$

Similarly, it is possible to express the rotor self-inductance of one rotor phase as:

$$L_r^- = L_{rl} + L_{rm}$$

and

$$\bar{M}_r = L_{rm}\cos\left(\frac{2\pi}{3}\right) = -\frac{L_{rm}}{2}$$

Also, from simple physical consideration of winding inductances, the magnetising inductance of the stator L_{Sm} is related to the maximum value of the mutual inductance between the stator and rotor, M_{sr} as follows:

$$L_{Sm} = \left(\frac{N_{SE}}{N_{rE}}\right) M_{Sr}$$

where, N_{SE} and N_{rE} are the effective number of stator turns and rotor turns per phase [as elaborated in Section 11.2]. The magnetising inductance of the rotor is similarly related to $\bar{M}_{Sr}$ as follows:

$$L_{rm} = \left(\frac{N_{rE}}{N_{SE}}\right) \bar{M}_{Sr}$$

and, it follows that:

$$M_{Sr}^2 = L_{Sm} L_{rm}$$

If the resultant 3-phase magnetising inductance, L_m—first introduced in Eq. (11.35) is used, where,

$$L_m = \frac{3}{2} \bar{M}_{Sr}$$

then, the total 3-phase stator inductance, L_S [also introduced for the first time in Eq. (11.35)], has the form:

$$L_S = \bar{L}_S - \bar{M}_S = L_{Sl} + L_{Sm} + \frac{1}{2} L_{Sm}$$
$$= L_{Sl} + \frac{3}{2} L_{Sm}$$

and the total 3-phase rotor inductance [introduced for the first time in Eq. (11.45)] could similarly be expressed as:

$$L_r = \bar{L}_r - \bar{M}_r = L_{rl} + L_{rm} + \frac{1}{2} L_{rm}$$
$$= L_{rl} + \frac{3}{2} L_{rm}$$

As was pointed out while discussing the unified theory of electrical machines in the earlier chapters of this book, considerable amount of simplification can be achieved if the 3-phase in the voltage equations could be replaced by corresponding 2-phase equivalents expressed in the same reference frames. Further simplifications can also be achieved by using other than the natural frames used to derive the voltage equations indicated in Eq. (11.108). Thus, expressing the rotor quantities (voltages, currents and flux-linkages) in their natural reference frames, if they are expressed in the stationary reference frame, the resulting voltage differential equations contain constant coefficients if the parameters of the machine are considered as constants.

We discuss the 2-phase slip-ring model again briefly, because of the advantages of these simplifications:

2-phase Slip-ring Model

While discussing a 2-phase slip-ring model of smooth air-gap machine, we reduce the number of phase variables of the 3-phase model, by assuming that there is no zero-sequence voltages and currents in the stator and rotor members. There is now only four voltage equations corresponding to the direct- and quadrature-axis stator and rotor voltage equations respectively. Physically, this amounts to using a 2-phase slip-ring model, shown in Figure 11.7, denoting the 2-axis stator windings (*SD* and *SQ*), and a 2-axis rotor winding ($r\alpha$, $r\beta$), such that the rotor-phase $r\alpha$ is displaced from the stator-phase *SD* by rotor angle θ_r.

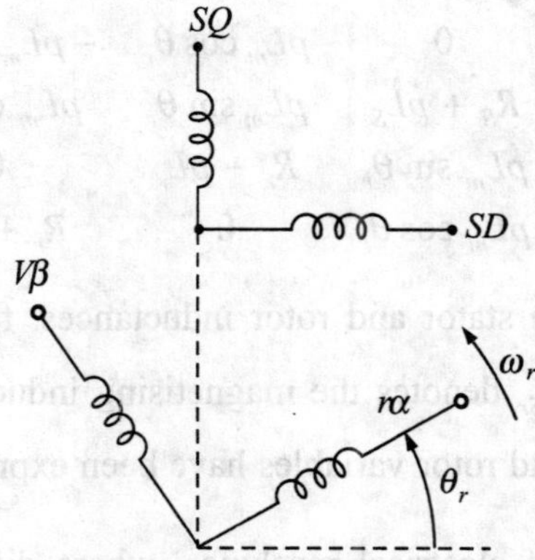

FIGURE 11.7 Schematic depiction of a 2-phase slip-ring model.

The relatioships connecting the 2-phase stator voltage and current components and the corresponding 3-phase components were shown in Eqs. (11.63) to (11.66), and are repeated here for completeness:

$$v_{SD} = \frac{2}{3}\left(V_{SA} - \frac{1}{2}V_{SB} - \frac{1}{2}V_{SC}\right) \quad ...(11.110)$$

$$v_{SQ} = \frac{(V_{SB} - V_{SC})}{\sqrt{3}} \quad ...(11.111)$$

and

$$i_{SD} = \frac{2}{3}\left(i_{SA} - \frac{1}{2}i_{SB} - \frac{1}{2}i_{SC}\right) \quad ...(11.112)$$

$$i_{SQ} = \frac{(i_{SB} - i_{SC})}{\sqrt{3}} \quad ...(11.113)$$

Also, similar transformations are effective for the rotor voltages and currents:

$$v_{r\alpha} = \frac{2}{3}\left(v_{ra} - \frac{1}{2}v_{rb} - \frac{1}{2}v_{rc}\right) \quad ...(11.114)$$

$$v_{r\beta} = \frac{(v_{rb} - v_{rc})}{\sqrt{3}} \quad ...(11.115)$$

and also,

$$i_{r\alpha} = \frac{2}{3}\left(i_{ra} - \frac{1}{2}i_{rb} - \frac{1}{2}i_{rc}\right) \quad ...(11.116)$$

$$i_{r\beta} = \frac{(i_{rb} - i_{rc})}{\sqrt{3}} \quad ...(11.117)$$

After suitable algebric manipulations using Eqs. (11.110) to (11.117), Eq. (11.108) derived for the natural frames for 3-phase smooth air-gap machine can be shown to have the shape shown in Eq. (11.118).

$$\begin{bmatrix} V_{SD} \\ V_{SQ} \\ V_{r\alpha} \\ V_{r\beta} \end{bmatrix} = \begin{bmatrix} R_S + pL_S & 0 & pL_m \cos\theta_r & -pL_m \sin\theta_r \\ 0 & R_S + pL_S & pL_m \sin\theta_r & pL_m \cos\theta_r \\ pL_m \cos\theta_r & pL_m \sin\theta_r & R_r + pL_r & 0 \\ -pL_m \sin\theta_r & pL_m \cos\theta_r & 0 & R_r + pL_r \end{bmatrix} \begin{bmatrix} i_{SD} \\ i_{SQ} \\ i_{r\alpha} \\ i_{r\beta} \end{bmatrix} \quad ...(11.118)$$

In expression, L_S and L_r are the stator and rotor inductances, first introduced in Eqs. (11.35) and (11.45). Further, $L_m = \frac{3}{2}\bar{M}_{Sr}$ denotes the magnetising inductance.

In Eq. (11.118), the stator and rotor variables have been expressed in their natural reference frames.

In case of unified theory of electrical machines, where different models can be derived by using matrix transformation, the 2-phase model described by voltage Eq. (11.118) can be obtained from its equivalent 3-phase counterpart shown by the voltage Eq. (11.108) by using the phase transformation matrix $[C_1]$. The inverse of the phase transformation matrix was defined in Eq. (11.69).

Thus, if Z_S is the impedance matrix of the 3-phase machine specified by the voltage equation (11.108), the impedance matrix for the 2-phase slip-ring model expounded by Eq. (11.118) could be derived by performing.

$$[Z'_S] = [C_t][Z_S][C]$$

where,

$$[C] = \text{diag}\ \{[C_{1t}];\ [C_1]\}$$

In the above relation, the suffix t attached to $[C_1]$ is the transpose.

11.7.2 Equivalent 2-Phase Commutator Model

If the stationary-axis stator quantities of the 2-axis slip-ring model remain unchanged, but the rotor operational parameters or the rotor voltages and currents, that is, $v_{r\alpha}$, $i_{r\alpha}$ and $v_{r\beta}$, $i_{r\beta}$ are transformed from the rotor reference frame to a new reference frame attached to the stator, we obtain 2-phase commutator model with pseudo-stationary rotor windings. The schematic of this machine is presented in Figure 11.8. On the stator, there are the direct and quadrature axis windings SD and SQ, and on the rotor, there are windings represented by rd and rq, as shown.

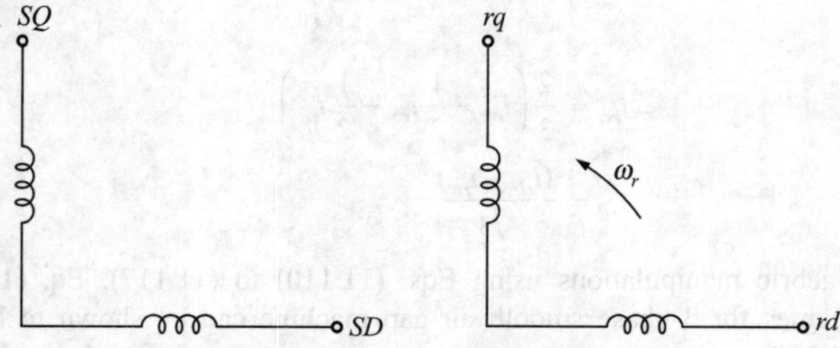

FIGURE 11.8 Schematic depiction of a 2-phase commutator model.

The following relations, described for the first time in Eqs. (11.40) and (11.74), offer transformations for the rotor voltages and currents respectively:

$$v_{r\alpha} = \cos\theta_r\, v_{rd} + \sin\theta_r v_{rq} \qquad \text{...(11.119)}$$

$$v_{r\beta} = -\sin\theta_r\, v_{rd} + \cos\theta_r\, v_{rq} \qquad \text{...(11.120)}$$

and, similarly:

$$i_{r\alpha} = \cos\theta_r\, i_{rd} + \sin\theta_r\, i_{rq} \qquad \text{...(11.121)}$$

$$i_{r\beta} = -\sin\theta_r i_{rd} + \cos\theta_r\, i_{rq} \qquad \text{...(11.122)}$$

Using the above transformations listed under Eqs. (11.119) to (11.122), we obtain the following transformed set of voltage [shown in Eq. (11.123)].

The above set of voltage relations are also referred to as the commutator primitive model voltage equations.

$$\begin{bmatrix} v_{SD} \\ v_{SQ} \\ v_{rd} \\ v_{rq} \end{bmatrix} = \begin{bmatrix} R_S + pL_S & 0 & pL_m & 0 \\ 0 & R_S + pL_S & 0 & pL_m \\ pL_m & \omega_r L_m & R_r + pL_r & \omega_r L_r \\ -\omega_r L_m & pL_m & -\omega_r L_r & R_r + pL_r \end{bmatrix} \begin{bmatrix} i_{SD} \\ i_{SQ} \\ i_{rd} \\ i_{rq} \end{bmatrix} \qquad \text{...(11.123)}$$

In case of the Unified Theory of Electrical Machines, the above model could be directly derived from the 2-phase slip-ring model shown in Eq. (11.118), by executing,

$$[Z''_S] = [C_t][Z'_S][C]$$

where,

$$[C] = \text{diag}\,\{[I_2];\ [C_2]\}$$

In the above relation, $[I_2]$ is a 2 × 2 identity matrix and the matrix $[C_2]^{-1}$ is defined in Eq. (11.40). We have arrived at identical results in Eq. (8.60) in Section 8.3 of Chapter 8 earlier, while assuming the direction of rotation of the rotor as clockwise based on physical considerations. There we have considered that in the pseudo-stationary rotor coils situated along the direct-axis, *rd*, as well as in the quadrature-axis rotor coil *rq*, there are voltages induced by transformer effect, that is, $p(Mi_{SD} + L_2\, i_{rd})$ for winding *rd*, and $p(Mi_{SQ} - L_2 i_{rq})$ for winding *rq*, as well as speed induced voltages, namely $\omega_r(M + L_2)$ for the winding *rd* and $-(M + L_2)\omega_r$ for the winding *rq*. Here *M* denotes the magnetising inductance and L_2 is the self-inductance of the rotor.

11.8 SPACE-PHASOR FORM OF THE EXPRESSION

In this section, the space-phasor form of voltage expressions for 3-phase as well as 2-phase, smooth air-gap configurations are elaborated. These relations have been developed with respect to the stationary reference frame as well as with regard to a general rotating reference frame, rotating at a general speed of ω_g.

11.8.1 Space-phasor Voltage Relations with Respect to a Stationary Reference Frame

The stator and rotor voltage differential equations of the smooth air-gap 3-phase configuration are presented in their natural reference frames in Eq. (11.108). If we use the definitions of the voltage space phasor enumareted in Eqs. (11.61) and (11.62), current space phasors contained in Eqs. (11.4) and (11.20), and flux-linkage space phasors elaborated by Eqs. (11.31) and (11.41), then the stator and rotor voltage equations defined under Eqs. (11.102) to (11.104) and (11.105) to (11.107), could be indicated in the following space phasor form:

$$\overline{v}_S = R_S\overline{i}_S + \frac{d\psi_S}{dt} \quad ...(11.124)$$

and

$$\overline{v}_r = R_r\overline{i}_r + \frac{d\overline{\psi}'_r}{dt} - j\omega_r\overline{\psi}'_r \quad ...(11.125)$$

In Eqs. (11.124) and (11.125), the stator and rotor space phasor flux-linkages, ψ_S and $\overline{\psi}'_r$ appear. These have been specified in terms of the machine impedances and the space-phasors of the stator and rotor currents, elaborated in Eqs. (11.35) and (11.51) respectively, while ω_r denotes the instantaneous angular speed of the rotor.

We record the definitions of all the space-phasors of 3-phase quantities for consolidation and overview. The space phasors of the stator voltages, currents and flux-linkages, with reference to stationary frame of reference attached to the stator, as introduced in Eqs. (11.61), (11.4), (11.31) and (11.35) are as follows:

$$\overline{v}_S = \frac{2}{3}\{v_{SA}(t) + av_{SB}(t) + a^2v_{SC}(t)\} = v_{SD} + jv_{SQ} \quad ...(11.126)$$

$$\overline{i}_S = \frac{2}{3}\{i_{SA}(t) + ai_{SB}(t) + a^2i_{SC}(t)\} = i_{SD} + ji_{SQ} \quad ...(11.127)$$

and

$$\left.\begin{aligned}\overline{\psi}_S &= \frac{2}{3}\{\psi_{SA}(t) + a\psi_{SB}(t) + a^2\psi_{SC}(t)\}\\ &= \psi_{SD} + j\psi_{SQ}\\ &= L_S\overline{i}_S + L_m\overline{i}'_r\end{aligned}\right\} \quad ...(11.128)$$

and similarly, the space phasors of the rotor voltages, currents and flux-linkages, corresponding to reference frame attached to the rotor member as per Eqs. (11.62), (11.20), (11.41) and (11.45) are:

$$\overline{v}_r = \frac{2}{3}\{v_{ra}(t) + av_{rb}(t) + a^2v_{rc}(t)\} = v_{r\alpha} + jv_{r\beta} \quad ...(11.129)$$

$$\overline{i}_r = \frac{2}{3}\{i_{ra}(t) + ai_{rb}(t) + a^2i_{rc}(t)\} = i_{r\alpha} + ji_{r\beta} \quad ...(11.130)$$

and

$$\left.\begin{aligned}\overline{\psi}_r &= \frac{2}{3}\{\psi_{ra}(t) + a\psi_{rb}(t) + a^2\psi_{rc}(t)\}\\ &= L_r\overline{i}_r + L_m\overline{i}'_S\\ &= L_r\overline{i}_r + L_m\overline{i}_S e^{j\theta_r}\\ &= \psi_{r\alpha} + j\psi_{r\beta}\end{aligned}\right\} \qquad ...(11.131)$$

The rotor quantities are attached to the reference frame fixed to the stator as per Eqs. (11.74), (11.39) and (11.51). Thus:

$$\left.\begin{aligned}\overline{v}'_r &= \overline{v}_r e^{j\theta_r}\\ &= v_{rd} + jv_{rq}\end{aligned}\right\} \qquad ...(11.132)$$

$$\left.\begin{aligned}\overline{i}'_r &= \overline{i}_r e^{j\theta_r}\\ &= i_{rd} + ji_{rq}\end{aligned}\right\} \qquad ...(11.133)$$

and

$$\left.\begin{aligned}\overline{\psi}'_r &= \overline{\psi}_r e^{j\theta_r}\\ &= L_r\overline{i}'_r + L_m\overline{i}_S\\ &= L_r\overline{i}_r e^{j\theta_r} + L_m\overline{i}_S\\ &= \psi_{rd} + j\psi_{rq}\end{aligned}\right\} \qquad ...(11.134)$$

Also, for completeness, the space phasors of the stator voltages, currents and flux-linkages, may be given with reference to the rotating reference frame attached to the rotor member, as indicated in Eqs. (11.73), (11.54) and (11.56), as follows:

$$\left.\begin{aligned}\overline{v}'_S &= \overline{v}_S e^{-j\theta_r}\\ &= v_{SD} + jv_{SQ}\end{aligned}\right\} \qquad ...(11.135)$$

$$\left.\begin{aligned}\overline{i}'_S &= \overline{i}_S e^{j\theta_r}\\ &= i_{SD} + ji_{SQ}\end{aligned}\right\} \qquad ...(11.136)$$

and

$$\left.\begin{aligned}\overline{\psi}'_S &= \overline{\psi}_S e^{-j\theta_r}\\ &= \psi_{SD} + j\psi_{SQ}\end{aligned}\right\} \qquad ...(11.137)$$

The first terms on the right hand side of Eqs. (11.124) and (11.125) are the space-phasor form of ohmic losses, and the second terms are the transformer emfs, the time derivative of the flux-linkage space-phasor of the stator and rotor respectively. The third term in Eq. (11.125), that is, the term $(-j\omega_r\overline{\psi}'_r)$ denotes rotational emf in the rotor, which contributes to electromechanical energy conversion.

If the flux-linkage space phasors specified in Eqs. (11.128), (11.131) and (11.134) are substituted into Eqs. (11.124) and (11.125), the space-phasor voltage expressions assume the form:

$$\bar{v}_S = R_S\bar{i}_S + \frac{d}{dt}(L_S\bar{i}_S) + \frac{d}{dt}(L_m\bar{i}'_r) \qquad ...(11.138)$$

and

$$\bar{v}'_r = R_r\bar{i}'_r + \frac{d}{dt}(L_r\bar{i}'_r) + \frac{d}{dt}(L_m\bar{i}_S) - j\omega_r(L_r\bar{i}'_r + L_m\bar{i}_s) \qquad ...(11.139)$$

The above relations expressed in matrix form becomes:

$$\begin{bmatrix}\bar{v}_S\\ \bar{v}'_r\end{bmatrix} = \begin{bmatrix}R_S & 0\\ 0 & R_r\end{bmatrix}\begin{bmatrix}\bar{i}_S\\ \bar{i}'_r\end{bmatrix} + \frac{d}{dt}\begin{bmatrix}L_S & L_m\\ L_m & L_r\end{bmatrix}\begin{bmatrix}\bar{i}_S\\ \bar{i}'_r\end{bmatrix} - j\omega_r\begin{bmatrix}0 & 0\\ L_m & L_r\end{bmatrix}\begin{bmatrix}\bar{i}_S\\ \bar{i}'_r\end{bmatrix} \qquad ...(11.140)$$

If we compare Eq. (11.140) with Eq. (11.108), we observe that in Eq. (11.108) while the phase variable voltages and currents appeared, in Eq. (11.140), the space-phasors are presented. If these space-phasors are now resolved into their real and imaginary-axis components using Eqs. (11.126), (11.127), (11.132) and (11.133), Eq. (11.140) reduces to the corresponding 2-phase commutator model shown in Eq. (11.123).

In other words, if the space-phasor voltage relations are established with respect to a stationary reference frame, they lead to the relation of the commutator model, without having to execute matrix transformation. The ability to develop models of machines in different reference frames without using matrix transformations is thus a definite advantage of space-phasors over the unified matrix theory of electrical machines. Furthermore, using Eqs. (11.132) and (11.133), we can express:

$$\bar{v}_r = \bar{v}'_r e^{-j\theta_r}$$

and

$$\bar{i}_r = \bar{i}'_r e^{-j\theta_r}$$

and the rotor quantities can be expressed in the reference frame attached to the rotor, which makes Eq. (11.140) as follows:

$$\begin{bmatrix}\bar{v}_S\\ \bar{v}_r\end{bmatrix} = \begin{bmatrix}R_S & 0\\ 0 & R_r\end{bmatrix}\begin{bmatrix}\bar{i}_S\\ \bar{i}_r\end{bmatrix} + \frac{d}{dt}\begin{bmatrix}L_S & L_m e^{j\theta_r}\\ L_m e^{-j\theta_r} & L_r\end{bmatrix}\begin{bmatrix}\bar{i}_S\\ \bar{i}_r\end{bmatrix} \qquad ...(11.141)$$

If all the space-phasor quantities are expressed in terms of their real-and imaginary-axis components, using Eqs. (11.126), (11.127), (11.129) and (11.130), Eq. (11.141) yields Eq. (11.118). This corresponds to the 2-phase slip-ring model. Thus, another model can therefore be derived from the space-phasor expressions without taking recourse to matrix transformations. We have, however, to note that the procedure leads to 'hidden' complex transformations in the form of the presence of terms $e^{j\theta_r}$ and $e^{-j\theta_r}$.

11.8.2 Space-phasor Voltage Relations in a General Reference Frame

We propose to take up the development of general voltage space-spasor relations with respect to a general reference frame, that rotates at a general speed ω_g.

Following the analysis presented earlier in Eq. (11.136) and in Figure 11.4, the space-phasor of the stator currents with reference to the rotor frame is given by:

$$\overline{i_S'} = \overline{i_S} e^{-j\theta_g}$$

where, $\overline{i_S}$ is the stator current space phasor with respect to the stator reference frame. Similarly it follows from Eq. (11.133) that the stator space phasor of the rotor currents in the stationary reference frame is denoted by:

$$\overline{i_r'} = \overline{i_r} e^{j\theta_r}$$

where $\overline{i_r}$ is the space phasor of the rotor currents with respect to the rotor reference frame with direct- and quadrature-axis, if x and y rotating at a general instantaneous speed, $\omega_g = \frac{d}{dt}(\theta_g)$ as shown in Figure 11.9 are used, where θ_g is the angle between the direct-axis of the stationary reference frame *SD* attached to the stator and the real-axis x of the general reference frame, then the following relation defines the stator current space phasor attached to the general reference frame:

$$\overline{i_{Sg}} = \overline{i_S} e^{-j\theta_g} = i_{Sx} + j i_{Sy} \quad \text{...(11.142)}$$

If we substitute $\theta_g = \theta_r$, we obtain Eq. (11.136), and this gives the space phasor of the stator currents in the rotor reference frame.

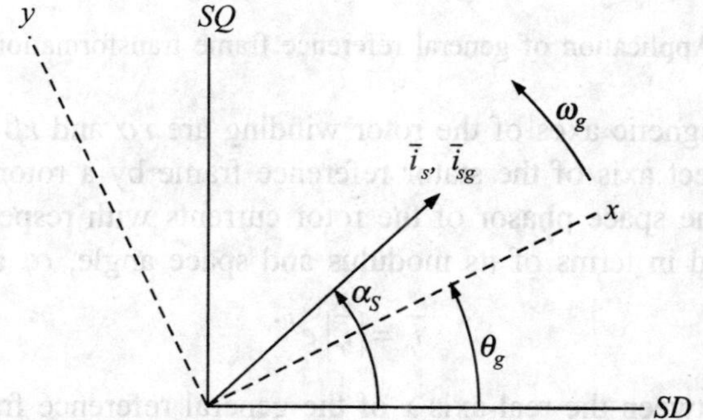

FIGURE 11.9 Application of general reference frame transformation of stator quantities.

Equation (11.142) can be proved mathematically by considering that from Figure 11.9 the stator reference frame,

$$\overline{i_{Sg}} = |\overline{i_S}| \, e^{j\alpha_S}$$

and the general reference frame,

$$\overline{i_{Sg}} = |\overline{i_S}| \, e^{j(\alpha_S - \theta_g)}$$

so that,
$$\bar{i}_{Sg} = |\bar{i}_S| e^{j\alpha_S} e^{-j\theta_g}$$
$$= \bar{i}_S e^{-j\theta_g}$$

The stator voltage and flux-linkage space phasors could be obtained similarly, with reference to the general reference frame as follows:

$$\bar{v}_{Sg} = \bar{v}_S e^{-j\theta_g} = v_{Sx} + jv_{Sy} \qquad \text{...(11.143)}$$

and
$$\bar{\psi}_{Sg} = \bar{\psi}_S e^{-j\theta_g} = \psi_{Sx} + j\psi_{Sy} \qquad \text{...(11.144)}$$

where, $\bar{v}_S$ and $\bar{\psi}_S$ are the space phasors of the stator voltages and the stator flux-linkages respectively with reference to the stationary reference frame attached to the stator.

Similar consideration is also applicable for specifying space phasors for the rotor voltages, currents and flux-linkages. In Figure 11.10, three reference frames are shown—the reference frame attached to the rotor, the reference frame fixed to the stator, and the general reference frame.

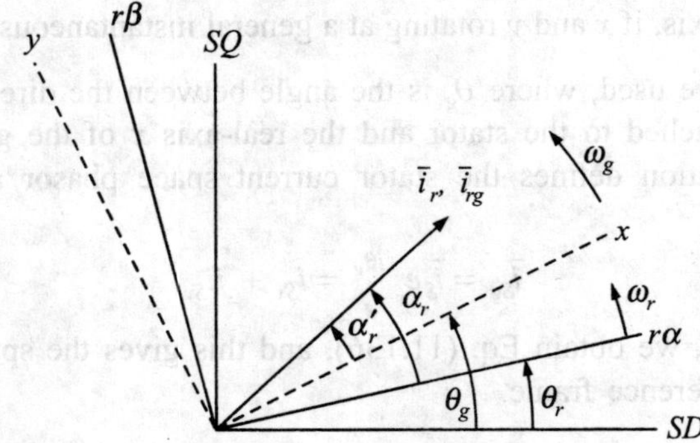

FIGURE 11.10 Application of general reference frame transformation of rotor quantities.

In the diagram, the magnetic axes of the rotor winding are $r\alpha$ and $r\beta$ and the rotor axis $r\alpha$ is displaced from the direct axis of the stator reference frame by a rotor angle θ_r.

From the figure, the space phasor of the rotor currents with respect to the rotor reference frame can be expressed in terms of its modulus and space angle, α_r as follows:

$$\bar{i}_r = |\bar{i}_r| e^{j\alpha_r}$$

However, the angle between the real-axis x of the general reference frame and the real axis of the reference frame rotating with the rotor (that is, $r\alpha$) is $(\theta_g - \theta_r)$, in the general reference frame, the space phasor of the rotor currents can be denoted by,

$$\bar{i}_{rg} = |\bar{i}_r| e^{j\alpha'_r}$$

where
$$\alpha'_r = \alpha_r - (\theta_g - \theta_r)$$

Hence,

$$\left.\begin{aligned} \bar{i}_{rg} &= |\bar{i}_r| e^{j\alpha_r} e^{-j(\theta_g - \theta_r)} \\ &= \bar{i}_r e^{-j(\theta_g - \theta_r)} \\ &= i_{rx} + j i_{ry} \end{aligned}\right\} \quad \text{...(11.145)}$$

with respect to the reference frame fixed to the stator, $\theta_g = 0$, and

$$\bar{i}_{rg} = \bar{i}_r' = \bar{i}_r e^{j\theta_r}$$

as was shown in Eq. (11.133).

Similarly, the space phasors of the rotor voltages and the rotor flux-linkages with reference to the general reference frame can be denoted by:

$$\bar{v}_{rg} = \bar{v}_r e^{-j(\theta_g - \theta_r)} = v_{rx} + j v_{ry} \quad \text{...(11.146)}$$

and

$$\bar{\psi}_{rg} = \bar{\psi}_r e^{-j(\theta_g - \theta_r)} = \psi_{rx} + j\psi_{ry} \quad \text{...(11.147)}$$

If we substitute Eqs. (11.142) to (11.147) to Eqs. (11.124), (11.125), (11.128) and (11.131), we get the following stator and rotor space-phasor voltage equations with respect to the general reference frame:

$$\bar{v}_{Sg} = R_S \bar{i}_{Sg} + \frac{d}{dt}\bar{\psi}_{Sg} + j\omega_g \bar{\psi}_{Sg} \quad \text{...(11.148)}$$

$$\bar{v}_{rg} = R_r \bar{i}_{rg} + \frac{d}{dt}\bar{\psi}_{rg} + j(\omega_g - \omega_r)\bar{\psi}_{rg} \quad \text{...(11.149)}$$

In the above relations, the stator and rotor flux-linkages in the general reference frame are expressed in terms of the stator and rotor current space-phasors as follows:

$$\bar{\psi}_{Sg} = L_S \bar{i}_{Sg} + L_m \bar{i}_{rg} \quad \text{...(11.150)}$$

and

$$\bar{\psi}_{rg} = L_r \bar{i}_{rg} + L_m \bar{i}_{Sg} \quad \text{...(11.151)}$$

By combining Eqs. (11.148) to (11.151), we arrive at:

$$\bar{v}_{Sg} = R_S \bar{i}_{Sg} + \frac{d}{dt}(L_S \bar{i}_{Sg}) + \frac{d}{dt}(L_m \bar{i}_{rg}) + j\omega_g (L_s \bar{i}_{Sg} + L_m \bar{i}_{rg}) \quad \text{...(11.152)}$$

and

$$\bar{v}_{rg} = R_r \bar{i}_{rg} + \frac{d}{dt}(L_r \bar{i}_{rg}) + \frac{d}{dt}(L_m \bar{i}_{Sg}) + j(\omega_g - \omega_r)(L_r \bar{i}_{rg} + L_m \bar{i}_{Sg}) \quad \text{...(11.153)}$$

which could be represented in matrix form as follows:

$$\begin{bmatrix} \bar{v}_{Sg} \\ \bar{v}_{rg} \end{bmatrix} = \begin{bmatrix} R_S & 0 \\ 0 & R_r \end{bmatrix}\begin{bmatrix} \bar{i}_{Sg} \\ \bar{i}_{rg} \end{bmatrix} + \frac{d}{dt}\begin{bmatrix} L_S & L_m \\ L_m & L_r \end{bmatrix}\begin{bmatrix} \bar{i}_{Sg} \\ \bar{i}_{rg} \end{bmatrix} + j\omega_g \begin{bmatrix} L_S & L_m \\ L_m & L_r \end{bmatrix}\begin{bmatrix} \bar{i}_{Sg} \\ \bar{i}_{rg} \end{bmatrix} - j\omega_r \begin{bmatrix} 0 & 0 \\ L_m & L_r \end{bmatrix}\begin{bmatrix} i_{Sg} \\ \bar{i}_{rg} \end{bmatrix}$$

...(11.154)

The right-hand side of Eq. (11.153) denotes voltage terms due to ohmic losses, together with transformer emfs (in the second and third terms) and voltages due to speed action of the general reference frame plus the true rotational voltages. The last component is only present in the rotor equation, and it is this term

$$-j\omega_r \overline{\psi}_{rg} = -j\omega_r (L_r \overline{i}_{rg} + L_m \overline{i}_{Sg})$$

which contributes to electromechanical energy convertion.

We can also use the rotational voltage to obtain the expression for the electromagnetic torque as will be seen in the next section.

We could express all space-phasor quantities in terms of their real- and imaginary-axis components, by using Eqs. (11.142), (11.143), (11.145), (11.146) and (11.154), to achieve the following 2-phase form.

$$\begin{bmatrix} v_{Sx} \\ v_{Sy} \\ v_{rx} \\ v_{ry} \end{bmatrix} = \begin{bmatrix} R_s + pL_S & -\omega_g L_S & pL_m & -\omega_g L_m \\ \omega_g L_S & R_S + pL_S & \omega_g L_m & pL_m \\ pL_m & -(\omega_g - \omega_r) L_m & R_r + pL_r & -(\omega_g - \omega_r) L_r \\ (\omega_g - \omega_r) L_m & pL_m & (\omega_g - \omega_r) L_r & R_r + pL_r \end{bmatrix} \begin{bmatrix} i_{Sx} \\ i_{Sy} \\ i_{rx} \\ i_{ry} \end{bmatrix} \qquad \text{...(11.155)}$$

where, p is the Heaviside's differential operator $\left(\dfrac{d}{dt}\right)$. Equation (11.155) is reduced to Eq. (11.123) for $\omega_g = 0$, as at this condition, we achieve the 2-phase commutator model. Also when $\omega_g = \omega_1$, the synchronous speed, and for an induction motor $\omega_1 - \omega_r = s\omega_1$, s defining the slip, the 2-axis equation of the induction motor in a synchronously rotating reference frame, is given by:

$$\begin{bmatrix} v_{Sx} \\ v_{Sy} \\ v_{rx} \\ v_{ry} \end{bmatrix} = \begin{bmatrix} R_S + pL_S & -\omega_1 L_S & pL_m & -\omega_1 L_m \\ \omega_1 L_S & R_S + pL_S & \omega_1 L_m & pL_m \\ pL_m & -s\omega_1 L_m & R_r + pL_r & -s\omega_1 L_r \\ s\omega_1 L_m & pL_m & s\omega_1 L_r & R_r + pL_r \end{bmatrix} \begin{bmatrix} i_{Sx} \\ i_{Sy} \\ i_{rx} \\ i_{ry} \end{bmatrix} \qquad \text{...(11.156)}$$

Here, v_{Sx}, v_{Sy} and i_{Sx}, i_{Sy} are the 2-phase components of the stator voltages and currents in the synchronously rotating reference frame. The rotor voltages and current components in the above reference frame are $v_{rx} = v_{ry} = 0$ and i_{rx}, i_{ry}.

The above voltage equations are valid for both transient as well as steady-state conditions of operation however, for the transient conditions, we have to consider also the equation of motion shown in Eq. (11.209), which is repeated for completeness:

$$T_e - T_L = J\frac{d\omega_r}{dt} + D\omega_r \qquad \text{...(11.157)}$$

where, T_L is the load torque and T_e is the electro-magnetic torque.

For a machine with P pole-pairs, the electromagnetic torque may be expressed in accordance with Eq. (11.99), in terms of the stator and rotor space phasors in their 2-axis components with respect to the general reference frame which was specified by Eqs. (11.142) and (11.145), as follows:

$$T_e = -\frac{3}{2} PL_m \bar{i}_{Sg} \times \bar{i}_{rg}$$

$$= -\frac{3}{2} PL_m \{i_{Sx} i_{ry} - i_{Sy} i_{rx}\} \qquad ...(11.158)$$

In Eq. (11.157), the rotor speed could also be expressed as follows:

$$\omega_r = \frac{d\theta_r}{dt} \qquad ...(11.159)$$

θ_r being the rotor angle.

In equation (11.98), $\bar{\psi}_r'$ and $\bar{i}_r'$ are the rotor flux-linkage and rotor current space phasors with respect to the stationary reference frame attached to the stator member. By considering $\theta_g = 0$ and scrutinizing Eqs. (11.145) and (11.147), their values are:

$$\bar{i}_{rg} = \bar{i}_r e^{-j(\theta_g - \theta_t)} = \bar{i}_r e^{-j\theta_g} e^{j\theta_r}$$

and

$$\bar{\psi}_{rg} = \bar{\psi}_r e^{-j(\theta_g - \theta_r)} = \bar{\psi}_r e^{-j\theta_g} e^{j\theta_r}$$

Hence, the value of the electromagnetic torque, using Eq. (11.98), becomes,

$$T_e = \frac{3}{2} \bar{\psi}_r' \times \bar{i}_r'$$

$$= -\frac{3}{2} \psi_{rg} e^{j\theta_g} \times \bar{i}_{rg} e^{j\theta_g}$$

$$= -\frac{3}{2} \bar{\psi}_{rg} \times \bar{i}_{rg} \qquad ...(11.160)$$

In Eq. (11.151), $\bar{\psi}_{rg}$ was specified, and substitution of Eq. (11.151) in Eq. (11.160) leads to

$$T_e = -\frac{3}{2} L_m \bar{i}_{Sg} \times \bar{i}_{rg}$$

for a 2-pole configuration.

Derivation of expression could also be arrived at on the technique applied for the Unified Matrix Theory.

With $[G]$ as the torque matrix and $[i]$ as the column vectorial representation in the general reference frame, the electromagnetic torque, T_e, is shown to be:

$$T_e = \frac{3}{2} P[i_t^*][G][i] \qquad ...(11.161)$$

where the suffix t denotes the transpose of the column matrix of current and * denotes the conjugate. Thus, the value of $[G]$ could be expressed as:

$$[G] = \begin{bmatrix} 0 & 0 & 0 & 0 \\ 0 & 0 & 0 & 0 \\ 0 & L_m & 0 & L_r \\ -L_m & 0 & -L_r & 0 \end{bmatrix} \qquad ...(11.162)$$

And since in Eq. (11.156), $s\omega_1 = \omega_1 - \omega_r$ and matrix $[G]$ contains terms multiplied by ω_r, we can rewrite Eqs. (11.161) and (11.162) as follows:

$$T_e = \frac{3}{2} P(i_{rx}\psi_{ry} - i_{ry}\psi_{rx}) \qquad ...(11.163)$$

When expressed in the form of a vector product, it takes the form:

$$T_e = -\frac{3}{2} P \,\bar{\psi}_{rg} \times \bar{i}_{rg}$$

which agrees with Eq. (11.160) for a 2-pole machine (P = 1).

11.9 ELECTROMAGNETIC TORQUE IN SMOOTH AIR-GAP MACHINES

We dedicate this section to the development of various expressions for electromagnetic torque through the use of space-phasor quantities like voltages, currents and flux-linkages with respect to alternative reference frames. When special reference frames are considered as attached to the rotor flux, the stator-flux, or the magnetising flux-linkage space phasor, the expression for the electromagnetic torque is found to bear strong resemblance with that of a separately excited dc machine.

11.9.1 Application of the General Reference Frame

We had developed expressions for the electromagnetic torque in Section 11.6.2, wherein we developed Eqs. (11.98) to (11.101), taking the stationary reference frame attached to the stator as reference frame.

For a clear physical understanding of the processes involved, it is first necessary that Eq. (11.149) of voltage equations with reference to the general reference frame be used directly to derive a general expression for the electromagnetic torque. We note from a scrutiny of Eq. (11.149) that the space-phasor form of the rotational emf in the rotor winding is given by:

$$-j\omega_r \bar{\psi}_{rg} = -j\omega_r (\psi_{rx} + j\psi_{ry})$$

The real-axis component of this emf is:

$$v_{\text{rot}\,x} = \omega_r \psi_{ry}$$

and the imaginary-axis, rotational-axis emf is:

$$v_{\text{rot}\,y} = -\omega_r \psi_{rx}$$

so that the mechanical power could be expressed as follows:

$$P_{\text{mech}} = \frac{3}{2}(v_{\text{rot}\,x} i_{rx} + v_{\text{rot}\,y}\, i_{ry})$$

$$= \frac{3}{2}\omega_r(\psi_{ry}i_{rx} - \psi_{rx}i_{ry})$$

$$= -\frac{3}{2}\omega_r\,\overline{\psi}_{rg} \times \overline{i}_{rg}$$

When this expression of power is divided by the rotor speed ω_r, it results in an expression for the electromagnetic torque, in agreement with Eq. (11.160).

Thus, the expression for the electromagnetic torque is similar in all reference frames. For a machine with P pole pairs, Eq. (11.160) assumes the following form:

$$\left.\begin{aligned} T_e &= -\frac{3}{2}P\overline{\psi}_r' \times \overline{i}_r' \\ &= -\frac{3}{2}P\,\overline{\psi}_{rg} \times \overline{i}_{rg} \end{aligned}\right\} \quad \text{...(11.164)}$$

which clearly indicates that torque is produced due to interaction of the rotor flux-linkage space-phasor and the rotor current space-phasor.

By extending a similar argument, using Eq. (11.99), we could write:

$$\left.\begin{aligned} T_e &= -\frac{3}{2}P\,L_m\overline{i}_S \times \overline{i}_r' \\ &= -\frac{3}{2}P\,L_m\overline{i}_{Sg} \times i_{rg} \end{aligned}\right\} \quad \text{...(11.165)}$$

This shows that electromagnetic torque is proportional to the cross product of stator and rotor space phasors. It is also possible to derive Eq. (11.165), by substituting Eq. (11.151) in Eq. (11.164). Similarly, we could have used Eq. (11.101) to obtain:

$$T_e = -\frac{3}{2}P\left(\frac{L_m}{L_S}\right)\psi_{Sg} \times \overline{i}_{rg} \quad \text{...(11.166)}$$

The above relation could be obtained by manipulating Eqs. (11.150) and (11.165). If we drop the negative sign in Eq. (11.164) according to Newton's third law and write as:

$$T_e = \frac{3}{2}P\overline{\psi}_{Sg} \times \overline{i}_{Sg} \quad \text{...(11.167)}$$

there will be many more ways of providing mathematical proof of Eq. (11.167).

Using Eq. (11.165):

$$T_e = \frac{3}{2}P\,L_m\,\overline{i}_{rg} \times \overline{i}_{Sg} \quad \text{...(11.168)}$$

This could be expanded as:

$$\left.\begin{aligned} T_e &= \frac{3}{2}P\,(L_S\,\overline{i}_{Sg} + L_m\overline{i}_{rg}) \times \overline{i}_{Sg} \\ &= \frac{3}{2}P\,\overline{\psi}_{Sg} \times \overline{i}_{Sg} \end{aligned}\right\} \quad \text{...(11.169)}$$

It is also possible to split the total stator flux-linkage into sum of leakage flux-linkage ($\bar{\psi}_{S\text{lg}}$) and magnetising flux-linkage ($\bar{\psi}_{mg}$) using Eq. (11.150).

$$\bar{\psi}_{Sg} = \bar{\psi}_{S\text{lg}} + \bar{\psi}_{mg} \quad \text{...(11.170)}$$

where,

$$\bar{\psi}_{S\text{lg}} = L_{Sl}\bar{i}_{Sg} \quad \text{...(11.171)}$$

is the space-phasor for the stator leakage flux linkages in the general reference frame, and L_{Sl} is the leakage inductance of the stator winding. If the stator and rotor windings have equal effective turns, then

$$\bar{\psi}_{mg} = L_m\bar{i}_{mg} \quad \text{...(11.172)}$$

where, $\bar{\psi}_{mg}$ is the space phasor of the magnetising flux-linkages in the general reference frame and $\bar{i}_{mg}$ denotes the magnetising current space phasor in the general reference frame,

$$\bar{i}_{mg} = \bar{i}_{Sg} + \bar{i}_{rg} \quad \text{...(11.173)}$$

Substitution of Eqs. (11.170) to (11.173) in Eq. (11.169) produces:

$$\left.\begin{aligned} T_e &= \frac{3}{2}P(L_{Sl}\bar{i}_{Sg} + L_m\bar{i}_{mg}) \times \bar{i}_{Sg} \\ &= \frac{3}{2}PL_m\bar{i}_{mg} \times \bar{i}_{Sg} \\ &= \frac{3}{2}P\bar{\psi}_{mg} \times \bar{i}_{Sg} \end{aligned}\right\} \quad \text{...(11.174)}$$

The above expressions indicate that the electromagnetic torque is expressed as the vector (or cross) product of the magnetising current and the stator current space phasors. It also establishes that the stator leakage fluxes do not contribute to electromagnetic torque production as is physically justified.

Resolution of all space phasors into their real (or x) and imaginary (or y) axes components, leads to development of electromagnetic torque in terms of 2-axis components, and it is possible to obtain expressions for the electromagnetic torque in terms of 2-axis components of the general reference frame. Using Eqs. (11.174) with

$$\psi_{mg} = \psi_{mx} + j\psi_{my}$$

and

$$i_{Sg} = i_{Sx} + ji_{Sy}$$

we can write:

$$T_e = \frac{3}{2}P(\psi_{mx}i_{Sy} - \psi_{my}i_{Sx}) \quad \text{...(11.175)}$$

If we expand Eq. (11.168) and incorporate Eq. (11.151) in it, we obtain expression of electromagnetic torque in the form:

$$T_e = \frac{3}{2} P L_m \bar{i}_{rg} \times \bar{i}_{Sg} = \frac{3}{2} P \frac{L_m}{L_r} \bar{i}_{Sg} \, (L_r \bar{i}_{rg} \times \bar{i}_{Sg})$$

$$= \frac{3}{2} P \frac{L_m}{L_r} (L_r \, \bar{i}_{rg} + L_m \bar{i}_{Sg}) \times \bar{i}_{Sg}$$

$$= \frac{3}{2} P \frac{L_m}{L_r} \bar{\psi}_{rg} \times \bar{i}_{Sg} \qquad \text{...(11.176)}$$

And by resolution into direct-axis (or x) and quadrature axis (or y) components of the general reference frame and using:

$$\bar{\psi}_{rg} = \psi_{rx} + j\psi_{ry}$$

and

$$\bar{i}_{Sg} = i_{Sx} + j i_{Sy}$$

we get:

$$T_e = \frac{3}{2} P \frac{L_m}{L_r} (\psi_{rx} i_{Sy} - \psi_{ry} i_{Sx}) \qquad \text{...(11.177)}$$

In Eq. (11.177), ψ_{rx} and ψ_{ry} are the 2-axis components of the rotor flux-linkage space phasors with respect to the general reference frame, which, using Eq. (11.151), could be expressed as follows:

$$\psi_{rx} = L_r i_{rx} + L_m i_{Sx}$$

and

$$\psi_{ry} = L_r i_{ry} + L_m i_{Sy}$$

11.9.2 Application of Special Reference Frames

One of the aims of applying vector control to ac motors was identified as the ability of applying of control schemes leading to good dynamic performance similar to those derived in case of dc machines, as was pointed out in the initial part of Section 11.1.6.

Under transient conditions, we should achieve the electromagnetic torque in a smooth air-gap machine to be proportional to the product of a flux-producing current component and a torque producing current component, such that they are in space quadrature.

There are three similar ways of achieving the above objective:

1. By use of reference frames to the magnetising flux-linkage space phasor (This being equivalent to fixing the reference frame to the magnetising current-flux phasor.)
2. By use of reference frames fixed to the rotor flux-linkage flux phasor
3. By use of reference frames attached to the stator flux-linkage flux phasor

to derive expression for the electromagnetic torque in a form that makes possible independent control of torque and flux producing current components.

11.9.3 Application of Special Reference Frame Attached to the Magnetizing Flux-linkage Space Phasor

An expression for the electromagnetic torque is derived in terms of the modulus of the magnetising flux-linkage space phasor and the quadrature-axis stator current component, expressed with respect to the reference frame attached to the magnetising flux-linkage space phasor.

A smooth air-gap ac machine having a symmetrical poly-phase winding and a short-circuited poly-phase rotor winding-like a 3-phase induction motor could be represented by a schematic diagram shown in Figure 11.11, where the 3-phase stator windings (*SA*, *SB* and *SC*) are replaced by its equivalent 2-phase stator configuration with windings (*SD* and *SQ*), together with a poly-phase short-circuited rotor winding arrangement.

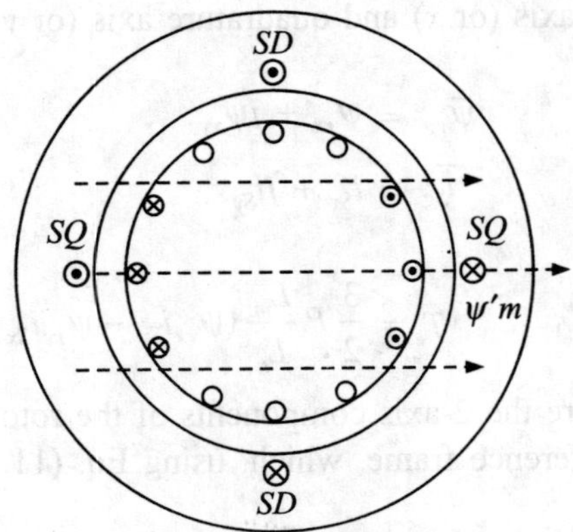

FIGURE 11.11 Schematic of a poly-phase induction machine with a 2-phase stator winding and a short-circuited rotor configuration.

Let us compare Figure 11.11 with Figure 11.5. This diagram was drawn and used to explain the mechanism of torque production in case of dc machines.

In the original diagram in Figure 11.5, we have considered the armature winding to be located on the rotor member, while the field system together with the compensating winding carrying the armature current on the stator. However, in case of dc machines, the commutator and brushes fix the spatial position of the armature reaction flux at rest along the quadrature axis, which is also the axis of the compensating winding. Therefore, from flux considerations alone we could have considered that in Figure 11.5, the armature winding was notionally located on the stator member with the armature coils behaving as pseudo-stationary coils with their effective axis along SQ, while the field coils were notionally positioned on the rotor with flux-axis effective along SD and the compensating winding also mounted on the rotor producing flux opposing the armature reaction flux along SQ axis. Hence, Figure 11.11 could notionally denote a dc machine in its modified form flux considerations.

The direct-axis stator current i_{SD} produces the magnetising flux linkage, ψ'_m. If the quadrature-axis current i_{SQ} is suddenly injected into the stator winding *SQ* (which now acts as the compensating winding of the dc machine equivalent), a rotor current i'_r shall flow in the rotor winding. At the first instnt of time (that is at $t = t_0$) the value of $i'_r = i_{rQ}$. Thus, space-phasor diagram of the magnetising flux linkage and stator and rotor currents shall be similar to that shown in Figure 11.6 for the dc machine. This is depicted in Figure 11.12, in which

$$\bar{i}_S = i_{SD} + ji_{SQ}$$

is the space phasor of the stator currents, and $\bar{i}'_r$ is the space phasor of the rotor currents in the stationary reference frame.

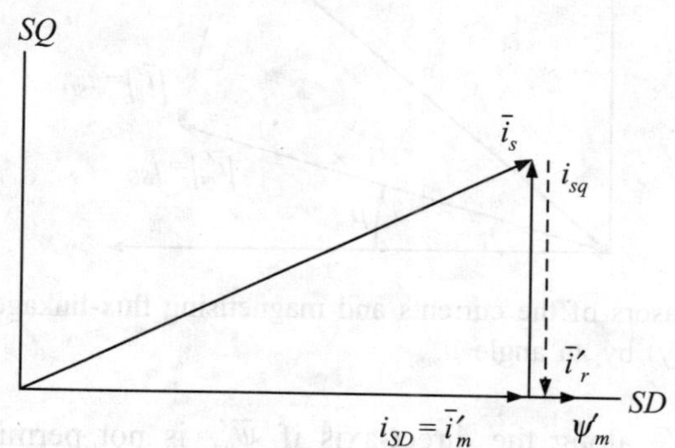

FIGURE 11.12 Space phasors of $\bar{i}_S$, $\bar{i}_r$ and $\bar{\psi}'_m$ corresponding to instant $t = t_0$.

If the stator and rotor windings are assumed to have the same number of effective turns, we observe from Eq. (11.173) that the sum of the stator and rotor current space phasors gives the magnetising current space phasor. Also, it follows from Eq. (11.172) that the magnetising flux-linkage space-phasor, with respect to the stationary reference frame is given by

$$\bar{\psi}'_m = L_m \bar{i}'_m$$

which is proportional to the magnetising current space phasor if the magnetising inductance L_m is constant and co-axial with direct-axis of the stationary reference frame.

However, for t greater than t_0, the space phasor diagram gets altered, since the field changes in order to induce rotor currents. If it is assumed that the rotor is locked in the same position as indicated in Figure 11.11, the new space-phasor diagram assumes the shape shown in Figure 11.13, with new changed values of magnetising flux and rotor current space-phasors. Thus, we see from Figure 11.13 that contrary to Figure 11.12 (or Figure 11.5, which refers to a dc machine), the magnetising flux-linkage space-phasor $\bar{\psi}'_m = L_m \bar{i}'_m$ is no longer co-axial with *SD*-axis of the stationary reference frame and is not in space quadrature with *SQ* axis of the same reference frame, but is rotated by an angle μ_m with respect to *SD* axis. Thus, $\bar{\psi}'_m$ is no longer along *SD* axis.

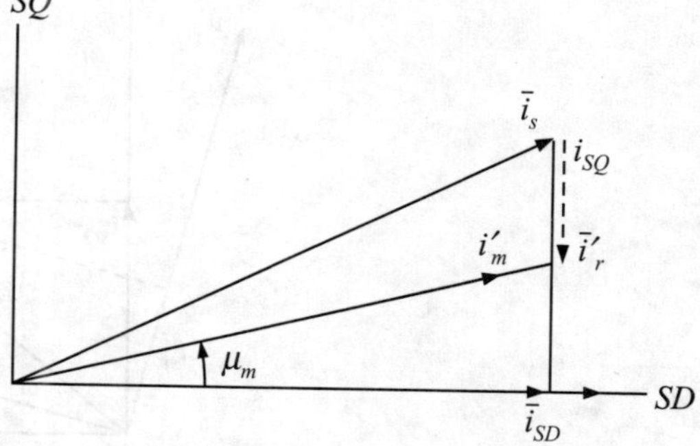

FIGURE 11.13 Space-phasors of $\bar{i}_S$, $\bar{i}'_r$ and ψ'_m at $t > t_0$, with rotor blocked.

If, however, there is a fictitious rotation of the stator through an angle μ_m in the counter-clockwise direction, the magnetising flux-linkage space phasor $\bar{\psi}'_m = L_m \bar{i}'_m$ again attains co-axial position with *SD*-axis and hence with the direct-axis stator current and is in space quadrature with the quadrature-axis stator current, as indicated in Figure 11.14—a condition similar to the condition indicated in Figure 11.5 for the dc machine.

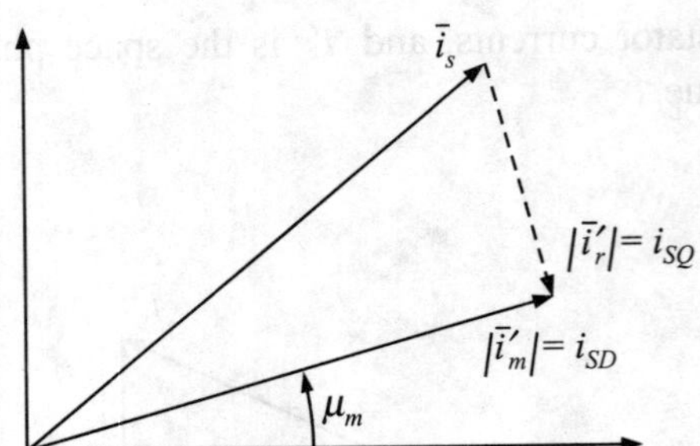

FIGURE 11.14 Space phasors of the currents and magnetising flux-linkage when the stator is rotated (notionally) by an angle μ_m.

In order to keep $\bar{\psi}'_m$ along the direct-axis if $\bar{\psi}'_m$ is not permitted to rotate, the stator has to be continuously rotated by the appropriate value of μ_m, until the required orientation is achieved. In reality, the stator is not rotated and the stator windings are held stationary to ensure separation of the stator currents into such two current components—one producing the magnetising flux-linkage and the other producing the electromagnetic torque. A special reference frame (with direct and quadrature axes nomenclatures of x and y) is used instead. There are many ways of choosing such a special reference frame.

One method to select special reference frame is to choose a reference frame that rotates at the speed of the magnetising flux-linkage space-phasor, such that its real axis (x), is always co-axial with the magnetising flux-linkage space-phasor. This is depicted in Figure 11.15.

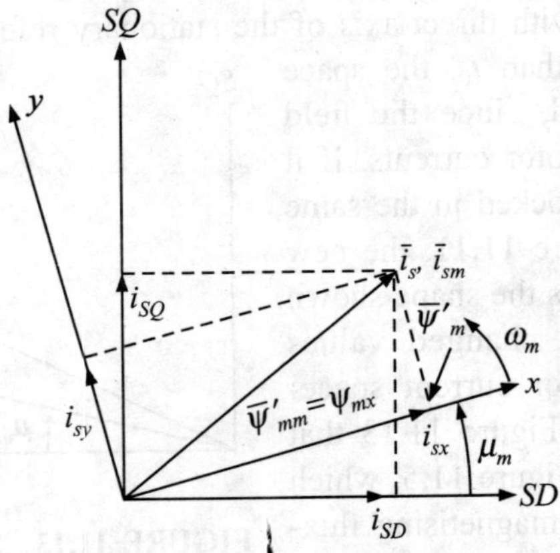

FIGURE 11.15 Relationship of stationary (SD, SQ) and spcial (x, y) reference frames.

We find from Figure 11.15 that if in the stationary reference frame (SD, SQ), the magnetising flux-linkage phasor is defined as:

$$\bar{\psi}'_m = |\bar{\psi}_m| \, e^{j\mu_m} = \psi_{mD} + j\psi_{mQ} \qquad ...(11.178)$$

where $|\bar{\psi}_m|$ and μ_m are the modulus and space angle respectively, of the magnetising space phasor, and ψ_{mD} and ψ_{mQ} are its two-axis components in the special reference frame, the space phasor of the stator currents with respect to the special reference frame rotating at speed